MATHEMATICAL STATISTICS

Other World Scientific Titles by the Author

Probability
Second Edition
ISBN: 978-981-98-1958-4
ISBN: 978-981-98-2047-4 (pbk)

MATHEMATICAL STATISTICS

LAWRENCE M. LEEMIS
William & Mary, USA

World Scientific
NEW JERSEY • LONDON • SINGAPORE • BEIJING • SHANGHAI • HONG KONG • TAIPEI • CHENNAI • TOKYO

Published by

World Scientific Publishing Co. Pte. Ltd.
5 Toh Tuck Link, Singapore 596224
USA office: 27 Warren Street, Suite 401-402, Hackensack, NJ 07601
UK office: 57 Shelton Street, Covent Garden, London WC2H 9HE

Library of Congress Control Number: 2026020427

British Library Cataloguing-in-Publication Data
A catalogue record for this book is available from the British Library.

MATHEMATICAL STATISTICS

ISBN 978-981-98-3292-7 (hardcover)
ISBN 978-981-98-3325-2 (paperback)

For any available supplementary material, please visit
https://www.worldscientific.com/worldscibooks/10.1142/14872#t=suppl

For Jill

Contents

Preface

I am pleased to have this new edition of *Mathematical Statistics* published by World Scientific Publishing Company. I am grateful for the opportunity to have world-wide distribution of the book. The pre-requisite for reading this book is a calculus-based class in probability, so this text follows another book titled *Probability* that has been published by World Scientific. I have always thought that there are two important directions that a student can go after taking a calculus-based probability class. The first direction is stochastic processes, which can be thought of as probability over time. The second direction is statistics, which can be thought of as probability applied to data. This book is aimed in that second direction. The material in this book is appropriate as the second class in a mathematical statistics sequence of classes. To aid in the transition from probability to statistics, the book includes many examples that build toward the 300-plus exercises at the end of each chapter. This example-driven approach extends to Monte Carlo simulation support of analytic solutions to examples and exercises, which is a unique feature of the textbook.

As with most mathematical statistics texts, point estimation, interval estimation, and hypothesis testing form the core of the material presented. The material chosen is that which I would like to have a student see upon their first exposure to mathematical statistics. One central theme is that statistics, such as the sample mean $\bar{X}$ and the sample variance S^2, are random variables that estimate their associated population analogs, which are the population mean μ and the population variance σ^2, which are constants. Furthermore, these statistics have probability distributions which allow us to assess their precision.

The R language is used throughout the text for graphics, computation, and Monte Carlo simulation. R can be downloaded for free at `r-project.org`. Most of the R code also works in S-Plus. There is not always a perfect match between the code for generating a graph and the graph that appears in the text in order to keep the number of lines of code to a minimum. Most of the figures were generated in S-Plus, so running the associated R code will not always provide a perfect match.

The Maple-based symbolic probability language APPL (A Probability Programming Language) has also been used throughout the text. My thanks goes to Andy Glen, Diane Evans, Billy Kaczynski, Jeff Yang, and Keith Webb for writing, maintaining, and extending the language. APPL is available from the author.

The text is organized into chapters and sections. When there are several topics within a section, they are set off by **boldface headings** to delineate the subsections. Definitions and theorems are boxed; examples are indented; proofs are terminated with a box □. Proofs are only included when they are instructive to the material being presented. Exercises are numbered sequentially at the end of each chapter. Computer code is set in `monospace font`, and is not punctuated. Indentation is used to indicate nesting in code. An index is included. Page numbers that are italicized in the index correspond to the primary source of information on a topic. Index entries set in `monospace font` are typically R functions. A solutions manual and teaching slides are available for instructors from the author.

Beginning a mathematical statistics textbook with a section on statistical graphics is unusual. This topic is presented first because (*a*) most students have not seen this material before, (*b*) it provides a transition from probability theory, which typically does not involve data, to statistics, which is highly data-centric, and (*c*) it serves as a reminder that although the data in the book is generally treated as the random variables $X_1, X_2, \ldots, X_n$, these values correspond to real-world values associated with some application. This section can be skipped with only a minimal loss of continuity.

Some data sets are used in the examples and exercises in the book. Many of these data sets are available on my home page.

My thanks goes to the colleagues and students who have caught errors and made suggestions in the various drafts of the text: John Delos, Tanujit Dey, Pam Farnham, Rex Kincaid, Samantha King, Rui Pereira, Yuxin Qin, Rebecca Rector, Mike Schilling, Heather Sasinowska, Amy Then, John Welch, Dave Williams, Diane Yamini (William & Mary), Jim Hartman, Bob Wooster (The College of Wooster), Yan Hao (Hobart and William Smith Colleges), Daniel Luckett (University of North Carolina), Barry Nelson (Northwestern), Ori Rosen (UTEP), Bruce Schmeiser (Purdue), Jason Wilson (Biola), and Andy Glen and Chris Weld (West Point). In addition, I am very grateful to Robert Lewis and Robert Jackson from the University of Alabama in Huntsville who proofread the final draft of this textbook and provided hundreds of edits which have improved the exposition substantially. Robert Jackson provided the basis for the maximum likelihood estimation example in Chapter 2 involving the sampling of motors from sheds. Ross Iaci has been very helpful with suggestions regarding notation. Guannan Wang has class-tested earlier versions of the text and made many helpful suggestions. Special thanks to Greg Hunt and his students for class-testing the first complete draft of the book, and particularly Cassie Chang who spotted numerous typographical errors. My thanks also goes to Lindsey Leemis for the handsome cover design and to Mark and Logan Leemis for their proofreading.

Rochelle Kronzek from World Scientific Publishing Company first contacted me regarding the publication of a sequence of books in probability and statistics. I am grateful for her help and persistence in getting this textbook over the finish line. I am also grateful to my production editors Angeline Husni and Tan Rok Ting for their help with converting the book to a smaller trim size and the cover design.

There are no references cited in the text for readability. The sources of materials in the four chapters are cited in the next four paragraphs.

Chapter 1 notes: Example 1.1 was adapted from Gelman (2011), "Tables as Graphs: The Ramanujan Principle," *Significance*, Volume 8, Issue 4, page 183. Further reading in statistical graphics can be found in the four-volume series by Edward R. Tufte: The Visual Display of Quantitative Information (1983), Envisioning Information (1990), Visual Explanations (1997), and Beautiful Evidence (2006), all published by Graphics Press. Other books in the area of statistical graphics include The Elements of Graphing Data (1985, Wadsworth) and Visualizing Data (1993, Hobart Press) by William S. Cleveland and Exploratory Data Analysis (1977, Addison–Wesley) by John W. Tukey. There are dozens of other books on statistical graphics that are also very well written. The graphical presentation of a basketball game in Example 1.2 was adapted from Westfall, P. (1990), "Graphical Presentation of a Basketball Game," *The American Statistician*, Volume 44, Number 4, pages 305–307. Gene Wojciechowski provided me with the play-by-play data from the Duke vs. Kentucky basketball game. The nursing data in Example 1.3 is from Konner, M., and Worthman, C. (1980), "Nursing Frequency, Gonadal Function and Birth Spacing Among !Kung Hunter-Gatherers," *Science*, Volume 207, Number 4432, pages 788–791. More information on Michelson's experiments to calculate the speed of light in Example 1.6 is given in Trosset, M. (2009), *An Introduction to Sta-*

tistical Inference and Its Applications with R, Chapman and Hall/CRC. The data for the population pyramid in Example 1.7 was obtained from `www.insee.fr`. The city weather data in Example 1.8 was obtained from `http://weather-warehouse.com`. The NFL regular season kickoff starting field position data given in Example 1.11 is from Horowitz, M. (2017), *nflscrapr: R Package for Scraping NFL Data* available at: `https://github.com/maksimhorowitz/nflscrapR`. More information concerning the construction of the graphic is given in Weld, C. and Leemis, L. (2019), "Mixed-Type Distribution Plots," *Information Visualization*, Volume 18, Number 3, 311–317. The sample Gini mean difference appears on page 241 of Kendall and Stuart (1958), *The Advanced Theory of Statistics, Volume 1*, Griffin. The data set of the maximum skull breadths is from Hand, D.J., Daly, F., Lunn, A.D., McConway, K.J., and Ostrowski, E. (1994), *Small Data Sets*, Chapman and Hall. The original data was published in Barnicot, N.A., and Brothwell, D.R. (1959), "The Evaluation of Metrical Data in the Comparison of Ancient and Modern Bones," in *Ciba Foundation Symposium on Medical Biology and Etruscan Origins*, Wolstenholme, G.E.W., and Connor, C.M., eds., pages 131–149, Little, Brown, and Company.

Chapter 2 notes: The ball bearing data set in Example 2.3 was first published in Lieblein, J. and Zelen, M. (1956), "Statistical Investigation of the Fatigue Life of Deep-Groove Ball Bearings," *Journal of Research of the National Bureau of Standards*, Volume 57, Number 5, pages 273–316. The initial estimate of the shape parameter in the Newton–Raphson procedure for finding the maximum likelihood estimates of the Weibull distribution parameters in Example 2.14 is from Menon, M.V. (1963), "Estimation of the Shape and Scale Parameters of the Weibull Distribution," *Technometrics*, Volume 5, Number 2, pages 175–182. A fixed-point algorithm for calculating the maximum likelihood estimators of the Weibull distribution parameters is given by Qiao, H., and Tsokos, C.P. (1994), "Parameter Estimation of the Weibull Probability Distribution, *Mathematics and Computers in Simulation*, Volume 37, Number 1, pages 47–55. Six regularity conditions associated with the information and the Cramér–Rao inequality are given in Chapter 6 of Hogg, R.E., McKean, J.W., and Craig, A.T. (2005), *Introduction to Mathematical Statistics*, 6th Edition, Prentice–Hall. Another proof of the Cramér–Rao inequality is given by Kagan, A. (2001), "Another Look at the Cramér–Rao Inequality," *The American Statistician*, Volume 55, Number 3, pages 211–212. The asymptotic distribution of the sample median for sampling from a normal population is given on page 327 of Hogg, R.E., McKean, J.W., and Craig, A.T. (2005), *Introduction to Mathematical Statistics*, 6th Edition, Prentice–Hall. Example 2.27 is adapted from page 378 of Hogg, R.E., McKean, J.W., and Craig, A.T. (2005), *Introduction to Mathematical Statistics*, 6th Edition, Prentice–Hall. A proof of the Fisher–Neyman factorization criteria in the case of sampling from a continuous population is given on page 377 of Hogg, R.E., McKean, J.W., and Craig, A.T. (2005), *Introduction to Mathematical Statistics*, 6th Edition, Prentice–Hall. The times between nuclear power plant accidents in Exercise 2.6 are given in Weld, C. (2019), Computational Graphics and Statistical Analysis: Mixed-Type Random Variables, Confidence Regions, and Golden Quantile Rank Sets, Ph.D. Dissertation, The College of William & Mary. The counts of the number of ticks appearing on sheep collected by Fisher in Exercise 2.19 are from page 156 of Hand, D.J., Daly, F., Lunn, A.D., McConway, K.J., and Ostrowski, E. (1994), *Small Data Sets*, Chapman and Hall.

Chapter 3 notes: The ten-question quiz appeared on page 71 of Russo, J.E., and Schoemaker, P.J.H. (1989), *Decision Traps: Ten Barriers to Brilliant Decision-Making and How to Overcome Them*, Doubleday. The answers are (1) 39 years, (2) 4187 miles, (3) 13 countries, (4) 39 books, (5) 2160 miles, (6) 390,000 pounds, (7) 1756, (8) 645 days, (9) 5959 miles, (10) 36,198 feet. More questions of this nature are given in Wang, X., Reich, N.G., and Horton, N.J. (2019), "Enriching Students' Conceptual Understanding of Confidence Intervals: An Interactive Trivia-Based Activity," *The American Statistician*, Volume 73, Number 1, pages 50–55. An excellent comprehensive overview of interval estimation is in Meeker, W.Q., Hahn, G.J., and Escobar, L.A. (2017), *Statis-*

tical Intervals: A Guide for Practitioners and Researchers, Second Edition, Wiley. Plotting (l, u) pairs as in Example 3.4 was devised by Kang, K., and Schmeiser, B. (1990), "Graphical Methods for Evaluating and Comparing Confidence-Interval Procedures," *Operations Research*, Volume 38, Number 3, pages 546–553. The Boeing 720 air conditioning failure data in Example 3.7 was from plane number 7913 from Proschan, F. (1963), "Theoretical Explanation of Observed Decreasing Failure Rate," *Technometrics*, Volume 5, Number 3, pages 375–383. The IQ scores in Example 3.9 are from page 33 of Moore, D., and McCabe, G. (1999), *Introduction to the Practice of Statistics*, Third Edition, Freeman. The tomato yield data in Example 3.11 is from page 316 of Hand, D.J., Daly, F., Lunn, A.D., McConway, K.J., and Ostrowski, E. (1994), *Small Data Sets*, Chapman and Hall. The original data set is from Box, G.E.P., Hunter, W.G., and Hunter, J.S. (1978), *Statistics for Experimenters*, John Wiley and Sons. The Clopper–Pearson confidence interval introduced in Example 3.13 first appeared in Clopper, C.J., Pearson, E.S. (1934), "The Use of Confidence or Fiducial Limits Illustrated in the Case of the Binomial," *Biometrika*, Volume 26, Number 4, pages 404–413. The reference to a lady tasting tea in Example 3.13 is from Salsburg, D. (2001), *A Lady Tasting Tea: How Statistics Revolutionized Science in the Twentieth Century*, W.H. Freeman and Company. References for the Wald, Wilson–score, Jeffreys, Agresti–Coull, and arcsine confidence intervals for the Bernoulli parameter p are given in Park, H. and Leemis, L. (2019), "Ensemble Confidence Intervals for Binomial Proportions," *Statistics in Medicine*, Volume 38, Number 18, pages 3460–3475. A conservative confidence interval for the Bernoulli parameter p that has improved statistical properties is given by Blaker, H. (2000), "Confidence Curves and Improved Exact Confidence Intervals for Discrete Distributions," *Canadian Journal of Statistics*, Volume 28, Number 4, pages 783–798. The confidence interval associated with Welch's approximation illustrated in Example 3.15 is from Welch, B.L. (1947), "The Generalization of "Student's" Problem when Several Different Population Variances are Involved," *Biometrika*, Volume 34 (1–2), pages 28–35. His proposal was modified in Aspin, A. A. (1949), "Tables for Use in Comparisons Whose Accuracy Involves Two Variances Separately Estimated," *Biometrika*, Volume 36 (3–4), pages 290–296. The rat survival data used in the bootstrapping subsection is from page 11 of Efron, B., and Tibshirani, R. (1993), *An Introduction to the Bootstrap*, Chapman and Hall. The coin flipping illustration that was used to motivate the likelihood ratio statistic is adapted from a similar illustration in Rice, J.A. (2007), *Mathematical Statistics and Data Analysis*, Third Edition, Thompson and Brooks/Cole. The discussion of the "fill problem" at the end of the subsection on tolerance intervals is adapted from page 253 of Hogg, R.E., McKean, J.W., and Craig, A.T. (2005), *Introduction to Mathematical Statistics*, 6th Edition, Prentice–Hall. The discussion of the truck tank design illustration at the end of the subsection on tolerance intervals is adapted from page 194 of Vardeman, S.B. (1992), "What About the Other Intervals?" *The American Statistician*. Volume 46, Number 3, pages 193–197. A discussion of Bayesian methods from several points of view is given on pages 1–17 of the February 2013 issue (Volume 67, Number 1) of *The American Statistician*. Dr. James Wilson of North Carolina State University suggested the figure associated with the trapezoidal shaped integration regions in Example 3.41.

Chapter 4 notes: The small birthday data set in Example 4.10 is from *A Guide to the Memorials of Bruton Parish Church*, which is self-published and available from the church. The large birthday data set is from Phillips, D.P., Van Voorhees, C.A., Ruth, T.E. (1992), "The Birthday: Lifeline or Deadline?" *Psychosomatic Medicine*, Volume 54, Number 5, pages 532–542. A litany of problems with p-values, along with links to the American Statistical Association statement and an article on misinterpretation of p-values is given at `www.fharrell.com/post/pval-litany`. The finger tapping experiment data given in Example 4.15 is from page 40 of Hand, D.J., Daly, F., Lunn, A.D., McConway, K.J., and Ostrowski, E. (1994), *Small Data Sets*, Chapman and Hall. The original data set is from Draper, N.R., and Smith, H. (1981), *Applied Regression Analysis*, Second Edition, John Wiley and Sons. Only two of the three groups of students performing the finger tapping experiment were considered here. The equivalence between using confidence intervals and traditional hypothesis testing to conduct a statistical test in Section 4.5 is proved on pages 421–422 of Casella, G., and Berger, R.L. (2002), *Statistical Inference*, Second Edition, Duxbury. The Neyman–Pearson theorem was first given in Neyman, J., and Pearson, E.S. (1933), "On the Problem of the Most Efficient Tests of Statistical Hypotheses," *Philosophical Transactions of the Royal Society of London. Series A, Containing Papers of a Mathematical or Physical Character*, Volume 231, pages 289–337. The height-to-width data for the beaded rectangles sewn by the Shoshone American Indians to decorate their blankets, clothes, and leather goods in Example 4.34 are given on page 118 of Hand, D.J., Daly, F., Lunn, A.D., McConway, K.J., and Ostrowski, E. (1994), *Small Data Sets*, Chapman and Hall. The original data set is from Dubois, C. (1970), *Lowie's Selected Papers in Anthropology*, University of California Press.

Williamsburg, Virginia, U.S.A.

Larry Leemis
March 2026

Chapter 1

Random Sampling

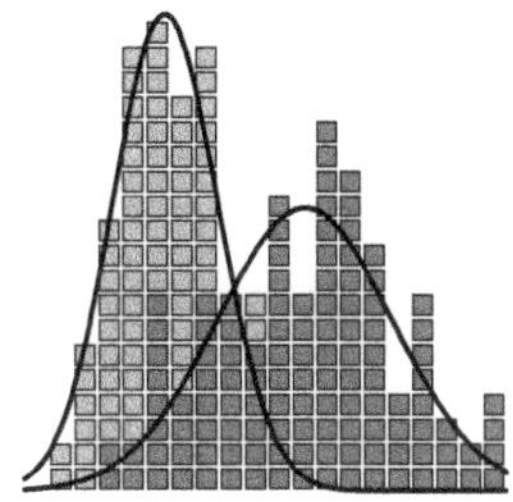

A *statistic* is a number that is calculated from a sample consisting of data values. One simple example of a statistic is the sample mean. The study of statistics concerns gathering data, analyzing data, drawing conclusions from data, and making decisions from data. Statisticians use mathematical techniques and algorithms to collect, display, and analyze data. Many of their methods rely on probability, so the study of statistics can be thought of as probability applied to data.

Sometimes, you just need to present data in a way that allows you to reach an obvious conclusion. In other cases, sophisticated mathematical models are required in order to draw a conclusion from a data set. We begin with the first case.

1.1 Statistical Graphics

Assuming that data values have been collected in a reasonable fashion (more detail will be provided later about what constitutes "reasonable"), statisticians face the challenge of how to present the data values in a fair, intuitive, and revealing fashion. The graphical display of a data set consists of tables and figures that highlight key features that address a relevant question of interest.

We begin with a simple example of placing a data set of just five observations in a table in order to compare the populations of five countries.

Example 1.1 A data set consisting of the populations of five countries in the year 2000 taken from the appropriate Wikipedia website is displayed in Table 1.1. Although it contains all of the information required to compare the populations, Table 1.1 is a dreadful presentation of this data set for the following reasons.

- The populations are placed horizontally. Although this takes up less vertical space on the page, it makes visual comparisons more difficult. Aligning the populations vertically on the decimal point is a better approach.

Country:	China	Indonesia	Montenegro	Serbia	U.S.A.
Pop.:	1,242,612,266	206,264,595	620,145	9,778,991	281,421,906

Table 1.1: Populations in 2000 (presented poorly).

- The countries are sorted alphabetically even though the interest is in comparing population sizes; it would have been more helpful to sort them by decreasing population size.
- The bland monospace font is hypnotizing for the reader and obscures rather than accentuates the population sizes.
- There are lots of lines separating the fields in the table; it is better to use fewer lines.
- Unnecessary extra ink (such as the colons after the row labels) should be avoided. Abbreviating the population label is not helpful.
- Using all of the digits on population is distracting and makes comparing the populations more difficult. Does the value in the ones digit really matter?

So how can this table be improved? We can improve Table 1.1 by addressing each of the six points above. Table 1.2 is a second attempt at displaying the five data values which, hopefully, you find to be a more intuitive way of presenting the data with the goal of comparing populations. One can easily and immediately see that China has the largest population with over one billion people, followed by the others in decreasing population size. The data values are arranged vertically, aligned on the decimal point, and sorted by population size; fonts have been altered, unnecessary ink has been removed, and distracting digits have been removed by presenting the populations in millions. The essence of the population data set is easily and quickly gleaned from Table 1.2 by using common-sense principles to redesign the table.

Country	**Population (millions)**
China	1,242.6
U.S.A.	281.4
Indonesia	206.3
Serbia	9.8
Montenegro	0.6

Table 1.2: Populations in 2000.

The display of data in an accurate, meaningful, and intuitive fashion is as much an art as it is a science. What constitutes a "good" table is a matter of preference. The two tables from Example 1.1 have illustrated a bad presentation and a good presentation. The key step in this process is placing yourself in the reader's position and thinking of ways to simplify the presentation of the data to illuminate the aspect of the data that is of interest.

Presenting data in tables, however, is not the only option. The graphical presentation of data provides a much more efficient way to convey the message a data set provides.

The practice of presenting data in graphical form is a field known as *statistical graphics*. One of the early pioneers in the field was William Playfair (1759–1823), who published the *Commercial and Political Atlas* in 1786. His atlas contained graphs that described imports and exports of England and Wales with their trading partners. Before the publication of Playfair's atlas, graphs were rare; after his publication, graphs began to appear with increasing regularity. More recent leaders in the field include William S. Cleveland, Edward R. Tufte, and John W. Tukey. Their books are listed in the preface and are recommended reading materials if this brief overview of statistical graphics given here sparks some interest on your part.

Contemplating how one should construct a graph in order to display quantitative information will help you think about data, which is at the core of statistical practice. Visualizing the data set via a graph is often the first view we get of a data set after the data is collected. It provides a first impression that often guides the next step, which might be using a statistical inference technique to draw a conclusion from the data. The graphs that will be produced in this section are drawn in the free statistical package called R, but there are many other tools available for constructing statistical graphics.

Do not prioritize fancy fonts, color, shading, or highlighting when it comes to statistical graphics. The best graphic is often a simple plot which conveys only the appropriate information in the data set. One overriding principle established by Tufte is to maximize the *data-ink ratio*. Make every bit of ink that you place on a graphic count. Use as little ink as possible to provide as much explanation as is necessary. Do not decorate your graphics with what has come to be known as *chartjunk*, which conveys no information, but is simply a misguided attempt to make the graphic more attractive. Display the data succinctly and clearly, bringing the key information contained in the data set to prominence in your graphic.

There are dozens of decisions that must be made when constructing a graph to display data. Human vision and perception considerations should drive all decisions. A sampling of the related questions includes the following.

- *Aspect ratio.* Is there a reason (by virtue of the meaning of the scales) that the plotting area should be a square? If not, is a tall thin graph (portrait orientation) more appropriate, or is a short wide graph (landscape orientation) more appropriate?

- *Axes.* Should axes be included? Does one of the variables naturally belong on the horizontal axis? Should axes be included on just the bottom and left sides of the plot, or should they be included on three or all four sides of the plot? Should the axes intersect or should there be a gap between them?

- *Scales.* Should the scales on the axes be linear? Should a logarithmic scale for an axis be used? Should a square root scale for an axis be used? Where should the scales begin and end? Is it helpful to have zero included on the scale? (This is not an option for a logarithmic scale.) Should a break be placed on a scale in order to include zero?

- *Axis labels.* Should the axis labels be in the same font style as the manuscript text? Should the axis labels be the same font size as the manuscript text? Should the labels be placed at the ends of the axes or in the center of the axes? Should the vertical axis label be displayed parallel to the axis or rotated clockwise 45° or 90° for easier reading?

- *Tick marks and tick labels.* Should the tick marks extend into the plotting area or out of the plotting area? How many tick marks should be included on each axis? Should all tick marks be the same size? How long should the tick marks be? Should all tick marks be labeled? Should the tick mark labels be in the same font style as the manuscript text? Should the tick mark labels be the same font size as the manuscript text? Should the tick mark labels for the vertical axis be displayed parallel to the axis or rotated clockwise 90° for easier reading?

- *Plotting area.* What symbol is appropriate for plotting a point? Is a legend necessary to describe the meaning of the plotted symbols? What should be done if two points fall on top of one another? Should points be connected with lines? Is placing text in the plotting area helpful? Should reference lines be included in the plotting area? If so, should they be solid, dotted, or dashed? Should the elements placed in the plotting area be black, gray, or colored?

Every graph that you construct must answer these questions. The first example of a statistical graphic comes from sports.

Example 1.2 *March Madness* occurs every spring when a 64-team single-elimination tournament is used to determine the best college basketball team in the United States. One game, which some consider to be the greatest basketball game ever played, occurred in the East Regional final game of the men's tournament on March 28, 1992 between the Blue Devils of Duke and the Wildcats of Kentucky. Duke prevailed 104–103 in an overtime victory, winning on a last-second shot by Christian Laettner. The game featured perfect shooting for Laettner (10 for 10 from the field and 10 for 10 free throws for 31 points), five lead changes in the last 31.5 seconds, and both teams shooting 63% from the field in the second half and overtime. ESPN sportswriter Gene Wojciechowski wrote a book about this game titled *The Last Great Game*. Table 1.3 contains the box score for those who played in the game.

Duke		Kentucky	
Christian Laettner	31	Jamal Mashburn	28
Bobby Hurley	22	Sean Woods	21
Thomas Hill	19	Dale Brown	18
Brian Davis	13	John Pelphrey	16
Grant Hill	11	Richie Farmer	9
Antonio Lang	4	Gimel Martinez	5
Cherokee Parks	4	Deron Feldhaus	5
Marty Clark	0	Aminu Timberlake	1
		Nehemiah Braddy	0
		Travis Ford	0
		Andre Riddick	0
Total	104	Total	103

Table 1.3: Duke vs. Kentucky box score.

The box score is helpful for knowing the final score and how points were distributed among the players, but it does not give you any indication of the dynamic, or time-dependent, aspect of the game. Was the game a see-saw battle with many lead changes, or did one team build a big lead and the other battled back? Did one team score a large portion of their points with three-point shots? The box score does not answer these questions; so, we must design a statistical graphic that includes the answers.

The first question concerns the axes on the graphic. One plot that might be of interest is scoring over time. We follow the standard practice of placing time on the horizontal axis. Now we move to the vertical axis. One measure of scoring which allows the viewer to easily capture the flow of the game is to plot the difference between the scores. The step function moves upward for a Duke score and downward for a Kentucky score; so whenever the difference lies above zero, Duke is leading. The fact that the basketball hoops are swapped at halftime is not particularly relevant.

Four more elements are added to the statistical graphic to help the viewer. First, a bit of shading helps highlight the area between the function and the horizontal line associated with a tied game, which helps the reader see how long a lead is held. Second, vertical lines are added at $t = 20$ minutes (halftime) and $t = 40$ minutes (end of regulation). Third, small dots are added at each scoring time. Finally, vertical axes on both the left and right sides of the graphic make the differences in scores easier to determine.

After some experimentation, the landscape orientation of the plotting area looked the best; the final graphic is shown in Figure 1.1. Several conclusions can be quickly drawn from this statistical graphic that are not apparent from the box score.

- Kentucky scored first with a 3-point shot.
- Kentucky built a lead of 8 points early in the first half.
- Duke lead by 5 points at halftime.
- The two longest scoring droughts occurred early in the second half.
- Duke built their lead to 12 midway through the second half.
- Kentucky brought the game back to within 1 point with a 9 point scoring run.
- There was a scoring frenzy at the very end of the game, which is often the case in a close game because of deliberate fouling, which stops the clock.
- Duke won the game on a buzzer shot.

The small dots on the plot allow the viewer to see the difference between two consecutive successful free throws (such as those by Duke midway through the overtime period) and a two point shot (such as that by Kentucky just prior to the successful free throws). The statistical graphic can be enhanced by labeling the times of key events (such as a key player fouling out) that might influence the momentum of the game.

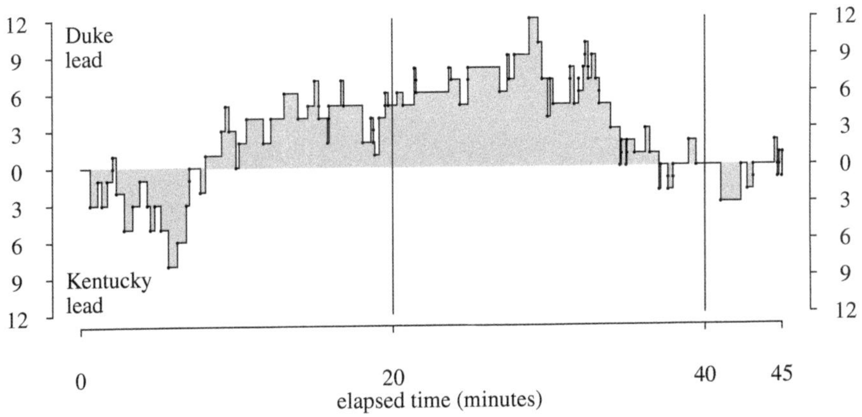

Figure 1.1: Duke vs. Kentucky, March 28, 1992.

The statistical graphic for the scoring in a basketball game allows one to glean the flow of the scoring for an entire basketball game in a glance. Would this type of graphic work for all sports? Consider football, where 1, 2, 3, or 6 points can be scored at a time. You could still produce the same statistical graphic for the scoring as we did for the basketball game, but football has other strategic elements (for example, field position) that are not captured with the single graphic alone.

The basketball graphic portrays quantitative variables on both the horizontal and vertical axes. Occasions arise when one of the variables is quantitative, but the other is qualitative, as is the case in the next example. This example also illustrates the case in which a statistical graphic is employed to help solve a mystery.

Example 1.3 The !Kung hunter-gatherers of Botswana and Namibia have long intervals between births, typically between 3 and 4 years, despite being a noncontracepting and nonabstinent population. Speculations linked the birth spacing to nutritional infertility, because the !Kung diet is sometimes low in calories, but no direct data had been collected to support this hypothesis.

Harvard anthropologists Melvin Konner and Carol Worthman investigated the unusually long birth spacings. Figure 1.2 shows one daylight cycle of interaction between a mother and her 14-day old son. As in the previous statistical graphics, the horizontal axis is time, which runs over the daytime interactions from 7:30 AM to 7:30 PM. The dependent variable here is not quantitative—it is one the following states: nursing, sleeping, holding, and crying. Furthermore, some of these states can occur simultaneously. For this reason, the states are labeled on the left using a text string and their durations are indicated by bars. Gaps between the bars imply that the state is not occurring. Nursing is placed at the top because it plays a central role in the conclusion that was drawn by Konner and Worthman.

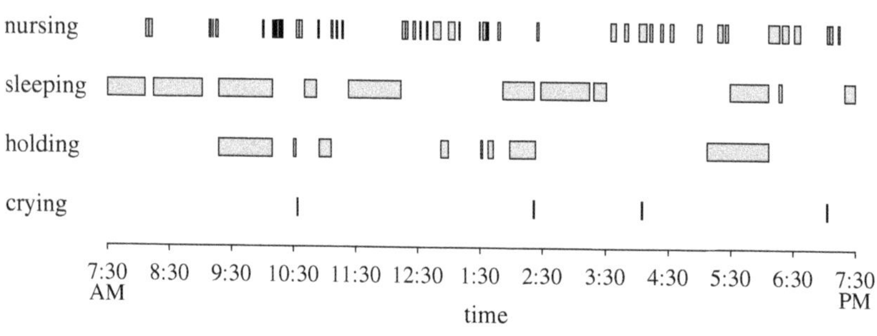

Figure 1.2: !Kung mother and baby daytime interactions.

The first thing that jumps out from the data (particularly to a Western mother) is the high frequency of nursing bouts. During the daytime hours when recordings were collected, there were 46 such bouts. This corresponds to a nursing bout every 15 minutes on average for this baby. There was a nursing bout every 13 minutes on average for all of the babies that they observed. Blood samples were also drawn daily on the nursing mothers. Konner and Worthman used the following logic to reject the conclusion that the low-calorie diet alone produced the long birth spacings: (1) nursing results in the release of the hormone prolactin, (2) prolactin has a half-life in the plasma of 10 to 30 minutes, (3) prolactin has an antigonadotrophic effect, which means that the mother will be less fertile if the prolactin level is high enough. So these frequent nursing bouts result in an elevated level of prolactin that results in the infertility of the mother. It is only late into the second year of life, when the baby's separations from the mother are longer as the baby spends more time playing, that the mother once again becomes fertile. The investigators used the statistical graphic, blood tests, and some biochemistry to draw their conclusion.

The statistical graphics in the previous two examples have had time on the horizontal axis. The next example considers a plot that shows the relationship between two categorical variables.

Example 1.4 All of the statistical graphics presented thus far have been prepared in the R language, which is open-source software. In this example, the R code required to produce the statistical graphic is presented, which only requires a single line of code. R has a built-in data set named `HairEyeColor`, which gives counts of the hair and eye color of 592 statistics students at the University of Delaware. The hair color is classified into four levels: black, brown, red, and blond. The eye color is also classified into four levels: brown, blue, hazel, and green. (The gender of the students was also collected, but will be ignored in the statistical graphic created here.) The hair color and eye color are known to statisticians as *categorical variables.* One way to investigate the relationship between hair color and eye color is a *mosaic plot*, which can be used to visualize the relationship between two or more variables. The single R command given below produces a mosaic plot of the hair and eye color data.

```
mosaicplot(~ Hair + Eye, data = HairEyeColor)
```

The mosaic plot is shown in Figure 1.3. The four hair colors are depicted horizontally; the four eye colors are depicted vertically. The areas of each rectangle are proportional to the counts of each combination of hair and eye color. For example, there are only 5 students in the statistics class with black hair and green eyes, so that rectangle has the smallest area. At the other extreme, there are 119 students with brown hair and brown

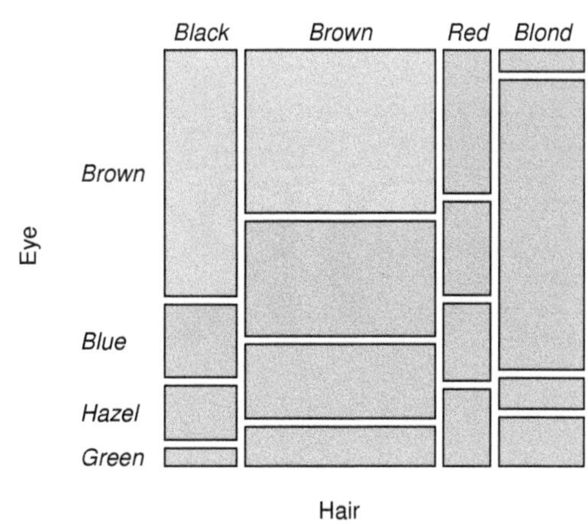

Figure 1.3: Mosaic plot of hair and eye color.

eyes, so that rectangle has the largest area. Clearly, based on the mosaic plot, hair color and eye color are related. If someone has blond hair, their eyes are most likely blue. If someone has black hair, their eyes are most likely brown.

Some statistical graphics don't require axes. *Word clouds*, which require no axes at all, are useful for showing frequencies of words in a text. They can also be used to quickly compare relative sizes of populations, relative number of search engine words, etc.

Example 1.5 A *word cloud* (also known as a wordle, tag cloud, or weighted list) uses the frequency of interesting words to give the viewer a quick visual overview of the content of a book, article, speech, or document. The words are packed as densely as possible into the cloud. Rotating some of the words so that they appear vertically allows the words to be packed more densely. The font size used on the three word clouds that follow is proportional to the frequency of the occurrence of the word in each text. I have arbitrarily chosen the top 50 "interesting" words in the three word clouds illustrated here. (This is a subjective process, but I left out words like "the," "is," "at," and "may.")

The Bible is divided into two parts: the Old Testament and the New Testament. I have used the 39 books for the Old Testament and the 27 books for the New Testament that are in common use by the Catholic, Protestant, Greek Orthodox, and Russian Orthodox churches, translated into the King James Version. The two different versions of the word "Lord" were counted together, making it the most frequent word in the text. The word cloud in Figure 1.4 contains the 50 most frequent interesting words that appear. By just viewing this graphic for a few seconds, it is easy to see what is emphasized in the text.

Figure 1.4: Word cloud of the Bible.

As a second contrasting example, Figure 1.5 contains a word cloud of the top 50 most frequent interesting words in William Shakespeare's play *Hamlet*. Several words appear (for example, "Denmark," "England") that could not have appeared in the first word cloud.

Figure 1.5: Word cloud of William Shakespeare's *Hamlet*.

The previous two word clouds are in stark contrast to the one for Mark Twain's *The Adventures of Huckleberry Finn* in Figure 1.6. Several of the words (for example, "reckon," "warn't," "ain't," "dey," "knowed") were common in the deep south at the time the book was written. The words "raft," "canoe," "river," and "woods" reflect the book's setting along the Mississippi River.

My hope is that these examples have conveyed the value of well-constructed statistical graphics. Their value is enormous in terms of both taking a first look at a data set for a statistician and communicating the information in a data set to non-statisticians. The following example includes

Figure 1.6: Word cloud of Mark Twain's *The Adventures of Huckleberry Finn*.

the R code necessary to produce some elementary statistical graphics that can be applied to a single data set of n values drawn from a univariate population. In this case the data set consists of $n = 100$ experimental estimates of the speed of light.

Example 1.6 We now know that the speed of light in a vacuum, oftentimes denoted by the constant c in Einstein's famous $E = mc^2$ formula, is 299,792.458 kilometers per second. The speed of light is slightly slower for light traveling through air—approximately 90 kilometers per second slower based on the refractive index of air. The speed of light in air also depends on the temperature and pressure of the air, so it is difficult to pin down one value. Before modern science emerged, many believed that the speed of light was infinite, meaning that light transmitted instantaneously. Galileo Galilei was one of the first to believe that the speed of light was finite. The first effort to determine the exact value of c was performed by Danish astronomer Ole Rømer in 1676, using the difference in the periods of Jupiter's innermost moon when the earth was approaching and receding from Jupiter. Christiaan Huygens combined Rømer's observation with an estimate of the diameter of the Earth's orbit to produce the first estimate of c, which was low by 26%. In 1879, Albert Michelson conducted experiments using a device with a rotating mirror (called an interferometer) to estimate the speed of light in air. These experiments resulted in the data set shown in Table 1.4. Each data value is the amount in excess of 299,000 kilometers per second. Each data value given is the average of ten experiments conducted by Michelson. The order of observation is given row-wise.

850	740	900	1070	930	850	950	980	980	880
1000	980	930	650	760	810	1000	1000	960	960
960	940	960	940	880	800	850	880	900	840
830	790	810	880	880	830	800	790	760	800
880	880	880	860	720	720	620	860	970	950
880	910	850	870	840	840	850	840	840	840
890	810	810	820	800	770	760	740	750	760
910	920	890	860	880	720	840	850	850	780
890	840	780	810	760	810	790	810	820	850
870	870	810	740	810	940	950	800	810	870

Table 1.4: Estimates of the speed of light in air (add 299,000 km/sec to each value).

The data displayed in Table 1.4 are rather unwieldy. Since the *order* that the observations were collected is not of particular interest, perhaps the data is better presented in sorted order, as in Table 1.5. This display is much more helpful to the reader. Clearly, the observations range from 299,620 kilometers per second to 300,070 kilometers per second, with the majority of the data values falling between 299,800 kilometers per second and 299,900 kilometers per second.

Although this second table is an improvement over the first, it is still rather difficult for the reader to intuit the *shape* of the probability distribution associated with this data set, even by spending significant time staring at the table of sorted observations. One crude, text-based graphic that can help determine the shape of the probability distribution is known as a *stem-and-leaf plot*. These plots were popularized in the 1970s, when monospace fonts (all numbers and letters use the same amount of horizontal space `like this`) were common on computer terminals and hard copy. A vertical line separates the

620	650	720	720	720	740	740	740	750	760
760	760	760	760	770	780	780	790	790	790
800	800	800	800	800	810	810	810	810	810
810	810	810	810	810	820	820	830	830	840
840	840	840	840	840	840	840	850	850	850
850	850	850	850	850	860	860	860	870	870
870	870	880	880	880	880	880	880	880	880
880	880	890	890	890	900	900	910	910	920
930	930	940	940	940	950	950	950	960	960
960	960	970	980	980	980	1000	1000	1000	1070

Table 1.5: Estimates of the speed of light in air (add 299,000 km/sec to each value).

stem values, which fall to the left of the line, and the leaf values, which fall to the right of the line. The first step in constructing a stem-and-leaf plot is to sort the data in ascending order, as in Table 1.5. The next step is to determine the meaning of the stem values. For the speed of light data, this will be the leftmost digit of the data values from Table 1.5 for the first 96 sorted observations, then 10 for the last four observations. The last step is to create a leaf entry for each data value. For the speed of light data set, we ignore the rightmost digit for each data value (because it is always zero) and use the tens digit of each entry as the leaf. The results of this process for the speed of light data are shown in Figure 1.7. There must be the same amount of horizontal space allocated to each leaf value for the shape of the probability distribution to be meaningful. The stem-and-leaf plot is one of the few statistical graphics in which the data set can be reconstructed from the plot. The stem-and-leaf plot can be viewed as an estimate of the probability density function of Michelson's observations as follows. Rotate your book 90° counterclockwise and look at the leaf values that are above the (now horizontal) line. These leaf values form a crude digital histogram by displaying the various cell frequencies. Not surprisingly, the speed of light estimates have a bell-shaped distribution, centered around the sample mean $\bar{x} = 852$, which is slightly higher than the known population mean of approximately $\mu = 792$. The spread of the probability distribution associated with the observations is partially explained by the error in the measuring device used in the experiment in the late 1800s.

The following two lines of code in R are used to produce a stem-and-leaf plot similar to the one in Figure 1.7.

```
x = MASS::michelson$Speed
stem(x)
```

6	25
7	222444566666788999
8	00000111111111122334444444455555555666777788888888889999
9	001123344455566667888
10	0007

Figure 1.7: Stem-and-leaf plot of the estimates of the speed of light in air.

The data set is stored in a data frame named `michelson` in the `MASS` package and is transferred to the vector `x` of length $n = 100$ by the first statement. Figure 1.8 shows the stem-and-leaf plot produced by the `stem` function. The `stem` function decided that the five cells in the stem-and-leaf plot plotted by hand in Figure 1.7 were not adequate, and decided internally to plot ten cells instead. I added a legend on the left that describes the meaning of the smallest and largest data values. This common practice helps the viewer interpret the values in the plot. Typing `help(stem)` at the command prompt in R gives the options associated with the `stem` command.

6 | 2 = 299,620 km/sec	6	2
	6	5
	7	222444
	7	566666788999
	8	000001111111111223344444444
	8	555555566677778888888888999
	9	0011233444
	9	55566667888
	10	000
10 | 7 = 300,070 km/sec	10	7

Figure 1.8: Another stem-and-leaf plot of the estimates of the speed of light in air.

Another common statistical graphic that captures the shape of the probability distribution of a data set is the *histogram*. Compared with the stem-and-leaf plot, the histogram is a bit more aesthetically pleasing and allows greater flexibility in choosing the cells that contain the data values. Unlike the stem-and-leaf plot, however, you are not able to recreate the data values from a histogram. The R statements

```
x = MASS::michelson$Speed
hist(x)
```

produce a histogram of the data values, which is plotted in Figure 1.9. The units on

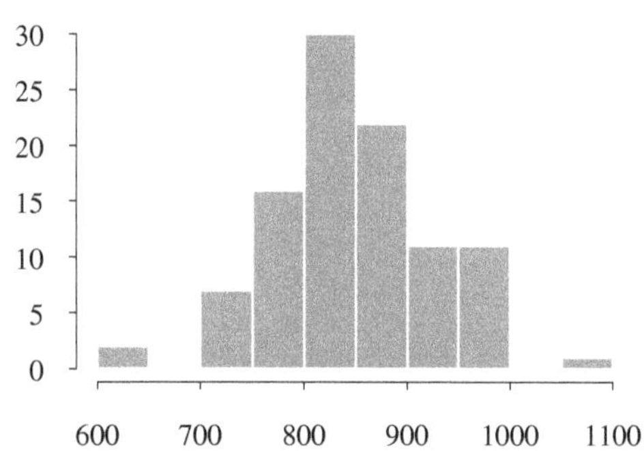

Figure 1.9: Histogram of speed of light estimates (km/sec over 290,000) in air.

the horizontal axis are the speeds in excess of 290,000 kilometers per second and the units on the vertical axis are the counts of observations falling into the cells $(600, 650]$, $(650, 700]$, ..., $(1050, 1100]$. The units on the vertical axis are the number of observations falling in each bin. Since the histogram is the statistical analog of the probability mass function or probability density function $f(x)$, some statisticians prefer to alter the vertical axis units so that the area under the histogram is one, just as it is for $f(x)$. Simply add the `probability = TRUE` option in the call to `hist` to alter the vertical scale in this fashion. This allows the analyst to superimpose a hypothesized probability density function on top of such a histogram to illustrate a potential population probability distribution that might have produced such a data set.

The main strength of the histogram is that it is one of the better ways of assessing the shape of a probability distribution. This shape can lead to a short list of probability models for the population distribution. Histograms also have several weaknesses. The first weakness concerns the arbitrary grouping of observations into cells. Choosing the number of cells and cell boundaries for a histogram is rather important for the following reasons.

- Choosing too few cells can mask important features of the data set. The default number of cells was chosen by R in Figure 1.9. Sturges's rule suggests using $1 + \log_2 n$ cells.
- Choosing too many cells can highlight the natural *random sampling variability* (that is, the chance fluctuations in the data associated with a finite sample size) rather than the shape of the parent probability distribution.

Even if you choose the right number of cells, shifting the cells slightly to the left or the right can cause subtle or even dramatic differences in the shape of the histogram. For the aesthetics of the histogram, it is preferable to have round numbers as the cell boundaries. The number of cells should increase with the sample size because random sampling variability is less pronounced for larger data sets.

The second weakness associated with histograms is that they are notoriously bad at comparing multiple populations; they do not stack well. One exception to this is the case of two populations, where the two histograms can be placed side-by-side. This case is illustrated in the following example.

Example 1.7 Demography is the statistical study of human populations. One aspect of human populations that can be summarized by two histograms is the age distribution for a particular sub-population. A *population pyramid* or *age structure diagram* consists of two histograms rotated 90° and placed side-by-side. These diagrams illustrate the longevity, birth rate, and probability distribution of the ages of the population by gender, race, etc. When two population pyramids are compared for the same sub-population at two different points in time, they reveal population dynamics due to various factors such as medical advances or immigration.

Figure 1.10 contains a population pyramid for France on January 1, 1960 using data from `www.insee.fr`. There are 100 ages plotted on the vertical axis, and corresponding male populations on the left and female populations on the right, in thousands, on the horizontal axis. The most pronounced features of this statistical graphic are the nearly-symmetric dents in the populations that achieve their lowest level at ages 19 and 44. These dents cannot be attributed to random sampling variability because the sample size is so large. So what caused the dents?

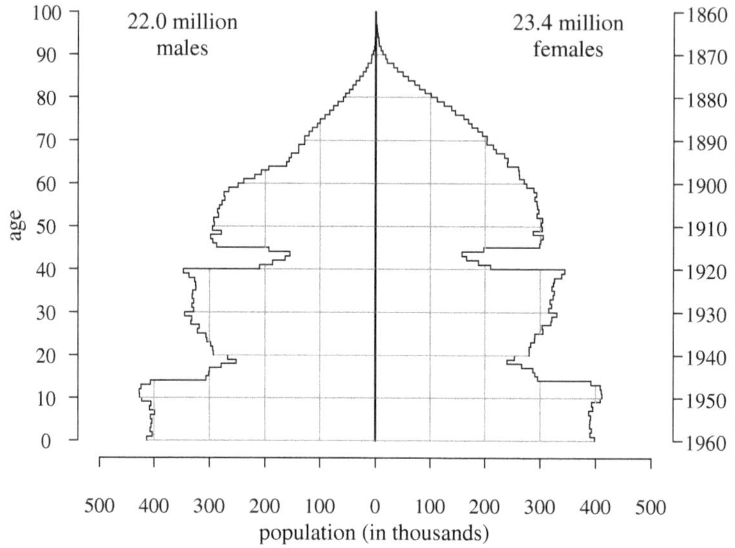

Figure 1.10: Population pyramid for France, January 1, 1960.

Using the birth year on the right-hand vertical scale, it can be concluded that the tragic effects of two world wars fought on French soil are the cause of the dents in the population pyramid. The durations of World War I (1914–1918) and World War II (1939–1945) coincide with the dents. There was a decreased birth rate during the wars, and a post-war baby boom after each war. Both male and female births are affected equally by the decline and subsequent bump in the birthrate. Looking more closely at the elderly population at the top of the graphic also indicates that there are significantly more elderly women than men. Could this be due to the increased longevity that women have over men, or are other factors at play? France sustained 1.7 million casualties in World War I and 600,000 casualties in World War II. These casualties would also account for some of the difference between the male and female population sizes. If you would like to see population pyramids for all countries for any year between 1950 and 2100, visit `populationpyramid.net`.

Population pyramids are useful ways of summarizing the age distribution of a population. Some extensions are listed below.

- Although a matter of taste, the gray grid lines inside of the pyramid could be removed, giving more emphasis to the shape of the two histograms.
- Labels are often added to a population pyramid to explain features of the pyramid (for example, dents).
- Population pyramids can be viewed over time, giving a dynamic sense as to how population is changing over time. In the case of the population pyramid of France on January 1, 1960, the dents would float upward as time advances.

- The two sides of the pyramid are not limited to male and female populations. The left side could be for left-handed individuals and the right side could be for right-handed individuals.
- Likewise, the vertical axis need not be age for constructing a pyramid of this type. The vertical axis could be SAT scores for a cohort of students applying to a particular university.
- The horizontal axis on most population pyramids is the population size, as in the previous example. But this need not be universally true. If the histogram on the left corresponds to the population (both male and female) in Europe, and the histogram on the right corresponds to the population (both male and female) in New Zealand, then it makes sense to use percent of population on the horizontal axis in order to assess the difference between the two age distributions.

Histograms are not an appropriate vehicle for comparing more than two populations simultaneously. A *box plot* (also known as the *box and whisker plot*) is a convenient statistical graphic for comparing multiple data sets simultaneously. Five numbers are used to summarize a data set in a box plot:

- the sample minimum (the smallest observation),
- the first quartile (the sample 25th percentile),
- the second quartile (the sample 50th percentile, also known as the sample median),
- the third quartile (the sample 75th percentile),
- the sample maximum (the largest observation).

Like the stem-and-leaf plot and the histogram, a box plot is "nonparametric" or "distribution-free" in the sense that no assumptions are made about the population distribution from which the data was drawn. Box plots can be drawn horizontally or vertically. A scale is typically included near the box plot. Returning to the speed of light data set, a box plot can be drawn with the R commands

```
x = MASS::michelson$Speed
boxplot(x)
```

which display the vertically-oriented box plot shown in Figure 1.11. The vertical axis displayed is the data values in km/sec in excess of 290,000 km/sec.

The difference between the third quartile and the first quartile, which captures the middle 50% of the data, is known as the *interquartile range* and is often abbreviated IQR. The interquartile range is a measure of the *dispersion, variability*, or *spread* of the data values. In a box plot, this is the height of the box when the box plot is arranged vertically as in Figure 1.11.

The symmetry of the data set is also apparent from the box plot. If the sample median (the middle line in the box) is about the average of the two ends of the box and is also about the average of the two extreme observations, then it is reasonable to conclude that the data set was drawn from a nearly symmetric distribution, and therefore the population skewness of the population probability distribution is approximately zero.

The box plot described here is the most common, but there are several variations on box plots. Although the ends of the boxes are universally the first and third quartiles, the whiskers that extend from the box do not always extend to the extremes. One common practice is to let the whiskers extend to a certain sample percentile, then include data values beyond this percentile with dots. Another practice is to place a notch in the box around the sample median value to indicate its

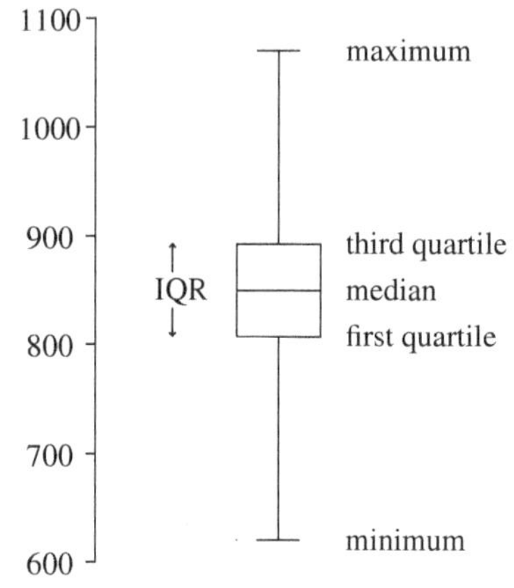

Figure 1.11: Box plot of the estimates of the speed of light in air.

precision. Still another practice is to let the width of the box reflect the sample size. More detail is given on most websites that describe box plots.

The real value of a box plot is not in just describing a single data set, as in Figure 1.11, but rather in comparing two or more data sets. The following example illustrates such a comparison.

Example 1.8 The average daily maximum temperature (in degrees Fahrenheit) for three U.S. cities (Monterey, California; Portland, Oregon; New York City, New York) in the year 2000 is given in Table 1.6.

	Jan	Feb	Mar	Apr	May	Jun	Jul	Aug	Sep	Oct	Nov	Dec
Monterey	58.4	61.0	61.7	65.5	66.1	68.8	64.8	68.2	73.8	65.7	61.1	62.3
Portland	45.1	50.2	53.9	64.1	66.0	76.4	78.4	78.6	73.9	63.2	49.5	45.5
New York	37.9	43.7	54.9	58.2	71.6	78.8	79.0	78.6	72.9	64.4	50.9	37.2

Table 1.6: Average maximum temperature in three cities in 2000.

Plotting the temperatures for the three cities over time would be cluttered because the curves would intersect one another at several points in time. A more elegant statistical graphic is to compare box plots of the data for the three cities. Although the time dependency is lost, the extreme values and quartiles are easily compared. The R statements

```
m = c(58.4, 61.0, 61.7, 65.5, 66.1, 68.8, 64.8, 68.2, 73.8, 65.7, 61.1, 62.3)
p = c(45.1, 50.2, 53.9, 64.1, 66.0, 76.4, 78.4, 78.6, 73.9, 63.2, 49.5, 45.5)
n = c(37.9, 43.7, 54.9, 58.2, 71.6, 78.8, 79.0, 78.6, 72.9, 64.4, 50.9, 37.2)
boxplot(m, p, n)
```

produce a plot similar to that shown in Figure 1.12, where the three box plots are oriented horizontally. Since there are $n = 12$ data values for each city, the sample median

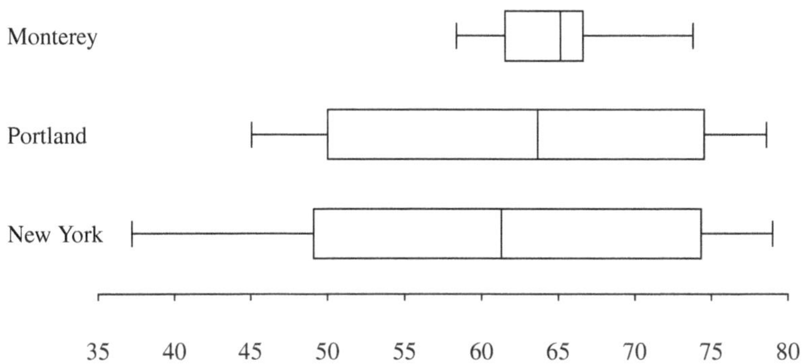

Figure 1.12: Box plots for the monthly average high temperature for three U.S. cities.

is calculated by averaging the two sorted middle values. One immediate conclusion that can be drawn from the box plots is that the sample medians are all quite close. The difference between the highest sample median (Monterey) and the lowest sample median (New York City) is less than four degrees. Even though the central tendency is nearly the same, the variability is drastically different. Monterey, California has one of the world's best climates, as reflected by both the top box plot and the associated housing prices. Portland is next in terms of variability, followed by New York City, whose residents endure the harshest winters and hottest summers of the three cities. What is the cause of the difference in variability? The earth's rotation gives cities on the west coast the advantage of warmer winters and cooler summers because of the damping effect on the temperature of the air that passes over the Pacific Ocean. The effect of the Atlantic Ocean on temperatures in New York City is minimal.

Box plots provide a way of comparing multiple probability distributions simultaneously. They also provide a way of describing a probability distribution that avoids the binning of data (that is, placing data values into cells) that is present in histograms. There is another statistical graphic that avoids binning data. The *empirical cumulative distribution function* is the statistical analog of the cumulative distribution function. The empirical cumulative distribution function is a step function with upward steps of height $1/n$ at each data value.

Another way of thinking about why the empirical cumulative distribution function is a reasonable estimate of $F(x)$ is as follows. If you just had the data values $x_1, x_2, \ldots, x_n$ and wanted to generate a "best guess" for $f(x)$, one option is to create an empirical probability mass function $\hat{f}(x)$ that has mass $1/n$ at each data value. (Statisticians use the hat, or caret, above f to indicate that $\hat{f}(x)$ is an estimator of $f(x)$.) This works fine if each of the data values is unique. If there are d values that are tied, then there will be a mass value of d/n at the tied value. Now what would the empirical cumulative distribution function $\hat{F}(x)$ associated with this empirical probability mass function look like? It would be exactly the one described above. It would have an upward step of height $1/n$ at each unique data value and an upward step of height d/n when there are d tied data values.

The good news about the empirical cumulative distribution function is that no binning is required, which means that an empirical cumulative distribution function is unique for a particular data set. The bad news is that its shape is not quite as distinctive as the histogram.

Example 1.9 Returning to Michelson's $n = 100$ estimates of the speed of light measured in excess of 290,000 km/sec, the empirical cumulative distribution function can be plotted with the R function `plot.ecdf` as shown in the code below.

```
x = MASS::michelson$Speed
plot.ecdf(x, verticals = TRUE, pch = "")
```

The empirical cumulative distribution function is plotted in Figure 1.13. There are several options for displaying the empirical cumulative distribution function, and the one displayed here has the vertical risers included on the steps. Some prefer these risers left off because the step function, after all, is still a function. This is largely a matter of personal taste.

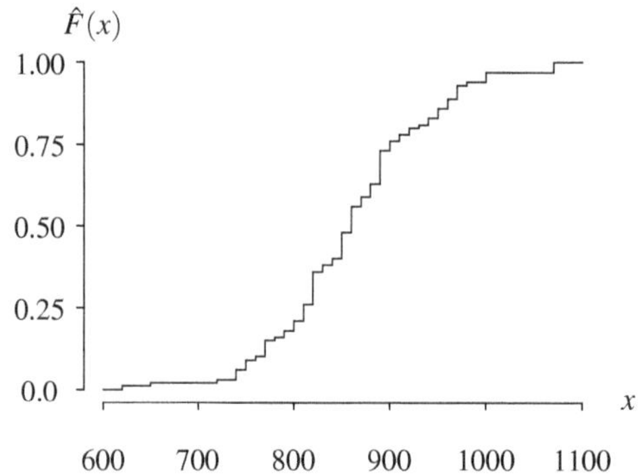

Figure 1.13: Empirical cumulative distribution function for the speed of light data.

Similar to overlaying a probability density function on top of a histogram, one often overlays a theoretical cumulative distribution function on top of an empirical cumulative distribution function. This will be illustrated in the next chapter.

This next example is drawn from business and finance. It illustrates the benefit of a logarithmic scale for displaying certain types of data sets.

Example 1.10 The Dow Jones Industrial Average (DJIA), also known as the Dow 30, was devised by Charles Dow and was initiated on May 26, 1896. The average bears Dow's name and that of statistician and business associate Edward Jones. The DJIA is the average stock price of 30 U.S.-based, publicly traded companies, adjusted for stock splits and the swapping of companies in and out of the average so that it adequately reflects the composition of the domestic stock market. These adjustments are made by altering the average's denominator for historical continuity, which is now much less than 30. The value of the denominator is given every day in the *Wall Street Journal*.

The evolution of the DJIA is not a true reflection of the yield of the 30 stocks because two important factors are not incorporated into the average. First, the average does not factor in dividends that are paid by some of the 30 stocks. Second, the average does not factor in inflation, which erodes the true return that a stock investment provides.

If dividends were factored in, the DJIA would be much higher than it is presently; if inflation were factored in, the DJIA would be much lower than it is presently.

This example develops statistical graphics associated with the DJIA that illustrate various ways to view its evolution over time. The first is a plot of the average annual DJIA closing values during the 20th century. This plot is generated with the R code given below.

```
x = 1901:2000
y = scan("djia")
plot(x, y, type = "l")
```

The file `djia` contains the 100 annual average closing values. The resulting graph is shown in Figure 1.14. Data sets of this nature in which a response variable is plotted over time are known as *time series*. Most economics and statistics departments at universities offer classes titled "time series analysis" in which probabilistic models are developed for describing a time series.

The DJIA had a sample mean closing value of 69.52 during 1901 and a sample mean closing value of 10731.15 during 2000. The linear vertical scale that is used in Figure 1.14 obscures most of the variability of the DJIA during the first half of the century. The graph can be made more meaningful by using a logarithmic scale on the vertical axis. This is accomplished by adding the `log = "y"` parameter to the `plot` command in the R code, resulting in the graph shown in Figure 1.15. Labels have been added to help highlight events that might have influenced the DJIA.

The stock market crash in October of 1929 that initiated the Great Depression is much more pronounced in Figure 1.15. The DJIA had peaked with a close of 381.20 on

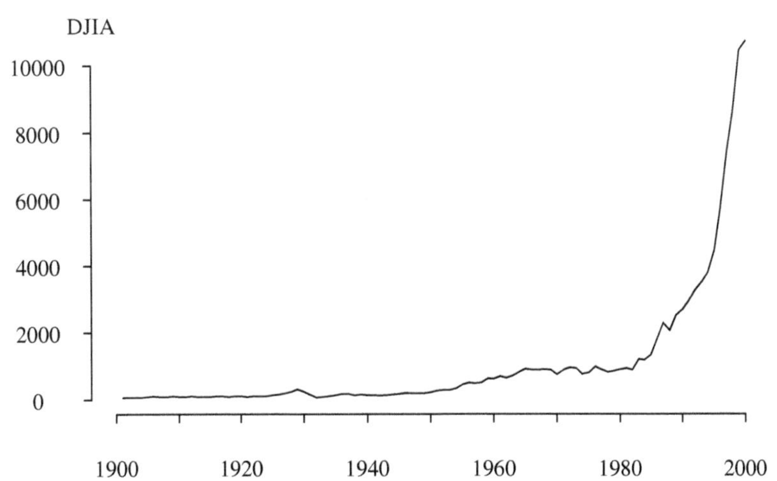

Figure 1.14: Dow Jones Industrial Average, 1901–2000.

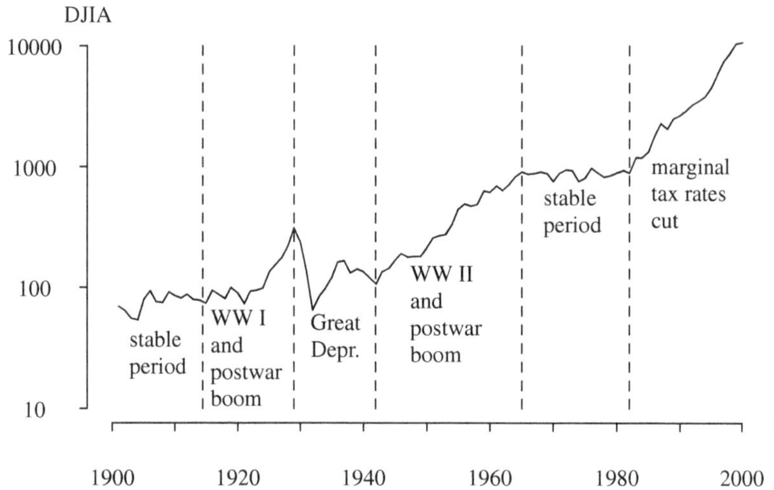

Figure 1.15: Dow Jones Industrial Average, 1901–2000.

September 3, 1929. The market bottomed out on July 8, 1932 when it closed at 41.20, which corresponds to a loss of almost 90%. Each of the two World Wars fought during the twentieth century was followed by a sustained bull market in the DJIA. The top marginal income tax rate was lowered from 70% to 28% and the federal budget was brought into balance in the 1980s and 1990s, resulting in a prolonged growth in the DJIA.

The next statistical graphic provides more detail associated with the DJIA during the first year of the twenty-first century. The DJIA closed on December 29, 2000, the last trading day of the century, at 10787.99. The DJIA closed on December 31, 2001 at 10021.57. This 7.1% decline is in part due to the terrorist attacks on the U.S. on September 11, 2001. The 7.1% decline is an average for the 30 stocks. Some performed better and some performed worse. The stocks that comprised the DJIA during 2001 are given in Table 1.7. Their ticker symbols are given in parentheses. General Electric is the only company that was in the original DJIA from its inception in 1886. A statistical graphic can be devised that captures the following five pieces of information about the 30 stocks comprising the DJIA from December 29, 2000 to December 31, 2001:

- the stock ticker symbol,
- the market sector,
- the absolute market capitalization (by including a legend),
- the relative market capitalization,
- the one-year performance.

This graphic is shown in Figure 1.16. There are 30 rectangles for each of the 30 stocks in the DJIA. The area of each rectangle is a monotonically increasing function of the

Alcoa (AA)	Allied Signal (ALD)	American Express (AXP)
AT&T (T)	Boeing (BA)	Caterpillar (CAT)
Citigroup (C)	Coca-Cola (KO)	DuPont (DD)
Exxon (XOM)	General Electric (GE)	General Motors (GM)
Hewlett-Packard (HPQ)	Home Depot (HD)	Intel (INTC)
IBM (IBM)	International Paper (IP)	Johnson & Johnson (JNJ)
J.P. Morgan (JPM)	Kodak (EK)	McDonald's (MCD)
Merck (MRK)	Microsoft (MSFT)	3M (MMM)
Philip Morris (PM)	Procter & Gamble (PG)	SBC Communications (SBC)
United Technologies (UTX)	WalMart (WMT)	Walt Disney (DIS)

Table 1.7: Dow Jones Industrial Average Companies in 2001.

market capitalization (that is, the value of the publicly-traded shares of stock). The scale of the market capitalization is seen in the legend in the lower-right-hand corner. Thinner lines separate individual stocks. Thicker lines separate the stocks by sector, and the sector labels are given outside of the large rectangle in italics. Each of the 30 rectangles contains a ticker symbol to identify the stock and the performance during 2001 as a percentage. Of the 30 stocks, 11 stocks increased throughout the year and 19 stocks decreased. The energy sector was the strongest throughout 2001; the financial sector was the weakest throughout 2001. Rectangular-shaped plots of this nature have generally replaced the more traditional pie diagram because of their ability to easily capture additional information in the smaller rectangles.

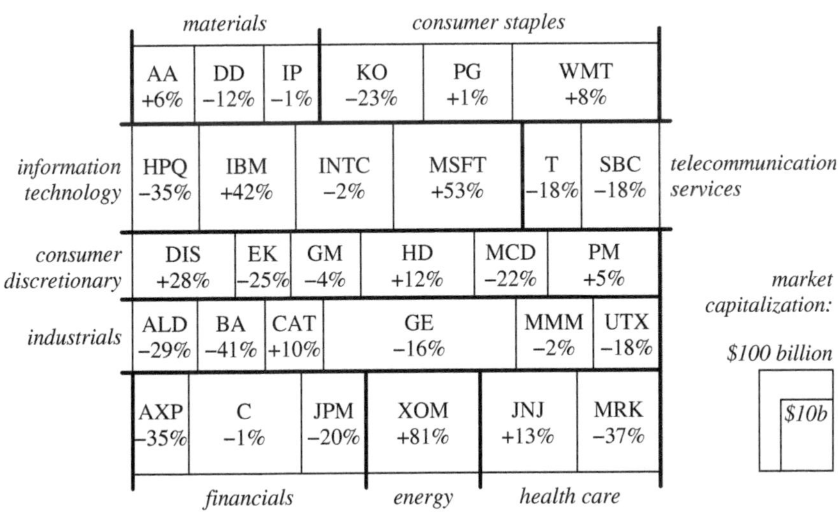

Figure 1.16: Dow Jones Industrial Average component stocks performance in 2001.

The applications of statistics span a wide range of disciplines. So far we have encountered data sets associated with comparing population sizes, showing the dynamics associated with a basketball game, solving a mystery concerning birth spacings, using a mosaic plot to visualize the relationship between hair and eye color, displaying word frequencies in a book, analyzing estimates of the speed of light in air, comparing the age distribution of men and women in France, comparing weather data for three U.S. cities, and displaying stock market data. This section ends with one final example that concerns the display of the estimate of a mixed discrete–continuous probability distribution. Mixed discrete–continuous random variables occur, for example, in queueing (the waiting time for a server), meteorology (the total rainfall in one day), and reliability (the lifetime of a product). In all three of these examples, there is a non-zero probability that the random variable will assume a value of zero, which accounts for a discrete portion of the probability distribution.

Example 1.11 A play-by-play account of the 2016 National Football League (NFL) regular-season games contains the field position for $n = 2593$ kickoffs. Kickoffs are further subdivided into those returned in the field of play and those with discrete categorical outcomes (safeties, touchbacks, out-of-bounds, and touchdowns). When a run-back is attempted by the kickoff returner and the return is concluded in the field of play (usually by tackling the kickoff returner but occasionally by a fumble recovery), the resulting field position is a continuous random variable with a support ranging from 0 to 100, measured as the distance from the return team's end zone. The categorical outcomes are associated with a discrete random variable with four mass values (0, 25, 40, and 100). Table 1.8 highlights the division between the continuous (1047 observations) and discrete (1546 observations) portions of the probability distribution and shows the frequency of the various outcomes during the 2016 season.

Type	Category	Starting field position	Frequency	Probability
Continuous	Returned in the field of play	$(0, 100)$	1047	$\frac{1047}{2593} \cong 0.404$
Discrete	End zone (return team)	0	3	$\frac{3}{2593} \cong 0.001$
	Touchback	25	1518	$\frac{1518}{2593} \cong 0.585$
	Out-of-bounds	40	18	$\frac{18}{2593} \cong 0.007$
	End zone (kicking team)	100	7	$\frac{7}{2593} \cong 0.003$
		Total:	2593	$\frac{2593}{2593} = 1.000$

Table 1.8: NFL 2016 regular-season kickoff starting field positions.

Let the random variable X be the starting field position following a NFL kickoff during the 2016 season measured in yards from the return team's end zone (regardless of whether a turn-over occurs). We wish to construct a statistical graphic that captures an estimate of the probability distribution of the starting field position X from the $n = 2593$ data values. Since the probability distribution of X is a mixed discrete–continuous random variable, a reasonable estimate of the contribution of the two parts of the probability distribution is the finite mixture

$$\hat{f}(x) = \frac{1047}{2593}\hat{f}_C(x) + \frac{1546}{2593}\hat{f}_D(x),$$

where $\hat{f}_C(x)$ and $\hat{f}_D(x)$ are

$$\hat{f}_C(x) = \begin{cases} \text{kernel density function} \\ \text{of outcomes returned} \\ \text{in the field-of-play} \end{cases} \quad \text{and} \quad \hat{f}_D(x) = \begin{cases} 3/1546 & x = 0 \\ 1518/1546 & x = 25 \\ 18/1546 & x = 40 \\ 7/1546 & x = 100 \end{cases}$$

and the hats denote estimators. Figure 1.17 displays the estimator for the 2016 NFL data. Even though a histogram would have worked perfectly fine for the estimate of the continuous portion of X, using a *kernel density function* emphasizes the continuous nature of the spotting of the ball on the field of play. Two different vertical scales (the scale for the continuous portion is on the left, labeled PDF for probability density function, and the scale for the discrete portion is on the right, labeled PMF for probability mass function) were necessary to avoid having the continuous portion crunched down to the horizontal axis. The scales were selected to reflect the approximately 40/60 split between the continuous and discrete portions of the probability distribution. A square root scale on the discrete axis makes it easier to differentiate between the three discrete-but-unlikely outcomes. Two distinct modes are evident in the continuous portion. The first—with a mode at the 22 yard-line—represents distances most often attained before a returner is tackled. The second—just past mid-field—is a consequence of the 54 onside kick attempts during the 2016 season.

Well-designed statistical graphics are of benefit to statisticians and non-statisticians alike. They display aspects of a data set that are often difficult to see by viewing the raw data or by viewing the data in tables. Statistical graphics are capable of displaying multiple variables simultaneously, and it is often apparent from a display how the variables are related to one another.

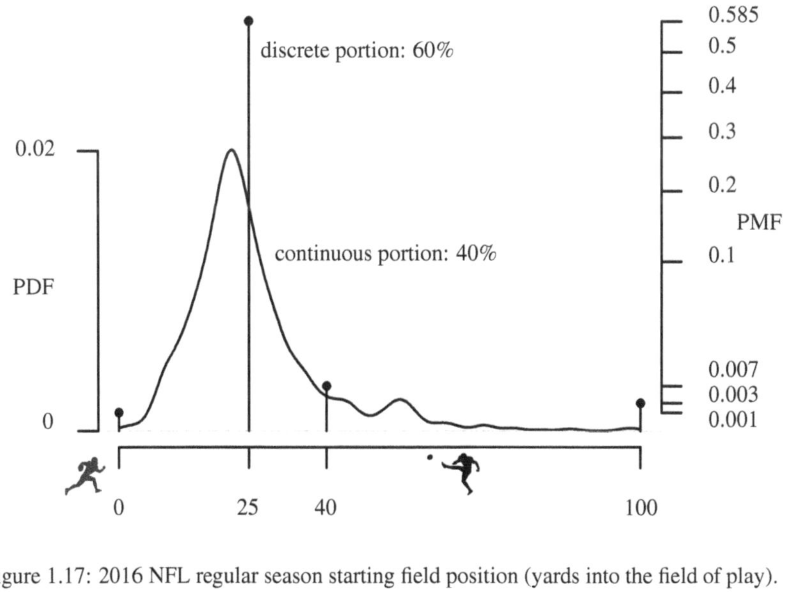

Figure 1.17: 2016 NFL regular season starting field position (yards into the field of play).

1.2 Random Sampling, Statistics, and Sampling Distributions

Statistical graphics give you a toolbox of techniques that are useful for visually summarizing a data set. Unfortunately, two knowledgeable people might draw opposite conclusions from a well-designed statistical graphic. This begs for a mathematical framework, which draws from probability theory, that can be used to remove personal opinion from the process. Two examples of questions that can be addressed by statistics are given below.

- How long, on average, does a particular brand of 60-watt light bulbs last under commonly-encountered environmental conditions?

- Should radiation or chemotherapy or both be used to treat a particular cancer?

There are lots of loose ends that have not been determined with these questions. Are the light bulbs burned continuously? Are the light bulbs used in an optimal environment (like a living room) or a high temperature or high vibration environment (like a rocket launch)? Are all people having the cancer women with diabetes, high blood pressure, and low cholesterol of approximately the same age? In order to answer the statistical questions, it is important to first pin down the details associated with the setting.

Statistical models used to describe data behave in a similar manner to mathematical models in economics, chemistry, or physics. A good model will be a close match to what is actually observed. A bell-shaped histogram, for example, is evidence that a normal population probability model could be an appropriate modeling assumption. The histogram would provide even more evidence, however, if there were $n = 1000$ observations producing the bell-shaped histogram rather than only $n = 20$ observations.

We have thus far avoided the question of how data values are collected. Now is the time to address that question. There are many different sampling mechanisms that can be used to cull the data. To keep the mathematics simple initially, we assume that *univariate data* is being collected on n subjects. This leads to the following definition of a random sample.

Definition 1.1 Let $X_1, X_2, \ldots, X_n$ be mutually independent random variables, each with the same but possibly unknown probability distribution described by $f_X(x)$. Realizations of the random variables $X_1, X_2, \ldots, X_n$ constitute a *random sample*.

There are several loose ends associated with the definition of a random sample that are outlined below.

- The integer n is known as the *sample size*.

- Some textbook authors refer to a random sample as a *simple random sample* (SRS).

- Mutually independent random variables, each with the same probability distribution, are often described using the abbreviation *iid* for "independent and identically distributed."

- Definition 1.1 implies that the joint distribution of the random sample can be found by

$$f(x_1, x_2, \ldots, x_n) = f_X(x_1)f_X(x_2)\ldots f_X(x_n).$$

- Definition 1.1 applies equally well to random sampling from discrete populations, continuous populations, and mixed discrete–continuous populations.

- The random sampling of vectors, that is, multivariate random variables, is considered in an advanced course.

- The assumption of mutual independence implies that each value must be sampled in a manner so that it is not influenced by any of the other values.

- Taking a "random sample" in industrial applications oftentimes requires significant effort. If a farmer delivers a truckload of potatoes to a potato chip company, the n potatoes sampled for quality should be selected from random positions in the truck, perhaps generated by a random number generator.

- Beware of selection bias. The famous headline "Dewey Defeats Truman" from the *Chicago Tribune* after the 1948 U.S. Presidential election was partially based on polling. If these polls were conducted by phone and more Republicans had phones than Democrats, then the estimate of the probability of Dewey defeating Truman would be biased.

- Beware also of response bias. Questions like "Do you use LSD?" or "Did you cheat on the French exam?" might not yield an honest response, which would lead to a biased estimator of the associated probabilities.

- Count the costs associated with collecting data. Sometimes a simple survey question can produce a data value cheaply. On the other hand, automobile safety data might involve crashing a vehicle into a wall to obtain data. *Destructive testing* destroys a test unit; *nondestructive testing* retains a test unit.

- Many statisticians follow the convention that

  $$X_1, X_2, \ldots, X_n$$

  are the data values in the abstract—they are random variables and hence described by uppercase letters. Realizations of these random variables that assume specific numerical values, however, are denoted by

  $$x_1, x_2, \ldots, x_n.$$

  This convention will be followed in this text.

- The data values $x_1, x_2, \ldots, x_n$ are known as a "data set."

Once a random sample has been collected, there are two potential next steps. The first is to construct one or more statistical graphics, as introduced in the previous section. The second is to compute one or more *statistics*, which are defined next.

Definition 1.2 Consider the data values $X_1, X_2, \ldots, X_n$. A *statistic* is some function of the data values that does not depend on any unknown parameter(s).

The key to this definition of a statistic is that no unknown parameters are involved. So, for example, the expressions

$$\frac{X_1 + X_2 + \cdots + X_n}{n} \qquad n \cdot X_{(1)} \qquad \frac{X_{(1)} + X_{(n)}}{2} \qquad \prod_{i=1}^{n} X_i$$

are all statistics because they do not involve any unknown parameters. (Recall that the order statistic $X_{(1)}$ is the smallest member of a data set and the order statistic $X_{(n)}$ is the largest member of a data set.) On the other hand, the expressions

$$\frac{\bar{X}-\mu}{\sigma/\sqrt{n}} \qquad \theta \cdot X_{(1)} \qquad \frac{\bar{X}}{\mu} \qquad \frac{X_7-\mu}{S/\sqrt{n}}$$

are not statistics because they involve the unknown parameters μ, σ, and θ.

This is an appropriate time to delineate the difference between probability and statistics. Figure 1.18 contains two ovals. The larger oval on the left represents the *population*. As one particular instance, the population might consist of the weight, in pounds, of every person in the world. The smaller oval on the right represents a *sample* of n values taken from the population. Using the current instance, the sample might be the weights of n people sampled at random and without replacement from the population. Populations and samples are fundamentally different entities:

- the population in the left-hand oval is often described by *parameters*, such as the population mean μ and the population variance σ^2, which are fixed constants;
- the sample in the right-hand oval can be described by *statistics*, such as the sample mean $\bar{X}$ and the sample variance S^2, which are random variables that vary from one sample to the next.

The arrow that points to the right represents the application of probability theory. Here is a typical probability problem:

> Ten people crowd into a small elevator. Their weights, in pounds, are mutually independent and identically distributed normal random variables, each with population mean $\mu = 140$ pounds and population standard deviation $\sigma = 30$ pounds. What is the probability that the elevator capacity of 1500 pounds is exceeded?

For this particular probability problem, the information about the population probability distribution, which is $N(140, 900)$, is known. The question asks about a sample statistic $X_1 + X_2 + \cdots + X_{10}$,

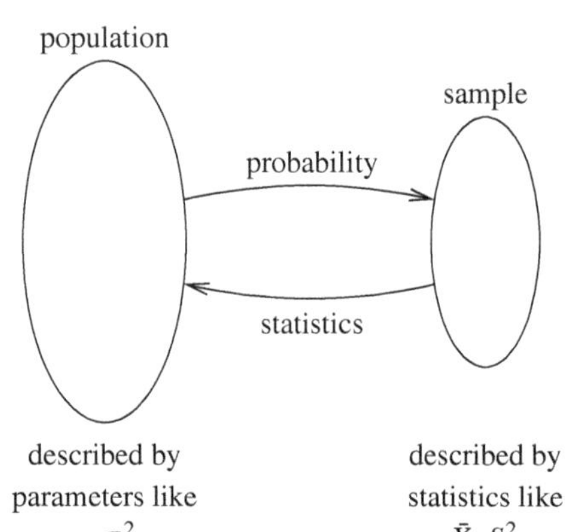

Figure 1.18: The difference between probability and statistics.

where X_i is the weight of the ith person on the elevator. More specifically, the question asks for $P(X_1 + X_2 + \cdots + X_{10} > 1500)$. Referring back to Figure 1.18, knowledge about the probability distribution of the population is used to answer a question concerning a statistic calculated from a sample. The arrow that points to the left represents the application of statistical theory. Here is a typical statistics problem:

> Ten people crowd into a small elevator. Their weights are $x_1 = 220$, $x_2 = 107$, $x_3 = 155$, $\ldots$, $x_{10} = 129$ pounds. If nothing is known about the population probability distribution of the weights, give estimates of the population mean μ and the population variance σ^2, along with some indication of the precision of the estimates.

In this particular setting, *nothing* is known about the population of weights. Questions are being asked about the population based only on the ten data values. Referring back to Figure 1.18, the data values in the sample are being used to answer questions about the population. This process is often referred to as *statistical inference* because a conclusion is being inferred about the population based on the sample. In some settings, it might be reasonable to assume that the ten values constitute a random sample.

The interest in computing a statistic is oftentimes to gain information about some unknown parameter. For example,

- the sample mean $\bar{X}$ is often used to estimate the population mean μ,
- the sample median M is often used to estimate the population median $x_{0.5}$,
- the sample variance S^2 is often used to estimate the population variance σ^2,
- the sample proportion $\hat{p}$ is often used to estimate the population proportion p.

An important distinction should be made between statistics and the quantities that they are estimating: the statistics (like $\bar{X}$ and S) are random variables, but the parameters that they are estimating (like μ and σ) are constants, which are typically unknown. Statistics take on different values from one sample to the next; population parameters assume just a single value. The sample mean $\bar{X}$ and the sample variance S^2 are formally defined in the next two sections.

Since a statistic is a random variable, its probability distribution is often of interest. The probability distribution of a statistic is called its *sampling distribution*. The following three examples concern a single random experiment—rolling a fair die five times—and highlight the probability distribution of three different statistics that can be gleaned from the five data values. The sample size $n = 5$ is rather small, but the simplicity of this setting allows us to calculate exact sampling distributions of the statistics.

Example 1.12 Consider the random variables X_1, X_2, X_3, X_4, X_5, which are the outcomes of five rolls of a fair die. What is the sampling distribution of the statistic $\bar{X}$, the sample mean?

The way that the data is collected (rolling a fair die five times) indicates that the five random variables are mutually independent, so the five values constitute a random sample. The first step in finding the probability distribution of the sample mean $\bar{X}$ is to determine the support (possible values) of the random variable $\bar{X}$. Since the numerator of

$$\bar{X} = \frac{X_1 + X_2 + X_3 + X_4 + X_5}{5}$$

can assume the values 5, 6, ..., 30, the support of $\bar{X}$ is

$$\mathcal{A} = \left\{ x \,\middle|\, x = 1, \frac{6}{5}, \frac{7}{5}, \ldots, \frac{29}{5}, 6 \right\}.$$

The next step is to determine the probabilities associated with each element in the support. Begin with $\bar{X} = 1$. There is only one way to achieve a sample mean of 1, which is the outcome $(X_1, X_2, X_3, X_4, X_5) = (1, 1, 1, 1, 1)$, so

$$P(\bar{X} = 1) = \frac{1}{6^5}.$$

Now consider $\bar{X} = 6/5$. This value for $\bar{X}$ can only be achieved with 4 ones and a single 2, for example the outcome $(X_1, X_2, X_3, X_4, X_5) = (1, 2, 1, 1, 1)$. Since the 2 can occur on any one of the five rolls,

$$P\left(\bar{X} = \frac{6}{5}\right) = \frac{5}{6^5}.$$

Next consider $\bar{X} = 7/5$. There are two ways to achieve this sample mean: 4 ones and a single 3, or 3 ones and 2 twos. So the probability that $\bar{X} = 7/5$ is

$$P\left(\bar{X} = \frac{7}{5}\right) = \frac{5}{6^5} + \frac{10}{6^5} = \frac{15}{6^5}$$

because there are $\binom{5}{2} = 10$ ways to place the 2 twos in the sequence of five rolls. One way to proceed is to continue in this fashion, which is a mind-numbing exercise. A much more efficient way to proceed is to use the following APPL code to calculate the probability mass function of $\bar{X}$.

```
X := UniformDiscreteRV(1, 6);
Y := ConvolutionIID(X, 5);
g := [[x -> x / 5], [5, 30]];
Xbar := Transform(Y, g);
```

This code returns the symmetric probability mass function for $\bar{X}$ as

$$f_{\bar{X}}(x) = \begin{cases} 1/7776 & x = 1 \\ 5/7776 & x = 6/5 \\ 15/7776 & x = 7/5 \\ 35/7776 & x = 8/5 \\ \vdots & \vdots \\ 1/7776 & x = 6. \end{cases}$$

The probability mass function is plotted in Figure 1.19. Even though the sample size of $n = 5$ is quite small, the effects of the central limit theorem are already being seen in the somewhat bell-shaped probability mass function.

In a statistical setting, you typically get only one instance of the statistic $\bar{X}$. For example, if one rolls a fair die five times, they might get

$$x_1 = 2, x_2 = 6, x_3 = 1, x_4 = 3, x_5 = 2,$$

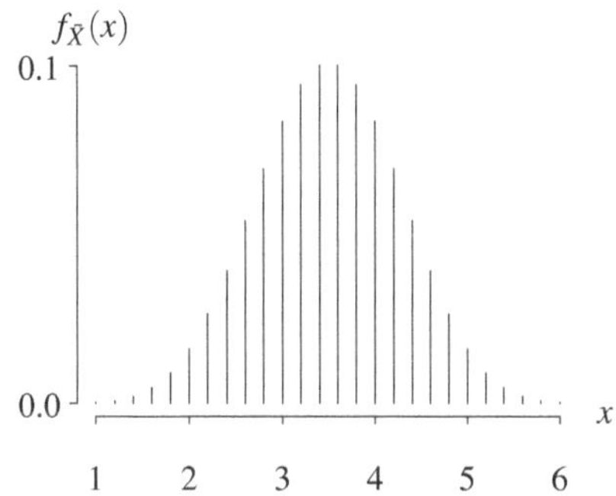

Figure 1.19: Sampling distribution of the sample mean.

which corresponds to $\bar{x} = 14/5 = 2.8$. Knowing where $\bar{x} = 2.8$ falls in the sampling distribution of $\bar{X}$ can be helpful in drawing conclusions concerning these particular rolls of the fair dice.

Here is one particular instance. Let's say a dice manufacturer claims that their dice are fair. Your friend, on the other hand, claims that the dice are loaded and producing too many sixes. As a budding statistician, you decide to purchase a die and roll it five times. If the results of your random experiment are

$$x_1 = 6, x_2 = 6, x_3 = 6, x_4 = 6, x_5 = 6,$$

you would certainly side with your friend. But could the die indeed have been fair as the manufacturer claimed? Possibly, but your work on determining the sampling distribution of $\bar{X}$ under the assumption that the manufacturer is telling the truth indicates that the outcome that you achieved occurs only one time in 7776, making the manufacturer's claim seem rather dubious. We may reject the "null" hypothesis that the die is fair because the likelihood of observing all sixes is extremely small if the die were indeed fair.

As a thought experiment, what would be our conclusion if we rolled all fours? It is for situations like this in which different statistics become valuable tools to support or refute various types of hypotheses.

The next example considers that same random experiment, rolling a fair die five times, but this time uses a different test statistic.

Example 1.13 Consider again the random variables X_1, X_2, X_3, X_4, X_5, which are the outcomes of five rolls of a fair die. What is the sampling distribution of the statistic $X_{(5)} = \max\{X_1, X_2, \ldots, X_5\}$?

The fifth order statistic satisfies the definition of a statistic because it is a function of the data alone and does not involve any unknown parameters. As before, the first step in determining the sampling distribution of the statistic is to determine its support. The largest of five rolls of a fair die has support

$$\mathcal{A} = \{x \,|\, x = 1, 2, 3, 4, 5, 6\}.$$

The next step is to assign probabilities to each of the six values in the support. The only way to obtain a maximum of one is to roll 5 ones, so

$$P\left(X_{(5)}=1\right)=P(X_1=X_2=X_3=X_4=X_5=1)=\frac{1}{6^5}.$$

There are multiple ways for the largest value rolled to be a two. One way is to roll all twos (and there is only one way to do so) and the other ways are various sequences of ones and twos. Using combinations to count all of the possibilities associated with the largest outcome being two gives

$$P\left(X_{(5)}=2\right)=\frac{\binom{5}{5}+\binom{5}{4}+\binom{5}{3}+\binom{5}{2}+\binom{5}{1}}{6^5}=\frac{1+5+10+10+5}{6^5}=\frac{31}{7776}.$$

One can continue in this fashion or use the APPL code given below to calculate the probability mass function of $X_{(5)}$.

```
X := UniformDiscreteRV(1, 6);
Y := MaximumIID(X, 5);
```

This code returns the probability mass function for $X_{(5)}$ as

$$f_{X_{(5)}}(x)=\begin{cases}1/7776 & x=1\\ 31/7776 & x=2\\ 211/7776 & x=3\\ 781/7776 & x=4\\ 2101/7776 & x=5\\ 4651/7776 & x=6.\end{cases}$$

This probability mass function is plotted in Figure 1.20. Not surprisingly, the most likely maximum is $X_{(5)}=6$.

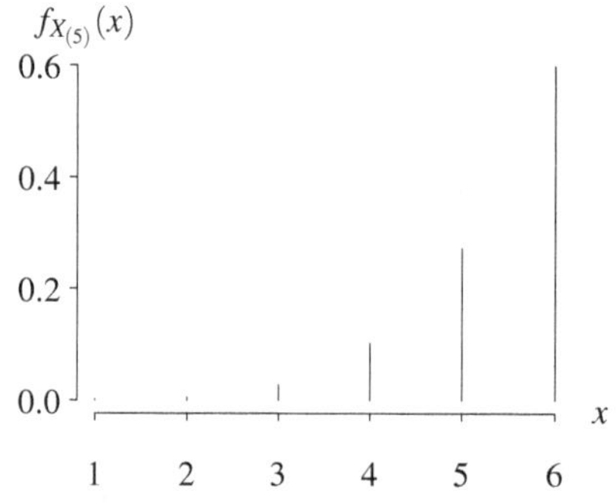

Figure 1.20: Sampling distribution of the sample maximum.

The previous two examples have illustrated two different statistics, the sample mean and the sample maximum, associated with the same random experiment. The next example introduces a new, and somewhat more obscure statistic, the *sample range*, which is used in a field known as *statistical quality control*.

Example 1.14 Consider once again the random variables X_1, X_2, X_3, X_4, X_5, which are the outcomes of five rolls of a fair die. What is the sampling distribution of the statistic $R = \max\{X_1, X_2, \ldots, X_5\} - \min\{X_1, X_2, \ldots, X_5\}$?

The sample range $R = X_{(5)} - X_{(1)}$ satisfies the definition of a statistic given in Definition 1.2 because it is a function of the data only and does not involve any unknown parameters. As before, the first step in determining the sampling distribution of the statistic is to determine its support. The difference between the largest outcome of the five rolls and the smallest outcome of the five rolls has support

$$\mathcal{A} = \{x \mid x = 0, 1, 2, 3, 4, 5\}.$$

The next step is to assign probabilities to each of the six values in the support. The only way to obtain a sample range of $R = 0$ is to roll five identical values, so

$$P(R = 0) = P(X_1 = X_2 = X_3 = X_4 = X_5) = \frac{6}{6^5}.$$

There are multiple ways to obtain a sample range of $R = 1$. Examples include the outcomes $(X_1, X_2, X_3, X_4, X_5) = (1, 1, 1, 2, 1)$ and $(X_1, X_2, X_3, X_4, X_5) = (4, 5, 4, 5, 4)$. Using combinations to count all of the possibilities associated with a sample range of $R = 1$, we obtain

$$P(R = 1) = 5 \cdot \frac{\binom{5}{4} + \binom{5}{3} + \binom{5}{2} + \binom{5}{1}}{6^5} = \frac{150}{7776}.$$

One can continue in this fashion or use the APPL code given below to calculate the probability mass function of R.

```
X := UniformDiscreteRV(1, 6);
R := RangeStat(X, 5);
```

This code returns the probability mass function for R as

$$f_R(x) = \begin{cases} 6/7776 & x = 0 \\ 150/7776 & x = 1 \\ 720/7776 & x = 2 \\ 1710/7776 & x = 3 \\ 2640/7776 & x = 4 \\ 2550/7776 & x = 5. \end{cases}$$

Some might prefer the notation $f_R(r)$ for this probability mass function, but we use $f_R(x)$ to have a consistent index x with the previous two examples. This probability mass function is plotted in Figure 1.21.

The previous three examples have illustrated three statistics and their associated sampling distributions. One key insight here is that when a random experiment is conducted, we get to calculate just one instance of the test statistic. The importance of the sampling distribution is to let you know whether the value of the statistic that you observe is common or rare. For example, if you roll a fair die five times and get all ones, then the three statistics give

$$\bar{x} = 1 \qquad x_{(5)} = 1 \qquad r = 0.$$

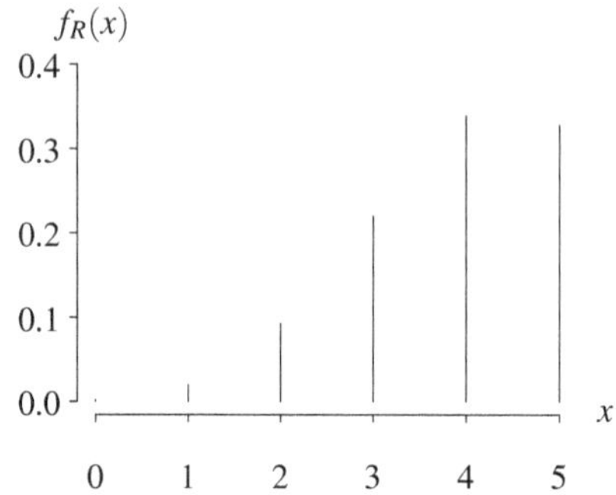

Figure 1.21: Sampling distribution of the sample range.

Looking at the three sampling distributions from the previous three examples, the values of these particular three statistics correspond to very unlikely events because the sampling distributions tell us

$$P(\bar{X}=1)=\frac{1}{7776} \qquad P\left(X_{(5)}=1\right)=\frac{1}{7776} \qquad P(R=0)=\frac{6}{7776}.$$

Different statistics are used to detect different types of "rare" events. For example, if you roll a fair die five times and get all sixes, then the three statistics are

$$\bar{x}=6 \qquad x_{(5)}=6 \qquad r=0.$$

Looking at the three sampling distributions from the previous three examples, $\bar{x}=6$ and $r=0$ are extraordinarily unlikely events, but $x_{(5)}=6$ occurs quite often. The sampling distributions tell us that

$$P(\bar{X}=6)=\frac{1}{7776} \qquad P\left(X_{(5)}=6\right)=\frac{4651}{7776} \qquad P(R=0)=\frac{6}{7776}.$$

The take-away message here is that certain types of statistics can be selected to detect one particular type of rarity over another. Finally, to consider a sequence of rolls that is a bit more mainstream, the rolls $(x_1, x_2, x_3, x_4, x_5)=(1, 4, 5, 2, 4)$ result in the three statistics

$$\bar{x}=3.2 \qquad x_{(5)}=5 \qquad r=4.$$

Looking at the graphs of the three sampling distributions from the previous three examples, none of these statistics point to this particular outcome as particularly rare because

$$P(\bar{X}=3.2)=\frac{735}{7776}\cong 0.09 \qquad P\left(X_{(5)}=5\right)=\frac{2101}{7776}\cong 0.27 \qquad P(R=4)=\frac{2640}{7776}\cong 0.34.$$

The next two sections introduce two broad classes of statistics that arise in many statistical problems. The first class consists of statistics that reflect the central tendency of the population probability distribution. The second class consists of statistics that reflect the dispersion of the population probability distribution. Having estimates of the central tendency and the dispersion is important because they allow a statistician to quantify both the center of the population probability distribution and how far from that center one can expect random variables to stray.

1.3 Estimating Central Tendency

As indicated in the previous section, statistics can be defined to estimate certain characteristics of a population distribution. One aspect of a population that is nearly always of interest is the central tendency of the population distribution. Several statistics that reflect this central tendency are formally defined in this section. We begin with the sample mean.

Sample mean

The *sample mean* is the most intuitive measure of central tendency. People naturally average data values in order to get a sense of the center of a probability distribution.

Definition 1.3 Let $x_1, x_2, \ldots, x_n$ be experimental values associated with the random variables $X_1, X_2, \ldots, X_n$. The *sample mean* is

$$\bar{X} = \frac{1}{n}\sum_{i=1}^{n} X_i.$$

As indicated previously, $\bar{X}$ is used in the abstract when there are no specific data values. When specific data values have been collected, the lower case version $\bar{x}$ is used to denote the sample mean. The sample mean is sometimes called the *sample arithmetic mean*.

Example 1.15 Ten kindergarten children from ten different families are polled to find the number of children that are in their family. The resulting values, $x_1, x_2, \ldots, x_{10}$ are

$$3, 1, 5, 1, 3, 2, 1, 1, 3, 2.$$

Calculate the sample mean.

The sample mean is

$$\bar{x} = \frac{3+1+5+1+3+2+1+1+3+2}{10} = \frac{22}{10} = 2.2 \text{ children.}$$

This calculation is straightforward and can be conducted in R with the statements given below.

```
x = c(3, 1, 5, 1, 3, 2, 1, 1, 3, 2)
mean(x)
```

Of course, polling ten different children would most likely result in a different sample mean. The experimental sample mean $\bar{x}$ given above is one instance from the sampling distribution of the random variable $\bar{X}$.

Another way of thinking about a sample mean is to consider it to be a special case of a *weighted average*, in which each of the data values is given a weight of $1/n$. If the data values constitute a random sample, then there is no reason to give more weight to one value over another. Returning to the kindergarten sibling data from the previous example, the sample mean could be written as

$$\bar{x} = \frac{3+1+5+1+3+2+1+1+3+2}{10} = 1 \cdot \frac{4}{10} + 2 \cdot \frac{2}{10} + 3 \cdot \frac{3}{10} + 5 \cdot \frac{1}{10}.$$

This way of thinking emphasizes the fact that the sample mean is a weighted average, where the weights reflect the relative frequency of a particular data value. Compare the expression on the far right with the formula for the population mean $E[X]$ for a discrete probability distribution from probability theory:

$$E[X] = \sum_{\mathcal{A}} x f(x),$$

where $\mathcal{A}$ is the support and $f(x)$ is the probability mass function. The weights $4/10$, $2/10$, $3/10$, and $1/10$ play the role of $f(x)$ from probability theory.

There is still another way to think about the sample mean. In order to develop this formulation, the notion of an empirical probability distribution must be defined.

Definition 1.4 Let $x_1, x_2, \ldots, x_n$ be experimental values associated with the random variables $X_1, X_2, \ldots, X_n$. The *empirical probability distribution* associated with $x_1, x_2, \ldots, x_n$ is the discrete probability distribution defined by assigning probability $1/n$ to each x_i value.

This empirical probability distribution can be expressed as either an *empirical probability mass function*, denoted by $\hat{f}(x)$, or an *empirical cumulative distribution function*, denoted by $\hat{F}(x)$, which are defined next. The empirical cumulative distribution function was introduced in the statistical graphics section as a way to avoid binning observations into cells when constructing a histogram.

Definition 1.5 Let $x_1, x_2, \ldots, x_n$ be experimental values associated with the random variables $X_1, X_2, \ldots, X_n$. The *empirical probability mass function* associated with $x_1, x_2, \ldots, x_n$ is

$$\hat{f}(x) = \frac{\text{number of } x_i \text{ equal to } x}{n}.$$

The *empirical cumulative distribution function* associated with $x_1, x_2, \ldots, x_n$ is

$$\hat{F}(x) = \frac{\text{number of } x_i \text{ less than or equal to } x}{n}.$$

The empirical probability distribution, regardless of whether it is expressed in either of its equivalent forms as $\hat{f}(x)$ or $\hat{F}(x)$, is our best guess for the population probability distribution based on the data values $x_1, x_2, \ldots, x_n$.

Let's return to the discussion of the sample mean. The empirical probability distribution associated with the data set has a population mean, which is typically called the "plug-in estimator of the population mean." Using the formula for the population mean from probability, the formula for the plug-in estimator of the population mean is

$$\hat{\mu} = \sum_{\mathcal{A}} x \hat{f}(x) = \frac{1}{n} \sum_{i=1}^{n} x_i,$$

where $\mathcal{A}$ is the support of the population distribution. This is, once again, the formula for the sample mean.

So, regardless of whether you simply use the defining formula, think of the sample mean as a weighted average, or use the plug-in estimator of the population mean, the same value results for the sample mean.

Since the sample mean $\bar{X}$ is a random variable, we can calculate its sampling distribution. This sampling distribution depends on the population probability distribution from which the data values are drawn. The two examples that follow consider the sampling distribution of the sample mean for observations drawn from a discrete population and a continuous population.

Example 1.16 Let $X_1, X_2, \ldots, X_n$ be a random sample from a Poisson(λ) distribution, where λ is a positive unknown parameter. What is the sampling distribution of $\bar{X}$?

One could easily envision a real-world scenario in which averaging observations sampled from a Poisson population could occur, for example,

- averaging the number of customers that arrive to a drive-up window at a fast food restaurant during the lunch hour for five consecutive weekdays,
- averaging the number of potholes per mile on a particular stretch of highway, and
- averaging the number of web hits per day at a popular website during February.

The first step in finding the probability distribution of the sample mean $\bar{X}$ is to determine the support of the random variable $\bar{X}$. The Poisson population distribution has support on the nonnegative integers, so the numerator of

$$\bar{X} = \frac{X_1 + X_2 + \cdots + X_n}{n}$$

can also assume the values $0, 1, 2, \ldots$. Therefore, the support of $\bar{X}$ is

$$\mathcal{A} = \left\{ x \,\middle|\, x = 0, \frac{1}{n}, \frac{2}{n}, \ldots \right\}.$$

The next step is to determine the probabilities associated with each element in the support. The random variable X_i has probability mass function

$$f_{X_i}(x) = \frac{\lambda^x e^{-\lambda}}{x!} \qquad x = 0, 1, 2, \ldots$$

for $i = 1, 2, \ldots, n$. The numerator in the sample mean

$$\bar{X} = \frac{X_1 + X_2 + \cdots + X_n}{n}$$

consists of the sum of mutually independent and identically distributed Poisson(λ) random variables because $X_1, X_2, \ldots, X_n$ is a random sample. Using a result from probability theory that can be proved by the moment generating function technique, the numerator $X_1 + X_2 + \cdots + X_n$ is Poisson($n\lambda$) with probability mass function

$$f_{X_1+X_2+\cdots+X_n}(x) = \frac{(n\lambda)^x e^{-n\lambda}}{x!} \qquad x = 0, 1, 2, \ldots.$$

Finally, dividing the numerator of $\bar{X}$ by n gives the probability mass function

$$f_{\bar{X}}(x) = \frac{(n\lambda)^{nx} e^{-n\lambda}}{(nx)!} \qquad x = 0, \frac{1}{n}, \frac{2}{n}, \ldots$$

by the transformation technique.

Now consider the sampling distribution of $\bar{X}$ for a particular sample size n and a particular population mean λ, say $n = 5$ and $\lambda = 2$. Figure 1.22 is a graph of the probability mass function of the first seven support values of the population from which the data values are drawn, that is, a Poisson(2) distribution. The graph of $f_X(x)$ continues to

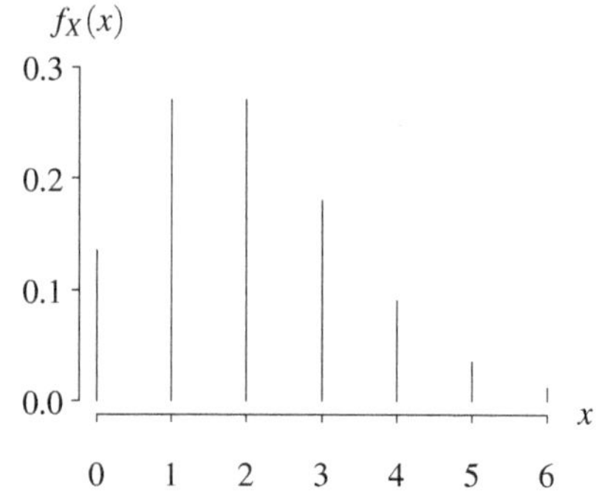

Figure 1.22: Population probability mass function.

decline as x increases. Using the formula for $f_{\bar{X}}(x)$, the probability mass function of $\bar{X}$ is

$$f_{\bar{X}}(x) = \frac{10^{5x}e^{-10}}{(5x)!} \qquad x = 0, \frac{1}{5}, \frac{2}{5}, \ldots .$$

Figure 1.23 is a graph of the probability mass function of the sampling distribution of the statistic $\bar{X}$ when $n = 5$ and $\lambda = 2$. The horizontal scales are identical, but the vertical scales differ on the two graphs. There are four observations that can be made concerning these two probability mass functions:

- Both the probability distribution of X_i and the probability distribution of $\bar{X}$ have the same expected value: $E[X_i] = E[\bar{X}] = 2$ in this particular example. As will be seen subsequently, this result, stated more generally as $E[\bar{X}] = \mu$, is true for any population distribution that has a finite population mean.
- The population variance of the sampling distribution of $\bar{X}$ is less than the population variance of the population distribution. Although X_i and $\bar{X}$ have the same

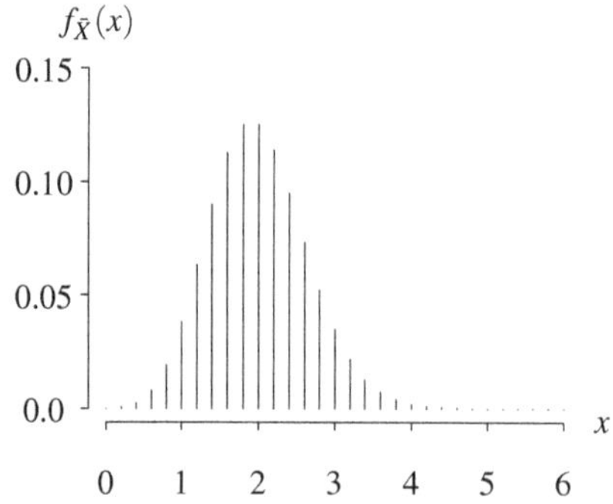

Figure 1.23: Probability mass function (sampling distribution) of the sample mean.

population mean, averaging the $n = 5$ observations decreases the population variance of $\bar{X}$ relative to X_i.

- The support of $\bar{X}$ is finer than the support of X_i. The data values can assume the values $0, 1, 2, \ldots$, but the sample mean can assume the values $0, 1/5, 2/5, \ldots$.
- Even though the sample size is only $n = 5$, the central limit theorem is evident in the distribution of $\bar{X}$ as it has more of a bell shape than the population distribution. The limiting distribution of $\bar{X}$ in this example is normal.

The sampling distribution of $\bar{X}$ when the data values are drawn from a continuous population is determined in a similar fashion, as illustrated in the next example.

Example 1.17 Let $X_1, X_2, \ldots, X_n$ be a random sample from a gamma(λ, κ) distribution, where λ and κ are positive unknown scale and shape parameters. Find the sampling distribution of $\bar{X}$. Also find $P(\bar{X} < 2)$ for a sample size of $n = 4$ when $\lambda = 1$ and $\kappa = 3$.

The probability density function of X_i sampled from a gamma(λ, κ) population is

$$f_{X_i}(x) = \frac{\lambda^\kappa x^{\kappa-1} e^{-\lambda x}}{\Gamma(\kappa)} \qquad x > 0$$

for $i = 1, 2, \ldots, n$. The corresponding moment generating function is

$$M_{X_i}(t) = \left(\frac{\lambda}{\lambda - t}\right)^\kappa \qquad t < \lambda$$

for $i = 1, 2, \ldots, n$. Since the observations are a random sample, $X_1, X_2, \ldots, X_n$ are mutually independent and identically distributed random variables. Hence, the moment generating function of $\bar{X}$ is

$$\begin{aligned} M_{\bar{X}}(t) &= E\left[e^{t\bar{X}}\right] \\ &= E\left[e^{t(X_1+X_2+\cdots+X_n)/n}\right] \\ &= M_{X_1+X_2+\cdots+X_n}(t/n) \\ &= M_{X_1}(t/n)M_{X_2}(t/n)\ldots M_{X_n}(t/n) \\ &= \left(\frac{\lambda}{\lambda - t/n}\right)^\kappa \left(\frac{\lambda}{\lambda - t/n}\right)^\kappa \cdots \left(\frac{\lambda}{\lambda - t/n}\right)^\kappa \\ &= \left(\frac{n\lambda}{n\lambda - t}\right)^{n\kappa} \qquad t < n\lambda. \end{aligned}$$

This moment generating function can be recognized as that of a gamma($n\lambda$, $n\kappa$) random variable.

Figure 1.24 is a plot of the population probability density function for $\lambda = 1$ and $\kappa = 3$. The $n = 4$ data values are sampled from this population probability distribution. Since $\bar{X} \sim \text{gamma}(n\lambda, n\kappa)$, $\lambda = 1$, $\kappa = 3$, and $n = 4$,

$$\bar{X} \sim \text{gamma}(4, 12).$$

Figure 1.25 contains a plot of the probability density function of the sample mean $\bar{X} \sim \text{gamma}(4, 12)$. The horizontal scales in Figures 1.24 and 1.25 are identical, but the vertical scales differ on the two graphs. The same effect as in the previous example (when the random sampling was from a Poisson population) takes place here:

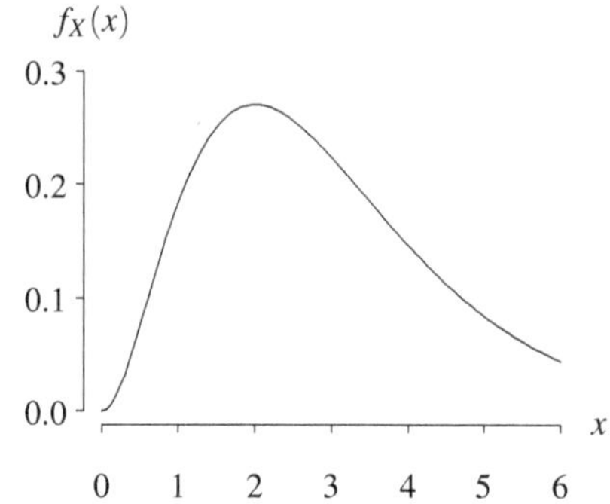

Figure 1.24: Population probability density function.

- The population probability distribution and the probability distribution of $\bar{X}$ have the same central value, which in this case is $E[X_i] = E[\bar{X}] = 3$.
- The probability distribution of $\bar{X}$ has a smaller population variance than the population probability distribution.
- The probability distribution of $\bar{X}$ looks more bell-shaped than the population probability distribution because of the central limit theorem. The probability density function of $\bar{X}$ is nearly symmetric. The limiting distribution of $\bar{X}$ is normal.

The final part of the question is to determine the probability that the sample mean is less than 2 for sample size $n = 4$ and population parameters $\lambda = 1$ and $\kappa = 3$. One way to calculate this probability is to integrate the probability density function over the appropriate limits. Since $\bar{X} \sim$ gamma(4, 12),

$$P(\bar{X} < 2) = \int_0^2 \frac{(n\lambda)^{n\kappa} x^{n\kappa-1} e^{-n\lambda x}}{\Gamma(n\kappa)} dx = \int_0^2 \frac{4^{12} x^{11} e^{-4x}}{\Gamma(12)} dx.$$

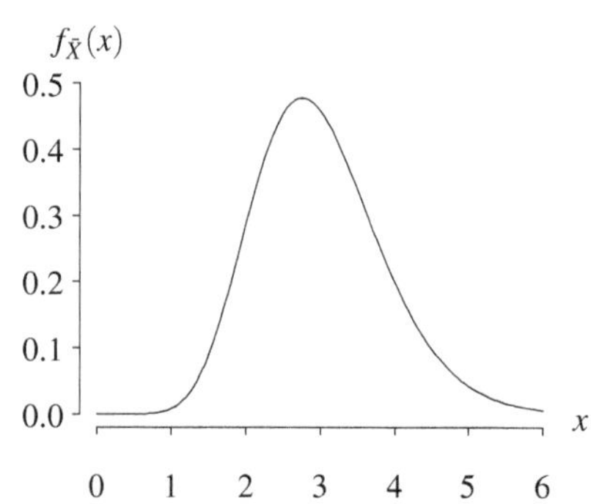

Figure 1.25: Probability density function (sampling distribution) of the sample mean.

This integral can be computed by hand using integration by parts repeatedly or can be calculated using a computer algebra system, giving the required probability as

$$P(\bar{X} < 2) = 1 - \frac{412782941}{155925} e^{-8} \cong 0.1119.$$

R can also be used to calculate the probability that the sample mean is less than 2. Using the `pgamma` function, which returns the cumulative distribution function of a gamma random variable, the single statement

```
pgamma(2, 12, 4)
```

also returns $P(\bar{X} < 2) \cong 0.1119$. Notice that R switches the order of the parameters as arguments relative to the convention gamma(λ, κ) used here.

Finally, to determine whether the derivation and associated numerical value are correct, a Monte Carlo simulation experiment can be conducted to estimate the probability that the sample mean is less than 2. The following R code generates one million sample means and prints the fraction of those sample means that are less than 2.

```
nrep  = 1000000
count = 0
for (i in 1:nrep) {
  xbar = mean(rgamma(4, 3, 1))
  if (xbar < 2) count = count + 1
}
print(count / nrep)
```

After a call to `set.seed(3)` to initialize the random number stream, five runs of this simulation yield the following estimates of $P(\bar{X} < 2)$:

0.1118 0.1119 0.1120 0.1117 0.1125.

Since these values hover about the analytic value $P(\bar{X} < 2) \cong 0.1119$, the Monte Carlo simulation supports our analytic solution.

The two previous examples have shown that the probability distribution of the sample mean depends on the probability distribution associated with the population. Every population probability distribution that the data values are drawn from requires a separate derivation—some simple and others quite intricate—to determine the probability distribution of $\bar{X}$. One piece of good news is that the expected value of $\bar{X}$ and the population variance of $\bar{X}$ can be computed with the same formulas for practically all probability distributions. Assuming that $X_1, X_2, \ldots, X_n$ constitute a random sample from some population distribution (discrete or continuous) with finite population mean μ and finite population variance σ^2, then the sample mean $\bar{X}$ has expected value

$$E[\bar{X}] = E\left[\frac{1}{n}\sum_{i=1}^{n} X_i\right] = \frac{1}{n}\sum_{i=1}^{n} E[X_i] = \frac{1}{n}(n\mu) = \mu$$

and population variance

$$V[\bar{X}] = V\left[\frac{1}{n}\sum_{i=1}^{n} X_i\right] = \frac{1}{n^2}\sum_{i=1}^{n} V[X_i] = \frac{1}{n^2}\left(n\sigma^2\right) = \frac{\sigma^2}{n}.$$

The first of these equations indicates that the sample mean $\bar{X}$ is on target for estimating the population mean μ. So the first equation addresses the *accuracy* of $\bar{X}$ in estimating μ. Since the expected value of the sample mean is the population mean, statisticians say that $\bar{X}$ is an *unbiased estimator* of μ. (Unbiased estimators will be presented formally in Chapter 2.) The second of these equations indicates that the variability of the sample mean $\bar{X}$ decreases as n increases. So the second equation addresses the *precision* of $\bar{X}$ in estimating μ. The sample mean $\bar{X}$ is often said to be a more *precise* estimator of μ as the sample size increases. This constitutes a derivation of the following result.

Theorem 1.1 Let $X_1, X_2, \ldots, X_n$ be a random sample from a population distribution with finite population mean μ and finite population variance σ^2. The sample mean $\bar{X}$ has population mean

$$E[\bar{X}] = \mu$$

and population variance

$$V[\bar{X}] = \frac{\sigma^2}{n}.$$

Having a good understanding of the behavior of $\bar{X}$ is important when drawing conclusions based on sample means, as illustrated in the next example.

Example 1.18 The six states with the highest age-adjusted incidence of kidney cancer in the United States in the years 2012–2016 are shown in Figure 1.26 using data from the Centers for Disease Control website. If you happen to be reading this book and live in one of those six states, you might be grabbing your belly right now and thinking "Oh no, I am living in a kill zone, I need to move away!" But where should you move? Figure 1.27 shows the six states with the lowest age-adjusted incidence of kidney cancer in the United States in the years 2012–2016. It seems like any one of these states would provide a much more hospitable home for your kidneys.

It is important to establish that the kidney cancer incidence rate is indeed a sample mean $\bar{X}$. Think of the kidney cancer status of each resident of a state as a Bernoulli ran-

Figure 1.26: Six states with the highest incidence of kidney cancer in 2012–2016.

Figure 1.27: Six states with the lowest incidence of kidney cancer in 2012–2016.

dom variable X_i, where $X_i = 0$ corresponds to not being diagnosed with kidney cancer in the years 2012–2016 and $X_i = 1$ corresponds to being diagnosed with kidney cancer in the years 2012–2016, for $i = 1, 2, \ldots, n$, where n is the population of the state. The sample mean $\bar{X}$ gives the estimated probability or incidence rate of kidney cancer for a particular state. Epidemiologists typically express the incidence rate for a rare cancer in terms of the number of cases per 100,000 in order to avoid writing too many leading zeros. This change of scale does not change the fact that the incidence rate behaves like an average from a statistical point of view.

So what is going on here? Are the states with the high kidney cancer incidence rates really less safe, or are we being tricked by random sampling variability? Another statistical graphic can lend some insight. The kidney cancer annual incidence rates for the states are plotted on the vertical axis against the population on the horizontal axis (which uses a logarithmic scale) in Figure 1.28. The weighted annual incidence rate

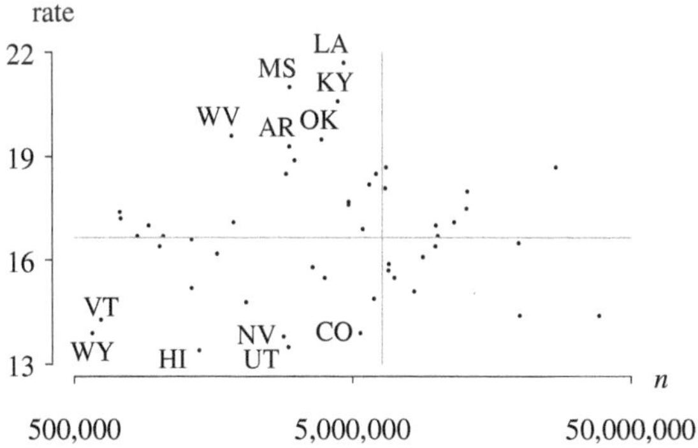

Figure 1.28: Population versus annual kidney incidence rate per 100,000 in 2012–2016.

for the entire U.S. over the five-year period is 16.7 incidences per 100,000 population, indicated by the horizontal line on the plot. The average population of a state over the five-year period is 6.4 million residents, indicated by the vertical line on the plot. The six states with the highest and lowest kidney cancer incidence rates in 2012–2016 are identified on the plot. One curiosity that appears immediately is that the states with high and low incidence rates tend to, on average, be smaller states. The most populous states don't show up on either list.

So the fact that smaller states tend to appear more often on the list of states with high and low kidney cancer incidence rates brings us back to the equation $V\left[\bar{X}\right] = \sigma^2/n$. Smaller states have a smaller value of n and are thus more susceptible to random sampling variability of $\bar{X}$ and are thus more likely to show up on the high incidence rate and low incidence rate lists. So it is quite possible that the six states with the highest and lowest kidney cancer incidence rates are really no more or less risky than any others. Looking at the data for subsequent years would help confirm whether or not there is a pattern developing, or if the results are simply due to random sampling variability of $\bar{X}$.

Professor Howard Wainer refers to $V\left[\bar{X}\right] = \sigma^2/n$ as a "dangerous equation" in his book *Picturing the Uncertain World.* He also considers kidney cancer incidence rates, but this time by county rather than state. Some counties have just a few hundred residents, so having no kidney cancers gives them a kidney cancer incidence rate of zero, which could potentially wrongly classify them as "safe." Having just a single kidney cancer in a small county, however, could potentially wrongly classify them as "unsafe." A large county with millions of residents will almost never be classified as safe or unsafe because of the n in the denominator of $V\left[\bar{X}\right] = \sigma^2/n$. He cites several other examples where the lack of knowledge concerning the effect of n on the population variance of $\bar{X}$ "has led to billions of dollars of loss over centuries, yielding untold hardship."

To visualize the effect of n on the sampling distribution of $\bar{X}$ in a Monte Carlo framework, consider a random sample of n data values drawn from a $N(10, 1)$ population. Ten values of $\bar{x}$ are plotted for $n = 2, 4, 8, 16, 32, 64$ in Figure 1.29, which has a logarithmic horizontal axis. The R code to generate and plot the sample means follows. The `plot` function sets up the axes, and the `points` function plots the points generated by

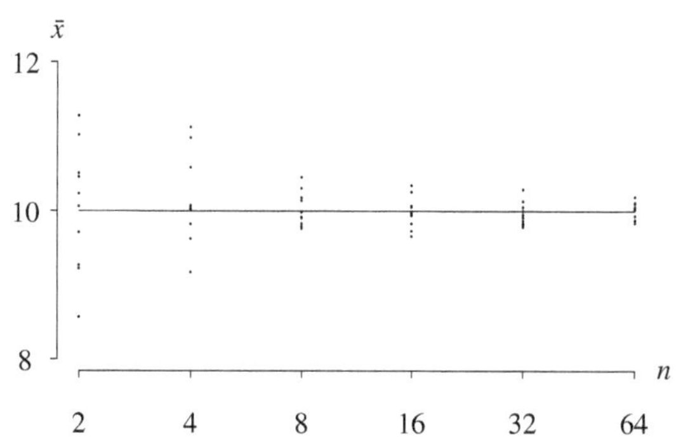

Figure 1.29: Monte Carlo experiment plotting $\bar{x}$ for several values of n.

rnorm, which generates random variates from a normal population. The two take-away messages are immediate: (a) since $\bar{X}$ is an unbiased estimate of μ, all averages have an expected value of $\mu = 10$, and (b) extreme values of the sample mean will occur at the smaller sample sizes because the sample mean has a larger population variance for smaller values of n.

```
set.seed(8)
plot(c(2, 64), c(10, 10), type = "l", xlim = c(2, 64),
                                       ylim = c(8, 12), log = "x")
nrep = 10
for (n in c(2, 4, 8, 16, 32, 64)) {
  for (j in 1:nrep) {
    xbar = mean(rnorm(n, 10, 1))
    points(n, xbar)
  }
}
```

The previous example concerned cancer incidence rates, but could apply equally well to any number of settings. Here are three examples.

- If a small high school has stellar average SAT scores, it could be that the high school is particularly good or it could be that the high school is particularly small.
- If a small hospital has an unusually high infection rate for patients, it could be that the hospital is careless with respect to sanitation or it could be that the hospital is particularly small.
- If your friend invests in just two stocks and brags about his average annual return, it could be that he is a brilliant investor or it could be that the number of stocks he invested in is particularly small.

The key take-away point of the previous example involving kidney cancer rates, stated in two equivalent fashions, is that

- the sample mean $\bar{X}$ is a more precise estimator of μ for larger sample sizes,
- small sample sizes can yield more extreme values of $\bar{X}$ than large sample sizes.

One last point should be made about the equation $V[\bar{X}] = \sigma^2/n$ from Theorem 1.1. Taking the positive square root of both sides of this equation results in

$$\sigma_{\bar{X}} = \frac{\sigma}{\sqrt{n}},$$

which can be read as "the standard error of the sample mean is the ratio of the population standard deviation to the square root of n." The term "standard error" here is synonymous with standard deviation. This equation implies that if you want to halve the standard deviation of $\bar{X}$, you must quadruple the sample size n. Even more extreme, if you want to decrease the standard deviation of $\bar{X}$ by a factor of 10, you must collect 100 times as many data values. The $\sqrt{n}$ will appear in the denominator of many expressions throughout this book and it will cause problems when lots of precision is required and data values are expensive to collect. This relationship between n and $\sigma_{\bar{X}}$ is an instance of the *law of diminishing returns*.

As a final example to conclude this subsection on the sample mean, consider the effect of sampling *dependent* observations. The mutual independence assumption from the previous examples, which was helpful in determining the sampling distribution of the sample mean, is lost because we are no longer dealing with a random sample. The formula for $\bar{X}$ remains the same, but the calculation of the probability distribution of $\bar{X}$ is more difficult. The next example considers the case of $n = 2$ dependent observations.

Example 1.19 Consider an experiment that consists of sampling a person at random from the community and asking them the following two (somewhat personal) questions:

1. Are you a medical doctor?
2. Is your annual salary greater than $100,000?

The responses to these $n = 2$ questions are positively correlated because medical doctors tend to have higher salaries than the general population. Let X_1 be 0 for "no" and 1 for "yes" to the first question. Likewise, let X_2 be 0 for "no" and 1 for "yes" to the second question. The responses to the questions have now been defined as the dependent random variables X_1 and X_2. The sample mean is

$$\bar{X} = \frac{X_1 + X_2}{2}.$$

Table 1.9 contains the joint probability mass function of the random variables X_1 and X_2, where p_1, p_2, p_3, and p_4 are unknown probabilities that sum to one. What is the probability mass function of $\bar{X}$?

x_1 \ x_2	0	1
0	p_1	p_2
1	p_3	p_4

Table 1.9: Joint probability mass function for X_1 and X_2.

The dependence between X_1 and X_2 indicates that the probabilities for each of the possible values for $\bar{X}$ needs to be assigned to the appropriate component probabilities. Since the sample size of $n = 2$ is so small, there are only three different values for $\bar{X}$:

- $\bar{X} = 0$, which corresponds to $X_1 = 0$ and $X_2 = 0$,
- $\bar{X} = 1/2$, which corresponds to $X_1 = 0$ and $X_2 = 1$, or $X_1 = 1$ and $X_2 = 0$,
- $\bar{X} = 1$, which corresponds to $X_1 = 1$ and $X_2 = 1$.

So the sampling distribution of $\bar{X}$ in this case is described by the probability mass function

$$f_{\bar{X}}(x) = \begin{cases} p_1 & x = 0 \\ p_2 + p_3 & x = 1/2 \\ p_4 & x = 1. \end{cases}$$

The sample mean is the most common statistical measure of central tendency. The sample median is considered next.

Sample median

The sample mean is the "gold standard" in terms of estimating the central tendency of a population probability distribution and is used in a vast majority of applications in which central tendency is of interest. Occasions arise, however, when the *sample median* is a better measure of central tendency.

> **Definition 1.6** Let $x_1, x_2, \ldots, x_n$ be experimental values associated with the random variables $X_1, X_2, \ldots, X_n$. The *sample median* is
>
> $$M = \begin{cases} X_{((n+1)/2)} & n \text{ odd} \\ \left(X_{(n/2)} + X_{(n/2+1)}\right)/2 & n \text{ even,} \end{cases}$$
>
> where $X_{(1)}, X_{(2)}, \ldots, X_{(n)}$ are the order statistics (the data values sorted into ascending order).

If n is odd, the sample median is just the middle sorted value; if n is even, the sample median is the average of the two middle sorted values.

Economists frequently use the sample median, rather than the sample mean, when reporting statistics concerning certain economic measures, such as incomes or house prices. To see why this is the case, consider a small M.S. program in operations research that graduates just $n = 5$ students in one particular academic year. The students assume positions in industry and report the following annual salaries:

$$\$71{,}000 \qquad \$65{,}000 \qquad \$74{,}000 \qquad \$194{,}000 \qquad \$73{,}000.$$

Now which would be a more accurate way to report the salary data in a recruiting brochure for the new class of operations researchers: use the sample mean $\bar{x} = \$95{,}400$ or use the sample median $m = \$73{,}000$ as the measure of central tendency? The student who graduated and took a salary of \$194,000 might have joined a family business or had a lucrative overseas offering. The other four salaries are fairly tightly clustered around the sample median $m = \$73{,}000$. The one high salary is a rarity, so it can either be considered an outlier or it can be an observation from a very long right-hand tail of the population probability distribution. In either case, reporting the sample median is the appropriate statistic to go in the brochure for next year. It gives the students the most accurate assessment of what their salary will be when they finish the M.S. program.

Determining the sampling distribution of the sample median can vary from simple to very complex. The two examples that follow span the two extremes.

Example 1.20 Let $X_1, X_2, \ldots, X_9$ be a random sample from a $U(0, 1)$ distribution. Find the sampling distribution of the sample median.

Unlike the three examples associated with determining the sampling distribution of the sample mean, this time the population distribution does not have any unknown parameters. The probability density function of X_i drawn from a $U(0, 1)$ population is

$$f_{X_i}(x) = 1 \qquad 0 < x < 1$$

for $i = 1, 2, \ldots, 9$. The corresponding cumulative distribution function on the support of X_i is

$$F_{X_i}(x) = x \qquad 0 < x < 1$$

for $i = 1, 2, \ldots, 9$. The observations are mutually independent and identically distributed random variables because they constitute a random sample. So the distribution of the sample median M, which is $X_{(5)}$ because $n = 9$ is odd, can be found using the formula for the probability density function of the kth order statistic drawn from a continuous population,

$$f_{X_{(k)}}(x) = \frac{n!}{(k-1)!(n-k)!}[F(x)]^{k-1} f(x)[1-F(x)]^{n-k} \qquad a < x < b$$

for $k = 1, 2, \ldots, n$, where a and b are the lower and upper limits of the support of the population probability distribution. Applying this formula to our sample of $n = 9$ observations from a $U(0, 1)$ population gives

$$f_M(x) = \frac{9!}{(5-1)!(9-5)!} x^{5-1} \cdot 1 \cdot (1-x)^{9-5} = 630x^4(1-x)^4 \qquad 0 < x < 1.$$

Notice that this probability density function is symmetric about $x = 1/2$. A Monte Carlo simulation experiment can be used to support our analytic work. The following R code generates 100 sample medians from 100 samples of size $n = 9$ drawn from a $U(0, 1)$ population distribution, plots a histogram, and overlays the histogram with the sampling distribution of the sample median derived above.

```
nrep = 100
medians = numeric(nrep)
for (i in 1:nrep) {
  x = runif(9)
  medians[i] = median(x)
}
hist(medians, probability = TRUE)
curve(630 * x ^ 4 * (1 - x) ^ 4, 0, 1, add = TRUE)
```

Executing this code after a call to `set.seed(7)` yields the graph shown in Figure 1.30.

For both the analytic values represented by the curve and the sample values represented by the histogram, the effect of choosing the fifth largest of the nine values is to push the probability distribution away from the extremes at 0 and 1 toward the center of the distribution at $1/2$. But are the histogram and the curve close enough to support our analytic work? The problem illustrated here is that we chose only `nrep = 100` replications of the simulation experiment, resulting in a rather noisy histogram. Random

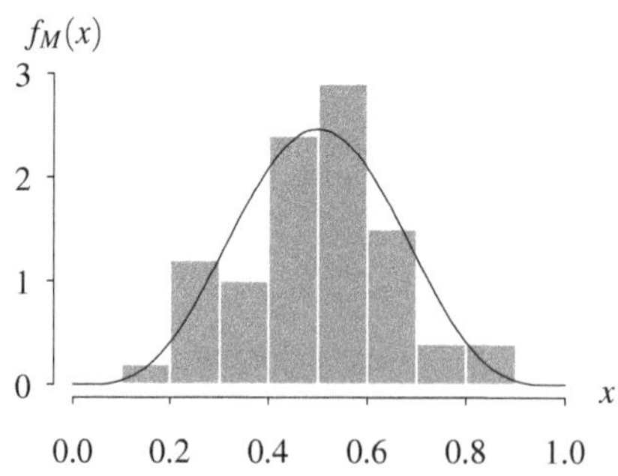

Figure 1.30: Sampling distribution of the sample median (100 replications).

sampling variability applies to Monte Carlo simulation as well as to collecting data. Figure 1.31 uses the same code, but this time with `nrep = 200000`. R chooses more cells for the histogram because of the larger number of replications; the histogram is much smoother this time. We now achieve a good match between the sampling distribution of M and its estimate via Monte Carlo simulation. This time our analytic work is supported by the simulation. The bell shape of the sampling distribution of M is not due to the central limit theorem, but rather due to the choice of the middle order statistic from a symmetric population probability distribution. Now that the sampling distribution of the sample median has been derived and supported by Monte Carlo simulation, it is often of value to know the expected value and population variance of the statistic of interest. The APPL statements below calculate the probability density function of the sample median M and its expected value and its population variance.

```
X := UniformRV(0, 1);
M := OrderStat(X, 9, 5);
Mean(M);
Variance(M);
```

The statements yield

$$E[M] = \frac{1}{2} \qquad \text{and} \qquad V[M] = \frac{1}{44}.$$

Notice that the expected value of the sample median equals the population median (this is $1/2$ by inspection because of the symmetry of the $U(0, 1)$ distribution). This is a good property for an estimator to have because the estimator is "on target" for estimating the population quantity. This property will be defined carefully in the next chapter, but for now $E[M] = E\left[X_{(5)}\right] = 1/2$ is stated in words as "the sample median is an unbiased estimator of the population median." The population variance of M is an indication of how far the sample median might stray from its target. We want the population variance of M to be as small as possible. One way to decrease the population variance of M is to increase the sample size n.

So the sample median seems like a reasonable estimator of the population median. But for this particular population distribution, the $U(0, 1)$ distribution, the population

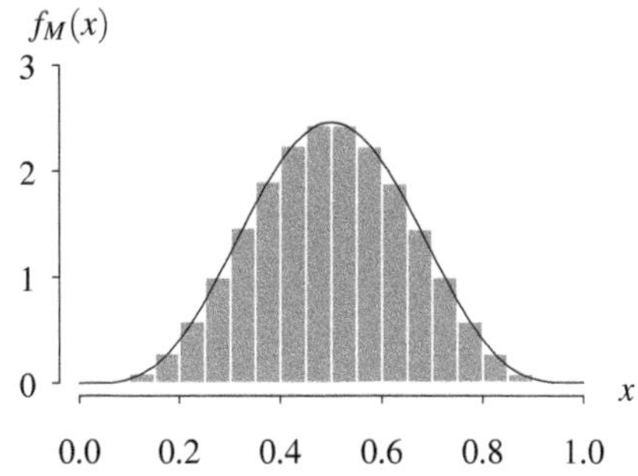

Figure 1.31: Sampling distribution of the sample median (200,000 replications).

median and the population mean both equal $1/2$. Would it be better to use the sample mean $\bar{X}$ to estimate the population median? One way to pin down the choice is to consider the population variance of the estimates. From above, the population variance of the sample median is

$$V[M] = \frac{1}{44},$$

but the population variance of the sample mean $\bar{X}$ by Theorem 1.1 is

$$V[\bar{X}] = \sigma_{\bar{X}}^2 = \frac{\sigma_X^2}{n} = \frac{1/12}{9} = \frac{1}{108}.$$

So the sample mean is more tightly clustered about the population median of $1/2$ than the sample median, and is therefore the preferred estimator of the population median for this particular symmetric population distribution. This will not be the case in general.

The previous example had two factors which made the analytic work tractable: an odd value for n and a particularly simple population distribution. In the next example, we remove both of those advantages and see the extra work associated with an even n and a more complicated population distribution.

Example 1.21 Let $X_1, X_2, \ldots, X_6$ be a random sample from a population having probability density function

$$f(x) = 2x \qquad 0 < x < 1.$$

Find the sampling distribution of the sample median.

As in the previous example, the population distribution does not involve any parameters. In contrast to the previous example, there are two complicating factors at play in this question: the *even* sample size $n = 6$ and the *slightly* more complicated population distribution. The even sample size implies that two adjacent order statistics will be averaged in order to arrive at the sample median. As you will see, these two extra factors create lots of extra work in deriving the sampling distribution of the sample median. The problem provides a good review, however, of the joint distribution of order statistics and the transformation technique. The random variable X_i has probability density function

$$f_{X_i}(x) = 2x \qquad 0 < x < 1$$

for $i = 1, 2, \ldots, 6$. The associated cumulative distribution function of X_i on its support is

$$F_{X_i}(x) = \int_0^x 2w\,dw = \left[w^2\right]_0^x = x^2 \qquad 0 < x < 1$$

for $i = 1, 2, \ldots, 6$. Since n is even, the sample median is calculated by averaging $X_{(3)}$ and $X_{(4)}$. Unfortunately, $X_{(3)}$ and $X_{(4)}$ are dependent random variables. So we begin the process of finding the probability density function of the sample median by finding the joint probability density function of $X_{(3)}$ and $X_{(4)}$. Using the same heuristic argument that gave us the probability density function of a single order statistic drawn from a continuous population, the joint probability density function of two order statistics $X_{(i)}$ and $X_{(j)}$, which is $f_{X_{(i)},X_{(j)}}(x_{(i)}, x_{(j)})$ for $i < j$, is given by the expression

$$\frac{n!}{(i-1)!(j-i-1)!(n-j)!}\left[F\left(x_{(i)}\right)\right]^{i-1} f\left(x_{(i)}\right)\left[F\left(x_{(j)}\right) - F\left(x_{(i)}\right)\right]^{j-i-1} f\left(x_{(j)}\right)\left[1 - F\left(x_{(j)}\right)\right]^{n-j}$$

for $a < x_{(i)} < x_{(j)} < b$, where a and b are the lower and upper bounds on the support of the population distribution. Applying this formula to the population distribution described here, the joint probability density function of $X_{(3)}$ and $X_{(4)}$ is

$$f_{X_{(3)},X_{(4)}}\left(x_{(3)},x_{(4)}\right) = \frac{6!}{2!\,0!\,2!}\left[x_{(3)}^2\right]^2 2x_{(3)}\left[x_{(4)}^2 - x_{(3)}^2\right]^0 2x_{(4)}\left[1-x_{(4)}^2\right]^2$$

for $0 < x_{(3)} < x_{(4)} < 1$. This simplifies to

$$f_{X_{(3)},X_{(4)}}\left(x_{(3)},x_{(4)}\right) = 720x_{(3)}^5 x_{(4)}\left(1-x_{(4)}^2\right)^2 \qquad 0 < x_{(3)} < x_{(4)} < 1.$$

Now the sample median by Definition 1.6 is the average of $X_{(3)}$ and $X_{(4)}$, that is,

$$M = \frac{X_{(3)} + X_{(4)}}{2}.$$

To determine the probability density function of the sample median, we will use the transformation technique, which requires a "dummy" transformation. So the transformation consists of $Y_1 = g_1(X_{(3)}, X_{(4)}) = (X_{(3)} + X_{(4)})/2$, the sample median, and the dummy transformation $Y_2 = g_2(X_{(3)}, X_{(4)}) = (X_{(4)} - X_{(3)})/2$ because these particular functions can be solved in closed form for $X_{(3)}$ and $X_{(4)}$ and the associated Jacobian is tractable. The fact that this is a *linear* transformation ensures that the Jacobian is nonzero. The transformation

$$y_1 = g_1\left(x_{(3)},x_{(4)}\right) = \frac{x_{(3)}+x_{(4)}}{2} \qquad \text{and} \qquad y_2 = g_2\left(x_{(3)},x_{(4)}\right) = \frac{x_{(4)}-x_{(3)}}{2}$$

is illustrated in Figure 1.32 and is a bivariate one-to-one transformation from

$$\mathcal{A} = \left\{\left(x_{(3)},x_{(4)}\right) \mid 0 < x_{(3)} < x_{(4)} < 1\right\}$$

to

$$\mathcal{B} = \left\{(y_1, y_2) \mid y_2 > 0,\, y_2 < y_1 < 1 - y_2\right\}.$$

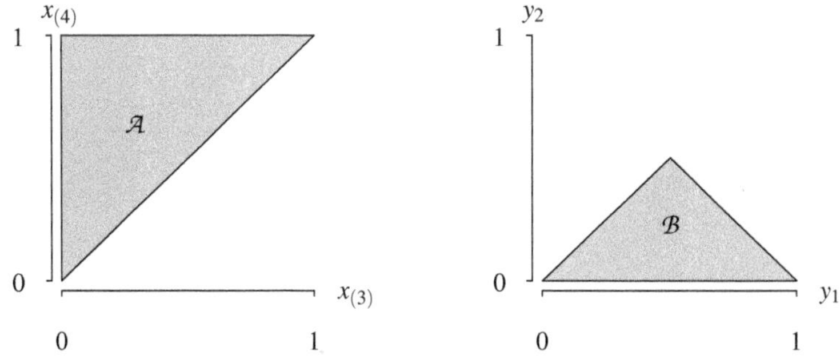

Figure 1.32: The support of $X_{(3)}$ and $X_{(4)}$ and the support of Y_1 and Y_2.

These functions can be solved in closed form for $x_{(3)}$ and $x_{(4)}$ as

$$x_{(3)} = g_1^{-1}(y_1, y_2) = y_1 - y_2 \qquad \text{and} \qquad x_{(4)} = g_2^{-1}(y_1, y_2) = y_1 + y_2$$

with associated Jacobian

$$J = \begin{vmatrix} 1 & -1 \\ 1 & 1 \end{vmatrix} = 2.$$

Applying the transformation technique, the joint probability density function of the random variables Y_1 and Y_2 is

$$f_{Y_1,Y_2}(y_1, y_2) = 720(y_1 - y_2)^5 (y_1 + y_2)\left[1 - (y_1 + y_2)^2\right]^2 |2| \qquad (y_1, y_2) \in \mathcal{B},$$

which simplifies to

$$f_{Y_1,Y_2}(y_1, y_2) = 1440(y_1 - y_2)^5 (y_1 + y_2)\left[1 - (y_1 + y_2)^2\right]^2 \qquad (y_1, y_2) \in \mathcal{B}.$$

Integrating y_2 out of the joint probability density function gives the marginal distribution of Y_1, which is

$$f_{Y_1}(y_1) = \begin{cases} \displaystyle\int_0^{y_1} 1440(y_1 - y_2)^5 (y_1 + y_2)\left(1 - (y_1 + y_2)^2\right)^2 dy_2 & 0 < y_1 < 1/2 \\ \displaystyle\int_0^{1-y_1} 1440(y_1 - y_2)^5 (y_1 + y_2)\left(1 - (y_1 + y_2)^2\right)^2 dy_2 & 1/2 < y_1 < 1, \end{cases}$$

or

$$f_{Y_1}(y_1) = \begin{cases} \frac{40960}{77}y_1^{11} - \frac{5200}{7}y_1^9 + \frac{1920}{7}y_1^7 & 0 < y_1 < 1/2 \\ -\frac{40960}{77}y_1^{11} + \frac{15280}{7}y_1^9 - \frac{28800}{7}y_1^7 + 7680y_1^5 - \\ \quad \frac{61440}{7}y_1^4 + 4800y_1^3 - \frac{10240}{7}y_1^2 + 240y_1 - \frac{1280}{77} & 1/2 < y_1 < 1, \end{cases}$$

which is the probability density function of the sample median. Replacing Y_1 with the sample median M, this could also be written as $f_M(x)$, which is a probability density function defined on the support $0 < x < 1$. This derivation had so many opportunities for a mathematical error that it is probably worthwhile conducting a Monte Carlo simulation check of this sampling distribution of the sample median. The following R code generates 200,000 sample medians for a sample of size $n = 6$ drawn from a population distribution with cumulative distribution function

$$F_X(x) = \begin{cases} 0 & x \le 0 \\ x^2 & 0 < x < 1 \\ 1 & x \ge 1. \end{cases}$$

The inverse cumulative distribution function is

$$F_X^{-1}(u) = \sqrt{u} \qquad 0 < u < 1,$$

so random variates are generated via

$$x \leftarrow \sqrt{u},$$

where u is a random number, that is, a realization of a $U(0, 1)$ random variable. The code that follows places a sorted sample of $n = 6$ values into the R vector `x`, then averages the two middle values to arrive at a sample median `m`.

```
nrep = 200000
m = numeric(nrep)
for (i in 1:nrep) {
  x = sort(sqrt(runif(6)))
  m[i] = (x[3] + x[4]) / 2
}
hist(m, probability = TRUE)
xx = seq(0, 0.5, by = 0.01)
yy = 40960 * xx ^ 11 / 77 - 5200 * xx ^ 9 / 7 + 1920 * xx ^ 7 / 7
lines(xx, yy, type = "l")
xx = seq(0.5, 1, by = 0.01)
yy = - 40960 * xx ^ 11 / 77 + 15280 * xx ^ 9 / 7
     - 28800 * xx ^ 7 / 7 + 7680 * xx  ^ 5 - 61440 * xx ^ 4 / 7
     + 4800 * xx ^ 3 - 10240 * xx ^ 2 / 7 + 240 * xx - 1280 / 77
lines(xx, yy, type = "l")
```

After a call to `set.seed(3)` to set the random number seed, the R code plots the histogram of the simulated values stored in the vector `m` in Figure 1.33, along with the theoretical probability density function of the sample median from the derivation. Because the population distribution is skewed to the left, the sampling distribution of the sample median is also skewed to the left.

Finally, how does the expected value of the sample median compare to the population median? The population median can be found by solving

$$F_X(x_{0.5}) = x_{0.5}^2 = \frac{1}{2}$$

for the population median $x_{0.5}$. This results in

$$x_{0.5} = \frac{1}{\sqrt{2}} \cong 0.7071.$$

The expected value of the sample median is

$$E[Y_1] = \int_0^1 y_1 f_{Y_1}(y_1)\,dy_1 = \frac{160}{231} \cong 0.6926.$$

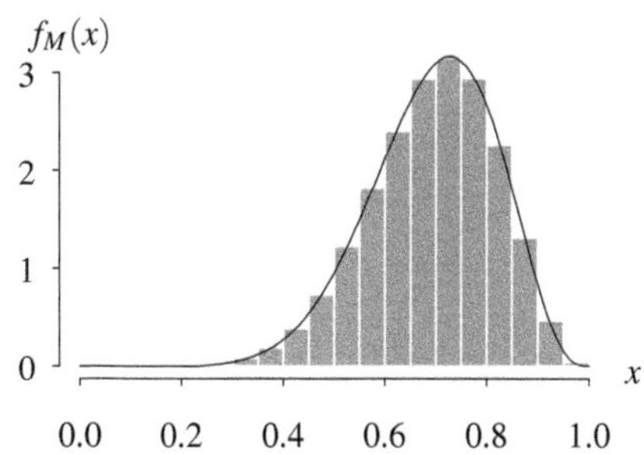

Figure 1.33: Sampling distribution of the sample median (200,000 replications).

The expected value of the sample median just slightly underestimates the population median. It is off target, aiming low by a small amount. Using the jargon of the statistician, "the sample median is a biased estimator of the population median." The sample median is a biased estimator of the population median because the population distribution is nonsymmetric.

Other measures of central tendency

The sample mean and the sample median will cover over 99% of the applications requiring an estimate of central tendency that will arise in practice. There are three other statistics that reflect the central tendency of a population probability distribution that are introduced now for completeness. They are

- the sample geometric mean,
- the sample harmonic mean, and
- the sample quadratic mean.

Definition 1.7 Let $x_1, x_2, \ldots, x_n$ be experimental values associated with the positive random variables $X_1, X_2, \ldots, X_n$. The *sample geometric mean* is

$$G = \left(\prod_{i=1}^{n} X_i \right)^{1/n}.$$

The sample geometric mean has a geometric interpretation. Consider the rectangle on the left in Figure 1.34 with side lengths x_1 and x_2 (that is, $n = 2$ in Definition 1.7). The square to the right of the rectangle has the same area as the rectangle and has side lengths that equal the sample geometric mean $g = \sqrt{x_1 x_2}$. Next consider the rectangular solid on the left in Figure 1.35 with side lengths x_1, x_2, and x_3 (that is, $n = 3$ in Definition 1.7). The cube to the right of the rectangular solid has the same volume as the rectangular solid. The cube has side lengths that equal the sample geometric mean $g = \sqrt[3]{x_1 x_2 x_3}$. This progression continues for larger values of the sample size n, but, of course, no pictures can be drawn.

There is another geometric interpretation of the sample geometric mean that can be applied in film and video. The *aspect ratio* of a rectangular image is the ratio of the width to the height of the rectangle, often expressed in the form *a*:*b*. For example,

- 35 millimeter film has an aspect ratio of 1.5:1,

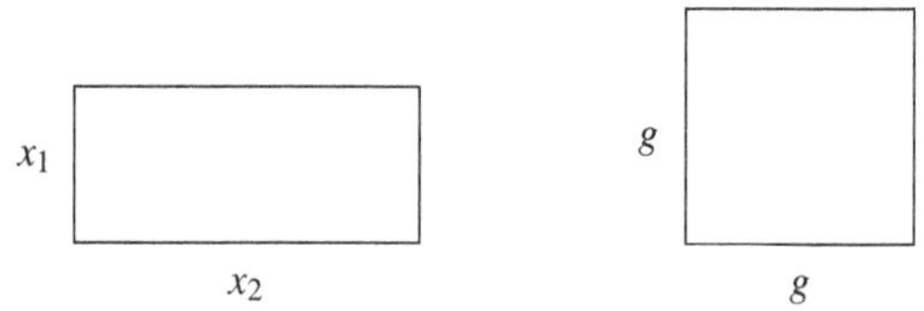

Figure 1.34: Equal-area geometric interpretation of the sample geometric mean for $n = 2$.

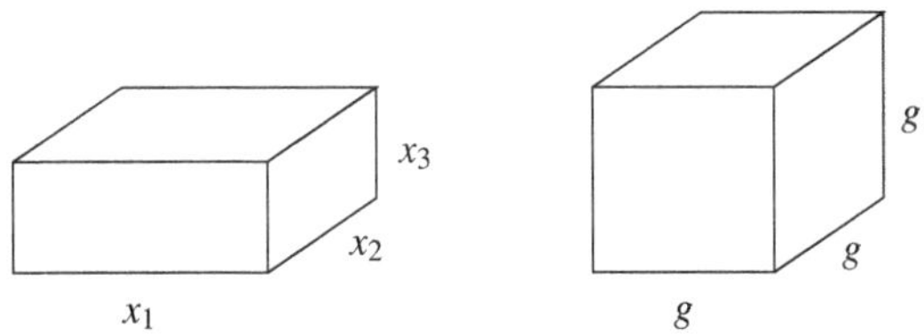

Figure 1.35: Equal-volume geometric interpretation of the sample geometric mean for $n = 3$.

- credit cards measure 85.6 mm by 54 mm, so their aspect ratio is about 1.59:1,
- a rectangle possessing the "golden ratio" has aspect ratio $\frac{1+\sqrt{5}}{2}$:1, or about 1.62:1, and
- the high-definition video standard aspect ratio is 16:9 or $1.\bar{7}$:1.

The sample geometric mean is an appropriate way to average the aspect ratios of two rectangles. More specifically, consider the two rectangles in Figure 1.36 that have (a) equal areas, (b) identical centers, (c) longer sides oriented horizontally, and (d) different aspect ratios. A smaller rectangle, which is formed by the intersection points of the two larger rectangles, has an aspect ratio which is the sample geometric mean of the aspect ratios of the two larger rectangles. For example, if the wider rectangle has width 8 and height 3 (area 24), then its aspect ratio is 8:3, or $8/3$. If the narrower rectangle has width 6 and height 4 (area 24), then its aspect ratio is 3:2, or $3/2$. The internal rectangle, which is a compromise between the two rectangles, will have an aspect ratio which is the sample geometric mean of the aspect ratios of the two larger rectangles:

$$g = \sqrt{\frac{8}{3} \cdot \frac{3}{2}} = \sqrt{4} = 2.$$

Since the dimensions of the smaller rectangle are width 6 and height 3, the minimum of the widths and the minimum of the heights, its aspect ratio is indeed given by 2:1, matching the value given by the sample geometric mean.

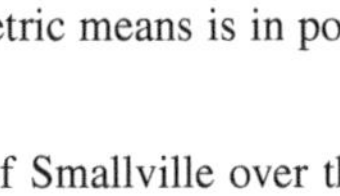

Figure 1.36: Two equal-area rectangles with different aspect ratios.

Another application area for geometric means is in population growth, as illustrated in the following example.

Example 1.22 The population of Smallville over the span of four annual censuses is 1000, 1600, 2000, and 3300. Find the sample geometric mean of the growth multipliers from year-to-year and explain why it is preferred to the sample arithmetic mean.

The three growth percentages between the censuses are

$$\frac{1600-1000}{1000} = 60\% \qquad \frac{2000-1600}{1600} = 25\% \qquad \frac{3300-2000}{2000} = 65\%.$$

The associated three growth multipliers between the censuses are

$$1.60 \qquad 1.25 \qquad 1.65.$$

First, consider the sample arithmetic mean of the $n = 3$ growth multipliers, which is

$$\bar{x} = \frac{1.60 + 1.25 + 1.65}{3} = \frac{4.50}{3} = 1.50.$$

If this multiplier is applied to the original population of 1000 residents of Smallville from the first census, one would expect to have $1000 \cdot 1.5^3 = 3375$ residents upon the last census. This is 75 more than the actual count. Alternatively, consider the sample geometric mean of the $n = 3$ growth multipliers, which is

$$g = \sqrt[3]{(1.60)(1.25)(1.65)} = \frac{\sqrt[3]{3300}}{10} \cong 1.4888.$$

If one uses the growth multiplier g on the original 1000 people in Smallville from the first census, one would expect to have $1000 \cdot 1.4888^3 = 3300$ residents upon the last census, which is exactly what we observe. In problems of this nature, the sample geometric mean outperforms the sample arithmetic mean in terms of identifying the appropriate mean growth multiplier.

Just the look of the formula for the sample geometric mean from Definition 1.7 tells us that determining its sampling distribution is going to be difficult. An example of finding that sampling distribution for one particular population probability distribution is given next.

Example 1.23 Let $X_1, X_2, \ldots, X_n$ be a random sample from a continuous population probability distribution having probability density function

$$f_X(x) = \lambda x^{-(\lambda+1)} \qquad x > 1,$$

where λ is a positive parameter. Find the sampling distribution of the sample geometric mean.

Before addressing the specific population probability distribution given in the problem, let's begin the derivation generically using the cumulative distribution function technique to determine the distribution of G:

$$\begin{aligned} F_G(g) &= P(G \le g) \\ &= P\left(\left(\prod_{i=1}^{n} X_i\right)^{1/n} \le g\right) \\ &= P\left(\prod_{i=1}^{n} X_i \le g^n\right) \\ &= P\left(\ln\left(\prod_{i=1}^{n} X_i\right) \le \ln g^n\right) \\ &= P\left(\sum_{i=1}^{n} \ln X_i \le n \ln g\right). \end{aligned}$$

Where the problem goes from here depends on whether the probability distribution of $\sum_{i=1}^{n} \ln X_i$ has a tractable cumulative distribution function. Consider the distribution of $\ln X_i$, the logarithm of the ith data value. Dropping the subscript i, the problem states that the probability density function of X is

$$f_X(x) = \lambda x^{-(\lambda+1)} \qquad x > 1.$$

The associated population cumulative distribution function on its support is

$$F_X(x) = \int_1^x \lambda w^{-\lambda-1}\,dw = \left[-w^{-\lambda}\right]_1^x = 1 - x^{-\lambda} \qquad x > 1.$$

Using the cumulative distribution function technique, the cumulative distribution function of $Y = \ln X$ is

$$\begin{aligned} F_Y(y) &= P(Y \le y) \\ &= P(\ln X \le y) \\ &= P(X \le e^y) \\ &= F_X(e^y) \\ &= 1 - (e^y)^{-\lambda} \\ &= 1 - e^{-\lambda y} \qquad y > 0. \end{aligned}$$

This is *very* good news. This cumulative distribution function is recognized as that of an exponential(λ) random variable. So $Y = \ln X$ is an exponential random variable. But our interest is in the probability distribution of $\sum_{i=1}^{n} \ln X_i$. Since the sum of n mutually independent and identically distributed exponential(λ) random variables is Erlang(λ, n) via a result from probability theory proved using the moment generating function technique, we can continue with our derivation of the cumulative distribution function of G because the cumulative distribution function of an Erlang random variable can be written as a summation. So

$$\begin{aligned} F_G(g) &= P(G \le g) \\ &= P\left(\sum_{i=1}^{n} \ln X_i \le n \ln g\right) \\ &= P\big(\text{Erlang}(\lambda, n) \le n \ln g\big) \\ &= 1 - \sum_{y=0}^{n-1} \frac{(\lambda n \ln g)^y e^{-\lambda n \ln g}}{y!} \qquad g > 1. \end{aligned}$$

A Monte Carlo simulation experiment can be used to support the claim that this is indeed the cumulative distribution function of the sample geometric mean G for particular values of n and λ, say $n = 6$ and $\lambda = 2$. Random variates from the population probability distribution can be generated via inversion. Since the population cumulative distribution function on its support is

$$F_X(x) = 1 - x^{-\lambda} \qquad x > 1,$$

the inverse cumulative distribution function is

$$F_X^{-1}(u) = (1-u)^{-1/\lambda} \qquad 0 < u < 1,$$

where u is a random number, that is, a $U(0, 1)$ random variate, which is a realization of a $U(0, 1)$ random variable. So random variates are generated via inversion by

$$x \leftarrow (1-u)^{-1/\lambda}.$$

The R code segment below shows how to generate `nrep = 200` sample geometric means, each from a sample size of $n = 6$ observations drawn from the population distribution with $\lambda = 2$.

```
nrep   = 200
n      = 6
lambda = 2
g      = numeric(nrep)
for (i in 1:nrep) {
  x = (1 - runif(n)) ^ (-1 / lambda)
  g[i] = prod(x) ^ (1 / n)
}
```

The resulting empirical cumulative distribution function for the sample geometric means generated by Monte Carlo and the cumulative distribution function of G calculated analytically on its support as

$$F_G(g) = 1 - \sum_{y=0}^{5} \frac{(12 \ln g)^y e^{-12 \ln g}}{y!} \qquad g > 1$$

are plotted in Figure 1.37. The empirical cumulative distribution function $\hat{F}_G(g)$ associated with the 200 sample geometric means is a step function with 200 tiny upward steps at each generated sample geometric mean. The analytic cumulative distribution function $F_G(g)$ is the smooth curve. The two are nearly identical, so our Monte Carlo simulation supports the analytic derivation. For much larger values of `nrep`, the step function and the smooth curve are indistinguishable, lending strong evidence to the validity of the analytic result.

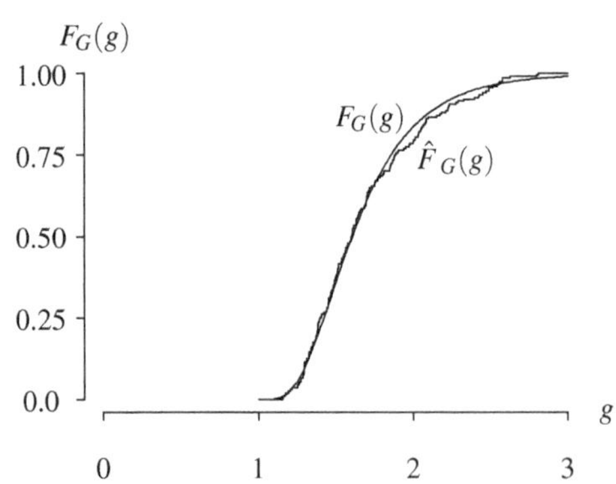

Figure 1.37: Cumulative distribution functions for the sample geometric mean.

This concludes the discussion of the sample geometric mean. The sample harmonic mean, which is defined next, appears in applications in which *rates* or *ratios* are being averaged.

Definition 1.8 Let $x_1, x_2, \ldots, x_n$ be experimental values associated with the positive random variables $X_1, X_2, \ldots, X_n$. The *sample harmonic mean* is

$$H = \frac{1}{\frac{1}{n}\sum_{i=1}^{n}\frac{1}{X_i}} = \frac{n}{\sum_{i=1}^{n}X_i^{-1}}.$$

The sample harmonic mean is the reciprocal of the arithmetic mean of the reciprocals of the data values. The following list contains applications of the sample harmonic mean. In all cases, the example illustrates a sample size of $n = 2$, but the thinking generalizes to larger sample sizes.

- If you bicycle down a hill at $x_1 = 30$ miles per hour, then bicycle up the same hill at $x_2 = 20$ miles per hour, what is your average speed? The arithmetic sample mean $\bar{x} = 25$ miles per hour is not the average speed. Rates do not average in this fashion. The sample harmonic mean

  $$h = \frac{2}{1/30 + 1/20} = \frac{2}{50/600} = 24 \text{ miles per hour}$$

  gives the appropriate average speed. To see that this is the case, rewrite the "rate times time equals distance" formula as "time equals distance over rate." Letting d be the distance in one direction, t_1 be the time going downhill, and t_2 be the time going uphill, the times going downhill and uphill are

  $$t_1 = \frac{d}{30} \qquad \text{and} \qquad t_2 = \frac{d}{20}.$$

  Since the total distance traveled is $2d$ and the total time is $t_1 + t_2$, the average rate is

  $$\frac{2d}{t_1 + t_2} = \frac{2d}{d/30 + d/20} = \frac{2}{1/30 + 1/20} = 24 \text{ miles per hour,}$$

  which is just the sample harmonic mean. Traveling the entire distance at 24 miles per hour will take the same amount of time as the original circuit.

- In electrical engineering, placing two resistors in parallel with resistance $x_1 = 30$ ohms and $x_2 = 20$ ohms has an equivalent effect to placing two identical resistors in parallel with resistances $h = 24$ ohms, which is the sample harmonic mean.

- If one person can mow the lawn in $x_1 = 30$ minutes and a second person can mow the lawn in $x_2 = 20$ minutes, the total time to mow the lawn when the two work together is half of the sample harmonic mean, which is $h/2 = 12$ minutes.

- If one car gets $x_1 = 30$ miles per gallon and a second car gets $x_2 = 20$ miles per gallon, then the average number of miles per gallon is the sample arithmetic mean $\bar{x} = 25$ miles per gallon. If, however, you want the average number of gallons per mile, the reciprocal of the units, use the reciprocal of the sample harmonic mean: $1/h = 1/24$ gallon per mile.

The sample arithmetic mean, sample geometric mean, and sample harmonic mean together comprise what are known as the *Pythagorean means*. The next example illustrates how to find the sampling distribution of the sample harmonic mean.

Example 1.24 Let $X_1, X_2, \ldots, X_n$ be a random sample from a continuous population with probability density function

$$f_X(x) = \frac{1}{\Gamma(\kappa)} \lambda^{\kappa} x^{-\kappa-1} e^{-\lambda/x} \qquad x > 0,$$

where λ and κ are positive unknown parameters. Find the sampling distribution of the sample harmonic mean

$$H = \frac{1}{\frac{1}{n}\sum_{i=1}^{n} \frac{1}{X_i}}.$$

Juggling all of the reciprocals in the formula for the sample harmonic mean is going to require a significant number of transformations to arrive at the sampling distribution of the sample harmonic mean. The three steps in the process are

- find the distribution of the reciprocals of the data values: $Y_i = 1/X_i$, $i = 1, 2, \ldots, n$,
- find the distribution of the sum of the reciprocals of the data values $Y = \sum_{i=1}^{n} Y_i$,
- find the distribution of the sample harmonic mean $H = n/Y$.

The ith data value X_i has probability density function

$$f_{X_i}(x) = \frac{1}{\Gamma(\kappa)} \lambda^{\kappa} x^{-\kappa-1} e^{-\lambda/x} \qquad x > 0$$

for $i = 1, 2, \ldots, n$. The first step in the process is to find the distribution of $Y_i = 1/X_i$; we use the transformation technique. Temporarily dropping the subscript i, the transformation $y = g(x) = 1/x$ is a one-to-one transformation from

$$\mathcal{A} = \{x \,|\, x > 0\}$$

to

$$\mathcal{B} = \{y \,|\, y > 0\}$$

with closed-form inverse $x = g^{-1}(y) = 1/y$ and Jacobian $\frac{dx}{dy} = -y^{-2}$. So the probability density function of Y_i is

$$f_{Y_i}(y) = f_{X_i}\left(g^{-1}(y)\right) \left|\frac{dx}{dy}\right| \qquad y \in \mathcal{B}$$

for $i = 1, 2, \ldots, n$. For this particular transformation,

$$f_{Y_i}(y) = \frac{1}{\Gamma(\kappa)} \lambda^{\kappa} y^{\kappa+1} e^{-\lambda y} \left|-y^{-2}\right| \qquad y > 0$$

for $i = 1, 2, \ldots, n$. This reduces to

$$f_{Y_i}(y) = \frac{1}{\Gamma(\kappa)} \lambda^{\kappa} y^{\kappa-1} e^{-\lambda y} \qquad y > 0$$

for $i = 1, 2, \ldots, n$. This probability density function is recognized as that of a gamma(λ, κ) random variable, which is good news for the next step.

Next we want to find the distribution of the sum of the reciprocals of the data values, that is,

$$Y = Y_1 + Y_2 + \cdots + Y_n.$$

Since $Y_i \sim$ gamma(λ, κ) for $i = 1, 2, \ldots, n$, the moment generating function of Y_i is

$$M_{Y_i}(t) = \left(\frac{\lambda}{\lambda - t}\right)^{\kappa} \qquad t < \lambda$$

for $i = 1, 2, \ldots, n$. Since $Y_1, Y_2, \ldots, Y_n$ are mutually independent and identically distributed random variables, the moment generating function of $Y_1 + Y_2 + \cdots + Y_n$ is

$$\begin{aligned} M_{Y_1+Y_2+\cdots+Y_n}(t) &= M_{Y_1}(t)M_{Y_2}(t)\ldots M_{Y_n}(t) \\ &= \left(\frac{\lambda}{\lambda - t}\right)^{\kappa}\left(\frac{\lambda}{\lambda - t}\right)^{\kappa}\cdots\left(\frac{\lambda}{\lambda - t}\right)^{\kappa} \\ &= \left(\frac{\lambda}{\lambda - t}\right)^{n\kappa} \qquad t < \lambda. \end{aligned}$$

This is recognized as the moment generating function of a gamma(λ, $n\kappa$) random variable. So the probability density function of Y is

$$f_Y(y) = \frac{1}{\Gamma(n\kappa)}\lambda^{n\kappa}y^{n\kappa-1}e^{-\lambda y} \qquad y > 0.$$

The last step is to find the probability density function of the sample harmonic mean $H = n/Y$. As before, we use the transformation technique. The transformation $h = g(y) = n/y$ is a one-to-one transformation from

$$\mathcal{A} = \{y \,|\, y > 0\}$$

to

$$\mathcal{B} = \{h \,|\, h > 0\}$$

with inverse $y = g^{-1}(h) = n/h$ and Jacobian $\frac{dy}{dh} = -nh^{-2}$. So the probability density function of H is

$$f_H(h) = f_Y\left(g^{-1}(h)\right)\left|\frac{dy}{dh}\right| \qquad h \in \mathcal{B}.$$

For this particular transformation,

$$f_H(h) = \frac{1}{\Gamma(n\kappa)}\lambda^{n\kappa}(n/h)^{n\kappa-1}e^{-n\lambda/h}\left|-\frac{n}{h^2}\right| \qquad h > 0,$$

which reduces to

$$f_H(h) = \frac{1}{\Gamma(n\kappa)}(n\lambda)^{n\kappa}h^{-n\kappa-1}e^{-n\lambda/h} \qquad h > 0.$$

This probability density function of the sample harmonic mean looks quite similar in form to the population probability density function. The population probability density function is known as the *inverse gamma distribution*. The result that has been derived here is as follows: if $X_1, X_2, \ldots, X_n$ is a random sample from an inverse gamma(λ, κ) distribution, then the sample harmonic mean has an inverse gamma($n\lambda$, $n\kappa$) distribution.

The sampling distribution of the sample harmonic mean was calculated for a random sample in the previous example. The next example considers the case of random variables that are mutually independent, but not identically distributed. The inverse gamma distribution is used once again in order to borrow some of the mathematics from the previous example.

Example 1.25 Let the mutually independent random variables X_1, X_2, X_3 represent the times, in hours, that it takes Huey, Dewey, and Louie to mow the lawn individually. The probability density functions of the three random variables is given by

$$f_{X_i}(x) = \frac{1}{\Gamma(\kappa_i)}\lambda^{\kappa_i}x^{-\kappa_i-1}e^{-\lambda/x} \qquad x > 0$$

for $i = 1, 2, 3$, where λ, κ_1, κ_2, κ_3 are positive unknown parameters.

(a) Find the probability density function of the time it takes for them to mow the lawn together, assuming no interference or loss of efficiency.

(b) Plot the probability density functions of the individual times to mow the lawn and the probability density function of the time to mow the lawn when they work together when $\lambda = 5.7$, $\kappa_1 = 3.7$, $\kappa_2 = 3.0$, $\kappa_3 = 2.3$.

(c) Conduct a Monte Carlo simulation experiment that supports the plot from part (b).

(a) The time for each of the boys to mow the lawn individually has an inverse gamma distribution. The three steps to find the probability distribution of the sample harmonic mean are similar to the previous example:

- find the distribution of the reciprocals of the data values: $Y_i = 1/X_i$, $i = 1, 2, 3$,
- find the distribution of the sum of the reciprocals of the X_i values $Y = \sum_{i=1}^{3} Y_i$,
- find the distribution of the time to mow the lawn $T = 1/Y$.

Using the result from the previous question, $Y_i \sim \text{gamma}(\lambda, \kappa_i)$ for $i = 1, 2, 3$. The next step is to find the distribution of the sum of the reciprocals of the data values, that is,

$$Y = Y_1 + Y_2 + Y_3.$$

The moment generating function of Y_i is

$$M_{Y_i}(t) = \left(\frac{\lambda}{\lambda - t}\right)^{\kappa_i} \qquad t < \lambda$$

for $i = 1, 2, 3$. Since Y_1, Y_2, Y_3 are mutually independent random variables, the moment generating function of $Y_1 + Y_2 + Y_3$ is

$$\begin{aligned} M_{Y_1+Y_2+Y_3}(t) &= M_{Y_1}(t)M_{Y_2}(t)M_{Y_3}(t) \\ &= \left(\frac{\lambda}{\lambda - t}\right)^{\kappa_1}\left(\frac{\lambda}{\lambda - t}\right)^{\kappa_2}\left(\frac{\lambda}{\lambda - t}\right)^{\kappa_3} \\ &= \left(\frac{\lambda}{\lambda - t}\right)^{\kappa_1+\kappa_2+\kappa_3} \qquad t < \lambda. \end{aligned}$$

This is recognized as the moment generating function of a $\text{gamma}(\lambda, \kappa_1 + \kappa_2 + \kappa_3)$ random variable. So the probability density function of Y is

$$f_Y(y) = \frac{1}{\Gamma(\kappa_1 + \kappa_2 + \kappa_3)}\lambda^{\kappa_1+\kappa_2+\kappa_3}y^{\kappa_1+\kappa_2+\kappa_3-1}e^{-\lambda y} \qquad y > 0.$$

The last step is to find the probability distribution of the time it takes the three boys to mow the lawn together $T = 1/Y$. The transformation $t = g(y) = 1/y$ is a one-to-one transformation from

$$\mathcal{A} = \{y \,|\, y > 0\}$$

to

$$\mathcal{B} = \{t \,|\, t > 0\},$$

with closed-form inverse $y = g^{-1}(t) = 1/t$ and Jacobian $\frac{dy}{dt} = -t^{-2}$. So the probability density function of T is

$$f_T(t) = f_Y\left(g^{-1}(t)\right)\left|\frac{dy}{dt}\right| \qquad t \in \mathcal{B}.$$

For this particular transformation,

$$f_T(t) = \frac{1}{\Gamma(\kappa_1+\kappa_2+\kappa_3)}\lambda^{\kappa_1+\kappa_2+\kappa_3}t^{-\kappa_1-\kappa_2-\kappa_3+1}e^{-\lambda/t}\left|-\frac{1}{t^2}\right| \qquad t > 0.$$

This reduces to

$$f_T(t) = \frac{1}{\Gamma(\kappa_1+\kappa_2+\kappa_3)}\lambda^{\kappa_1+\kappa_2+\kappa_3}t^{-\kappa_1-\kappa_2-\kappa_3-1}e^{-\lambda/t} \qquad t > 0,$$

which can be recognized as the probability density function of an inverse gamma random variable with parameters λ and $\kappa_1+\kappa_2+\kappa_3$.

(b) The probability density functions of the individual times to mow the lawn and the probability density function of the time to mow the lawn when they work together when $\lambda = 5.7$, $\kappa_1 = 3.7$, $\kappa_2 = 3.0$, $\kappa_3 = 2.3$ are plotted in Figure 1.38.

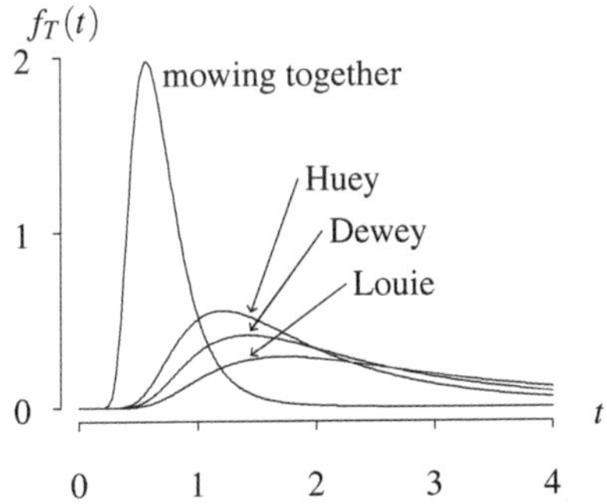

Figure 1.38: Probability density functions of mowing times.

(c) Using the R random variate generator `rgamma` to generate random gamma variates, the following code generates 200,000 lawn mowing times when the three boys work together, then plots a histogram. The histogram and analytic probability density function $f_T(t)$ are displayed in Figure 1.39. Their similar shapes support the analytic derivation.

```
nrep = 200000
x1 = 1 / rgamma(nrep, 3.7, 5.7)
x2 = 1 / rgamma(nrep, 3.0, 5.7)
x3 = 1 / rgamma(nrep, 2.3, 5.7)
tt = 1 / (1 / x1 + 1 / x2 + 1 / x3)
hist(tt, probability = TRUE)
```

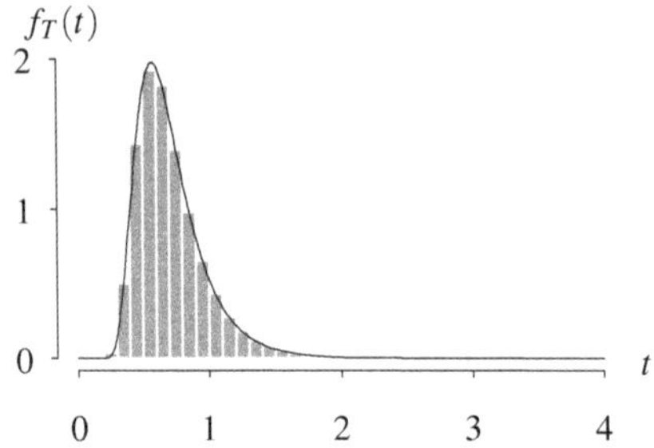

Figure 1.39: Monte Carlo simulation and analytic results.

The final measure of central tendency, known as the *sample quadratic mean*, also goes by the name root mean square, and is often abbreviated with RMS. It is useful for averaging the magnitude of a quantity that varies with time, which makes it popular in electrical engineering (for example, alternating current).

Definition 1.9 Let $x_1, x_2, \ldots, x_n$ be experimental values associated with the random variables $X_1, X_2, \ldots, X_n$. The *sample quadratic mean* is

$$Q = \sqrt{\frac{1}{n}\sum_{i=1}^{n} X_i^2}.$$

The sample quadratic mean can be used for sampling a function that varies over time that might contain negative and positive values, such as a sinusoidal function. Averaging the squares of the observations under the square root emphasizes that the sample quadratic mean is measuring the average magnitude of the data values. The next example considers the sampling distribution of the sample quadratic mean for a random sample.

Example 1.26 Let $X_1, X_2, \ldots, X_n$ be a random sample from a continuous population with probability density function

$$f_X(x) = \sqrt{\frac{2}{\pi}}\, e^{-x^2/2} \qquad x > 0.$$

Find the sampling distribution of the sample quadratic mean

$$Q = \sqrt{\frac{1}{n}\sum_{i=1}^{n} X_i^2}.$$

The population probability density function is that of a standard normal distribution truncated on the left at 0. This distribution is known as the *chi distribution* with one degree of freedom. Rather than give all of the details associated with the derivation of Q, this time we just outline the steps and leave the reader to fill in the details.

- The ith data value X_i has probability density function

  $$f_{X_i}(x) = \sqrt{\frac{2}{\pi}}\, e^{-x^2/2} \qquad x > 0$$

  for $i = 1, 2, \ldots, n$.

- Using the transformation technique, $Y_i = X_i^2$ has probability density function

  $$f_{Y_i}(y) = \frac{1}{\sqrt{2\pi}} y^{-1/2} e^{-y/2} \qquad y > 0$$

  for $i = 1, 2, \ldots, n$, which is recognized as the chi-square distribution with one degree of freedom.

- Using the moment generating function technique, the sum of n mutually independent $\chi^2(1)$ random variables has the $\chi^2(n)$ distribution, so

  $$Y = Y_1 + Y_2 + \cdots + Y_n = X_1^2 + X_2^2 + \cdots + X_n^2 \sim \chi^2(n),$$

  with associated probability density function

  $$f_Y(y) = \frac{1}{2^{n/2}\Gamma(n/2)} y^{n/2-1} e^{-y/2} \qquad y > 0.$$

- Using the transformation technique, the sample quadratic mean $Q = \sqrt{Y/n}$ has probability density function

  $$f_Q(q) = \frac{1}{2^{n/2-1}\Gamma(n/2)} n^{n/2} q^{n-1} e^{-nq^2/2} \qquad q > 0.$$

This probability density function is plotted for the population distribution ($n = 1$) and sample sizes $n = 2$ and $n = 12$ in Figure 1.40. As was the case with the other sample means, the population variance of the sampling distribution of the sample quadratic mean decreases as n increases.

We have now surveyed five measures of central tendency associated with a data set $X_1, X_2, \ldots, X_n$, which are:

- the sample mean $\bar{X}$,
- the sample median M,
- the sample geometric mean G,
- the sample harmonic mean H, and
- the sample quadratic mean Q.

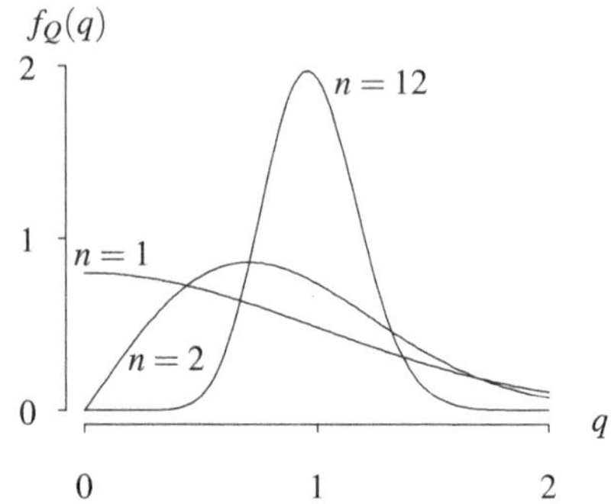

Figure 1.40: Sampling distribution of the sample quadratic mean.

The sample mean, also known as the sample arithmetic mean, is used in far more applications in statistics than any of the others. We end this section by listing two properties associated with the four sample means $\bar{X}$, G, H, and Q.

- If all of the data values are identical to some positive constant c, then all of the various sample means assume the same value, that is,

$$x_1 = x_2 = \cdots = x_n = c \qquad \Rightarrow \qquad \bar{x} = g = h = q = c.$$

- The four sample means satisfy the inequality

$$\min\{x_1, x_2, \ldots, x_n\} \le h \le g \le \bar{x} \le q \le \max\{x_1, x_2, \ldots, x_n\}$$

  for positive data values $x_1, x_2, \ldots, x_n$. Equality holds if and only if $x_1 = x_2 = \cdots = x_n$.

The R function `mean` computes the sample arithmetic mean. The R code below defines the functions `gmean`, `hmean`, and `qmean` that calculate the sample geometric mean, harmonic mean, and quadratic mean, respectively.

```
gmean = function(x) prod(x) ^ (1 / length(x))
hmean = function(x) 1 / mean(1 / x)
qmean = function(x) sqrt(mean(x ^ 2))
```

When the data set of $n = 10$ observations:

$$3, 1, 5, 1, 3, 2, 1, 1, 3, 2,$$

from Example 1.15 is used as an argument to the four functions, the code

```
x = c(3, 1, 5, 1, 3, 2, 1, 1, 3, 2)
min(x)
hmean(x)
gmean(x)
mean(x)
qmean(x)
max(x)
```

results in an ordering consistent with the inequality concerning the measures of central tendency:

$$x_{(1)} = 1 \qquad h = 1.61 \qquad g = 1.88 \qquad \bar{x} = 2.2 \qquad q = 2.53 \qquad x_{(10)} = 5.$$

1.4 Estimating Dispersion

Central tendency is just one metric associated with a population probability distribution. Measuring how far a random variable might stray from a central value is a second key aspect of a population probability distribution. This section introduces statistics that statisticians use to measure dispersion about a central value from n data values. We begin with the most important measure of dispersion: the sample variance.

Sample variance

One way to begin to develop a statistical analog of the population variance is to return to the empirical distribution. Recall from Definition 1.4 that the empirical distribution was developed by assigning a probability of $1/n$ to each data value. This allowed us to see that the "plug in" estimator of the population mean was the population mean of this empirical distribution. We will now determine the plug-in estimator of the population variance.

As in the previous two sections, let $x_1, x_2, \ldots, x_n$ be experimental values associated with the random variables $X_1, X_2, \ldots, X_n$. We will apply the formula for the population variance from probability theory

$$\sigma^2 = V[X] = E\left[(X-\mu)^2\right]$$

to the empirical probability distribution associated with the data values. Since the empirical probability distribution is a discrete probability distribution, the formula for the population variance becomes

$$\sigma^2 = V[X] = E\left[(X-\mu)^2\right] = \sum_{\mathcal{A}} (x-\mu)^2 f(x),$$

where $\mathcal{A}$ is the support of the random variable X. Since $f(x_i) = 1/n$, for $i = 1, 2, \ldots, n$ and μ can be replaced by its plug-in estimator $\bar{x}$, the plug-in estimator of the population variance from the empirical distribution becomes

$$\hat{\sigma}^2 = \frac{1}{n} \sum_{i=1}^{n} (x_i - \bar{x})^2.$$

At first glance, this estimator seems quite reasonable. It is the arithmetic average of the squared deviations of each data value from the sample mean. One flaw will become apparent through the following thought experiment.

Let's say you go to Mars and observe the height of $n = 1$ Martian, who is exactly 2 feet tall. So the estimate of the population mean is just the sample mean: $\bar{x} = 2$. Now you would like to estimate the population variance σ^2. The summation in the plug-in estimator of the population variance has only a single term, which is zero because $x_1 = \bar{x}$, so $\hat{\sigma}^2 = 0$. But is a population variance estimate of zero appropriate here? A population variance of zero implies that *all* Martians are exactly 2 feet tall. All Martians *might* indeed be 2 feet tall or they might range from 1 foot tall to 10 feet tall. You just cannot tell by observing the height of a single Martian. In fact, you cannot estimate the population variance at all when $n = 1$. So the definition of the sample variance given next is undefined when $n = 1$, and this is accomplished by dividing the sum of squares by $n - 1$ rather than n.

Definition 1.10 Let $x_1, x_2, \ldots, x_n$ be experimental values associated with the random variables $X_1, X_2, \ldots, X_n$. The *sample variance* is

$$S^2 = \frac{1}{n-1} \sum_{i=1}^{n} \left(X_i - \bar{X}\right)^2.$$

The positive square root of the sample variance is the *sample standard deviation S*.

The $n-1$ in the denominator is counter-intuitive to most people who encounter this version of the sample variance for the first time. Here are three comments on the unusual denominator.

- As mentioned earlier, a sample size of $n = 1$ means that S^2 is undefined. This is consistent with the notion that the population variance cannot be estimated from a single observation.

- The summation $\sum_{i=1}^{n} \left(X_i - \bar{X}\right)^2$ is a statistic with $n-1$ "degrees of freedom." In many application areas, statisticians divide statistics by their degrees of freedom. So dividing by $n-1$ here is consistent with that practice. One degree of freedom is lost by using $\bar{X}$ rather than μ in the terms of the summation.

- Perhaps more important than the first two comments, the sample variance is an unbiased estimator of the population variance, that is,

$$E\left[S^2\right] = \sigma^2$$

for any population with finite population mean and finite population variance. This result will be formally stated and proved in Section 2.4. The important take-away here is that the sample variance S^2 is "on target" for estimating σ^2. Using anything other than $n-1$ in the denominator of the formula for S^2 would result in a sample variance that is not aimed at the population variance σ^2.

Computing the sample variance is not a trivial matter for large data sets. The "defining formula" given in Definition 1.10 leads to a two-pass algorithm for computing s^2, in that the data must be passed over twice in order to compute s^2: once to compute $\bar{x}$, then a second time to compute $\sum_{i=1}^{n}(x_i - \bar{x})^2$. This procedure is inefficient computationally. Some simple algebra can lead to a one-pass algorithm:

$$s^2 = \frac{1}{n-1} \sum_{i=1}^{n} (x_i - \bar{x})^2 = \frac{1}{n-1} \left[\sum_{i=1}^{n} x_i^2 - n\bar{x}^2\right] = \frac{1}{n-1} \left[\sum_{i=1}^{n} x_i^2 - \frac{1}{n}\left(\sum_{i=1}^{n} x_i\right)^2\right].$$

In the one-pass algorithm, the data is passed over just once, computing the running values of $\sum_{i=1}^{n} x_i^2$ and $\sum_{i=1}^{n} x_i$.

Even with the one-pass algorithm, problems still arise. For huge data sets that must be processed on a computer using floating-point arithmetic, there is the possibility of round-off and overflow errors. There is additional risk when the data values themselves are large numbers. A dozen or so algorithms have been designed to decrease the possibility of these issues having an impact on the computed value of s^2.

Analogous to the definition of the population standard deviation in probability theory, the positive square root of the sample variance is the sample standard deviation. The units of the sample standard deviation are inherited from the units of the data values. The sample variance is computed in R with the `var` function and the standard deviation is computed with the `sd` function, as illustrated next for a small sample.

Example 1.27 Consider again the data set of $n = 10$ observations from Example 1.15, where kindergarteners were polled concerning the number of children in their family. The data values $x_1, x_2, \ldots, x_{10}$ are

$$3, 1, 5, 1, 3, 2, 1, 1, 3, 2.$$

Calculate the sample variance s^2 and the sample standard deviation s.

The sample mean is

$$\bar{x} = \frac{3+1+5+1+3+2+1+1+3+2}{10} = \frac{22}{10} = 2.2 \text{ children.}$$

Implementing the two-pass algorithm with the defining formula from Definition 1.10, the sample variance is

$$\begin{aligned} s^2 &= \frac{1}{9}\sum_{i=1}^{10}(x_i - \bar{x})^2 \\ &= \frac{1}{9}\left[(3-2.2)^2 + (1-2.2)^2 + (5-2.2)^2 + \cdots + (2-2.2)^2\right] \\ &= \frac{1}{9} \cdot 15.6 \\ &\cong 1.73. \end{aligned}$$

The units on the sample variance are the rather awkward-sounding "children squared." The one-pass algorithm calculates

$$\sum_{i=1}^{10} x_i = 22 \qquad \text{and} \qquad \sum_{i=1}^{10} x_i^2 = 64$$

on a single pass through the data set, then calculates

$$\begin{aligned} s^2 &= \frac{1}{9}\left[\sum_{i=1}^{10} x_i^2 - 10\bar{x}^2\right] \\ &= \frac{1}{9}\left[64 - 10 \cdot 2.2^2\right] \\ &= \frac{1}{9}\left[64 - 48.4\right] \\ &= \frac{1}{9} \cdot 15.6 \\ &\cong 1.73. \end{aligned}$$

The sample variance is the second moment of the data about $\bar{X}$, analogous to the moment of inertia in physics. For large data sets, a statistical package is typically used to calculate the sample mean and the sample variance. For this particular data set, the R statements

```
x = c(3, 1, 5, 1, 3, 2, 1, 1, 3, 2)
mean(x)
var(x)
```

return

$$\bar{x} = 2.2 \qquad\qquad s^2 \cong 1.73.$$

If you are interested in the plug-in estimator of the population variance, which divides the sum of squares in the definition by n rather than $n-1$, use

```
(length(x) - 1) * var(x) / length(x)
```

Finally, the sample standard deviation is just the square root of the sample variance

$$s \cong \sqrt{1.73} \cong 1.32 \text{ children},$$

and can be calculated in R with `sd(x)`.

The next example is concerned with finding the expected value of the sample standard deviation for a sample size of $n = 2$. As you will see, calculating the *value* of s is much easier than determining the *probability distribution* of S.

Example 1.28 Let X_1, X_2 be a random sample of size $n = 2$. Find the expected value of the standard deviation, $E[S]$, when the sampling is from a

(a) Bernoulli(p) population,

(b) $U(a, b)$ population.

The sample variance reduces to a mathematically tractable expression when $n = 2$:

$$\begin{aligned} S^2 &= (X_1 - \bar{X})^2 + (X_2 - \bar{X})^2 \\ &= \left(X_1 - \frac{X_1 + X_2}{2}\right)^2 + \left(X_2 - \frac{X_1 + X_2}{2}\right)^2 \\ &= \left(\frac{X_1 - X_2}{2}\right)^2 + \left(\frac{X_2 - X_1}{2}\right)^2 \\ &= \frac{(X_1 - X_2)^2}{2}. \end{aligned}$$

So the sample standard deviation can be expressed in terms of the order statistics as

$$S = \frac{|X_1 - X_2|}{\sqrt{2}} = \frac{X_{(2)} - X_{(1)}}{\sqrt{2}},$$

and the expected value of the sample standard deviation is

$$E[S] = \frac{1}{\sqrt{2}} E\left[X_{(2)} - X_{(1)}\right] = \frac{1}{\sqrt{2}} \left(E\left[X_{(2)}\right] - E\left[X_{(1)}\right]\right).$$

(a) Consider the case of random sampling from a Bernoulli(p) population. The population probability mass function is

$$f_X(x) = \begin{cases} 1-p & x = 0 \\ p & x = 1 \end{cases}$$

for $0 < p < 1$. Since X_1 and X_2 constitute a random sample, the joint probability mass function of X_1 and X_2 is the product of the marginal probability mass functions $f_{X_1}(x)$ and $f_{X_2}(x)$ and given in Table 1.10.

x_1 \ x_2	0	1
0	$(1-p)^2$	$p(1-p)$
1	$p(1-p)$	p^2

Table 1.10: Joint probability mass function for X_1 and X_2.

Since $P\left(X_{(2)} - X_{(1)} = 0\right) = (1-p)^2 + p^2$ and $P\left(X_{(2)} - X_{(1)} = 1\right) = 2p(1-p)$ from Table 1.10, the expected value of the sample standard deviation is

$$E[S] = \frac{1}{\sqrt{2}} E\left[X_{(2)} - X_{(1)}\right] = \frac{1}{\sqrt{2}} \cdot 2p(1-p) = \sqrt{2}p(1-p)$$

for $0 < p < 1$. This expected value should be compared to the population standard deviation of a Bernoulli(p) random variable, which is $\sigma = \sqrt{p(1-p)}$. The expected standard deviation underestimates the population standard deviation because

$$\sqrt{2}p(1-p) < \sqrt{p(1-p)}$$

for all values of p on the interval $0 < p < 1$. So in this case the sample standard deviation is a *biased estimator* of the population standard deviation, missing low on average. This result can be supported by a Monte Carlo simulation experiment for a particular value of p, say $p = 1/3$. The R code given below generates 900,000 samples of size $n = 2$ from the Bernoulli distribution, then calculates the sample standard deviation of the two values. The 900,000 standard deviations are then averaged.

```
nrep  = 900000
p     = 1 / 3
total = 0
x1    = rbinom(nrep, 1, p)
x2    = rbinom(nrep, 1, p)
for (i in 1:nrep) total = total + sqrt(var(c(x1[i], x2[i])))
print(total / nrep)
```

After a call to `set.seed(3)`, five runs of this simulation yield

0.3147 0.3140 0.3143 0.3144 0.3148.

These values hover about the theoretical value

$$E[S] = \sqrt{2} \cdot \frac{1}{3} \cdot \frac{2}{3} = \frac{2\sqrt{2}}{9} \cong 0.3143,$$

so the Monte Carlo simulation supports the analytic result. The expected value of the sample standard deviation falls below the population standard deviation, which, for $p = 1/3$, is

$$\sigma = \sqrt{p(1-p)} = \sqrt{\frac{1}{3} \cdot \frac{2}{3}} = \frac{\sqrt{2}}{3} \cong 0.4714.$$

(b) Now consider the case of random sampling from a $U(a, b)$ population. The population probability density function is

$$f_X(x) = \frac{1}{b-a} \qquad a < x < b.$$

The associated cumulative distribution function on its support is

$$F_X(x) = \int_a^x \frac{1}{b-a}\,dw = \left[\frac{w}{b-a}\right]_a^x = \frac{x-a}{b-a} \qquad a < x < b.$$

In order to find the probability density functions of $X_{(1)}$ and $X_{(2)}$, we could use the result for the probability density function of the kth order statistic from a continuous population, which is

$$f_{X_{(k)}}(x) = \frac{n!}{(k-1)!(n-k)!}[F(x)]^{k-1} f(x)[1-F(x)]^{n-k}$$

for $a < x < b$ and $k = 1, 2, \ldots, n$, where a and b are the lower and upper limits of the support of the population probability distribution. After simplification, these marginal probability density functions for $n = 2$, $k = 1$, and $k = 2$ are

$$f_{X_{(1)}}(x) = \frac{2(b-x)}{(b-a)^2} \qquad a < x < b$$

and

$$f_{X_{(2)}}(x) = \frac{2(x-a)}{(b-a)^2} \qquad a < x < b.$$

The associated expected values are

$$E\left[X_{(1)}\right] = \frac{2a+b}{3} \qquad \text{and} \qquad E\left[X_{(2)}\right] = \frac{a+2b}{3}.$$

So the expected value of the sample standard deviation is

$$E[S] = \frac{1}{\sqrt{2}}\left(E\left[X_{(2)}\right] - E\left[X_{(1)}\right]\right) = \frac{1}{\sqrt{2}}\left(\frac{a+2b}{3} - \frac{2a+b}{3}\right) = \frac{b-a}{3\sqrt{2}}.$$

This should be compared with the population standard deviation

$$\sigma = \frac{b-a}{\sqrt{12}} = \frac{b-a}{2\sqrt{3}}.$$

The sample standard deviation underestimates the population standard deviation for any fixed values of a and b. Alternatively, the following APPL code does all of the necessary integrations to find $E[S]$.

```
X  := UniformRV(a, b);
X1 := OrderStat(X, 2, 1);
X2 := OrderStat(X, 2, 2);
(Mean(X2) - Mean(X1)) / sqrt(2);
```

The mathematics worked out well for $n = 2$ observations. For sample sizes $n > 2$, however, the mathematics do not work out as cleanly.

The sample variance is the gold standard for estimating dispersion. Its primary downside is that it is tedious to calculate using pencil and paper. For someone working in an industrial setting, the few minutes required to calculate the value of S^2 might be onerous. A much easier statistic to calculate, the sample range, is a second statistic that can be calculated from data that reflects the dispersion of a probability distribution.

Sample range

Particularly for a small data set, the *sample range* can be calculated by hand in just a few seconds, which makes it an important measure of dispersion of the population probability distribution in fields like quality control.

Definition 1.11 Let $x_1, x_2, \ldots, x_n$ be experimental values associated with the random variables $X_1, X_2, \ldots, X_n$. The *sample range* is

$$R = X_{(n)} - X_{(1)}.$$

The sample range is just the largest observation minus the smallest observation. Calculating the value of R from a data set is easy. Determining the sampling distribution of R is typically nontrivial. The next result gives a formula for the probability density function of the sample range when a random sample is drawn from a continuous population.

Theorem 1.2 Let $X_1, X_2, \ldots, X_n$ be a random sample from a continuous population with probability density function $f_X(x)$ and cumulative distribution function $F_X(x)$ and support on the real numbers. The probability density function of the sample range $R = X_{(n)} - X_{(1)}$ is

$$f_R(r) = n(n-1)\int_{-\infty}^{\infty} f_X(x)\left[F_X(r+x) - F_X(x)\right]^{n-2} f_X(r+x)\,dx \qquad r > 0.$$

Proof Using a result concerning order statistics, the joint probability density function of the two order statistics $X_{(i)}$ and $X_{(j)}$, which is $f_{X_{(i)},X_{(j)}}\left(x_{(i)}, x_{(j)}\right)$ for $i < j$, is given by the expression

$$\frac{n!}{(i-1)!(j-i-1)!(n-j)!}\left[F\left(x_{(i)}\right)\right]^{i-1} f\left(x_{(i)}\right)\left[F\left(x_{(j)}\right) - F\left(x_{(i)}\right)\right]^{j-i-1} f\left(x_{(j)}\right)\left[1 - F\left(x_{(j)}\right)\right]^{n-j},$$

where $f(x)$ and $F(x)$ are the population probability density function and cumulative distribution function defined on the real numbers. So the joint probability density function of $X_{(1)}$ and $X_{(n)}$ is

$$f_{X_{(1)},X_{(n)}}\left(x_{(1)}, x_{(n)}\right) = \frac{n!}{(n-2)!} f_X\left(x_{(1)}\right)\left[F_X\left(x_{(n)}\right) - F_X\left(x_{(1)}\right)\right]^{n-2} f_X\left(x_{(n)}\right)$$

on the support $-\infty < x_{(1)} < x_{(n)} < \infty$. We now use the transformation technique to find the distribution of the sample range $Y_1 = X_{(n)} - X_{(1)}$ by using the dummy transformation $Y_2 = X_{(1)}$. The transformation

$$y_1 = g_1\left(x_{(1)}, x_{(n)}\right) = x_{(n)} - x_{(1)} \qquad \text{and} \qquad y_2 = g_2\left(x_{(1)}, x_{(n)}\right) = x_{(1)}$$

is a bivariate one-to-one transformation from the support of $X_{(1)}$ and $X_{(n)}$, which is

$$\mathcal{A} = \left\{ \left(x_{(1)}, x_{(n)} \right) \middle| x_{(1)} < x_{(n)} \right\},$$

to the support of Y_1 and Y_2, which is

$$\mathcal{B} = \{ (y_1, y_2) \,|\, y_1 > 0, -\infty < y_2 < \infty \}.$$

The functions g_1 and g_2 can be solved in closed form for $x_{(1)}$ and $x_{(n)}$ as

$$x_{(1)} = g_1^{-1}(y_1, y_2) = y_2 \qquad \text{and} \qquad x_{(n)} = g_2^{-1}(y_1, y_2) = y_1 + y_2$$

with associated Jacobian

$$J = \begin{vmatrix} 0 & 1 \\ 1 & 1 \end{vmatrix} = -1.$$

Therefore, using the transformation technique, the joint probability density function of $Y_1 = g_1(X_{(1)}, X_{(n)}) = X_{(n)} - X_{(1)}$ and $Y_2 = g_2(X_{(1)}, X_{(n)}) = X_{(1)}$ is

$$\begin{aligned} f_{Y_1,Y_2}(y_1, y_2) &= f_{X_{(1)},X_{(n)}}\left(g_1^{-1}(y_1, y_2), g_2^{-1}(y_1, y_2)\right) |J| \\ &= n(n-1) f_X(y_2) \left[F_X(y_1 + y_2) - F_X(y_2)\right]^{n-2} f_X(y_1 + y_2) \end{aligned}$$

on the support $y_1 > 0$, $-\infty < y_2 < \infty$. Integrating out the dummy variable y_2 gives the probability density function of the sample range as

$$f_{Y_1}(y_1) = n(n-1) \int_{-\infty}^{\infty} f_X(y_2) \left[F_X(y_1 + y_2) - F_X(y_2)\right]^{n-2} f_X(y_1 + y_2)\, dy_2 \qquad y_1 > 0.$$

Replacing Y_1 with the sample range R, the index y_1 with r, and the index y_2 with x yields the desired result. □

Theorem 1.2 is stated for population distributions with infinite support. For population distributions with finite support, the result should be re-derived with the appropriate adjustments made to the integration limits. Unfortunately, the result expresses the probability density function of the sample range as an integral. This might work out well for some distributions, but not for others. The next example derives the sampling distribution of the sample range for observations drawn from an exponential population.

Example 1.29 Let $X_1, X_2, \ldots, X_n$ be a random sample from an exponential(λ) distribution. Find the sampling distribution of the sample range $R = X_{(n)} - X_{(1)}$.

The population probability density function is

$$f_X(x) = \lambda e^{-\lambda x} \qquad x > 0.$$

The population cumulative distribution function is

$$F_X(x) = \begin{cases} 0 & x \le 0 \\ 1 - e^{-\lambda x} & x > 0. \end{cases}$$

Theorem 1.2 gives the probability density function of the sample range as

$$\begin{aligned} f_R(r) &= n(n-1) \int_{-\infty}^{\infty} f_X(x) \left[F_X(r + x) - F_X(x)\right]^{n-2} f_X(r + x)\, dx \\ &= n(n-1) \int_0^{\infty} \lambda e^{-\lambda x} \left[e^{-\lambda x} - e^{-\lambda(r+x)}\right]^{n-2} \lambda e^{-\lambda(r+x)}\, dx \qquad r > 0. \end{aligned}$$

Leaving out the integration details, this simplifies to

$$f_R(r) = \lambda(n-1)e^{-\lambda r}\left(1-e^{-\lambda r}\right)^{n-2} \qquad r > 0$$

for $n = 2, 3, \ldots$. This probability density function is associated with the cumulative distribution function

$$F_R(r) = \begin{cases} 0 & r \le 0 \\ \left(1-e^{-\lambda r}\right)^{n-1} & r > 0. \end{cases}$$

A plot of the cumulative distribution function of R for $\lambda = 1$ and $n = 2, 3, \ldots, 9$ is displayed in Figure 1.41. Consistent with intuition, increasing the sample size pushes the probability distribution of the sample range R to the right.

The following APPL statements also compute the cumulative distribution function of the sample range $R = X_{(n)} - X_{(1)}$.

```
assume(n, posint);
X := ExponentialRV(lambda);
R := RangeStat(X, n);
CDF(R);
```

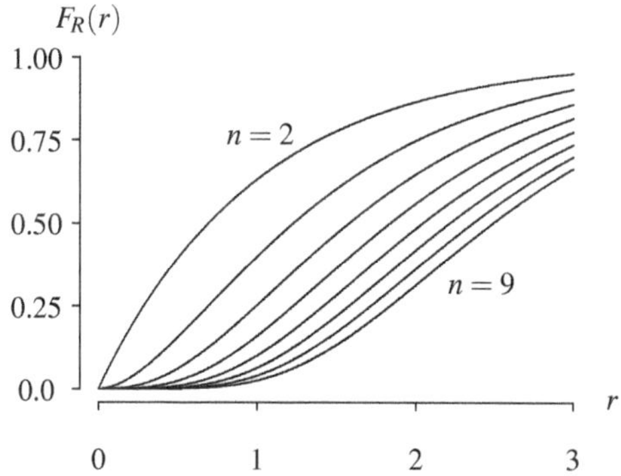

Figure 1.41: Cumulative distribution functions of the sample range ($\lambda = 1$ and $n = 2, 3, \ldots, 9$).

Other measures of dispersion

The sample variance (along with its positive square root, the sample standard deviation) and the sample range are the two most common measures of the dispersion of a data set. Another measure, introduced earlier in the chapter during the discussion on box plots, is the interquartile range (IQR), which is the estimate of the 75th percentile minus the estimate of the 25th percentile. This is an estimate of the range of the middle 50% of the data values.

Another measure of dispersion is the sample *mean absolute deviation*, which is a twist on the sample variance. The squaring that occurs in the formula for the sample variance is used to treat values that are *above* the sample mean and *below* the sample mean by the same amount equally.

But the squaring also tends to accentuate data values that are far away from the sample mean. This emphasizes outliers. The sample mean absolute deviation uses the absolute value operator, rather than squaring, to account for data values below and above the sample mean.

Definition 1.12 Let $x_1, x_2, \ldots, x_n$ be experimental values associated with the random variables $X_1, X_2, \ldots, X_n$. The *sample mean absolute deviation* (MAD) is

$$\frac{1}{n}\sum_{i=1}^{n}|X_i - \bar{X}|.$$

One final measure of dispersion is the sample Gini mean difference, which is the arithmetic mean of the absolute difference between every possible pair of data values. Since the sample size is n, there are $\binom{n}{2}$ possible pairs of data values.

Definition 1.13 Let $x_1, x_2, \ldots, x_n$ be experimental values associated with the random variables $X_1, X_2, \ldots, X_n$. The *sample Gini mean difference* is

$$\frac{1}{\binom{n}{2}}\sum_{j=2}^{n}\sum_{i=1}^{j-1}|X_i - X_j|.$$

You might have noticed that none of the examples so far this chapter have referred to a random sample drawn from a normal population. That is because the normal distribution arises so often in the practice of statistics that it is given its own section.

1.5 Sampling from Normal Populations

This section introduces eight results associated with random sampling from normally-distributed populations. These results are central to what is known as "classical statistics." The first result gives the distribution of $\bar{X}$ for a random sample from a normal population.

Theorem 1.3 If $X_1, X_2, \ldots, X_n$ is a random sample from a $N\left(\mu, \sigma^2\right)$ population, then

$$\bar{X} \sim N\left(\mu, \frac{\sigma^2}{n}\right).$$

Proof Consider the mutually independent random variables $X_1, X_2, \ldots, X_n$, where $X_i \sim N(\mu_i, \sigma_i^2)$, for $i = 1, 2, \ldots, n$, and nonzero real constants $a_1, a_2, \ldots, a_n$. From probability theory, the linear combination $Y = a_1X_1 + a_2X_2 + \cdots + a_nX_n$ is normally distributed with population mean

$$\mu_Y = a_1\mu_1 + a_2\mu_2 + \cdots + a_n\mu_n$$

and population variance

$$\sigma_Y^2 = a_1^2\sigma_1^2 + a_2^2\sigma_2^2 + \cdots + a_n^2\sigma_n^2.$$

The present result is a special case of this result from probability theory in which

$$\mu_1 = \mu_2 = \cdots = \mu_n = \mu,$$

$$\sigma_1^2 = \sigma_2^2 = \cdots = \sigma_n^2 = \sigma^2,$$

and

$$a_1 = a_2 = \cdots = a_n = \frac{1}{n}.$$ □

Sometimes the parameters for the population are written as μ_X and σ_X^2. Likewise, sometimes the parameters associated with the sample mean can be written as $\mu_{\bar{X}} = \mu_X$ and $\sigma_{\bar{X}}^2 = \sigma_X^2/n$.

The conclusion to Theorem 1.3 can also be written in normalized form, that is,

$$\frac{\bar{X} - \mu}{\sigma/\sqrt{n}} \sim N(0, 1).$$

In order to illustrate the application of this theorem, we return to the elevator problem posed earlier.

Example 1.30 Ten people crowd into a small elevator. Their weights, in pounds, are mutually independent and identically distributed $N(140, 900)$ random variables. What is the probability that the elevator capacity of 1500 pounds is exceeded?

Let $X_1, X_2, \ldots, X_{10}$ denote the weights of the ten people. The probability that the elevator capacity is exceeded is

$$P(X_1 + X_2 + \cdots + X_{10} > 1500).$$

Since Theorem 1.3 concerns the sample mean, this probability can be rewritten in terms of $\bar{X}$ by dividing both sides of the inequality by $n = 10$, which gives

$$P(\bar{X} > 150).$$

Since the population mean is $\mu = 140$ and the population standard deviation is $\sigma = 30$, Theorem 1.3 can now be invoked by normalizing the left-hand side of the inequality:

$$\begin{aligned} P(\bar{X} > 150) &= P\left(\frac{\bar{X} - 140}{30/\sqrt{10}} > \frac{150 - 140}{30/\sqrt{10}}\right) \\ &= P\left(Z > \frac{\sqrt{10}}{3}\right) \\ &\cong P(Z > 1.0541) \\ &\cong 0.1459. \end{aligned}$$

This probability can be calculated in R with the statement

```
1 - pnorm(1.0541)
```

or you can leave the normalizing to R and use

```
1 - pnorm(150, 140, 30 / sqrt(10))
```

Figure 1.42 shows the population distribution, which is $N(140, 900)$, and the sampling distribution of the sample mean $\bar{X}$, which is $N(140, 90)$. As in the previous examples, the effect of computing a sample mean is to maintain the same expected value, but decrease the population variance. The area under the probability density function of $\bar{X}$ to the right of 150, which is $P(\bar{X} > 150) \cong 0.1459$, is shaded. This is the probability that the elevator capacity is exceeded. The R commands

```
x  = 80:200
y1 = dnorm(x, 140, 30)
y2 = dnorm(x, 140, 30 / sqrt(10))
matplot(x, cbind(y1, y2), type = "l")
```

generate the graphs of the two probability density functions. The `matplot` function plots two curves on the same set of axes. The column bind function `cbind` pastes the two vectors together, treating each as a column.

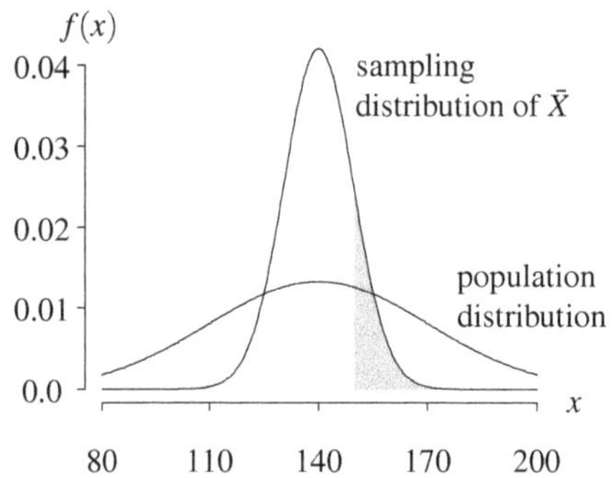

Figure 1.42: Population distribution and sampling distribution of $\bar{X}$.

The next result considers the sum of squares of mutually independent and identically distributed standardized normal random variables. This result brings the chi-square distribution into the picture when a random sample is drawn from a normal population.

Theorem 1.4 If $X_1, X_2, \ldots, X_n$ is a random sample from a $N(\mu, \sigma^2)$ population, then

$$\sum_{i=1}^{n} \left(\frac{X_i - \mu}{\sigma}\right)^2 \sim \chi^2(n).$$

Proof First, standardize the normal random variable X_i:

$$\frac{X_i - \mu}{\sigma} \sim N(0, 1)$$

for $i = 1, 2, \ldots, n$. Since the square of a standard normal random variable has the chi-square distribution with one degree of freedom,

$$\left(\frac{X_i - \mu}{\sigma}\right)^2 \sim \chi^2(1)$$

for $i = 1, 2, \ldots, n$. Finally, since the sum of n mutually independent chi-square random variables is also chi-square (with degrees of freedom equal to the sum of the degrees of freedom of the component chi-square random variables),

$$\sum_{i=1}^{n} \left(\frac{X_i - \mu}{\sigma}\right)^2 \sim \chi^2(n),$$

which proves the result. □

Example 1.31 Ten people crowd into a small elevator. Denote their mutually independent and identically distributed weights, which are $N(140, 900)$ random variables, by $X_1, X_2, \ldots, X_{10}$. Find a constant c such that

$$P\left(\sum_{i=1}^{10} \left(\frac{X_i - 140}{30}\right)^2 < c\right) = 0.99.$$

Applying Theorem 1.4, the random variable

$$\sum_{i=1}^{10} \left(\frac{X_i - 140}{30}\right)^2 \sim \chi^2(10).$$

Therefore, the probability in the problem statement is equivalent to

$$P\left(\sum_{i=1}^{10} \left(\frac{X_i - 140}{30}\right)^2 < c\right) = P\left(\chi^2(10) < c\right) = 0.99.$$

So the constant c is just the 99th percentile of a $\chi^2(10)$ random variable. The value of the constant c can be calculated in R using the `qchisq` function

```
qchisq(0.99, 10)
```

which returns $c \cong 23.2093$.

The next result concerning random sampling from a normal population is somewhat counter-intuitive: the sample mean and the sample variance are independent random variables. This result is counter-intuitive because the sample mean appears in the formula for the sample variance, so you would naturally expect them to be dependent.

Theorem 1.5 If $X_1, X_2, \ldots, X_n$ is a random sample from a $N\left(\mu, \sigma^2\right)$ population, then the sample mean $\bar{X}$ and the sample variance S^2 are independent random variables.

Proof Without loss of generality, assume that $\mu = 0$ and $\sigma^2 = 1$. The sample variance can be written as

$$\begin{aligned} S^2 &= \frac{1}{n-1}\sum_{i=1}^{n}\left(X_i - \bar{X}\right)^2 \\ &= \frac{1}{n-1}\left[\left(X_1 - \bar{X}\right)^2 + \sum_{i=2}^{n}\left(X_i - \bar{X}\right)^2\right] \\ &= \frac{1}{n-1}\left[\left(\sum_{i=2}^{n}\left(X_i - \bar{X}\right)\right)^2 + \sum_{i=2}^{n}\left(X_i - \bar{X}\right)^2\right] \end{aligned}$$

because

$$\sum_{i=1}^{n}\left(X_i - \bar{X}\right) = \sum_{i=1}^{n} X_i - n\bar{X} = 0$$

implies that

$$X_1 - \bar{X} = -\sum_{i=2}^{n}\left(X_i - \bar{X}\right).$$

Instead of writing S^2 in its usual form in terms of $X_1, X_2, \ldots, X_n$ from Definition 1.10, we have now written it in terms of the random variables $X_2 - \bar{X}, X_3 - \bar{X}, \ldots, X_n - \bar{X}$ and $\bar{X}$. The remainder of this proof shows that the random variables $X_2 - \bar{X}$, $X_3 - \bar{X}$, $\ldots$, $X_n - \bar{X}$ and $\bar{X}$ are independent. Since the random sample consists of mutually independent random variables, the joint probability density function of $X_1, X_2, \ldots, X_n$ is the product of the marginal probability density functions:

$$\begin{aligned} f(x_1, x_2, \ldots, x_n) &= f_{X_1}(x_1) \cdot f_{X_2}(x_2) \cdot \ldots \cdot f_{X_n}(x_n) \\ &= (2\pi)^{-1/2} e^{-x_1^2/2} \cdot (2\pi)^{-1/2} e^{-x_2^2/2} \cdot \ldots \cdot (2\pi)^{-1/2} e^{-x_n^2/2} \\ &= (2\pi)^{-n/2} e^{-(1/2)\sum_{i=1}^{n} x_i^2} \end{aligned}$$

for $-\infty < x_i < \infty$. Now consider the transformation $Y_1 = \bar{X}$ and $Y_i = X_i - \bar{X}$ for $i = 2, 3, \ldots, n$. Using the transformation technique, the transformation

$$\begin{aligned} y_1 &= g_1(x_1, x_2, \ldots, x_n) = \bar{x} \\ y_2 &= g_2(x_1, x_2, \ldots, x_n) = x_2 - \bar{x} \\ y_3 &= g_3(x_1, x_2, \ldots, x_n) = x_3 - \bar{x} \\ &\vdots \\ y_n &= g_n(x_1, x_2, \ldots, x_n) = x_n - \bar{x} \end{aligned}$$

is an n-variate transformation from the support of $X_1, X_2, \ldots, X_n$:

$$\mathcal{A} = \left\{(x_1, x_2, \ldots, x_n) \mid -\infty < x_1 < \infty, -\infty < x_2 < \infty, \ldots, -\infty < x_n < \infty\right\}$$

to the support of $Y_1, Y_2, \ldots, Y_n$:

$$\mathcal{B} = \left\{(y_1, y_2, \ldots, y_n) \mid -\infty < y_1 < \infty, -\infty < y_2 < \infty, \ldots, -\infty < y_n < \infty\right\}.$$

These functions can be solved in closed form for $x_1, x_2, \ldots, x_n$ as

$$\begin{aligned}
x_1 &= g_1^{-1}(y_1, y_2, \ldots, y_n) = y_1 - y_2 - \cdots - y_n \\
x_2 &= g_2^{-1}(y_1, y_2, \ldots, y_n) = y_1 + y_2 \\
x_3 &= g_3^{-1}(y_1, y_2, \ldots, y_n) = y_1 + y_3 \\
&\vdots \\
x_n &= g_n^{-1}(y_1, y_2, \ldots, y_n) = y_1 + y_n
\end{aligned}$$

with associated Jacobian

$$J = \begin{vmatrix} 1 & -1 & -1 & \cdots & -1 & -1 \\ 1 & 1 & 0 & \cdots & 0 & 0 \\ 1 & 0 & 1 & \cdots & 0 & 0 \\ \vdots & \vdots & \vdots & \ddots & \vdots & \vdots \\ 1 & 0 & 0 & \cdots & 1 & 0 \\ 1 & 0 & 0 & \cdots & 0 & 1 \end{vmatrix} = n.$$

So the joint probability density function of $Y_1, Y_2, \ldots, Y_n$ is

$$\begin{aligned}
f(y_1, y_2, \ldots, y_n) &= (2\pi)^{-n/2} e^{-(1/2)\left((y_1 - y_2 - \cdots - y_n)^2 + (y_1+y_2)^2 + (y_1+y_3)^2 + \cdots + (y_1+y_n)^2\right)} |n| \\
&= n(2\pi)^{-n/2} e^{-ny_1^2/2} \cdot e^{-(1/2)\left(y_2^2 + y_3^2 + \cdots + y_n^2 + (y_2+y_3+\cdots+y_n)^2\right)}
\end{aligned}$$

for $-\infty < y_i < \infty$, $i = 1, 2, \ldots, n$. Since the joint probability density function can be factored into a function of y_1 times a function of $y_2, y_3, \ldots, y_n$ and is defined on a product space, one can conclude that Y_1 is mutually independent of $Y_2, Y_3, \ldots, Y_n$ using a result from probability theory. Using the formula for S^2 from the beginning of the proof, this means that $\bar{X}$ and S^2 are independent random variables. □

Theorem 1.5 carries little utility on its own—but is needed to prove subsequent results. The next result is helpful in statistical inference involving the sample variance.

Theorem 1.6 If $X_1, X_2, \ldots, X_n$ is a random sample from a $N\left(\mu, \sigma^2\right)$ population, then

$$\frac{(n-1)S^2}{\sigma^2} \sim \chi^2(n-1).$$

Proof The first step is to rewrite the quantity of interest in the following fashion:

$$\frac{(n-1)S^2}{\sigma^2} = \frac{\sum_{i=1}^{n}(X_i - \bar{X})^2}{\sigma^2} = \sum_{i=1}^{n}\left(\frac{X_i - \bar{X}}{\sigma}\right)^2.$$

The second step is to subtract and add $\bar{X}$ in the sum of squares

$$\begin{aligned}\sum_{i=1}^{n}(X_i-\mu)^2 &= \sum_{i=1}^{n}(X_i-\bar{X}+\bar{X}-\mu)^2\\ &= \sum_{i=1}^{n}(X_i-\bar{X})^2+\sum_{i=1}^{n}(\bar{X}-\mu)^2+2\sum_{i=1}^{n}(X_i-\bar{X})(\bar{X}-\mu)\\ &= \sum_{i=1}^{n}(X_i-\bar{X})^2+n(\bar{X}-\mu)^2+2(\bar{X}-\mu)\sum_{i=1}^{n}(X_i-\bar{X})\\ &= \sum_{i=1}^{n}(X_i-\bar{X})^2+n(\bar{X}-\mu)^2\end{aligned}$$

because $\sum_{i=1}^{n}(X_i-\bar{X})=0$. The third step is to divide the three terms in this equation by σ^2, resulting in

$$\sum_{i=1}^{n}\left(\frac{X_i-\mu}{\sigma}\right)^2=\sum_{i=1}^{n}\left(\frac{X_i-\bar{X}}{\sigma}\right)^2+n\left(\frac{\bar{X}-\mu}{\sigma}\right)^2.$$

Finally, pulling n inside the far right-hand term and replacing the summation on the right-hand side of the equation with the quantity of interest,

$$\sum_{i=1}^{n}\left(\frac{X_i-\mu}{\sigma}\right)^2=\frac{(n-1)S^2}{\sigma^2}+\left(\frac{\bar{X}-\mu}{\sigma/\sqrt{n}}\right)^2.$$

The random variable on the left-hand side of this equation has the chi-square distribution with n degrees of freedom by Theorem 1.4. The two terms being summed on the right-hand side of the equation are independent because they are functions of two independent random variables, S^2 and $\bar{X}$, by Theorem 1.5. The second term on the right-hand side of the equation is the square of a standard normal random variable by Theorem 1.3, so it has the chi-square distribution with 1 degree of freedom. So, because the degrees of freedom add when summing independent chi-square random variables, $(n-1)S^2/\sigma^2$ has the chi-square distribution with $n-1$ degrees of freedom, which proves the result. □

Theorem 1.6 is used for determining probabilities associated with the sample variance S^2. The next example illustrates one such instance.

Example 1.32 Let $X_1, X_2, \ldots, X_6$ be a random sample from a $N(77, 4)$ population. The sample standard deviation associated with the sample is $s = 2.6$. Is this unusual?

At first this seems like an ill-posed question, but questions of this type arise quite often in statistics. Begin by looking at the big picture. The sample size $n = 6$ is fairly small, so we expect significant random sampling variability associated with the sampling distribution of the sample standard deviation S. The sample is drawn from a normal population with population mean $\mu = 77$ and population standard deviation $\sigma = 2$. The question is asking whether a sample standard deviation of $s = 2.6$ is unusual. A sample standard deviation that is significantly larger than the value of the population standard deviation does seem rather unusual, but we would like to quantify that with a probability statement. Calculating $P(S = 2.6)$ gets us nowhere. The population is continuous, which means that the probability distribution of the sample standard deviation S is also continuous, which means that $P(S = 2.6) = 0$.

But here is another approach. Since $s = 2.6$ is a *larger* sample standard deviation than we expect to get from a population with $\sigma = 2$, calculating the probability that S exceeds 2.6 would give us some sense of how unusual it is to get a sample standard deviation this large. But we don't know the distribution of S. Fortunately, Theorem 1.6 gives us the distribution of a function of S, namely $(n-1)S^2/\sigma^2$. So the desired probability can be calculated in the following fashion:

$$\begin{aligned} P(S > 2.6) &= P\left(S^2 > 6.76\right) \\ &= P\left(\frac{(6-1)S^2}{4} > \frac{(6-1)6.76}{4}\right) \\ &= P\left(\chi^2(5) > 8.45\right) \\ &\cong 0.1331, \end{aligned}$$

which can be calculated with the R statement

```
1 - pchisq(8.45, 5)
```

So this particular value of the sample standard deviation, namely $s = 2.6$, is at the 87th percentile of its distribution. Now we return to the original question "Is this unusual"? The answer is "not particularly." We run into people at the 87th percentile of the distribution of height, weight, athletic ability, and IQ quite frequently. It is not enough for us to classify such a person as "unusual." Likewise, a value of s that is at the 87th percentile of the probability distribution of S should not be considered unusual.

In the previous example, the sample size was small, resulting in enough random sampling variability so that we could not classify a larger-than-usual standard deviation as "unusual." We now rework the example with a larger sample size.

Example 1.33 Let $X_1, X_2, \ldots, X_{36}$ be a random sample from a $N(77, 4)$ population. The sample standard deviation associated with the sample is $s = 2.6$. Is this unusual?

The only element that has changed from the previous example is the sample size. The larger sample size of $n = 36$ means that there will be less random sampling variability. This means that the sample standard deviation S is less likely to stray from the population standard deviation $\sigma = 2$. The desired probability can be calculated in a similar fashion:

$$\begin{aligned} P(S > 2.6) &= P\left(S^2 > 6.76\right) \\ &= P\left(\frac{(36-1)S^2}{4} > \frac{(36-1)6.76}{4}\right) \\ &= P\left(\chi^2(35) > 59.15\right) \\ &\cong 0.006549. \end{aligned}$$

which can be calculated with the R statement

```
1 - pchisq(59.15, 35)
```

Now the sample standard deviation is beyond the 99th percentile of the distribution of S, so we would label this outcome as "unusual" because it is not very likely to observe a standard deviation of 2.6 or more. There are two explanations for what is going on. The first explanation is that we just happened to draw this one particular very unlikely sample that had an unusually large spread to the observations which lead to an unusually large sample standard deviation. We have observed a truly rare event. If this is the case, we are certainly tempted to draw another sample of $n = 36$ observations to see if a more reasonable sample standard deviation value occurs, that is, a value of s which is closer to $\sigma = 2$. The second explanation is that one or more of the assumptions in the problem statement are wrong. The population might not have a population standard deviation of $\sigma = 2$, for example, as we have assumed. It seems to be larger than that based on our observed sample standard deviation $s = 2.6$ from a sample of $n = 36$ observations. Alternatively, our assumption of sampling from a normally distributed population might be incorrect. One or more of these explanations could explain the low probability associated with $P(S > 2.6)$.

Theorem 1.3 stated that when $X_1, X_2, \ldots, X_n$ is a random sample from a $N\left(\mu, \sigma^2\right)$ population,

$$\frac{\bar{X} - \mu}{\sigma/\sqrt{n}} \sim N(0, 1).$$

In many practical problems, the population standard deviation σ is not known, but rather is estimated by the sample standard deviation S. So a natural question is whether the probability distribution of

$$\frac{\bar{X} - \mu}{S/\sqrt{n}}$$

is mathematically tractable. Fortunately, it is tractable, and the specifics are given by the following theorem. The t distribution was discovered by William Sealy Gosset, who published the work under the pen name "Student" because his employer, the Guinness Brewery, would not allow him to publish his discovery under his real name.

Theorem 1.7 If $X_1, X_2, \ldots, X_n$ is a random sample from a $N\left(\mu, \sigma^2\right)$ population, then

$$\frac{\bar{X} - \mu}{S/\sqrt{n}} \sim t(n-1).$$

Proof Using algebra to manipulate the fraction into a familiar form,

$$\frac{\bar{X} - \mu}{S/\sqrt{n}} = \frac{\frac{\bar{X}-\mu}{\sigma/\sqrt{n}}}{\frac{S/\sqrt{n}}{\sigma/\sqrt{n}}} = \frac{\frac{\bar{X}-\mu}{\sigma/\sqrt{n}}}{\sqrt{\frac{(n-1)S^2}{\sigma^2(n-1)}}} = \frac{Z}{\sqrt{\chi^2(n-1)/(n-1)}} \sim t(n-1).$$

The numerator Z has the standard normal distribution by Theorem 1.3. The numerator and denominator of the expression are independent because $\bar{X}$ and S^2 are independent by Theorem 1.5. The denominator includes a chi-square random variable with $n-1$ degrees of freedom by Theorem 1.6. Using a result from probability theory, the ratio of a standard normal random variable to the square root of an independent chi-square random variable with $n-1$ degrees of freedom divided by $n-1$ has the $t(n-1)$ distribution, which proves the result. □

Statisticians often encounter data values that are drawn from bell-shaped populations with unknown population mean μ and unknown population variance σ^2. In many settings, there is interest in drawing conclusions concerning μ. The next example illustrates one such setting.

Example 1.34 The origins of the Etruscan empire, which existed in modern-day Italy, are a mystery. One question of interest is whether Etruscans were native Italians or immigrants from elsewhere. To address this question, anthropologists measured the maximum head breadth, in millimeters, of the skulls of $n = 84$ adult male Etruscans. The data values are given in Table 1.11. The sample mean and sample standard deviation associated with this data set are $\bar{x} = 144$ and $s = 5.97$. Notice that the sample mean and the sample standard deviation are reported to three significant digits, consistent with the number of significant digits in the data. To report the sample mean to more digits, for example, as $\bar{x} = 143.7738$, implies that there is some meaning to the digits after the decimal point, which is not true in this case; the measuring device only gives the data to the nearest millimeter. On the other hand, when performing subsequent statistical calculations, all of these digits are carried forward in calculations. There is no purpose served in compounding error due to sampling variability with error due to rounding intermediate values.

141	148	132	138	154	142	150	146	155	158	150	140
147	148	144	150	149	145	149	158	143	141	144	144
126	140	144	142	141	140	145	135	147	146	141	136
140	146	142	137	148	154	137	139	143	140	131	143
141	149	148	135	148	152	143	144	141	143	147	146
150	132	142	142	143	153	149	146	149	138	142	149
142	137	134	144	146	147	140	142	140	137	152	145

Table 1.11: Maximum skull breadths, in millimeters, for $n = 84$ adult Etruscan males.

Clearly the breadth of a skull is a continuous random variable, even though the observations in Table 1.11 are given to finite precision. The histogram of the data values in Figure 1.43 is helpful in assessing the normality assumption. The histogram is bell-shaped, and the slight non-symmetry is due to the choice of round numbers for cell widths and cell boundaries rather than any non-symmetry in the data values. We will proceed under the assumption that the population is bell-shaped so that our theorems concerning sampling from normal populations can be applied. (Non-normal populations will be addressed later in the text.)

Can a statement be made about the precision of the sample mean $\bar{x}$ in estimating the population mean μ?

This is another example of a common question that often arises in statistics. Our sample mean $\bar{x} = 144$ is an estimate of the unknown population mean μ, but it is stated without any indication of its precision. Are we reasonably sure that μ is within 2 millimeters of 144? Are we reasonably sure that μ is within 10 millimeters of 144? We invoke Theorem 1.7 to help answer this question.

First of all, to quantify the notion of "reasonably sure" in the above statements, we arbitrarily set this probability as 0.9, which corresponds to being 90% sure. One way

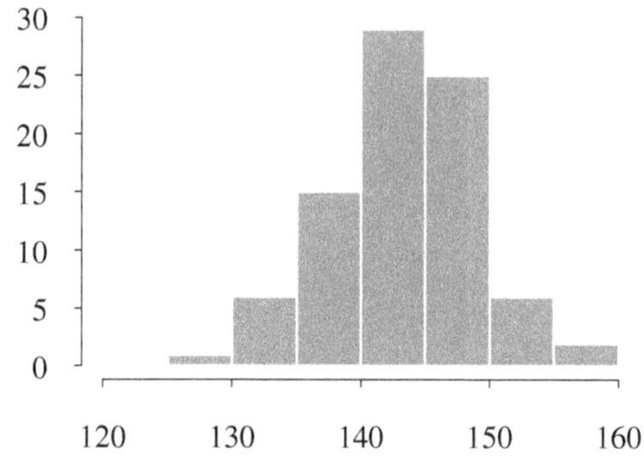

Figure 1.43: Histogram of the maximum skull breadth for $n = 84$ Etruscan males.

to determine the precision of $\bar{x}$ is to give an interval with lower bound L and upper bound U that contains the population mean μ with probability 0.9, that is,

$$P(L < \mu < U) = 0.9.$$

Statisticians call the interval $L < \mu < U$ a "90% confidence interval for μ." The uppercase letters L and U are used for the endpoints of this interval because they will be random variables that are functions of the data. Theorem 1.7 states that

$$\frac{\bar{X} - \mu}{S/\sqrt{n}} \sim t(n-1)$$

when sampling from a normal population. Since $n = 84$ skulls were measured, the statistic has the t distribution with 83 degrees of freedom. Figure 1.44 shows the probability density function of the $t(83)$ distribution on the interval $(-3, 3)$. Two special values are identified on the horizontal axis: the 5th and 95th percentiles. We introduce

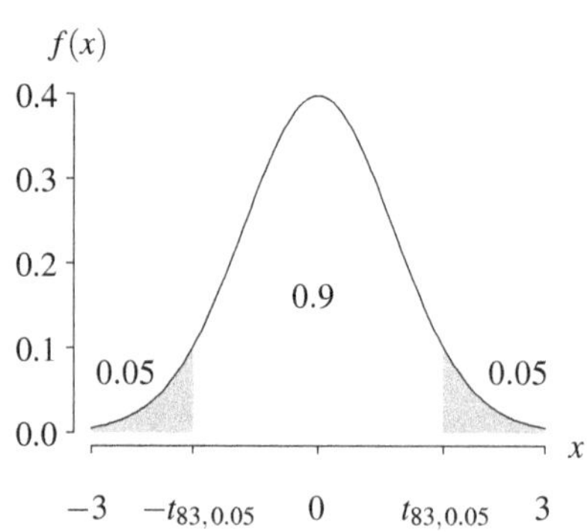

Figure 1.44: Probability density function of the t distribution with 83 degrees of freedom.

the notation

$$t_{83,0.05}$$

to denote the 95th percentile of a t distribution with 83 degrees of freedom. The first parameter in the subscript is the degrees of freedom and the second parameter in the subscript is the associated right-hand tail probability. (The use of subscripts for percentiles is commonplace; using the right-hand tail probability in the subscript is not universal.) This value can be calculated with the R statement

```
qt(0.95, 83)
```

which gives $t_{83,0.05} = 1.6634$. (Notice that R uses left-hand tail probabilities and swaps the order of the two arguments.) The large number of degrees of freedom associated with the t distribution means that this percentile is quite close to the associated standard normal percentile $z_{0.05} = 1.6449$. Since the probability density function for the t distribution is an even function, the left-hand percentile can be written as either

$$-t_{83,0.05} \qquad \text{or} \qquad t_{83,0.95},$$

which is $-t_{83,0.05} = -1.6634$. The area under the probability density function between -1.6634 and 1.6634 is 0.9, as indicated in Figure 1.44. By symmetry, the shaded area under the probability density function to the left of -1.6634 is 0.05; the shaded area under the probability density function to the right of 1.6634 is also 0.05. We are now in a position to make a probability statement concerning the statistic from Theorem 1.7. In our case, the generic expression

$$P\left(-t_{n-1,0.05} < \frac{\bar{X}-\mu}{S/\sqrt{n}} < t_{n-1,0.05}\right) = 0.9$$

becomes

$$P\left(-1.6634 < \frac{\bar{X}-\mu}{S/\sqrt{84}} < 1.6634\right) = 0.9.$$

The only work that remains is to manipulate the interval in the probability statement so that μ is isolated in the middle. Multiplying all three expressions in the inequality by $S/\sqrt{84}$, then subtracting $\bar{X}$ and negating gives

$$P\left(\bar{X} - 1.6634 \cdot \frac{S}{\sqrt{84}} < \mu < \bar{X} + 1.6634 \cdot \frac{S}{\sqrt{84}}\right) = 0.9.$$

Notice that the two endpoints of the interval are random variables because they are functions of the random variables $\bar{X}$ and S. Calculating the values of the endpoints of the interval for this particular data set gives

$$144 - 1.6634 \cdot \frac{5.97}{\sqrt{84}} < \mu < 144 + 1.6634 \cdot \frac{5.97}{\sqrt{84}},$$

which simplifies to the 90% confidence interval (reported to three significant digits to match the precision of the data values)

$$143 < \mu < 145.$$

We are 90% "confident" (the precise meaning of this word will be defined carefully in Chapter 3) that the population mean lies between 143 and 145. These calculations can be performed with the R statements given below. The $n = 84$ maximum skull breadths are stored in the file named `etruscan.d`.

```
x    = scan("etruscan.d")
n    = length(x)
xbar = mean(x)
sdev = sd(x)
crit = qt(0.95, n - 1)
l    = xbar - crit * sdev / sqrt(n)
u    = xbar + crit * sdev / sqrt(n)
print(c(l, u))
```

Since calculations of this type are performed routinely by statisticians, R has a built-in function called `t.test` that can save you a few key strokes:

```
x = scan("etruscan.d")
t.test(x, conf.level = 0.9)
```

These statements return the same interval $142.6902 < \mu < 144.8574$ as the previous code, but these values should be rounded to $143 < \mu < 145$ using the appropriate number of significant digits (three digits, in this case) when reported. The important conclusion here is that we are 90% confident that the sample mean $\bar{x} = 144$ is within one millimeter of the true mean maximum skull breadth based on our data set of $n = 84$ Etruscan skulls. The analysis might be somewhat compromised due to potential violation of the normality or independence assumptions.

The previous five results have dealt with a random sample drawn from a *single* normal population. Oftentimes, questions arise concerning the comparison of random samples from *two* normal populations. Two representative examples are given below.

- Is the population mean drying time associated with paint brand 1 less than the population mean drying time associated with paint brand 2?
- Is the population mean corn yield using fertilizer 1 greater than the population mean corn yield using fertilizer 2?

These questions assume that all other variables (for example, humidity for the paint drying and rainfall for the fertilizer) have been fixed. In the three remaining results concerning sampling from a normal population, we assume that we have two samples—one taken from one normal population and a second taken from another normal population. We use $X_1, X_2, \ldots, X_n$ to denote the first random sample and $Y_1, Y_2, \ldots, Y_m$ to denote the second random sample.

Theorem 1.8 If $X_1, X_2, \ldots, X_n$ is a random sample from a $N\left(\mu_X, \sigma_X^2\right)$ population and $Y_1, Y_2, \ldots, Y_m$ is a random sample (independent of the first sample) from a $N\left(\mu_Y, \sigma_Y^2\right)$ population, then

$$\bar{X} - \bar{Y} \sim N\left(\mu_X - \mu_Y, \frac{\sigma_X^2}{n} + \frac{\sigma_Y^2}{m}\right).$$

Proof Using a result from probability theory, we know that linear combinations of independent, normally distributed random variables are normally distributed, so the linear combination $\bar{X} - \bar{Y}$ is normally distributed. Furthermore,

$$E\left[\bar{X} - \bar{Y}\right] = E\left[\bar{X}\right] - E\left[\bar{Y}\right] = \mu_X - \mu_Y$$

and

$$V\left[\bar{X} - \bar{Y}\right] = V\left[\bar{X}\right] + V\left[\bar{Y}\right] = \frac{\sigma_X^2}{n} + \frac{\sigma_Y^2}{m}$$

by Theorem 1.3. Because $\bar{X} - \bar{Y}$ is normally distributed with the appropriate population mean and population variance, the result follows. □

There are several important aspects of Theorem 1.8 that are worth highlighting before working an example problem.

- The two sample sizes n and m are not necessarily equal.
- The two population means and the two population variances of the two normally distributed populations are not necessarily equal.
- By normalizing, the result in Theorem 1.8 could also have been stated as

$$\frac{\bar{X} - \bar{Y} - (\mu_X - \mu_Y)}{\sqrt{\frac{\sigma_X^2}{n} + \frac{\sigma_Y^2}{m}}} \sim N(0, 1).$$

- When the two population standard deviations are equal, that is, when $\sigma_X = \sigma_Y = \sigma$, the result in Theorem 1.8 reduces to

$$\frac{\bar{X} - \bar{Y} - (\mu_X - \mu_Y)}{\sigma\sqrt{\frac{1}{n} + \frac{1}{m}}} \sim N(0, 1).$$

Compare this corollary of the general result to the next theorem and some similarities will be apparent.

Example 1.35 Let $X_1, X_2, \ldots, X_6$ be a random sample from a $N(5, 4)$ population and Y_1, Y_2, Y_3 be a random sample (independent of the first sample) from a $N(7, 10)$ population. Find $P(\bar{X} > \bar{Y})$.

We expect a fairly small probability for $P(\bar{X} > \bar{Y}) = P(\bar{X} - \bar{Y} > 0)$ because the first population mean is $\mu_X = 5$ and the second population mean is $\mu_Y = 7$. The first sample mean $\bar{X}$ will occasionally exceed the second sample mean $\bar{Y}$ because of the small sample sizes ($n = 6$ and $m = 3$) and significant dispersion in the two populations ($\sigma_X^2 = 4$ and $\sigma_Y^2 = 10$). The probability density functions of the sampling distributions of $\bar{X}$ and $\bar{Y}$, that is,

$$\bar{X} \sim N\left(5, \frac{4}{6}\right) \qquad \text{and} \qquad \bar{Y} \sim N\left(7, \frac{10}{3}\right)$$

are plotted in Figure 1.45.

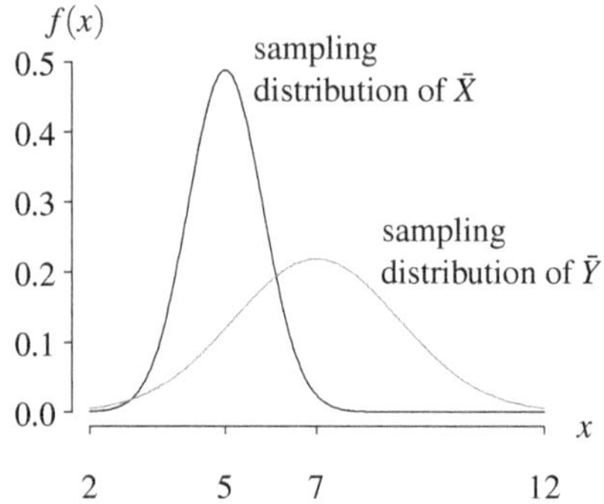

Figure 1.45: Sampling distributions of $\bar{X}$ and $\bar{Y}$.

Using Theorem 1.8,

$$\begin{aligned}
P(\bar{X} > \bar{Y}) &= P(\bar{X} - \bar{Y} > 0) \\
&= P\left(\frac{\bar{X} - \bar{Y} - (5-7)}{\sqrt{\frac{4}{6} + \frac{10}{3}}} > \frac{0 - (5-7)}{\sqrt{\frac{4}{6} + \frac{10}{3}}}\right) \\
&= P\left(Z > \frac{2}{\sqrt{4}}\right) \\
&= P(Z > 1) \\
&\cong 0.1587.
\end{aligned}$$

(The sample sizes and parameters were chosen so that the calculations would work out to round numbers.) This probability can be calculated with the R statement

```
1 - pnorm(1)
```

The following R code conducts a Monte Carlo simulation experiment with one million replications to estimate the probability that the first sample mean exceeds the second sample mean.

```
nrep  = 1000000
count = 0
for (i in 1:nrep) {
  x = rnorm(6, 5, sqrt(4))
  y = rnorm(3, 7, sqrt(10))
  if (mean(x) > mean(y)) count = count + 1
}
print(count / nrep)
```

After a call to `set.seed(3)` to initialize the random number stream, five runs of this simulation yields the following estimates for $P(\bar{X} > \bar{Y})$:

0.1587 0.1591 0.1582 0.1585 0.1583.

Since these values hover about the analytic value $P(\bar{X} > \bar{Y}) \cong 0.1587$, the Monte Carlo simulation supports the analytic solution.

Both σ_X and σ_Y are rarely known in statistical applications. This makes the application of Theorem 1.8 rare in most real-world statistical problems. The more common case is that σ_X and σ_Y need to be estimated from their statistical counterparts S_X and S_Y from data collected from the two populations. Theorem 1.9 addresses this scenario, with the additional restriction that the population variances of the two populations must be equal.

Theorem 1.9 If $X_1, X_2, \ldots, X_n$ is a random sample from a $N(\mu_X, \sigma^2)$ population and $Y_1, Y_2, \ldots, Y_m$ is a random sample (independent of the first sample) from a $N(\mu_Y, \sigma^2)$ population, then

$$\frac{\bar{X} - \bar{Y} - (\mu_X - \mu_Y)}{S_P\sqrt{\frac{1}{n} + \frac{1}{m}}} \sim t(n+m-2),$$

where

$$S_P^2 = \frac{(n-1)S_X^2 + (m-1)S_Y^2}{n+m-2}$$

is the *pooled sample variance.*

Proof Using algebra to manipulate the fraction into a familiar form,

$$\frac{\bar{X} - \bar{Y} - (\mu_X - \mu_Y)}{S_P\sqrt{\frac{1}{n} + \frac{1}{m}}} = \frac{\frac{\bar{X}-\bar{Y}-(\mu_X-\mu_Y)}{\sigma\sqrt{\frac{1}{n}+\frac{1}{m}}}}{\frac{S_P\sqrt{\frac{1}{n}+\frac{1}{m}}}{\sigma\sqrt{\frac{1}{n}+\frac{1}{m}}}} = \frac{\frac{\bar{X}-\bar{Y}-(\mu_X-\mu_Y)}{\sigma\sqrt{\frac{1}{n}+\frac{1}{m}}}}{\sqrt{\frac{\frac{(n-1)S_X^2}{\sigma^2}+\frac{(m-1)S_Y^2}{\sigma^2}}{n+m-2}}} = \frac{Z}{\sqrt{\frac{\chi^2(n+m-2)}{n+m-2}}}.$$

The numerator has the standard normal distribution by Theorem 1.8 in the special case in which the population variances are equal, which is assumed for this result. Theorem 1.6 is used to show that

$$\frac{(n-1)S_X^2}{\sigma^2} \sim \chi^2(n-1) \qquad \text{and} \qquad \frac{(m-1)S_Y^2}{\sigma^2} \sim \chi^2(m-1)$$

are independent chi-square random variables. Since the sum of independent chi-square random variables has the chi-square distribution using a result from probability theory,

$$\frac{(n-1)S_X^2}{\sigma^2} + \frac{(m-1)S_Y^2}{\sigma^2} \sim \chi^2(n+m-2).$$

So, the denominator in the expression is the square root of a chi-square random variable with $n+m-2$ degrees of freedom divided by its degrees of freedom. Finally, the ratio of a standard normal random variable divided by the square root of an independent chi-square random variable divided by its degrees of freedom has the t distribution with $n+m-2$ degrees of freedom using a result from probability theory. □

The primary purpose of Theorem 1.9 is to determine whether the difference between two sample means, that is, $\bar{x} - \bar{y}$, is "statistically significant," as illustrated in the next example.

Example 1.36 One of the purposes of measuring the maximum breadth of the $n = 84$ Etruscan skulls from Example 1.34 was to determine whether the Etruscans were native Italians or immigrants. To that end, an additional $m = 70$ modern Italian skulls were measured, as shown in Table 1.12. Were the Etruscans native Italians or immigrants?

133	138	130	138	134	127	128	138	136	131	126	120
124	132	132	125	139	127	133	136	121	131	125	130
129	125	136	131	132	127	129	132	116	134	125	128
139	132	130	132	128	139	135	133	128	130	130	143
144	137	140	136	135	126	139	131	133	138	133	137
140	130	137	134	130	148	135	138	135	138		

Table 1.12: Maximum skull breadths for $m = 70$ modern Italian adult males.

The first step in the analysis of these two data sets is to construct an appropriate statistical graphic. One such graphic is an adaptation of the population pyramid from Example 1.7. For that population pyramid, the population for each age was displayed for men and women. In this particular case, we want to adjust the pyramid so that the effect of the two sample sizes ($n = 84$ and $m = 70$) is not apparent. The pyramid in Figure 1.46 plots a histogram of the maximum skull breadth for the Etruscan skulls on the left and a histogram of the modern Italian skulls on the right. In this case, the line that is plotted for the histogram bins is the fraction of the observations falling into the bins. This way, the area under each of the two histograms is equal. The grid lines that were

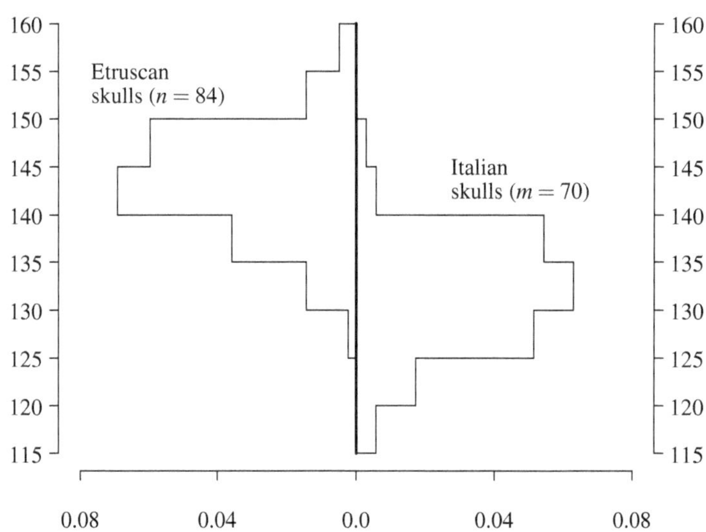

Figure 1.46: Normalized histograms of maximum skull breadth, in millimeters.

included in the French population pyramid have been suppressed in order to highlight the shape of the two histograms.

Two conclusions can be drawn from Figure 1.46. First, the Etruscan skulls are larger than the modern Italian skulls on average. This is verified by computing the sample means:

$$\bar{x} = 144 \text{ mm} \qquad \text{and} \qquad \bar{y} = 132 \text{ mm}.$$

Second, the population variances of the two data sets appear to be roughly equal. But can these two conclusions which have been drawn from the statistical graphic be due to sampling variability, or is there a pattern here? Is the difference of 12 millimeters between the sample means of the two data sets a matter of sampling variability, or is there a "statistically significant" difference between the two populations means? This is where Theorem 1.9 comes into play.

Assume for the moment that our visual inspection of the two histograms leads us to conclude that the two populations have equal population variances, that is, $\sigma_X^2 = \sigma_Y^2$, where X corresponds to the Etruscan skulls and Y corresponds to the modern Italian skulls. If that is the case, and we also assume that the data are random samples drawn from two independent normal populations (which is also plausible given the shape of the two histograms), then all of the assumptions are in place to invoke Theorem 1.9. One device that statisticians use is to begin by assuming that there is no difference between the population means of the two normal populations, that is, $\mu_X = \mu_Y$. The pooled sample variance is

$$s_p^2 = \frac{(n-1)s_X^2 + (m-1)s_Y^2}{n+m-2} = \frac{(84-1)35.6470 + (70-1)33.0619}{84+70-2} = 34.4735,$$

which is computed with the R statements

```
x = scan("etruscan.d")
y = scan("italian.d")
n = length(x)
m = length(y)
s = ((n - 1) * var(x) + (m - 1) * var(y)) / (n + m - 2)
```

This allows us to compute the value of the "test statistic" defined in Theorem 1.9, which is

$$\frac{\bar{X} - \bar{Y} - (\mu_X - \mu_Y)}{S_p\sqrt{\frac{1}{n} + \frac{1}{m}}} = \frac{143.7738 - 132.4429}{\sqrt{34.4735\left(\frac{1}{84} + \frac{1}{70}\right)}} = 11.9,$$

with $\mu_X = \mu_Y$, which is computed with the additional R statement

```
(mean(x) - mean(y)) / sqrt(s * (1 / n + 1/ m))
```

This statistic can be thought of as one value that is drawn from a t distribution with $n+m-2 = 84+70-2 = 152$ degrees of freedom. So the question now is whether this observation of a t random variable with 152 degrees of freedom is unusual. The t distribution with 152 degrees of freedom is nearly identical to a standard normal distribution because of the large number of degrees of freedom. So it is clear that the value of 11.9 is way out in the far right-hand tail of the distribution. In fact, it is so far out in the right-hand tail that we have strong evidence that our assumed hypothesis of equal means is false. There is overwhelming statistical evidence to support the hypothesis that the Etruscans immigrated to modern-day Italy. This does assume that skull size is not changing over time, which must be verified separately.

Calculating the t statistic arises so often in statistics that it is built into an R function named `t.test`. An additional flag must be added in order to tell R that the population variances associated with the two populations are assumed to be equal. The R code given below calculates the statistic 11.9 automatically.

```
x = scan("etruscan.d")
y = scan("italian.d")
t.test(x, y, var.equal = TRUE)
```

We based our conclusion that $\sigma_X = \sigma_Y$ in the previous example on a statistical graphic. This is never a good idea because the eye often improperly assesses the effect of the sample size and can be tricked. The next theorem gives the statistical basis for determining if the population variances for two normal populations are equal based on two data sets.

Theorem 1.10 If $X_1, X_2, \ldots, X_n$ is a random sample from a $N\left(\mu_X, \sigma_X^2\right)$ population and $Y_1, Y_2, \ldots, Y_m$ is a random sample (independent of the first sample) from a $N\left(\mu_Y, \sigma_Y^2\right)$ population, then

$$\frac{S_X^2/\sigma_X^2}{S_Y^2/\sigma_Y^2} \sim F(n-1, m-1).$$

Proof Using algebraic manipulations to get familiar expressions in the numerator and denominator,

$$\frac{S_X^2/\sigma_X^2}{S_Y^2/\sigma_Y^2} = \frac{\frac{(n-1)S_X^2}{\sigma^2}\Big/(n-1)}{\frac{(m-1)S_Y^2}{\sigma^2}\Big/(m-1)} = \frac{\chi^2(n-1)/(n-1)}{\chi^2(m-1)/(m-1)} \sim F(n-1, m-1).$$

The conclusion relies on Theorem 1.6 being applied to both the numerator and the denominator. Since the quotient of a chi-square random variable divided by its degrees of freedom and an independent chi-square random variable divided by its degrees of freedom has the F distribution using a well-known result from probability theory, the result is proved. □

The primary purpose of Theorem 1.10 is to determine whether the difference between two sample variances s_X^2 and s_Y^2 is "statistically significant." This is checked by considering the ratio of the two sample variances, as illustrated in the next example.

Example 1.37 In the previous example, the population variances of the skull sizes for the Etruscan and modern Italian populations were assumed to be equal based on an inspection of the histograms of the two data sets. Can Theorem 1.10 be used to draw a more formal conclusion about whether or not this is the case?

The sample variances for the two populations were computed as

$$s_X^2 = 35.6 \qquad \text{and} \qquad s_Y^2 = 33.1.$$

There is slightly more spread to the measurements made on the Etruscan skulls. The difference between these two sample variances seems slight. We would like to determine whether or not there is a statistically significant difference between the two sample variances. As before, we assume that the two population variances are equal, $\sigma_X^2 = \sigma_Y^2$, then compute the statistic from Theorem 1.10:

$$\frac{s_X^2}{s_Y^2} = \frac{35.6}{33.1} = 1.08.$$

This value can be viewed as one observation from an F distribution with $n-1 = 84-1 = 83$ and $m-1 = 70-1 = 69$ degrees of freedom. So the additional R statement

```
pf(var(x) / var(y), n - 1, m - 1)
```

returns 0.625, which indicates that the statistic 1.08 falls between the 62nd and 63rd percentiles of the appropriate sampling distribution. This is consistent with our assumption that the population variances are equal, so we can conclude that the assumption is not contradicted. We followed Theorem 1.10 by placing s_X^2 in the numerator and s_Y^2 in the denominator. If instead we use the reciprocal $s_Y^2/s_X^2 = 0.927$, this test statistic is less than 1, but not close enough to zero to conclude that the population variances are unequal. It is between the 37th and 38th percentiles of an F distribution with 69 and 83 degrees of freedom. An observed F random variable falling between the 37th and 38th percentile or between the 62nd and 63rd percentiles of the appropriate sampling distribution is not considered particularly unusual by statisticians.

This concludes this initial chapter on random sampling, which forms the foundation for the remainder of the book. Section 1.1 introduced statistical graphics, in which a carefully-designed graph helps to visualize a data set. Although these graphics can be very helpful as a first step in the analysis of a data set, it is oftentimes the case that a mathematical framework is necessary for drawing a conclusion regarding a data set.

Section 1.2 introduced the notion of a *random sample*, which is a sequence of n mutually independent and identically distributed observations drawn from a population distribution. These observations, also known as data values, are usually denoted by $X_1, X_2, \ldots, X_n$. The positive integer n is known as the *sample size*. A *statistic* is a function of the data values that does not depend on any unknown parameters. Statistics are functions of random variables, so they are also random variables. The probability distribution of a statistic is known as a *sampling distribution*. Statistics can be selected to reflect various aspects of interest in a statistical study, such as central tendency or dispersion, associated with the population distribution.

Section 1.3 introduced the following five statistics which reflect the central tendency of a population:

- the sample mean $\bar{X}$,
- the sample median M,
- the sample geometric mean G,
- the sample harmonic mean H, and
- the sample quadratic mean Q.

The sample mean and sample median are used as measures of central tendency in the vast majority of statistical applications. Upper-case letters, such as $\bar{X}$, M, G, H, and Q, are used when a statistic is being referred to in the abstract, and lower case letters, such as $\bar{x}$, m, g, h, and q, are used when a specific value of the statistic is being calculated for a particular data set $x_1, x_2, \ldots, x_n$.

Section 1.4 introduced the following five statistics which reflect the dispersion of a population:

- the sample variance S^2 (and its square root the sample standard deviation S),
- the sample range R,
- the sample interquartile range,
- the sample mean absolute deviation, and
- the sample Gini mean difference.

The sample variance and sample standard deviation are used as measures of dispersion in the vast majority of statistical applications.

Section 1.5 considered the special case of random sampling from a normal population. This framework is given a section of its own because (a) many real-world populations, such as heights of adult women, are normally distributed, and (b) certain mathematical properties of the normal distribution make for tractable statistical inference. The discussion revolved around eight results associated with random sampling from a normal population, summarized in Table 1.13. The first five results concerned a random sample of size n drawn from a single normal population; the last three concerned two independent random samples of sizes n and m drawn from two normal populations.

Theorem	Result
1.3	$\bar{X} \sim N\left(\mu, \dfrac{\sigma^2}{n}\right)$
1.4	$\displaystyle\sum_{i=1}^{n}\left(\frac{X_i-\mu}{\sigma}\right)^2 \sim \chi^2(n)$
1.5	$\bar{X}$ and S^2 are independent
1.6	$\dfrac{(n-1)S^2}{\sigma^2} \sim \chi^2(n-1)$
1.7	$\dfrac{\bar{X}-\mu}{S/\sqrt{n}} \sim t(n-1)$
1.8	$\bar{X}-\bar{Y} \sim N\left(\mu_X-\mu_Y, \dfrac{\sigma_X^2}{n}+\dfrac{\sigma_Y^2}{m}\right)$
1.9	$\dfrac{\bar{X}-\bar{Y}-(\mu_X-\mu_Y)}{S_P\sqrt{\frac{1}{n}+\frac{1}{m}}} \sim t(n+m-2)$
1.10	$\dfrac{S_X^2/\sigma_X^2}{S_Y^2/\sigma_Y^2} \sim F(n-1, m-1)$

Table 1.13: Results concerning random sampling from normal populations.

1.6 Exercises

1.1 The built-in R data set named `faithful` contains $n = 272$ waiting times between eruptions and the associated durations of the eruptions for the Old Faithful geyser in Yellowstone National Park. Draw a scatterplot of the eruption durations (on the horizontal axis) and the associated time between eruptions (on the vertical axis). What conclusions can be drawn?

1.2 The built-in R data set named `Titanic` contains data values concerning the fate of the passengers on the maiden voyage of the *RMS Titanic*, which struck an iceberg on April 14, 1912 and sunk. Draw a mosaic plot for plotting gender against survival status. What conclusions can be drawn?

1.3 The built-in R data set named `islands` contains the areas, in thousands of square miles, of the landmasses on earth that exceed 10,000 square miles. What is the largest landmass on earth? Draw a histogram of the areas of the landmasses. What conclusions can be drawn?

1.4 The built-in R data set named `InsectSprays` is a data frame that contains counts of insects in $n = 72$ agricultural experimental units which were treated with six different insecticides labeled A, B, C, D, E, and F. Draw side-by-side box plots of the counts for the six insecticides. What conclusions can be drawn?

1.5 The built-in R data set named `quakes` is a data frame that contains data on $n = 1000$ seismic events of magnitude 4.0 and above that occurred in a cube near Fiji beginning in 1964. Draw a histogram and empirical cumulative distribution function of the depth (in kilometers) of the 1000 seismic events. What conclusions can be drawn?

1.6 A bag contains 5 billiard balls numbered 1, 2, 3, 4, 5. A random sample of size $n = 3$ is drawn *without replacement* from the bag. What is the probability mass function of the sample median?

1.7 Consider the random discrete population described by the probability mass function

$$f_X(x) = x/10 \qquad x = 1, 2, 3, 4.$$

A sample of $n = 3$ observations is drawn at random and without replacement from this distribution. (After $x_1 = 2$, for example, is observed, the next observation is drawn from a conditional probability mass function that excludes 2.)

(a) Find the $P(X_{(2)} = 2)$ by hand.

(b) Find the $P(X_{(2)} = 2)$ using APPL.

(c) Verify the two results above using Monte Carlo simulation.

1.8 Let X_1, X_2, X_3, X_4 be a random sample from a $U(0, 5)$ population. Find a constant c such that $P\big(\max\{X_1, X_2, X_3, X_4\} > c\big) = 4/5$.

1.9 Let X_1, X_2, X_3 be a random sample from a $U(0, 1)$ population. Find the population correlation between $X_{(1)}$ and $X_{(3)}$.

1.10 Let X_1, X_2, X_3 be a random sample from a population with probability density function

$$f_X(x) = 2x \qquad 0 < x < 1.$$

Find $P\left(X_{(3)} > \frac{2}{3}\right)$.

1.11 Let X_1, X_2, X_3 be a random sample from a population with probability density function

$$f_X(x) = \theta x^{\theta-1} \qquad 0 < x < 1,$$

where θ is a positive constant. Give an expression for $P(X_{(3)} - X_{(2)} > c)$, where c is a constant satisfying $0 < c < 1$, and $X_{(1)}, X_{(2)}, X_{(3)}$ are the order statistics.

1.12 Consider the random discrete population described by the Zipf probability mass function

$$f(x) = \frac{1}{x^\alpha \sum_{i=1}^{k} (1/i)^\alpha} \qquad x = 1, 2, \ldots, k,$$

for some positive integer k and some nonnegative α. A sample of $n = 2$ observations is drawn at random and without replacement from this distribution. Find $E\left[X_{(1)}\right]$ to ten digits when $k = 6$ and $\alpha = 1/3$. You may use APPL.

1.13 A game of Yahtzee™ consists of 13 rounds, each beginning with the roll of five fair dice. Let $X_1, X_2, \ldots, X_{13}$ be the observed *maximums* of the number of spots showing on the initial rolls of each round.

(a) Use APPL to find the exact value of the population skewness of $\bar{X}$.

(b) Verify your solution to part (a) using Monte Carlo simulation.

1.14 Let $X_1, X_2, \ldots, X_n$ be a random sample from a $U(0, 1)$ population. These random variables partition the interval $(0, 1)$ into $n + 1$ segments given by the "gap" random variables

$$G_1 = X_{(1)},\ G_2 = X_{(2)} - X_{(1)},\ G_3 = X_{(3)} - X_{(2)},\ \ldots,\ G_n = X_{(n)} - X_{(n-1)},\ G_{n+1} = 1 - X_{(n)},$$

where $X_{(1)}, X_{(2)}, \ldots, X_{(n)}$ are the $U(0, 1)$ order statistics. Find the joint probability density function of the first n gaps $G_1, G_2, \ldots, G_n$. *Note:* The last gap, $G_{n+1} = 1 - X_{(n)}$, is not included here because it is determined by the first n gaps.

1.15 Let $X_1, X_2, \ldots, X_n$ be a random sample from a population with probability density function

$$f(x) = \lambda e^{-\lambda x} \qquad x > 0.$$

The corresponding order statistics are $X_{(1)}, X_{(2)}, \ldots, X_{(n)}$. The "gap" statistics are defined as $G_k = X_{(k)} - X_{(k-1)}$ for $k = 1, 2, \ldots, n$, where $X_{(0)}$ is defined to be zero. Find the distribution of G_k for $k = 1, 2, \ldots, n$ and show that the gap statistics are mutually independent random variables.

1.16 Let X_1 and X_2 be a random sample from a $U(0, 1)$ population. Find $P(X_1/X_2 > c)$, where c is a real positive constant.

1.17 Let X_1, X_2, and X_3 be mutually independent random variables with the following probability density functions:

$$f_{X_1}(x_1) = 1 \qquad 0 < x_1 < 1,$$
$$f_{X_2}(x_2) = 2x_2 \qquad 0 < x_2 < 1,$$

and

$$f_{X_3}(x_3) = 2(1 - x_3) \qquad 0 < x_3 < 1.$$

Find the population variance of the second order statistic $X_{(2)}$ by analytic methods, and support your result via a Monte Carlo simulation.

1.18 Let $X_1 \sim U(0, 1)$, $X_2 \sim U(0, 2)$, and $X_3 \sim U(0, 3)$ be mutually independent random variables.

(a) Find the probability density function and expected value of the second order statistic $X_{(2)}$. *Hint*: Use conditioning to find the probability density function.

(b) Use Monte Carlo simulation to support the expected value that you obtained in part (a).

1.19 Let X_1 and X_2 be a random sample from a population having probability mass function

$$f_X(x) = \begin{cases} 1/3 & x = 0 \\ 2/3 & x = 1. \end{cases}$$

(a) Find the probability mass function of the sample mean.

(b) Find the probability mass function of the sample median.

(c) Find the probability mass function of the sample geometric mean.

1.20 Let $X_1, X_2, \ldots, X_n$ be a random sample from a Bernoulli(p) population. What is the probability mass function of the sample geometric mean $G = (X_1 X_2 \ldots X_n)^{1/n}$?

1.21 Let $X_1 \sim \text{binomial}(10, 1/8)$, $X_2 \sim \text{binomial}(4, 1/8)$, $X_3 \sim \text{binomial}(6, 1/8)$ be mutually independent random variables. Find the exact value of $P\big(\bar{X} < 1.8\big)$.

1.22 Let X_1, X_2, X_3 be a random sample from a population with moment generating function

$$M_X(t) = \frac{2}{5}e^{3t} + \frac{3}{5}e^{6t} \qquad -\infty < t < \infty.$$

What is the value of the 75th percentile of the distribution of $\bar{X}$?

1.23 Let X_1 and X_2 be a random sample from a $U(0, 1)$ population. Find $P\big(\bar{X} > 0.8\big)$.

1.24 Let X_1, X_2, X_3, X_4 be a random sample drawn from a population with probability density function

$$f(x) = 2x \qquad 0 < x < 1.$$

Find the probability density function of the sample geometric mean.

1.25 Let X_1, X_2, X_3 be a random sample from a Bernoulli(p) population, where p is an unknown parameter satisfying $0 < p < 1$. Find the probability mass function of $\bar{X}$.

1.26 Let X_1 and X_2 have joint probability density function

$$f(x_1, x_2) = 2 \qquad 0 < x_1 < x_2 < 1.$$

Find the probability density function of $\bar{X}$.

1.27 Determine whether the following statement is true or false. The plug-in estimate of the sample median of a chi-square random variable with 7 degrees of freedom from a random sample of size $n = 100$ is correctly simulated with the following R command.

```
(rchisq(100, 7)[50] + rchisq(100, 7)[51]) / 2
```

1.28 The *bootstrapping* methodology resamples a data set of n values with replacement. For example, consider the (small) data set of $n = 3$ observations:

$$1, 2, 9.$$

If these three values are sampled three times with replacement, find the probability mass function of the sample mean of the three values sampled.

1.29 Let X_1, X_2, X_3 be a random sample from a Poisson(λ) population. Find the value of λ such that $P(\bar{X} = 4) = 6^{12}e^{-6}/12!$.

1.30 Let $X_1, X_2, \ldots, X_5$ be a random sample of adult heights, each rounded to the nearest inch. The sum of the sample mean and the sample median is 139 inches. The shortest person in the sample is 66 inches tall. If the shortest person in the sample is replaced by an even shorter person who is 61 inches tall, what will be the updated sum of the sample mean and the sample median?

1.31 Give an example of a data set in which the sample harmonic mean is undefined.

1.32 Let $X_1, X_2, \ldots, X_n$ be data values. Show that

$$H = \frac{n\prod_{i=1}^{n} X_i}{\sum_{i=1}^{n}(1/X_i)\prod_{j=1}^{n} X_j},$$

where H is the sample harmonic mean.

1.33 Let X_1 and X_2 be data values. Write an equation relating the sample mean $\bar{X}$, the sample geometric mean G, and the sample harmonic mean H.

1.34 Let $X_1, X_2, \ldots, X_n$ be data values. The *sample weighted harmonic mean* H_w is defined as

$$H_w = \frac{\sum_{i=1}^{n} w_i}{\sum_{i=1}^{n}(w_i/X_i)},$$

where $w_1, w_2, \ldots, w_n$ are the weights associated with the data values.

(a) Show that the sample weighted harmonic mean H_w reduces to the sample harmonic mean H when all of the weights are equal.

(b) Consider the calculation of the sample weighted harmonic mean of the price-to-earnings (P/E) ratio on an index consisting of two stocks. Stock A has a market capitalization of \$50 billion and annual earnings of \$5 billion, resulting in the P/E ratio

$$x_1 = \$50 \text{ billion}/\$5 \text{ billion} = 10.$$

Stock B has a market capitalization of \$1 billion and annual earnings of \$10 million, resulting in the P/E ratio

$$x_2 = \$1 \text{ billion}/\$10 \text{ million} = 100.$$

These two stocks are combined in an index that consists of 10% invested in Stock A ($w_1 = 0.1$) and 90% invested in Stock B ($w_2 = 0.9$). We desire to know a P/E ratio for the index. We expect that this P/E ratio will fall between 10 and 100, probably closer to 100 because of the weights. Calculate the ordinary sample weighted mean and the sample weighted harmonic mean and write a paragraph arguing for which of these two sample weighted means is appropriate for the P/E ratio of the index.

1.35 The data values x_1, x_2, x_3 result in sample mean $\bar{x} = 5$ and sample variance $s^2 = 4$. Give the updated values of the sample mean and sample variance if an additional data value $x_4 = 9$ is collected.

1.36 Let $X_1, X_2, \ldots, X_n$ be a random sample from a $N\left(\mu, \sigma^2\right)$ population. Find the probability density function of the sample variance S^2.

1.37 Let X_1, X_2, X_3 be a random sample from a population with probability mass function

$$f(x) = \begin{cases} 1/3 & x = -2 \\ 2/3 & x = 1. \end{cases}$$

(a) Find the probability mass function of the sample mean $\bar{X}$.

(b) Find the probability mass function of the sample quadratic mean Q.

(c) Find the probability mass function of the sample variance S^2.

1.38 Find the population variance of the sample variance of the independent random variables $X_1 \sim U(0, 1)$ and $X_2 \sim U(0, 2)$, where the sample variance is defined as

$$S^2 = \left(X_1 - \frac{X_1 + X_2}{2}\right)^2 + \left(X_2 - \frac{X_1 + X_2}{2}\right)^2.$$

You may use APPL. Support your result with a Monte Carlo simulation.

1.39 Let X_1 and X_2 be a random sample from a population having probability mass function

$$f_X(x) = \begin{cases} 1/3 & x = 0 \\ 2/3 & x = 1. \end{cases}$$

(a) Find the expected value of the sample range.

(b) Find the expected value of the sample standard deviation.

(c) Conduct a Monte Carlo simulation to assess the correctness of your solution to part (b).

1.40 Prove the shortcut formula for computing the sample variance:

$$s^2 = \frac{1}{n-1} \sum_{i=1}^{n} (x_i - \bar{x})^2 = \frac{1}{n-1} \left[\left(\sum_{i=1}^{n} x_i^2 \right) - n\bar{x}^2 \right].$$

1.41 Let $X_1, X_2, \ldots, X_n$ be a random sample from a $U(0, \theta)$ population.

(a) Find the probability density function of the sample range $R = X_{(n)} - X_{(1)}$. *Hint*: Make sure to re-derive the probability density function of the sample range which is given in Theorem 1.2.

(b) Find the cumulative distribution function of the sample range $R = X_{(n)} - X_{(1)}$.

(c) Find $E[R]$.

(d) Conduct a Monte Carlo simulation experiment to verify your expression for $E[R]$ in the case of $\theta = 3$ and $n = 5$.

1.42 Every morning, Ivan shakes a cup containing five fair dice, and rolls them. He then calculates the sample mean of the number of spots showing on the up faces of the five dice and writes this sample mean on a ledger pad. Ivan has been doing this for years, so he has thousands of sample means written on his ledger pad. Give an estimate of the sample variance of the numbers that Ivan has written down.

1.43 Let X_1 and X_2 be a random sample from a Bernoulli(p) population, where p is a parameter satisfying $0 < p < 1$. Find the probability mass function of the sample mean absolute deviation

$$\frac{1}{2}\left(\left|X_1 - \bar{X}\right| + \left|X_2 - \bar{X}\right|\right).$$

1.44 Let X_1 and X_2 be a random sample from a $U(0,\,\theta)$ population, where θ is a positive parameter. Find the probability density function of the sample mean absolute deviation

$$\frac{1}{2}\left(\left|X_1 - \bar{X}\right| + \left|X_2 - \bar{X}\right|\right).$$

1.45 Write the R functions `mad` and `gini`, each with a single argument `x` which is a vector containing data values, which calculate the sample mean absolute deviation and the sample Gini mean difference. Test your functions on the data set of $n = 10$ observations

$$3,\,1,\,5,\,1,\,3,\,2,\,1,\,1,\,3,\,2.$$

1.46 Let $X_1 \sim U(0,\,1)$, $X_2 \sim U(0,\,2)$, and $X_3 \sim U(0,\,3)$ be mutually independent random variables.

(a) Find the population variance of the third order statistic $X_{(3)}$.

(b) Find the probability density function of the sample mean $\bar{X}$. You may use APPL.

(c) Find the expected value of the sample variance, that is, $E[S^2]$.

1.47 For a random sample $X_1, X_2, \ldots, X_n$, the sample Gini mean difference is

$$G = \frac{1}{\binom{n}{2}} \sum_{j=2}^{n} \sum_{i=1}^{j-1} |X_i - X_j|.$$

The sample Gini mean difference is a measure of dispersion.

(a) What are the units of G? How many terms are there in the double summation that defines G? What is the range of G?

(b) Express G as a function of the order statistics $X_{(1)}, X_{(2)}, \ldots, X_{(n)}$ using only a single summation.

(c) When the sampling is from a $U(0,\,1)$ population, find $E[G]$.

1.48 If Z_1 and Z_2 are a random sample from a standard normal population, find

(a) $P\left(Z_1^2 + Z_2^2 < 1\right)$,

(b) an expression for the 100pth percentile of the distribution of $Z_1^2 + Z_2^2$ for $0 < p < 1$.

1.49 Let $X_1, X_2, \ldots, X_4$ be a random sample from a $N(\mu_X, \sigma_X^2)$ population and let $Y_1, Y_2, \ldots, Y_8$ be a random sample from a $N(\mu_Y, \sigma_Y^2)$ population, independent of the first sample. Find the 95th percentile of the distribution of

$$\frac{\sum_{i=1}^{4}(X_i - \bar{X})^2}{\sigma_X^2} + \frac{\sum_{i=1}^{8}(Y_i - \bar{Y})^2}{\sigma_Y^2}.$$

1.50 Let Z_1 and Z_2 be a random sample from a $N(0, 1)$ population. Find the population median of $Z_1^2 + Z_2^2$.

1.51 Let $X_1, X_2, \ldots, X_n$ be a random sample from a $N\left(\mu, \sigma^2\right)$ population and let $Y_1, Y_2, \ldots, Y_m$ be a random sample from a $N\left(\mu, \sigma^2\right)$ population, independent of the first sample. Find the population median of the distribution of $\bar{X} - \bar{Y}$.

1.52 Let $X_1, X_2, \ldots, X_n$ be a random sample from a $N\left(\mu, \sigma^2\right)$ population and let $Y_1, Y_2, \ldots, Y_m$ be a random sample from a $N\left(\mu, \sigma^2\right)$ population, independent of the first sample. Find the 95th percentile of the distribution of $\bar{X} - \bar{Y}$. *Hint*: The 95th percentile is a function of n and m.

1.53 Let $X_1, X_2, \ldots, X_{19}$ be a random sample of IQ scores drawn from a $N(100, 100)$ population.

(a) Find the probability that the sample mean exceeds 105.

(b) Find the probability that the sample standard deviation exceeds 12.

1.54 Let $X_1, X_2, \ldots, X_n$ be a random sample from a standard normal population. Find a sample size n such that

$$P\left(\sum_{i=1}^{n}(X_i - \bar{X})^2 > 3\right) \cong 0.7.$$

1.55 Let $X_1, X_2, \ldots, X_{18}$ be a random sample from a $N(\mu, \sigma^2)$ population, where μ and $\sigma^2 > 0$ are known parameters.

(a) Find a value c such that

$$P\left(\frac{\bar{X} - \mu}{\sigma/\sqrt{n}} < c\right) = 0.67.$$

(b) Find a value c such that

$$P\left(\frac{\bar{X} - \mu}{S/\sqrt{n}} < c\right) = 0.99.$$

1.56 Kent, Dreama, Preston, and Evelyn go out on a double date. The men's heights, in inches, are independent $N(68, 9)$ random variables. The women's heights, in inches, are independent $N(65, 4)$ random variables. Furthermore, the men's heights are independent of the women's heights.

(a) Find the probability that Dreama is taller than Kent.

(b) Find the probability that Dreama is taller than both Kent and Preston.

(c) Find the probability that the average of the women's heights exceeds the average of the men's heights.

(d) Set up, but do not evaluate, an expression for the probability that the taller woman is taller than the shorter man.

1.57 Find a mathematical relationship between $z_{\alpha/2}$ and $\chi^2_{1-\alpha,1}$ for $0 < \alpha < 1$.

1.58 Consider the F distribution when the degrees of freedom of the numerator and denominator are equal.

(a) Find the population median of this distribution.

(b) Find a mathematical relationship between the α and $1-\alpha$ percentiles of this distribution for $0 < \alpha < 0.5$.

1.59 The time to complete a pit stop at the Indianapolis 500 is normally distributed with population mean μ seconds and population standard deviation σ seconds.

(a) Assume that the Formula 1 crew knows that $\mu = 8$ and $\sigma = 2$ for their crew. If the Formula 1 driver makes three independent pit stops during the race, find the probability that the total pit stop time is less than 25 seconds. Provide the analytic solution and the R statements to solve this problem.

(b) If the Formula 2 crew does not know their population mean pit stop time, μ, how many observations n must they collect to be at least 99% certain that their sample mean is within one second of the true value? Assume that $\sigma = 2$. Provide the analytic solution and the R statements to solve this problem.

1.60 Classify each of the results below as "exact," "approximately true," or "false."

(a) If $X_1, X_2, \ldots, X_n$ denote a random sample from a population with finite population mean μ and finite population variance σ^2, then $E\left[\bar{X}\right] = \mu$.

(b) If $X_1, X_2, \ldots, X_n$ denote a random sample from a population with finite population mean μ and finite population variance σ^2, then $\bar{X} \sim N(\mu, \sigma^2/n)$ for all sample sizes n.

(c) If $X_1, X_2, \ldots, X_n$ denote a random sample from a normally distributed population with finite population mean μ and finite population variance σ^2, then

$$\frac{\bar{X} - \mu}{S/\sqrt{n}} \sim t(n-1)$$

for all sample sizes n.

(d) If $X_1, X_2, \ldots, X_n$ denote a random sample from a population with finite population mean μ and finite population variance σ^2, then $E\left[S^2\right] = \sigma^2$.

(e) If X is drawn from a $U(0, 2)$ population, then $P(X < 0.8) = 0.4$.

(f) If $X_1, X_2, \ldots, X_9$ denote a random sample from a $U(0, 2)$ population, then $P\left(\bar{X} < 0.8\right) = 0.4$.

(g) If $X_1, X_2, \ldots, X_{30}$ denote a random sample from a $U(0, 2)$ population, then

$$\frac{\bar{X} - 1}{1/\sqrt{90}} \sim N(0, 1).$$

1.61 Let $X_1, X_2, \ldots, X_9$ denote a random sample from a $N(4, 9)$ population. Let $Y_1, Y_2, \ldots, Y_{36}$ denote a random sample from a $N(5, 16)$ population that is independent of the first sample. Find $P\left(\bar{X} < \bar{Y}\right)$.

1.62 Let $X_1, X_2, \ldots, X_n$ denote a random sample from a $N(\mu, \sigma^2)$ population. Let the sample variance be defined by

$$S^2 = \frac{1}{n-1} \sum_{i=1}^{n} (X_i - \bar{X})^2.$$

Find $E[S]$.

1.63 Let X_1, X_2, X_3 be a random sample from a normal population with population mean $\mu = 65$ and population variance $\sigma^2 = 10$. Find $E\left[S^2\right]$ and $V\left[S^2\right]$, where

$$S^2 = \frac{1}{2}\left[\left(X_1 - \bar{X}\right)^2 + \left(X_2 - \bar{X}\right)^2 + \left(X_3 - \bar{X}\right)^2\right]$$

and

$$\bar{X} = \frac{X_1 + X_2 + X_3}{3}.$$

Hint: The expected value of a chi-square random variable is its degrees of freedom and the population variance of a chi-square random variable is twice its degrees of freedom.

1.64 Let $X_1, X_2, \ldots, X_n$ be a random sample from a normal population with population mean μ_X and population variance σ^2. Let $Y_1, Y_2, \ldots, Y_m$ be a random sample from a normal population with population mean μ_Y and population variance $4\sigma^2$. Assuming that the two random samples are independent, find the distribution of

$$\frac{4\sum_{i=1}^{n}\left(X_i - \bar{X}\right)^2 + \sum_{i=1}^{m}\left(Y_i - \bar{Y}\right)^2}{4\sigma^2}.$$

1.65 Let $X_1, X_2, \ldots, X_9$ be a random sample from a $N(68, 4)$ population. Let

$$S^2 = \frac{1}{8}\sum_{i=1}^{9}\left(X_i - \bar{X}\right)^2$$

be the sample variance. Find $V\left[S^2\right]$ and $P\left(S^2 > 8.93\right)$.

1.66 Francy collects a random sample of IQ scores $X_1, X_2, \ldots, X_9$, which are assumed to be drawn from a normal population with population mean $\mu = 100$ and population standard deviation $\sigma = 15$. Find the probability that the sample variance exceeds 565, that is, find $P\left(S^2 > 565\right)$.

1.67 Let $X_1, X_2, \ldots, X_{25}$ be a random sample from a normal population with population mean 100 and population variance 16. Find the probability that

(a) the sample mean exceeds 102,

(b) the sample variance exceeds 22.13.

1.68 Let X_1, X_2, X_3, X_4 be a random sample from the heights of adult males having a normal distribution with population mean 68 inches and population variance $\sigma_X^2 = 16$ square inches. Let $Y_1, Y_2, \ldots, Y_9$ be a random sample from the heights of adult females having a normal distribution with population mean 65 inches and population variance $\sigma_Y^2 = 9$ square inches. Find the probability that the average height of the women exceeds the average height of the men.

1.69 An automatic filling machine fills 16-ounce juice bottles with a normally distributed amount of juice with population mean 16 ounces and population standard deviation 0.1 ounces. Assume that the number of ounces in $n = 9$ bottles constitutes a random sample. Find

(a) the probability that the first bottle sampled contains less than 16.1 ounces,

(b) the probability that the sample mean is less than 16.1 ounces, and

(c) the probability that the sample standard deviation is less than 0.2 ounces.

1.70 Let $X_1, X_2, \ldots, X_n$ be a random sample from a $N\left(\mu_X, \sigma_X^2\right)$ population and let $Y_1, Y_2, \ldots, Y_m$ be a random sample from a $N\left(\mu_Y, \sigma_Y^2\right)$ population, independent of the first sample. Find the 95th percentile (to four decimal places) of the distribution of the following random variables:

(a) $\dfrac{X_1 - \mu_X}{\sigma_X}$,

(b) $\dfrac{X_1 - \mu_X}{S_X}$, when $n = 4$,

(c) $\left(\dfrac{X_1 - \mu_X}{\sigma_X}\right)^2$,

(d) $\displaystyle\sum_{i=1}^{n} \frac{X_i - \mu_X}{\sigma_X}$,

(e) X_1, when $\mu_X = 100$ and $\sigma_X = 10$,

(f) $\displaystyle\sum_{i=1}^{n} \left(\frac{X_i - \mu_X}{\sigma_X}\right)^2$, when $n = 17$,

(g) $\dfrac{\bar{X} - \mu_X}{\sigma_X / \sqrt{n}}$,

(h) $\dfrac{\bar{X} - \mu_X}{S_X / \sqrt{n}}$, when $n = 14$,

(i) $\dfrac{6S_X^2}{\sigma_X^2}$, when $n = 7$,

(j) $\dfrac{S_X^2}{S_Y^2}$, when $n = 8$, $m = 10$, and $\sigma_X = \sigma_Y$,

(k) $\dfrac{\bar{X} - \bar{Y} - (\mu_X - \mu_Y)}{\sigma\sqrt{\frac{1}{n} + \frac{1}{m}}}$, when $\sigma_X = \sigma_Y = \sigma$,

(l) $\left(\displaystyle\sum_{i=1}^{n} X_i^2\right) \Big/ \left(\displaystyle\sum_{i=1}^{m} Y_i^2\right)$, when $n = 12$, $m = 17$, $\mu_X = \mu_Y = 0$, and $\sigma_X = \sigma_Y$,

(m) $\dfrac{3\left(\bar{X} - \mu_X\right)^2}{\left(\bar{Y} - \mu_Y\right)^2}$, when $n = 18$, $m = 6$, and $\sigma_X = \sigma_Y$.

1.71 Let $X_1, X_2, \ldots, X_7$ be a random sample from a $N(\mu, 16)$ population. Find the value of the constant c that satisfies $P\left(S^2 < c\right) = 0.99$.

1.72 Let X_1, X_2, X_3, X_4 be a random sample from a $N\left(\mu, \sigma^2\right)$ population. Find $P\left(\bar{X} > \mu + 2S\right)$.

1.73 Let $X_1, X_2, \ldots, X_{100}$ be a random sample from a standard normal population.

(a) Calculate $P(X_{29} < 0.2)$.

(b) Calculate $P(\bar{X} < 0.2)$.

(c) Calculate $P(S^2 > 0.9)$.

1.74 Let X_1, X_2, X_3, X_4 be a random sample from a $N(10, 4)$ population. Let $Y_1, Y_2, \ldots, Y_{16}$ be a random sample from a $N(12, 1)$ population. Let

$$S_X^2 = \frac{1}{3} \sum_{i=1}^{4} (X_i - \bar{X})^2$$

and

$$S_Y^2 = \frac{1}{15} \sum_{i=1}^{16} (Y_i - \bar{Y})^2$$

be the sample variances of the two random samples. Assuming that the two random samples are independent, find $P(1 < S_X^2 / S_Y^2 < 4)$.

1.75 Let $X_1, X_2, \ldots, X_n$ and $Y_1, Y_2, \ldots, Y_n$ be two mutually independent random samples from two normal populations. If it is known that the population variance of the first population is two times the population variance of the second population, find the smallest sample size n required to assure that $P(S_X^2 > S_Y^2)$ is at least 0.99, where S_X^2 and S_Y^2 are the sample variances calculated from the two samples.

1.76 Let X_1, X_2, X_3, X_4 be a random sample from a standard normal population. What is the probability distribution (give the name of the distribution and the value of any parameter(s)) of

(a) $(X_1 - \bar{X})^2 + (X_2 - \bar{X})^2 + (X_3 - \bar{X})^2 + (X_4 - \bar{X})^2$,

(b) $\dfrac{(X_1 - X_2 + X_3 + X_4)^2}{4}$.

1.77 Let $X_1, X_2, \ldots, X_9$ be a random sample from a $N(\mu, \sigma^2)$ population. Write an R command to find $P(\bar{X} - \mu < S)$.

1.78 Let $X_1, X_2, \ldots, X_5$ be a random sample from a normal population with population mean $\mu = 17$ and standard deviation $\sigma = 3$. Write an R command to calculate the 95th percentile of S^2.

1.79 Let $X_1, X_2, \ldots, X_{25}$ and $Y_1, Y_2, \ldots, Y_{100}$ be independent random samples from normal populations with population means $\mu_X = 12$ and $\mu_Y = 10$ and population standard deviations $\sigma_X = 15$ and $\sigma_Y = 20$, respectively. Write a single R command to calculate $P(\bar{X} < \bar{Y})$.

1.80 Let $X_1, X_2, \ldots, X_5$ be a random sample from a $N(\mu_X, \sigma^2)$ population. Let $Y_1, Y_2, \ldots, Y_5$ be a random sample from a $N(\mu_Y, \sigma^2)$ population. Assuming that the two random samples are independent, find the probability density function of

$$\max\left\{\frac{S_X^2}{S_Y^2}, \frac{S_Y^2}{S_X^2}\right\}.$$

1.81 Let $X_1, X_2, \ldots, X_n$ be a random sample from a normal population with population mean 68 and population variance 9.

(a) Find the smallest n such that $P\big(67 < \bar{X} < 69\big) > 0.99$.

(b) Write a single R statement that returns the solution to part (a).

(c) If $n = 20$, find $P\left(S^2 < 12.886\right)$.

(d) Write a single R statement that returns the solution to part (c).

1.82 Let $X_1, X_2, \ldots, X_9$ be a random sample from a $N\left(\mu, \sigma^2\right)$ population. Let $Y_1, Y_2, \ldots, Y_5$ be a random sample from a $N\left(\mu, 3\sigma^2\right)$ population. Assuming that the two random samples are independent, write an R command to calculate $P\left(S_X^2 < 4S_Y^2\right)$.

1.83 Let $X_1 \sim N(\mu, 9)$ and $X_2 \sim N(\mu, 16)$ be independent random variables, where μ is a nonzero parameter. Find real-valued constants c_1 and c_2 such that

$$c_1X_1 + c_2X_2 \sim N(0, 1).$$

1.84 Let X_1 and X_2 have the bivariate normal distribution with parameters $\mu_{X_1} = \mu$, $\mu_{X_2} = \mu$, $\sigma_{X_1}^2 = 9$, $\sigma_{X_2}^2 = 16$, and $\rho = 2/3$, where μ is a nonzero parameter. Find real-valued constants c_1 and c_2 such that

$$E[c_1X_1 + c_2X_2] = 0 \qquad \text{and} \qquad V[c_1X_1 + c_2X_2] = 1.$$

With your choices of c_1 and c_2, will $c_1X_1 + c_2X_2$ be normally distributed?

1.85 Let $Z_1, Z_2, \ldots, Z_5$ be a random sample from a $N(0, 1)$ population. For what real-valued constant c does

$$c \cdot \frac{Z_1 + Z_2}{\sqrt{Z_3^2 + Z_4^2 + Z_5^2}}$$

have the t distribution?

1.86 Let $X_1, X_2, \ldots, X_n$ be a random sample from a $N\left(\mu, \sigma_X^2\right)$ population, where μ and $\sigma_X^2 > 0$ are unknown parameters. Likewise, let $Y_1, Y_2, \ldots, Y_n$ be a random sample from a $N\left(\mu, \sigma_Y^2\right)$ population, where μ and $\sigma_Y^2 > 0$ are unknown parameters. The two random samples are independent. The corresponding values in the two populations are paired and differenced in the following fashion: $Z_i = X_i - Y_i$, for $i = 1, 2, \ldots, n$. Let $\bar{Z}$ denote the sample mean of $Z_1, Z_2, \ldots, Z_n$. Let S_Z^2 denote the sample variance of $Z_1, Z_2, \ldots, Z_n$.

(a) What is the probability distribution of Z_1?

(b) What is the probability distribution of $\bar{Z}$?

(c) What is the probability distribution of $\sqrt{n}\bar{Z}/S_Z$?

Chapter 2

Point Estimation

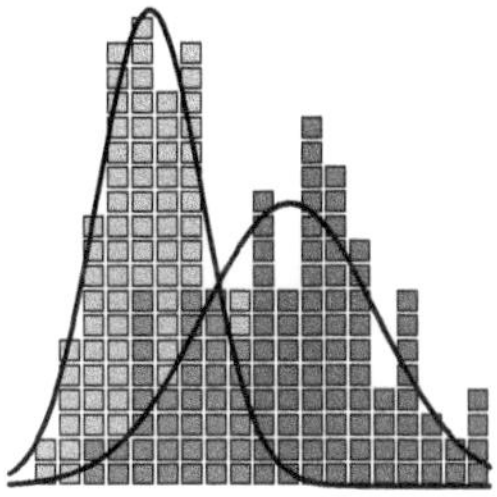

The previous chapter introduced sampling distributions of statistics for data drawn from various population probability distributions. This chapter raises the important question of how to estimate the parameters for a population parametric distribution from collected data values.

A statistic encapsulates information within a data set in a single number. Statistical inference typically uses these statistics to draw conclusions about the population from which the random sample was drawn. Drawing these inferences often involves fitting a parametric distribution, such as the normal distribution or the exponential distribution, to a data set. The process of estimating these unknown parameters from a data set is generally known as "point estimation." This chapter presents two techniques for determining point estimates of parameters, namely the *method of moments* and *maximum likelihood* techniques.

The general problem associated with point estimation can be stated as follows. Let $x_1, x_2, \ldots, x_n$ be the experimental values of the random variables $X_1, X_2, \ldots, X_n$, where n is the sample size. We want to find a statistic $\hat{\theta} = g(x_1, x_2, \ldots, x_n)$ such that $\hat{\theta}$ is a "good" point estimator of some unknown parameter θ. (Recall that the hat that is placed on top of θ is a common practice that signifies that $\hat{\theta}$ is an estimator of θ.) The Greek letter θ is popular with statisticians as a generic parameter. For a specific population distribution, it might be λ for a Poisson population, p for a Bernoulli population, etc. The specific properties used to determine whether an estimator is a good point estimator or a bad point estimator will also be surveyed in the chapter. In general, we want $\hat{\theta}$ to be close to θ with high probability.

2.1 Introduction

The parameters in a probability model, such as μ and σ for the normal distribution or λ for the Poisson distribution, are typically assumed to be known constants in probability theory. In a statistical setting, the parameters are typically unknown and need to be estimated from data.

The process begins by posing questions of interest to the experimenter. Once these questions have been established, data is collected by performing appropriate random experiments that address the questions of interest. Examples of such data include

- the lifetimes of toasters,
- the "social distance" between two strangers having a conversation,

- Olympic marathon winning times,
- odometer readings associated with the first oil change for a particular make of automobile,
- data pairs consisting of the number of days per year with a low temperature below freezing and the latitude of the location, or
- data pairs consisting of father's and son's adult heights collected by Sir Francis Galton.

For now, we limit the analysis to *univariate* data associated with a random experiment of interest.

What are some popular targets that statisticians would like to estimate? In addition to estimating parameters for a population distribution, statisticians are often interested in estimating

- the population mean μ,
- the population variance σ^2,
- the population probability of success p, for example, in a political poll or consumer survey,
- the difference between two population means $\mu_1 - \mu_2$, or
- the ratio of two population variances σ_1^2/σ_2^2.

It is not hard to imagine which statistics are appropriate for estimating these population parameters of interest. For example,

- the sample mean $\bar{X}$ estimates the population mean μ,
- the sample variance S^2 estimates the population variance σ^2, and
- the fraction of successes $\hat{p} = X/n$ estimates population probability of success p.

There is not just a single option, however, for estimating, say, the population mean μ. The following three statistics, for example, might be used to estimate the population mean in a particular setting:

$$\bar{X} \qquad X_7 \qquad \frac{X_{(1)} + X_{(n)}}{2}.$$

Is there a way to decide which of these statistics is the best one to use for estimating μ? This chapter addresses questions of this nature.

Why is it of interest to fit probability distributions to data? One possible answer is that the data is "bumpy" due to a small sample size and the associated random sampling variability, and fitting a parametric distribution smooths out the bumps. A second answer is that there might be some type of conclusion that needs to be drawn using evidence contained in the data set, and having a parametric distribution approximating the data set facilitates the inference procedure. A third answer is that a parametric probability model is necessary for a subsequent application, such as modeling a random element in a discrete-event simulation model.

The first step in the process of fitting probability distributions to data is to determine which of the probability distributions is an appropriate model for approximating the random variable associated with the random experiment of interest. Sometimes knowing information about the random experiment can help to exclude certain models, as in the two instances described below.

- If the random experiment involves counting, such as counting the number of ticks on a sheep or the number of daily hits on a website, then discrete distributions are candidates for a model, and the continuous distributions can be excluded.

- Conversely, if the random experiment records an inherently continuous random variable, such as the burning time of a light bulb prior to failure or the survival time after heart surgery, then continuous distributions are candidates for a model, and the discrete distributions can be excluded.

In addition, the support values might inform the list of potential probability distributions. In the case of the light bulb burning times described above, the normal distribution might be excluded because its support includes negative values. Probability distributions with entirely positive support, such as the exponential or Weibull distributions, would be candidates.

A helpful tool in identifying candidate probability distributions is the histogram. If the sample size n is moderate to large, the shape of the histogram can be compared to your inventory of known probability mass function or probability density function shapes, and this can be used to further limit the potential probability distributions for modeling the population distribution.

The second step in the process is to estimate the parameter(s) of a candidate probability distribution. Let θ denote a generic parameter to be estimated. Two estimation techniques will be described in this chapter. The first is the *method of moments* technique, in which an estimator $\hat{\theta}$ is chosen so that certain population and sample moments match. The second is the *maximum likelihood estimation* technique, in which an estimator $\hat{\theta}$ is chosen so that it is the most likely parameter value from the candidate distribution to have yielded the observed data set. The two methods do not necessarily yield the same estimator. The next two sections in this chapter survey the method of moments and the maximum likelihood estimation techniques.

The term *point estimator* describes $\hat{\theta}$ written in terms of $X_1, X_2, \ldots, X_n$, which is a random variable. The term *point estimate* describes $\hat{\theta}$ written in terms of the specific data values $x_1, x_2, \ldots, x_n$, which is a realization of the point estimator. This subtle difference in terminology is not universal, but is used by many authors of mathematical statistics textbooks.

The third step in the process is to assess how well the fitted probability distribution approximates the data. Although this third step will be done informally by displaying statistical graphics in this chapter, formal methods known as "goodness-of-fit tests" can be used to assess model adequacy.

This section ends with an elementary example that highlights some of the issues associated with the analysis of a data set.

Example 2.1 Let's say that you want to estimate the fraction of lefties in the world. One way to do so is to simply guess based on your experience in life thus far. Although this guess is your best estimate without additional information, you are concerned that it might not be accurate. So instead you proceed to initiate a statistical survey to answer the question.

One issue that must be resolved initially is: "what constitutes a lefty?" Some people write with their left hand, but kick a soccer ball with their right foot. Still others are ambidextrous. You must decide on a definition. You decide to survey people and determine whether they are lefties by observing which hand they use to fill out a short survey questionnaire. By observing the hand they use to fill out the survey you have eliminated the possibility of someone sabotaging your survey by lying about their handedness, a problem that statisticians have termed *response bias*.

A second issue that arises is that you want to estimate the fraction of lefties in the world, but are taking the survey in your hometown. Care must be taken to generalize appropriately. There might be customs that parents and educators take with small children in different parts of the world that influence their handedness.

Other important issues are (a) where the survey will be conducted, (b) when the survey will be conducted, and (c) how the survey will be conducted. These issues will be taken on in order. You decide to conduct the survey at the local shopping mall. This is perfectly fine as long as this particular shopping mall does not attract or repel lefties. You decide to conduct your survey on Saturday morning. This is also perfectly fine as long as lefties do not tend to sleep in or wake up early on Saturdays in a different proportion than righties. In terms of conducting the study, one would prefer to get a random sample $X_1, X_2, \ldots, X_n$ of binary observations, where $X_i = 1$ corresponds to success (surveying a lefty) and $X_i = 0$ corresponds to failure (surveying a righty). One aspect of a random sample is that the observations should be mutually independent random variables. If you survey identical twins or happen to encounter a gang of lefties, the independence assumption will be violated. So in order to achieve an approximately mutually independent set of observations, you decide to survey every fifth adult customer that comes off of an escalator at the mall who is willing to be surveyed.

The next step is to define a probability model for the data set. Since so much care has been taken in the sampling procedure, it is reasonable to assume that the data values can be modeled as mutually independent Bernoulli trials, with probability of success (being left-handed) p. In other words, the data values $X_1, X_2, \ldots, X_n$ are a random sample from a Bernoulli(p) population. Next, you would like a point estimator of p. The point estimator of p is a guess, informed by the data, of a likely value of the parameter p. A natural choice in this setting is

$$\hat{p} = \frac{X_1 + X_2 + \cdots + X_n}{n},$$

the estimated fraction of lefties. This point estimator is a random variable because the data values $X_1, X_2, \ldots, X_n$ are random variables. If $n = 100$ individuals are sampled, for instance, using the procedure described above, and 14 of those individuals were lefties, then $\hat{p} = 0.14$ is the point estimate. The observed fraction $\hat{p} = 0.14$ is an experimental or observed value of a random variable. If you return to the mall the following Saturday and survey 100 more individuals, it is likely you will arrive at a different value for $\hat{p}$. The probability distribution of $\hat{p}$ is known as the random sampling distribution of $\hat{p}$.

Although the formula for $\hat{p}$ is intuitive in the previous example, this is not the case in all settings. The focus of this chapter is to develop methods for determining point estimators and to consider their properties. We begin with the method of moments technique for finding point estimators.

2.2 Method of Moments

A desirable property of a fitted distribution is that population and sample moments should be approximately equal. Furthermore, we desire that the lowest-order moments of the population and sample be equal because they will give the closest match between the sample and the fitted distribution. This method for estimating parameters is known as "matching moments" or the "method of moments."

The technique is conceptually straightforward and proceeds as follows. Let k be the number of unknown parameters in the candidate probability distribution. Solve the equations

$$E\left[X^j\right] = \frac{1}{n}\sum_{i=1}^{n} x_i^j \qquad j = 1, 2, \ldots, k$$

for all unknown parameters, assuming all expectations exist, which gives the method of moments estimators of those parameters. For notational simplicity, the right-hand side of this equation can be written more compactly as m_j. For some distributions, this $k \times k$ set of simultaneous equations has a closed-form solution; for other distributions, this $k \times k$ set of simultaneous equations needs to be solved using numerical methods (for example, the Newton–Raphson procedure).

To most readers encountering the method of moments for the first time, the equation given above is troubling. The left-hand side of the equation is a constant: the jth population moment about the origin. The right-hand side of the equation is a random variable: the jth sample moment about the origin, which changes from one sample to the next. Equating a constant to a random variable here is simply a device that is used to get the first k population moments to match the first k sample moments for one particular choice of the parameters.

The first example illustrates the method of moments procedure for fitting data values to a uniform distribution in which the lower limit of the support is a known constant and the upper limit of the support is a positive unknown parameter to be estimated from the data.

Example 2.2 Let $X_1, X_2, \ldots, X_n$ be a random sample from a $U(0, \theta)$ population, where θ is a positive unknown parameter. Find the method of moments estimate of θ.

We assume that the preliminary work of collecting the data and drawing a histogram has already been completed and the conclusion is that the histogram is fairly flat with lower limit zero, so a $U(0, \theta)$ model is warranted. There is only a single ($k = 1$) parameter θ to estimate in this case. The population mean of a $U(0, \theta)$ random variable is $\mu = \theta/2$. So, equating the population mean (the first population moment) to the sample mean (the first sample moment) results in

$$\frac{\theta}{2} = \frac{1}{n}\sum_{i=1}^{n} x_i.$$

Solving for θ is simply a matter of multiplying both sides of this equation by 2, which gives the *point estimate*

$$\hat{\theta} = \frac{2}{n}\sum_{i=1}^{n} x_i,$$

which is twice the sample mean, that is, $\hat{\theta} = 2\bar{x}$. The data values $x_1, x_2, \ldots, x_n$ are written in lower case to signify that these are the collected realizations (experimental values) of the random variables $X_1, X_2, \ldots, X_n$. When we are interested in the *point estimator* written in terms of the random variables $X_1, X_2, \ldots, X_n$, we will use

$$\hat{\theta} = \frac{2}{n}\sum_{i=1}^{n} X_i.$$

There is a problem that arises with the method of moments estimate of θ for the $U(0, \theta)$ population distribution with certain data sets. Consider fitting the $U(0, \theta)$ distribution to the (tiny) data set of $n = 3$ observations (integers are used for easy computations) x_1, x_2, x_3, which are

$$1, 2, 9.$$

In this case, the sample mean is $\bar{x} = 4$, and the associated method of moments estimate is

$$\hat{\theta} = 8.$$

The problem here is that the fitted $U(0, 8)$ distribution is incapable of producing one of the data values, namely 9, because this particular data value falls outside of the support of the estimated probability distribution. This is certainly not a desirable feature of the fitted distribution. This problem will be overcome for the $U(0, \theta)$ distribution in the next section when maximum likelihood estimation is introduced.

The next example illustrating the method of moments estimation procedure also involves a single unknown parameter.

Example 2.3 Let $X_1, X_2, \ldots, X_n$ be a random sample from a population with probability density function

$$f(x) = \frac{1}{\theta}e^{-x/\theta} \qquad x > 0,$$

where θ is a positive unknown parameter. Find the method of moments estimate of θ.

The first step is to recognize that the distribution being fitted is an exponential distribution with population mean θ. We again assume that the preliminary work of collecting the data and drawing a histogram has already been completed. The conclusion is that the histogram approximates the shape of the probability density function of an exponential random variable to a reasonable degree, so we can proceed with parameter estimation by the method of moments. There is only a single ($k = 1$) unknown parameter θ to estimate in this case. The population mean of the exponential random variable is $\mu = \theta$. So, equating the population mean to the sample mean results in

$$\theta = \bar{x} = \frac{1}{n}\sum_{i=1}^{n} x_i.$$

This equation is already solved for θ, so the final remaining step is to simply place a hat on θ, yielding the point estimate

$$\hat{\theta} = \bar{x} = \frac{1}{n}\sum_{i=1}^{n} x_i.$$

For this population probability distribution, the population mean is estimated by the sample mean, which is intuitively appealing.

Now that the method of moments estimate has been determined, consider fitting an actual data set to the exponential distribution. To illustrate the point estimation process, fit the $n = 23$ ball bearing failure times (measured in 10^6 revolutions)

17.88	28.92	33.00	41.52	42.12	45.60	48.48	51.84
51.96	54.12	55.56	67.80	68.64	68.64	68.88	84.12
93.12	98.64	105.12	105.84	127.92	128.04	173.40	

to the exponential distribution using the method of moments technique. Assume that the analyst is lazy and does not first plot a histogram of the data to see if the data is likely to be drawn from an exponential population. Rather, the fitting of the exponential distribution is performed before the histogram is drawn. The method of moments estimate of θ is

$$\hat{\theta} = \frac{1}{n}\sum_{i=1}^{n} x_i = \frac{1}{23} \cdot 1661.16 \cong 72.22$$

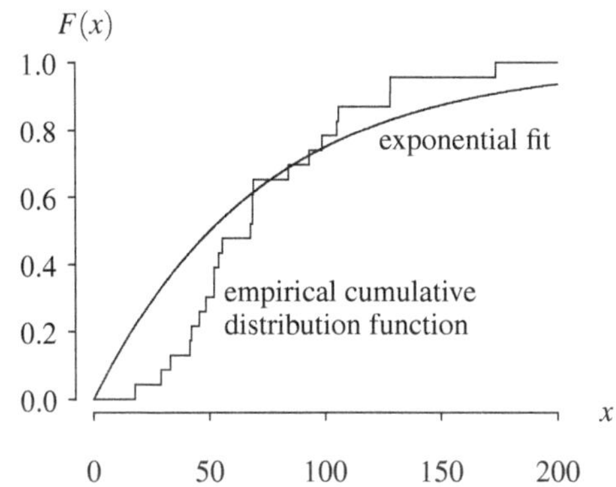

Figure 2.1: Empirical and fitted cumulative distribution functions for the ball bearing data.

million revolutions. One way to visually assess how well the fitted distribution approximates the data is to plot the empirical cumulative distribution function along with the fitted cumulative distribution function on its support:

$$F(x) = 1 - e^{-x/\hat{\theta}} \qquad x > 0.$$

These functions are plotted in Figure 2.1. In this case the fitted exponential distribution does *not* do an adequate job of approximating the data. The cumulative distribution function for the exponential distribution is simply not capable of bending in a fashion to approximate the empirical cumulative distribution function. The fact that the exponential distribution does not make a good model for the ball bearing lifetimes could also have been recognized by plotting a histogram of the data, as shown in Figure 2.2. The histogram shape is clearly not consistent with an exponential population because it is not monotone decreasing. Further evidence that an exponential distribution will not be a good model for ball bearing failure times can be made via the memoryless property.

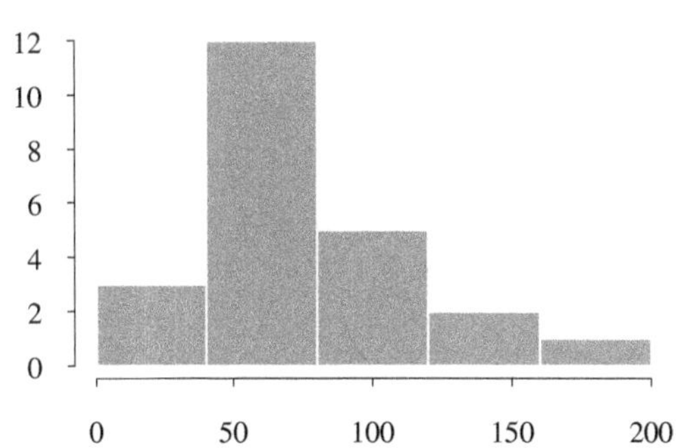

Figure 2.2: Histogram of the ball bearing failure times.

Recall that the exponential distribution is the only continuous distribution having the memoryless property. If ball bearing failure times have the memoryless property, then used ball bearings are as good as new ball bearings. Intuition indicates that this should not be the case, because the ball bearings are wearing out. The lesson to be learned from this analysis is that the exponential distribution can be easily fit to any data set of positive observations, so care must be taken to ensure that it is being applied appropriately. The next example illustrates that a two-parameter distribution, in particular the gamma distribution, is capable of more adequately modeling the ball bearing failure times.

The previous two examples involved fitting probability distributions with just a single ($k = 1$) parameter. The next example involves a parametric distribution with $k = 2$ parameters.

Example 2.4 Let $X_1, X_2, \ldots, X_n$ be a random sample from a gamma(λ, κ) population, where λ and κ are positive unknown parameters. Find the method of moments estimates of λ and κ. Fit the gamma distribution to the ball bearing failure times.

The histogram in Figure 2.2 provides evidence that the gamma distribution, among others, is a potential model for the ball bearing failure times. There are $k = 2$ unknown parameters to be estimated, namely λ and κ, so a 2×2 set of equations must be solved on the parameter space $\Omega = \{(\lambda, \kappa) \,|\, \lambda > 0, \kappa > 0\}$. The generic equations for the method of moments procedure are

$$E[X] = \frac{1}{n} \sum_{i=1}^{n} x_i$$

$$E\left[X^2\right] = \frac{1}{n} \sum_{i=1}^{n} x_i^2.$$

For a gamma(λ, κ) population, these equations become

$$\frac{\kappa}{\lambda} = m_1 = \frac{1}{n} \sum_{i=1}^{n} x_i$$

$$\frac{\kappa(\kappa+1)}{\lambda^2} = m_2 = \frac{1}{n} \sum_{i=1}^{n} x_i^2.$$

The value of $E\left[X^2\right]$ for the gamma distribution is found by the shortcut formula for the population variance: $V[X] = E\left[X^2\right] - E[X]^2$. Fortunately, this 2×2 set of equations has a closed-form solution that is easily calculated for a data set. Squaring both sides of the first equation and then dividing the first equation by the second equation gives

$$\frac{\kappa}{\kappa+1} = \frac{m_1^2}{m_2}$$

or

$$m_2 \kappa = m_1^2 \kappa + m_1^2.$$

Solving for κ in this equation, then plugging this estimate into the first original equation, the method of moments estimates are

$$\hat{\lambda} = \frac{m_1}{m_2 - m_1^2} \qquad \text{and} \qquad \hat{\kappa} = \frac{m_1^2}{m_2 - m_1^2}.$$

Returning to the ball bearing failure times for which the exponential distribution provided a dismal fit in the previous example, the following R statements can be used to calculate the method of moments estimates of λ and κ.

```
x   = c(17.88, 28.92, 33.00, 41.52, 42.12, 45.60, 48.48, 51.84, 51.96,
        54.12, 55.56, 67.80, 68.64, 68.64, 68.88, 84.12, 93.12, 98.64,
        105.12, 105.84, 127.92, 128.04, 173.40)
n   = length(x)
m1  = mean(x)
m2  = sum(x ^ 2) / n
lambda.hat = m1 / (m2 - m1 ^ 2)
kappa.hat  = m1 ^ 2 / (m2 - m1 ^ 2)
```

The method of moments point estimates for this particular data set are

$$\hat{\lambda} = 0.05373 \qquad \text{and} \qquad \hat{\kappa} = 3.8804.$$

The empirical cumulative distribution function and the fitted cumulative distribution function are plotted in Figure 2.3. These reveal that the gamma distribution fit is far superior to the exponential distribution fit to the ball bearing failure times. Even though the fit to the ball bearing failure times is vastly improved by the gamma distribution over the exponential distribution, there may be even better parametric models. The right-hand tails of all gamma random variables are asymptotically exponentially distributed, which is not consistent with the physics of failure associated with ball bearings. The ball bearings are likely to continue to wear away as they age. A continuous distribution with a lighter right-hand tail, such as the Weibull distribution with a shape parameter greater than one, might be a more appropriate population probability model.

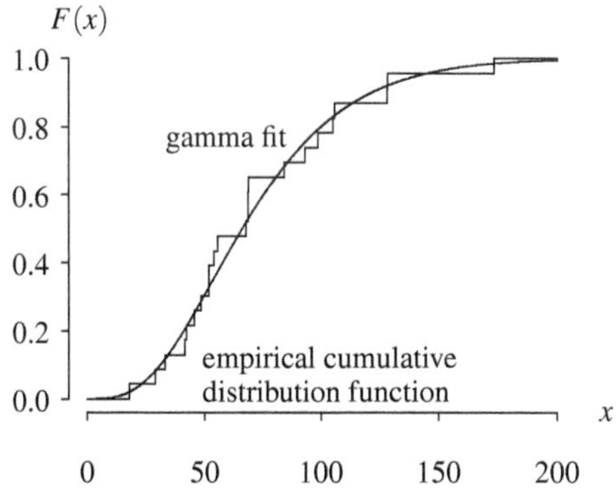

Figure 2.3: Empirical and fitted cumulative distribution functions for the ball bearing data.

The next example of the method of moments estimation technique for parameter estimation fits a normal distribution to a data set.

Example 2.5 Let $X_1, X_2, \ldots, X_n$ be a random sample from a $N\left(\mu, \sigma^2\right)$ population, for real-valued unknown parameter μ and positive unknown parameter σ^2. Find the method of moments estimates of μ and σ^2. Fit the normal distribution to Michelson's speed of light data from Example 1.6.

Assume that a histogram of the data reveals that a probability distribution with a bell-shaped probability density function is an appropriate probability model. We propose a normal model with its parameters estimated by the method of moments. There are $k = 2$ parameters to estimate, μ and σ^2, which are defined on the parameter space $\Omega = \left\{\left(\mu, \sigma^2\right) \mid -\infty < \mu < \infty, \sigma^2 > 0\right\}$. The generic set of 2×2 equations for the method of moments procedure are

$$E[X] = \frac{1}{n}\sum_{i=1}^{n} x_i$$

$$E\left[X^2\right] = \frac{1}{n}\sum_{i=1}^{n} x_i^2.$$

For a $N\left(\mu, \sigma^2\right)$ population, these equations become

$$\mu = m_1 = \frac{1}{n}\sum_{i=1}^{n} x_i$$

$$\sigma^2 + \mu^2 = m_2 = \frac{1}{n}\sum_{i=1}^{n} x_i^2$$

because $\sigma^2 = E\left[X^2\right] - \mu^2$ by the shortcut formula for computing the population variance. The first equation is already solved for μ, so the method of moments estimate of μ is

$$\hat{\mu} = m_1 = \frac{1}{n}\sum_{i=1}^{n} x_i,$$

which is the first sample moment about the origin. Estimating the population mean μ with the sample mean $\bar{X}$ is intuitively appealing. When this estimate is inserted into the second equation for μ and solved for σ^2, the resulting method of moments estimate of σ^2 is

$$\hat{\sigma}^2 = m_2 - m_1^2.$$

So the population variance σ^2 is estimated by $m_2 - m_1^2$, which, after some algebra, is

$$\hat{\sigma}^2 = \frac{1}{n}\sum_{i=1}^{n} x_i^2 - \bar{x}^2 = \frac{1}{n}\sum_{i=1}^{n} (x_i - \bar{x})^2.$$

This point estimate is the second sample moment about the sample mean. The method of moments estimate of the population variance, $\hat{\sigma}^2$, is equivalent to the sample variance with an n, rather than an $n-1$, in the denominator. This is a biased estimate of σ^2.

Michelson's estimates of the speed of light from Example 1.6 produced a bell-shaped histogram, so a reasonable next step might be to fit the normal distribution to the data. The resulting method of moments estimates are

$$\hat{\mu} = \bar{x} = 852.4 \qquad \text{and} \qquad \hat{\sigma}^2 = \frac{1}{n}\sum_{i=1}^{n} (x_i - \bar{x})^2 = 6180.24.$$

A plot of the histogram for the data set with the fitted normal probability density function superimposed can be generated with the following R code.

```
x = MASS::michelson$Speed
n = length(x)
hist(x, probability = TRUE)
xx = 600:1100
muhat = mean(x)
sighat = sqrt(sum((x - muhat) ^ 2) / n)
yy = dnorm(xx, muhat, sighat)
lines(xx, yy, type = "l")
```

The call to the function `hist` draws the histogram. The `probability = TRUE` argument scales the histogram so that the area under the histogram is one. The call to the function `lines` draws the fitted probability density function on top of the histogram. The statistical graphic is given in Figure 2.4. The fitted normal distribution effectively smooths out the random sampling variability and provides a reasonable estimate of the population probability distribution for Michelson's estimates. This, plus the fact that zero is so many standard-deviation units away from the center of this probability density function, makes the assumption of normality reasonable for this particular data set. The area under the fitted probability density function to the left of 0 is negligible for this fitted population distribution.

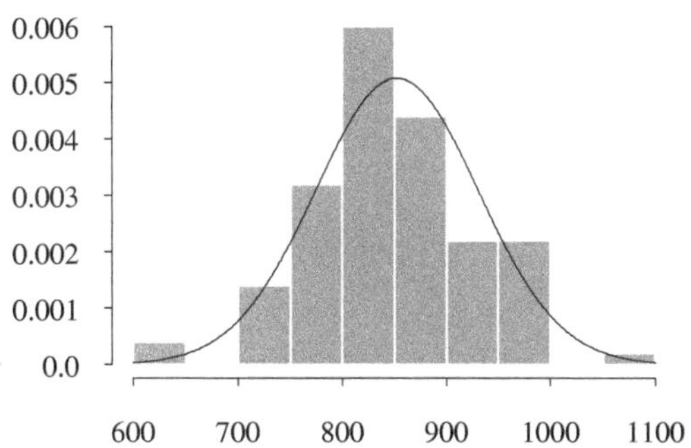

Figure 2.4: Histogram and fitted probability density function for Michelson's data.

The final example of the method of moments technique for estimating parameters illustrates another case in which problems arise with the method of moments estimates.

Example 2.6 Every morning before teaching his mathematical statistics class, Professor Glen loosens up by shooting free throws. He attempts the same number of free throws every morning and believes that each free throw is a mutually independent Bernoulli trial with identical probability of success p. The goal here is to estimate the number of shots he *attempts* every morning. Over the last ten days, he has kept a record of the number of free throws he has *made*:

$$8, 10, 7, 11, 5, 12, 8, 11, 8, 10.$$

Estimate the parameters associated with a distribution that models the number of shots that he attempts every morning. More specifically, estimate (*a*) the number of shots that he attempts every morning, and (*b*) the probability that he makes a single shot. Both parameters are fixed but unknown constants.

Assuming that each free throw is an independent Bernoulli trial, a binomial distribution is appropriate for modeling the number of shots that Professor Glen makes each morning. A notational conflict immediately arises. The variable n is used in statistics to denote the sample size, so the data set consists of $n = 10$ independent experimental values of a binomial random variable with unknown parameters. The variables used for the parameters in the binomial distribution, however, are typically denoted by n and p. We will re-name the binomial parameters to r and p in this example in order to avoid having n assume two roles. Since there are $k = 2$ unknown parameters r and p that need to be estimated, we need to match the first two moments on the parameter space $\Omega = \{(r, p) \mid r \in Z^+, 0 < p < 1\}$, where Z^+ is the set of positive integers. The 2×2 set of generic equations for the method of moments procedure are

$$E[X] = \frac{1}{n}\sum_{i=1}^{n} x_i$$

$$E\left[X^2\right] = \frac{1}{n}\sum_{i=1}^{n} x_i^2.$$

Since the population mean and variance of $X \sim \text{binomial}(r, p)$ are $E[X] = rp$ and $V[X] = rp(1-p)$, respectively, these equations become

$$rp = m_1 = \frac{1}{n}\sum_{i=1}^{n} x_i$$

$$rp(1-p) + r^2p^2 = m_2 = \frac{1}{n}\sum_{i=1}^{n} x_i^2,$$

because $\sigma^2 = E\left[X^2\right] - \mu^2$ by the shortcut formula for computing the population variance. After some algebra, the method of moments estimators of r and p are

$$\hat{r} = \frac{m_1^2}{m_1 - m_2 + m_1^2} \qquad \text{and} \qquad \hat{p} = \frac{m_1 - m_2 + m_1^2}{m_1}.$$

There are two problems associated with these two estimators. First, the value of $\hat{r}$ is not necessarily an integer, which is a requirement for the binomial distribution. Second, both of the estimators contain the expression

$$m_1 - m_2 + m_1^2 = \bar{x} - \frac{1}{n}\sum_{i=1}^{n} x_i^2 + \bar{x}^2.$$

If this quantity is zero or negative, then the parameter estimates lie outside of the parameter space.

The following R code computes the method of moments estimators from the $n = 10$ data values.

```
x = c(8, 10, 7, 11, 5, 12, 8, 11, 8, 10)
m1 = mean(x)
m2 = mean(x ^ 2)
r.hat = m1 ^ 2 / (m1 - m2 + m1 ^ 2)
p.hat = (m1 - m2 + m1 ^ 2) / m1
```

On the other hand, if you want to have Maple solve the method of moments equations, use the following code. (This code computes the parameter estimates as exact fractions.)

```
with(stats):
x  := [8, 10, 7, 11, 5, 12, 8, 11, 8, 10];
x2 := [64, 100, 49, 121, 25, 144, 64, 121, 64, 100];
m1 := describe[mean](x);
m2 := describe[mean](x2);
solve({r * p = m1, r * p * (1 - p) + r ^ 2 * p ^ 2 = m2}, {r, p});
```

The result of these calculations are $m_1 = 9$, $m_2 = 426/5$, and the point estimates of the binomial parameters are

$$\hat{r} = \frac{135}{8} = 16.875 \qquad \text{and} \qquad \hat{p} = \frac{8}{15} \cong 0.53.$$

So our best guess using the method of moments is that Professor Glen attempts about 17 free throws every morning and shoots about 53% in his bizarre morning ritual. This choice of parameters results in an *approximate* match in the first two sample and population moments.

The method of moments technique for estimating parameters has intuitive appeal because of the matched lower moments. The estimators do not have particularly appealing statistical properties, and that is why a second technique for estimating parameters, maximum likelihood estimation, is presented next.

2.3 Maximum Likelihood Estimation

Like the method of moments technique, the maximum likelihood estimation technique also has an intuitive appeal: the parameter estimates that it produces are the most likely ones to have generated the data set being analyzed. We begin by motivating the maximum likelihood estimation technique with a return to the question in Example 2.1 of estimating the fraction of lefties in the world.

Example 2.7 Let's say that you want to estimate the fraction of lefties in the world, but you are in a hurry and you want to do so with a sample size of only $n = 3$. If you are careful about designing your survey so as to achieve mutual independence in your three sample values, the data set corresponds to a random sample of three values X_1, X_2, X_3 from a Bernoulli(p) population, where p is the probability of being left-handed as before, which satisfies $0 < p < 1$. Thus, the population probability mass function of each individual observation is

$$f(x) = \begin{cases} 1-p & x = 0 \\ p & x = 1. \end{cases}$$

Now in a specific experiment, let's say that you observe the data values

$$x_1 = 0, x_2 = 1, x_3 = 0,$$

that is, you survey a righty, then a lefty, then a righty. What would be an appropriate estimate of p, the probability of being a lefty from this data set? One way to approach this question is to calculate the probability of the data set that was observed:

$$P(X_1 = 0, X_2 = 1, X_3 = 0) = P(X_1 = 0)P(X_2 = 1)P(X_3 = 0) = p(1-p)^2.$$

This function is known as the "likelihood function" and is denoted by

$$L(p) = p(1-p)^2$$

in this particular setting. The likelihood function is plotted in Figure 2.5 for $0 < p < 1$. One way to proceed is to find the p value that maximizes the probability of sampling the values that were observed. Either by inspecting this graph or by using calculus, it is clear that the likelihood function is maximized at $p = 1/3$, so this value is the maximum likelihood estimate $\hat{p} = 1/3$.

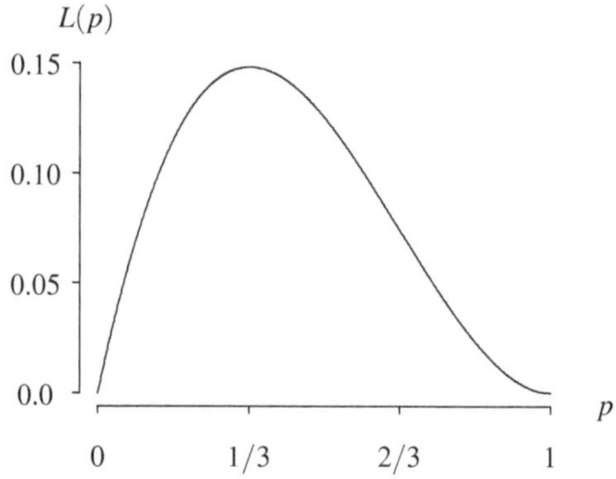

Figure 2.5: Likelihood function for $n = 3$ Bernoulli trials $x_1 = 0, x_2 = 1, x_3 = 0$.

We now generalize this thinking to an arbitrary sample size n and replace the parameter p with the generic population parameter θ associated with some population distribution. For a random sample $x_1, x_2, \ldots, x_n$ drawn from a probability distribution described by $f(x)$ with single unknown parameter θ or a vector of k unknown parameters $\boldsymbol{\theta}$, the *likelihood function* is

$$L(\theta) = L(\theta;\, x_1, x_2, \ldots, x_n) = \prod_{i=1}^{n} f(x_i).$$

Since the data values in a random sample are assumed to be realizations of independent and identically distributed random variables, the likelihood function $L(\theta)$ is also the joint probability mass function or probability density function of the associated random variables $X_1, X_2, \ldots, X_n$.

There are several facts about the likelihood function that are important. These facts are discussed in the list that follows in terms of a single unknown parameter θ; they generalize to a vector of k parameters $\boldsymbol{\theta} = (\theta_1, \theta_2, \ldots, \theta_k)'$.

- Some authors prefer the notation $L(\theta)$, while others prefer to write the likelihood function as $L(\theta; x_1, x_2, \ldots, x_n)$ to emphasize that it is a function of θ and the data values. For compactness, $L(\theta)$ will be used here.

- The *maximum likelihood estimator* $\hat{\theta}$ is the value of θ that maximizes $L(\theta)$ with respect to θ over the parameter space Ω. So a mathematical expression for the maximum likelihood estimator $\hat{\theta}$ of the unknown parameter θ is

$$\hat{\theta} = \underset{\Omega}{\operatorname{argmax}}\, L(\theta).$$

  In the ideal case, this simply involves solving $L'(\theta) = 0$ for θ, reducing the parameter estimation procedure to a calculus problem. As will be seen in the examples that follow, the procedure does not always evolve in this fashion for various reasons.

- The abbreviation MLE stands for the maximum likelihood estimator.

- In practice it is often easier to maximize the *log likelihood function*, $\ln L(\theta)$, rather than the likelihood function $L(\theta)$. The maximum likelihood estimators will be the same for maximizing either function because a function and its increasing monotonic transformation are maximized at the same value.

- The derivative of the log likelihood function $\frac{\partial \ln L(\theta)}{\partial \theta}$ is often called the *score*, which generalizes to the k-element *score vector* when there are several unknown parameters. The k elements of the score vector are the partial derivatives of the log likelihood function with respect to the k unknown parameters.

- When the second partial derivative of the likelihood function evaluated at the maximum likelihood estimator $\frac{\partial^2 L(\hat{\theta})}{\partial \theta^2} < 0$ or when the second partial derivative of the log likelihood function evaluated at the maximum likelihood estimator $\frac{\partial^2 \ln L(\hat{\theta})}{\partial \theta^2} < 0$, the maximum likelihood estimator $\hat{\theta}$ maximizes $L(\theta)$ or $\ln L(\theta)$ by the second derivative test from calculus. In the case of k unknown parameters, the k-element vector of maximum likelihood estimators $\hat{\boldsymbol{\theta}}$ maximizes $L(\boldsymbol{\theta})$, or, equivalently, $\ln L(\boldsymbol{\theta})$, when the Hessian matrix is positive definite at the maximum likelihood estimators.

- As will be seen in a subsequent example, maximum likelihood estimators are not necessarily unique.

- Maximum likelihood estimators have the *invariance property*: if $\hat{\theta}$ is the maximum likelihood estimator of θ that exists uniquely and h is a function, then $h(\hat{\theta})$ is the maximum likelihood estimator of $h(\theta)$.

- Although the mathematics required to find the maximum likelihood estimator of a parameter θ is often a straightforward calculus problem, the derivation is mathematically intractable for some distributions and numerical methods must be used to calculate $\hat{\theta}$.

There is an important property of point estimators that will be discussed in the examples that follow. When the expected value of a point estimator $\hat{\theta}$ is θ, that is, when $E\left[\hat{\theta}\right] = \theta$, the point estimator is known as an *unbiased estimator*. This is a good property for a point estimator to possess because it means that the estimator is, in some sense, aiming at the right target. A formal definition and more details associated with unbiased estimators are given in the next section. The unbiased property can be assessed for any point estimator, including a method of moments estimator.

Example 2.8 Let $X_1, X_2, \ldots, X_n$ be a random sample drawn from a population with probability density function

$$f(x) = \frac{1}{\theta} e^{-x/\theta} \qquad x > 0,$$

where θ is a positive unknown parameter. Find the maximum likelihood estimator of θ.

Once again, the population distribution is recognized as an exponential distribution with population mean θ. As before, we assume that a visual inspection of the histogram and/or theoretical considerations have revealed that the exponential distribution is an appropriate probability model for the data set, so we proceed with parameter estimation. The data values are denoted by $x_1, x_2, \ldots, x_n$. The likelihood function is

$$L(\theta) = \prod_{i=1}^{n} f(x_i) = \prod_{i=1}^{n} \frac{1}{\theta} e^{-x_i/\theta} = \frac{1}{\theta^n} e^{-\sum_{i=1}^{n} x_i/\theta}.$$

The log likelihood function is

$$\ln L(\theta) = -n \ln \theta - \frac{1}{\theta} \sum_{i=1}^{n} x_i.$$

The score is

$$\frac{\partial \ln L(\theta)}{\partial \theta} = -\frac{n}{\theta} + \frac{1}{\theta^2} \sum_{i=1}^{n} x_i.$$

When the score is equated to zero,

$$-\frac{n}{\theta} + \frac{1}{\theta^2} \sum_{i=1}^{n} x_i = 0,$$

and this equation is solved for θ, the maximum likelihood estimate of θ is

$$\hat{\theta} = \frac{1}{n} \sum_{i=1}^{n} x_i.$$

For data drawn from an exponential population, the point estimate via maximum likelihood estimation for the population mean is the sample mean. To see that the maximum likelihood estimator maximizes (rather than minimizes) the log likelihood function, a second derivative is taken:

$$\frac{\partial^2 \ln L(\theta)}{\partial \theta^2} = \frac{n}{\theta^2} - \frac{2}{\theta^3} \sum_{i=1}^{n} x_i.$$

When θ is replaced with the maximum likelihood estimator $\hat{\theta}$, this expression becomes

$$\left. \frac{\partial^2 \ln L(\theta)}{\partial \theta^2} \right|_{\theta=\hat{\theta}} = \frac{n}{\hat{\theta}^2} - \frac{2}{\hat{\theta}^3} \sum_{i=1}^{n} x_i = \frac{n^3}{\left(\sum_{i=1}^{n} x_i\right)^2} - \frac{2n^3}{\left(\sum_{i=1}^{n} x_i\right)^2} = -\frac{n^3}{\left(\sum_{i=1}^{n} x_i\right)^2}.$$

For data values drawn from a distribution with positive support (such as the exponential population in this example), this expression is always negative. This implies that $\hat{\theta}$ maximizes the log likelihood function, and therefore, $\hat{\theta}$ also maximizes the likelihood function. For this particular population, the maximum likelihood estimate happens to be identical to the method of moments estimate. This is not universally true, however.

As mentioned earlier, parameter estimators are random variables that have probability density functions. Switching the notation from the data values $x_1, x_2, \ldots, x_n$ to the associated random variables $X_1, X_2, \ldots, X_n$, the expected value of the maximum likelihood estimator is

$$E[\hat{\theta}] = E\left[\frac{1}{n}\sum_{i=1}^{n} X_i\right] = \frac{1}{n}\sum_{i=1}^{n} E[X_i] = \frac{1}{n}\sum_{i=1}^{n} \theta = \frac{1}{n}(n\theta) = \theta.$$

So the maximum likelihood estimator $\hat{\theta}$ is an unbiased estimator of θ. Having the expected value of the parameter estimate equal to the parameter itself is a highly desirable property for the estimate, because the estimator is "on target" or "unbiased" on average.

In order to appreciate the geometry associated with maximum likelihood estimation, consider the tiny data set with just $n = 4$ observations:

$$x_1 = 1.3, \qquad x_2 = 0.5, \qquad x_3 = 0.3, \qquad x_4 = 1.9.$$

Now consider all of the possible probability density functions for exponential populations, that is, all of the probability density functions of the form

$$f(x) = \frac{1}{\theta}e^{-x/\theta} \qquad x > 0$$

for some θ in the parameter space $\Omega = \{\theta \mid \theta > 0\}$. The θ value corresponding to the maximum likelihood estimate, which is

$$\hat{\theta} = \bar{x} = \frac{1.3 + 0.5 + 0.3 + 1.9}{4} = 1,$$

has the largest (maximum) product of the lengths of the vertical lines shown in Figure 2.6. Any other choice of θ would give a lower value for this product, which is also the value of the likelihood function $L(\theta)$. The data values are plotted as $\times$s on the horizontal axis in Figure 2.6, and the particular probability density function plotted is

$$f(x) = e^{-x} \qquad x > 0,$$

which is the probability density function associated with the maximum likelihood estimate $\hat{\theta} = 1$. The vertical lines connecting the data values to the probability density function have lengths $f(x_1)$, $f(x_2)$, $f(x_3)$, and $f(x_4)$. The product of these lengths is the value of the likelihood function for $\hat{\theta}$, which is

$$L\left(\hat{\theta}\right) = f(x_1) \cdot f(x_2) \cdot f(x_3) \cdot f(x_4) = e^{-x_1}e^{-x_2}e^{-x_3}e^{-x_4} = e^{-1.3}e^{-0.5}e^{-0.3}e^{-1.9} = e^{-4}.$$

In this sense, this particular choice of θ gives the exponential distribution that is most likely to have resulted in the observed data set. Hence, the name *maximum likelihood* is used to describe this parameter estimator.

The next example shows that the procedure for finding maximum likelihood estimates is essentially the same for a discrete population as it is for a continuous population.

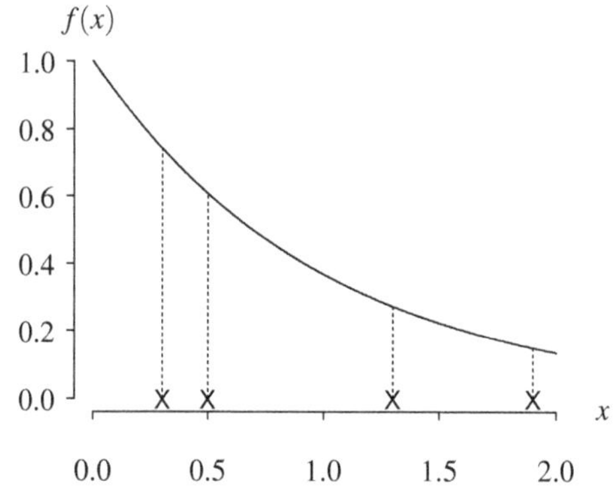

Figure 2.6: Geometry associated with the maximum likelihood estimator.

Example 2.9 Let $X_1, X_2, \ldots, X_n$ be a random sample from a Poisson(λ) population, where λ is a positive unknown parameter. Find the maximum likelihood estimator $\hat{\lambda}$.

To begin, a visual inspection of a histogram associated with a data set of values from a discrete population suggests that the Poisson distribution is an appropriate probability model for the data set in order to proceed with fitting via maximum likelihood. The probability mass function for the Poisson distribution is

$$f(x) = \frac{\lambda^x e^{-\lambda}}{x!} \qquad x = 0, 1, 2, \ldots,$$

where λ is an unknown parameter in the parameter space $\Omega = \{\lambda \,|\, \lambda > 0\}$. The likelihood function is

$$L(\lambda) = \prod_{i=1}^{n} f(x_i) = \prod_{i=1}^{n} \frac{\lambda^{x_i} e^{-\lambda}}{x_i!} = \frac{e^{-n\lambda} \lambda^{\sum_{i=1}^{n} x_i}}{\prod_{i=1}^{n} x_i!}.$$

The log likelihood function is

$$\ln L(\lambda) = -n\lambda + \left(\sum_{i=1}^{n} x_i \right) \ln \lambda - \sum_{i=1}^{n} \ln(x_i!)$$

and the score is

$$\frac{\partial \ln L(\lambda)}{\partial \lambda} = -n + \frac{1}{\lambda} \sum_{i=1}^{n} x_i.$$

When the score is equated to zero,

$$-n + \frac{1}{\lambda} \sum_{i=1}^{n} x_i = 0,$$

and when this equation is solved for λ, the maximum likelihood estimate of λ is

$$\hat{\lambda} = \frac{1}{n} \sum_{i=1}^{n} x_i.$$

As in the previous example, the sample mean is the point estimate of the population mean. The second partial derivative of the log likelihood function is

$$\frac{\partial^2 \ln L(\lambda)}{\partial \lambda^2} = -\frac{1}{\lambda^2} \sum_{i=1}^{n} x_i.$$

When λ is replaced by the maximum likelihood estimate, this expression becomes

$$\left. \frac{\partial^2 \ln L(\lambda)}{\partial \lambda^2} \right|_{\lambda=\hat{\lambda}} = -\frac{1}{\hat{\lambda}^2} \sum_{i=1}^{n} x_i = -\frac{n^2}{\sum_{i=1}^{n} x_i}.$$

Since this expression is always negative for data drawn from a Poisson population (except in the rare case in which all of the data values are zeros), the maximum likelihood estimate is associated with a local maximum of the log likelihood function.

Again switching the data values $x_1, x_2, \ldots, x_n$ to their associated random variables $X_1, X_2, \ldots, X_n$, the expected value of the maximum likelihood estimator is

$$E\left[\hat{\lambda}\right] = E\left[\frac{1}{n} \sum_{i=1}^{n} X_i\right] = \frac{1}{n} \sum_{i=1}^{n} E[X_i] = \frac{1}{n} \sum_{i=1}^{n} \lambda = \frac{1}{n}(n\lambda) = \lambda.$$

As was the case with the maximum likelihood estimator of the population mean for a random sample from an exponential population, the maximum likelihood estimator $\hat{\lambda}$ is an unbiased estimator of the unknown parameter λ.

Now consider fitting a Poisson distribution to a data set. The famous horse kick data set consists of annual deaths of Prussian cavalry soldiers due to horse kicks. There are $n = 200$ corps-years of data given in Table 2.1. This table shows that there are $n = 200$ data values, and they consist of 109 zeros, 65 ones, 22 twos, 3 threes, and 1 four. Since the maximum likelihood estimate of the Poisson distribution is the sample mean,

$$\hat{\lambda} = \frac{(109)(0) + (65)(1) + (22)(2) + (3)(3) + (1)(4)}{200} = \frac{122}{200} = 0.61$$

fatalities per corps-year. The R statements

```
x = 0:10
200 * dpois(x, 0.61)
```

can be used to add a third row to the table containing the data. Table 2.2 includes the fitted Poisson probabilities, rounded to the nearest tenth, that are predicted by the Poisson model. The spectacular agreement between the data and the fitted Poisson probability model allows us to conclude that the Poisson distribution is an appropriate model—horse kick deaths were likely to have been a Poisson process over time. The deaths appear to be occurring randomly over time, consistent with the assumptions

number of deaths per corps per year	0	1	2	3	4
number of observed values	109	65	22	3	1

Table 2.1: Observed horse kick deaths.

number of deaths per corps per year	0	1	2	3	4 or more
number of observed values	109	65	22	3	1
number of predicted values	108.7	66.3	20.2	4.1	0.7

Table 2.2: Observed and predicted horse kick deaths.

associated with a Poisson process concerning random events. This is sensible because the horses would be unlikely to have conspired against the soldiers in a systematic fashion.

Another geometric aspect of the maximum likelihood estimation technique can be seen by plotting the log likelihood function in a vicinity of the maximum likelihood estimate. Figure 2.7 shows the log likelihood function

$$\ln L(\lambda) = -n\lambda + \left(\sum_{i=1}^{n} x_i\right) \ln \lambda - \sum_{i=1}^{n} \ln(x_i!)$$

plotted in the vicinity of $\hat{\lambda} = 0.61$. Such a plot is generated by the R code given below.

```
x     = c(rep(0, 109), rep(1, 65), rep(2, 22), rep(3, 3), 4)
n     = length(x)
lam   = mean(x)
xlam  = seq(lam - 0.02, lam + 0.02, length = 200)
logl  = -n * xlam + sum(x) * log(xlam) - sum(log(factorial(x)))
plot(xlam, logl, type = "l")
```

Figure 2.7 shows that the log likelihood function achieves a local maximum at the point $\left(\hat{\lambda}, \ln L(\hat{\lambda})\right) = (0.61, -206.1067)$. The second coordinate is negative because it is the natural logarithm of a joint probability mass function, which typically assumes values between 0 and 1.

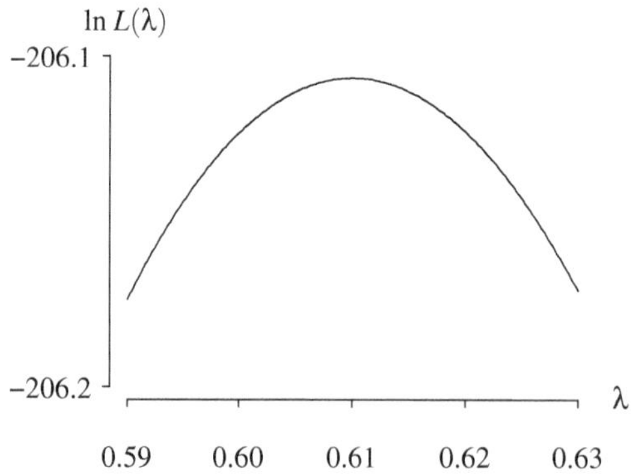

Figure 2.7: The log likelihood function near $\hat{\lambda} = 0.61$ for the horse kick data.

For the two previous examples involving the exponential and Poisson distributions, the maximum likelihood estimators have been found in a standard fashion using calculus. The next example illustrates a non-standard case.

Example 2.10 Let $X_1, X_2, \ldots, X_n$ be a random sample from a $U(0, \theta)$ population, where θ is a positive unknown parameter. Find the maximum likelihood estimator $\hat{\theta}$.

Once again, a histogram of the data values should be assessed prior to fitting a data set, and we assume that the histogram is fairly flat and the smallest data value is positive, so a $U(0, \theta)$ model might be appropriate. The population probability density function is

$$f(x) = \frac{1}{\theta} \qquad\qquad 0 < x < \theta,$$

where θ is an unknown parameter from the parameter space $\Omega = \{\theta \,|\, \theta > 0\}$. For the data values $x_1, x_2, \ldots, x_n$, the likelihood function is

$$L(\theta) = \prod_{i=1}^{n} f(x_i) = \left(\frac{1}{\theta}\right)^n$$

for $0 < x_i < \theta$, $i = 1, 2, \ldots, n$, and 0 otherwise. The log likelihood function is

$$\ln L(\theta) = -n \ln \theta.$$

The score is

$$\frac{\partial \ln L(\theta)}{\partial \theta} = -\frac{n}{\theta}.$$

Calculus is of no help here because $\ln L(\theta)$ is a monotonically decreasing function of the positive unknown parameter θ. But we can still proceed using the following logic. Since $L(\theta)$ is a decreasing function of θ, we want to choose the smallest possible θ. Only $\theta > x_{(n)}$ should be considered because any $\theta \le x_{(n)}$ would correspond to $L(\theta) = 0$. Figure 2.8 shows the geometry associated with the likelihood function for this problem. Since the constrained maximization problem includes only $\theta > x_{(n)}$, which is the solid portion of the curve in Figure 2.8, there is no maximum likelihood estimator. This is the first case we have encountered in which a maximum likelihood estimator does not exist. There are two possible workarounds in this case. First, the problem can be overcome by simply replacing the notion of *maximum* with *supremum*. The supremum of $L(\theta)$ is achieved at

$$\hat{\theta} = \max\{x_1, x_2, \ldots, x_n\} = x_{(n)}.$$

Second, if the support of the population distribution is changed from $\mathcal{A} = \{x \,|\, 0 < x < \theta\}$ to $\mathcal{A} = \{x \,|\, 0 \le x \le \theta\}$, then the constrained maximization problem is over $\theta \ge x_{(n)}$. In this case Figure 2.8 would have the open point occurring at $\theta = x_{(n)}$ replaced by a closed point. In this case, $L(\theta)$ also achieves a maximum at

$$\hat{\theta} = \max\{x_1, x_2, \ldots, x_n\} = x_{(n)},$$

so this is the maximum likelihood estimate. Since both workarounds are trivial extensions of the maximum likelihood estimation technique and the population distribution, the maximum likelihood estimate of θ for a $U(0, \theta)$ population is simply

$$\hat{\theta} = \max\{x_1, x_2, \ldots, x_n\} = x_{(n)}.$$

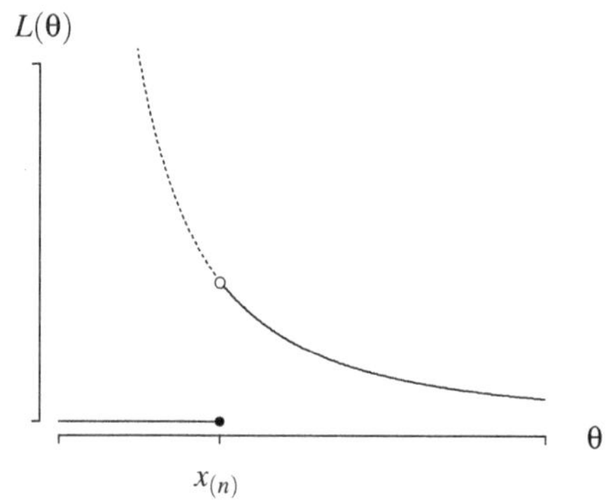

Figure 2.8: The likelihood function $L(\theta) = 1/\theta^n$ for $\theta > x_{(n)}$.

This is the first example that we have encountered in which the maximum likelihood estimate differs from the method of moments estimate from Example 2.2.

The maximum likelihood estimator successfully avoids the possibility of a fitted distribution that is incapable of producing the largest value in the data set. The maximum likelihood estimator, however, has problems of its own. Using an order statistic result, the expected value of the maximum likelihood estimator is

$$E\big[\hat{\theta}\big] = E\left[X_{(n)}\right] = \frac{n}{n+1}\theta < \theta,$$

so it can be concluded that $\hat{\theta}$ is a biased estimator of θ because $E\left[\hat{\theta}\right] \neq \theta$. This is an undesirable quality for this estimator because it implies that the estimator will consistently underestimate θ. This problem is most pronounced for small sample sizes. One solution to this problem is to modify the maximum likelihood estimator by multiplying it by an *unbiasing constant* to create a new point estimator. Using $(n+1)/n$ as an unbiasing constant, an unbiased estimate of θ is

$$\hat{\theta} = \frac{n+1}{n} \cdot x_{(n)}.$$

This new estimate is no longer the maximum likelihood estimate, however. The unbiased estimate is now applied to the tiny $(n = 3)$ data set:

$$1, 2, 9.$$

The unbiased estimate is

$$\hat{\theta} = \frac{3+1}{3} \cdot x_{(3)} = \frac{4}{3} \cdot 9 = 12.$$

So the best guess is that the data set 1, 2, 9 came from a $U(0,\, 12)$ population using the maximum likelihood estimate modified by an unbiasing constant.

Although the previous example has limited practical application, the discrete analog of this problem played an important role in World War II in what has become known as the *German tank problem*. The Allies wanted to estimate the number of German tanks. They approached the problem with conventional intelligence gathering methods, which estimated the monthly tank production to be 1400 tanks per month between June of 1940 and September of 1942. They also approached the problem with statistical sampling. The Allies were pleased to find sequential serial numbers beginning at 1 on captured tanks that could be used to estimate the total number of tanks. Using the methods described in the next example, the statistical estimate of tank production would be 256 tanks per month. After the war, German production was determined to be 255 tanks per month. The statisticians outperformed the spies.

Example 2.11 To frame the German tank problem as a statistical estimation problem, assume that the serial numbers are $1, 2, \ldots, \theta$, where the parameter θ is a positive unknown integer. The Allies capture n tanks with serial numbers $x_1, x_2, \ldots, x_n$. What is the maximum likelihood estimator of θ?

To begin with the simplest possible case, assume that the random sampling is performed with replacement (this is not a brilliant military strategy). The population probability mass function of the serial number of a single tank captured is

$$f(x) = \frac{1}{\theta} \qquad x = 1, 2, \ldots, \theta,$$

which is the discrete uniform distribution. Assuming that each serial number collected is a mutually independent random variable (a dubious assumption), the multiplication rule can be used to conclude that there are θ^n different samples, assuming order is considered. This includes, of course, n captures of the tank with serial number 1. The likelihood function in this case is the joint distribution of $X_1, X_2, \ldots, X_n$, which is

$$L(\theta) = f(x_1, x_2, \ldots, x_n) = \frac{1}{\theta^n}$$

for the θ^n possible permutations, with possible repeated values, of the data values $x_1, x_2, \ldots, x_n$ defined on the support $\mathcal{A} = \{x_i \,|\, x_i = 1, 2, \ldots, \theta\}$ for $i = 1, 2, \ldots, n$, and 0 otherwise. This likelihood function has the same functional form as that in the previous example. Since $1/\theta^n$ is again a decreasing function of θ, the maximum likelihood estimate is

$$\hat{\theta} = x_{(n)}$$

as in the previous example. So in this case the maximum likelihood estimate is the largest serial number observed.

Now consider the more realistic case of sampling without replacement. In this case, there are $\binom{\theta}{n}$ equally-likely samples that can be drawn from the population. The likelihood function is again the joint distribution of $X_1, X_2, \ldots, X_n$, which is

$$L(\theta) = f(x_1, x_2, \ldots, x_n) = \frac{1}{\binom{\theta}{n}}$$

for the $\binom{\theta}{n}$ possible combinations of $x_1, x_2, \ldots, x_n$ in the support, and $L(\theta) = 0$ otherwise. As in the case of sampling with replacement, the likelihood function is a decreasing function of θ, so the maximum likelihood estimate is again

$$\hat{\theta} = x_{(n)},$$

the largest serial number observed. The maximum likelihood estimate was originally presented as having appealing statistical properties. But this does not mean that they are high quality estimates in all settings. In this particular setting the maximum likelihood estimate is not a good estimate of θ. To illustrate, suppose that $n = 4$ tanks have been captured with serial numbers 155, 803, 243, and 606. The maximum likelihood estimate is $\hat{\theta} = x_{(4)} = 803$. Is 803 a good point estimate of the number of tanks? It is only correct if we just so happened to have the largest tank number in our sample. Otherwise it misses low. This is not an intuitively appealing estimate. The maximum likelihood estimate $\hat{\theta}$ is a *biased* point estimate of θ.

There must be a way to improve the point estimator so that it is an unbiased estimator of θ. One such heuristic method proceeds along the following lines. Begin with the assumption that the point estimator must be greater than or equal to the largest observed serial number $x_{(n)}$. Next, it is reasonable to assume that the maximum serial number should be increased by a small amount when n is large, and increased by a large amount when n is small. One such heuristic estimate is

$$\hat{\theta} = x_{(n)} + [\text{average gap in the set of captured serial numbers in sorted order}]$$

or, equivalently,

$$\hat{\theta} = x_{(n)} + \frac{[\text{number of known missing tank serial numbers}]}{n}.$$

Since there are $x_{(n)} - n$ known missing tank serial numbers, this becomes

$$\hat{\theta} = x_{(n)} + \frac{x_{(n)} - n}{n}.$$

For our previous set of $n = 4$ serial numbers, 155, 803, 243, and 606, this point estimate is

$$\hat{\theta} = 803 + \frac{803 - 4}{4} = 803 + 199.75 = 1002.75 \text{ tanks.}$$

We can use APPL to get a picture of how the maximum likelihood estimator and the heuristic estimator perform as point estimators. Begin with the maximum likelihood estimator. The APPL statements below calculate the probability mass function of the maximum likelihood estimator $\hat{\theta}$ when $n = 4$ tanks are sampled without replacement and the true number of tanks is $\theta = 100$. The optional fourth argument `"wo"` in `OrderStat` indicates that sampling is performed without replacement.

```
n := 4;
X := UniformDiscreteRV(1, 100);
T := OrderStat(X, n, n, "wo");
Mean(T);
Variance(T);
```

A plot of the probability mass function of $\hat{\theta}$, that is, the sampling distribution of $\hat{\theta}$, is given in Figure 2.9. The support of $\hat{\theta}$ ranges from 4 (which corresponds to capturing tanks with the serial numbers 1, 2, 3, and 4) to 100 (which corresponds to having tank number 100 in the sample). The population mean and variance of $\hat{\theta}$ are

$$E\left[\hat{\theta}\right] = \frac{404}{5} = 80.8 \qquad \text{and} \qquad V\left[\hat{\theta}\right] = \frac{6464}{25} = 258.56.$$

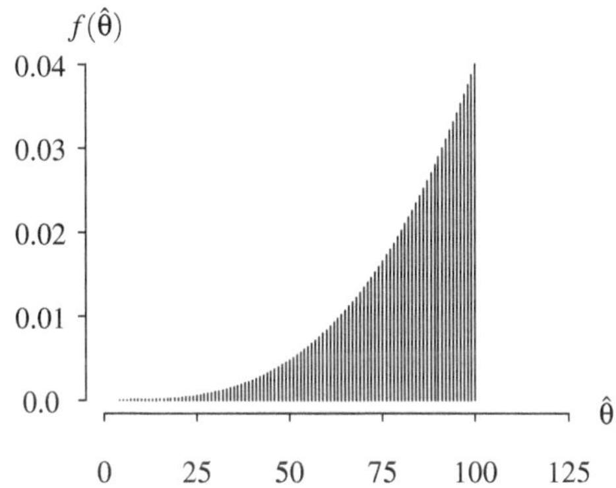

Figure 2.9: The probability mass function of $\hat{\theta} = X_{(4)}$ when $\theta = 100$.

The maximum likelihood estimator has a significant bias and is therefore a very poor point estimator of the number of tanks in the population.

Now consider the second point estimate

$$\hat{\theta} = x_{(n)} + \frac{x_{(n)} - n}{n}$$

when $n = 4$ of the $\theta = 100$ tanks are sampled without replacement. Rearranging the terms in the estimate, it can be seen that the second estimate is simply a linear transformation of the maximum likelihood estimate:

$$\hat{\theta} = \left(1 + \frac{1}{n}\right) x_{(n)} - 1 = \frac{5}{4} x_{(4)} - 1.$$

A graph of the probability mass function for the second estimator, that is, the sampling distribution of $\hat{\theta}$, is displayed in Figure 2.10. The support of $\hat{\theta}$ now ranges from 4 (which corresponds to capturing tanks with the serial numbers 1, 2, 3, and 4) to 124 (which corresponds to having tank number 100 in the sample), but in steps of $5/4$. The population mean and variance of $\hat{\theta}$ are

$$E\left[\hat{\theta}\right] = E\left[\frac{5}{4} X_{(4)} - 1\right] = \frac{5}{4} E\left[X_{(4)}\right] - 1 = \frac{5}{4} \cdot \frac{404}{5} - 1 = 100$$

and

$$V\left[\hat{\theta}\right] = V\left[\frac{5}{4} X_{(4)} - 1\right] = \frac{25}{16} V\left[X_{(4)}\right] = \frac{25}{16} \cdot \frac{6464}{25} = 404.$$

This estimator is unbiased, although it has a somewhat larger population variance than the maximum likelihood estimator. The graph of its probability mass function shows that it is strongly preferred to the maximum likelihood estimator. This example has shown that the maximum likelihood estimator can, in certain statistical estimation problems, be a very poor point estimator of an unknown parameter. Attention should be paid to its statistical properties.

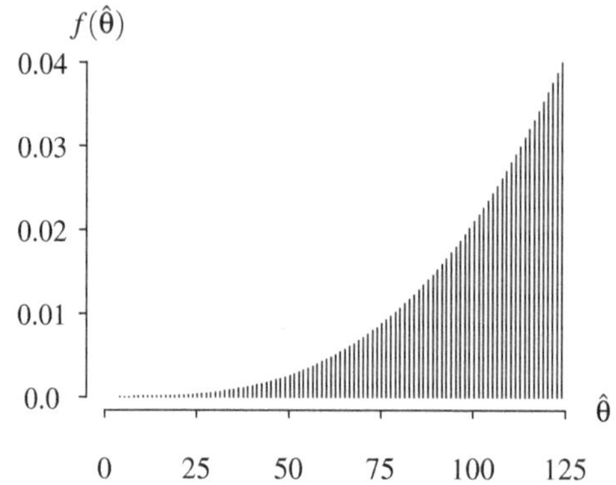

Figure 2.10: The probability mass function of $\hat{\theta} = \frac{5}{4}X_{(4)} - 1$ when $\theta = 100$.

The next example illustrates the application of maximum likelihood estimation to the setting of what is known in the field of quality control as *acceptance sampling*, which uses random sampling to determine whether to accept or reject a production lot of goods.

Example 2.12 Fifty motors are stored in sheds. There are ten motors stored in each of five sheds. The number of defective motors in each shed is given in Table 2.3. The setting is also displayed in Figure 2.11, where the rectangles depict the sheds, and the squares in the rectangles depict the motors. The defective motors are shaded. A quality control engineer selects a shed at random, then randomly samples and tests three motors from that shed. The engineer records x, the number of defective motors in the sample of three motors. Unfortunately, the shed number selected was not recorded. Find the maximum likelihood estimator of the shed number selected as a function of x.

shed number	1	2	3	4	5
number of defectives	0	1	2	3	4

Table 2.3: Number of defective motors in each shed.

This problem illustrates a common issue that arises in statistics: incomplete information. The fact that the shed number was not recorded complicates the analysis. The notation used to solve this problem is given below.

- Let the random variable X denote the random number of defectives in the sample.
- Let the constant x denote the observed number of defectives in the sample.
- Let the random variable S denote the random shed selected.
- Let the constant θ denote the shed actually selected.
- Let the random variable $\hat{\theta}$ denote the maximum likelihood estimator of the shed selected.

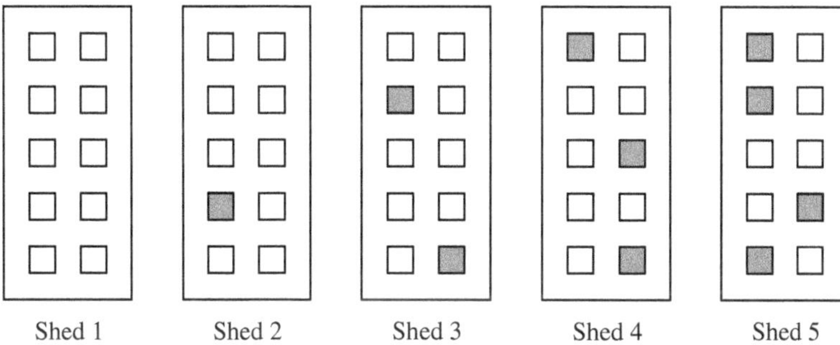

Figure 2.11: Ten motors stored in each of five sheds (defective motors shaded).

The rule of Bayes can be used to calculate the probabilities associated with selecting the various sheds. The probability that shed s was selected given that x defective motors were observed is

$$P(S=s\,|\,X=x)=\frac{P(X=x\,|\,S=s)P(S=s)}{\sum_{j=1}^{5}P(X=x\,|\,S=j)P(S=j)}$$

for $x=0,1,2,3$ and $s=1,2,3,4,5$. Because sampling from the selected shed is performed without replacement, the hypergeometric distribution is used to calculate these probabilities:

$$P(S=s\,|\,X=x)=\frac{\frac{\binom{s-1}{x}\binom{11-s}{3-x}}{\binom{10}{3}}\cdot\frac{1}{5}}{\sum_{j=1}^{5}\frac{\binom{j-1}{x}\binom{11-j}{3-x}}{\binom{10}{3}}\cdot\frac{1}{5}}=\frac{\binom{s-1}{x}\binom{11-s}{3-x}}{\sum_{j=1}^{5}\binom{j-1}{x}\binom{11-j}{3-x}}$$

for $x=0,1,2,3$ and $s=1,2,3,4,5$. The following R code can be used to calculate these probabilities. The `fractions` function in the `MASS` package can be used to obtain these probabilities as exact fractions.

```
for (x in 0:3) {
  for (s in 1:5) {
    top = choose(s - 1, x) * choose(11 - s, 3 - x);
    j   = 1:5
    bot = sum(choose(j - 1, x) * choose(11 - j, 3 - x))
    print(top / bot);
  }
}
```

The probabilities are displayed in Table 2.4. As expected, the sum of the entries in each row of the table is one. The largest element in each row has been set in **boldface**. Returning to the maximum likelihood estimation question in the problem statement, we would now like to determine the maximum likelihood estimate $\hat{\theta}$ for each value of x.

All that is necessary is to select the largest value in each row as the most likely shed selected for a given value of x. Using the boldface entries in Table 2.4, the maximum likelihood estimate of θ is

$$\hat{\theta} = \begin{cases} 1 & x = 0 \\ 4 & x = 1 \\ 5 & x = 2, 3. \end{cases}$$

The interpretation of the maximum likelihood estimate is as follows.

- If there are no defectives in the sample, then the sample was most likely to have been drawn from Shed 1.
- If there is one defective in the sample, then the sample was most likely to have been drawn from Shed 4.
- If there are two or three defectives in the sample, then the sample was most likely to have been drawn from Shed 5.

This maximum likelihood estimate is consistent with intuition. Consider the case of $x = 1$ defective, for example. Since one out of the three motors inspected is defective, the sample fraction of defective motors is $1/3$. The shed with the closest fraction of defective motors is Shed 4, in which $3/10$ of the motors are defective.

	$s=1$	$s=2$	$s=3$	$s=4$	$s=5$
$x=0$	**8/21**	4/15	8/45	1/9	4/63
$x=1$	0	36/215	56/215	**63/215**	12/43
$x=2$	0	0	8/65	21/65	**36/65**
$x=3$	0	0	0	1/5	**4/5**

Table 2.4: $P(S = s \mid X = x)$.

All of the examples of maximum likelihood estimation so far have only had a single ($k = 1$) parameter to estimate. When there are $k > 1$ unknown parameters $\theta_1, \theta_2, \ldots, \theta_k$, a unique solution of the $k \times k$ set of equations

$$\frac{\partial \ln L(\theta_1, \theta_2, \ldots, \theta_k)}{\partial \theta_i} = 0 \qquad i = 1, 2, \ldots, k$$

yields the maximum likelihood estimators $\hat{\theta}_1, \hat{\theta}_2, \ldots, \hat{\theta}_k$. The next example illustrates the technique for finding the maximum likelihood estimators of the $k = 2$ parameters in the normal distribution.

Example 2.13 Let $X_1, X_2, \ldots, X_n$ be a random sample from a normal population with probability density function

$$f(x) = \frac{1}{\sqrt{2\pi}\sigma} e^{-\frac{1}{2}\left(\frac{x-\mu}{\sigma}\right)^2} \qquad -\infty < x < \infty,$$

where μ and σ^2 are unknown parameters defined on the parameter space

$$\Omega = \left\{ (\mu, \sigma^2) \mid -\infty < \mu < \infty, \sigma^2 > 0 \right\}.$$

Find the maximum likelihood estimators of μ and σ^2.

As usual, it is assumed that a preliminary analysis of the shape of the histogram for the data set of interest is bell shaped, which indicates that a normal probability model might well be appropriate. In the derivation of the maximum likelihood estimators below, the estimation of σ^2 is performed by treating the population variance σ^2 as a unit, rather than working with the population standard deviation σ. This is done to simplify the mathematics. The maximum likelihood estimates of μ and σ^2 can be found by maximizing the likelihood function

$$L(\mu, \sigma^2) = \prod_{i=1}^{n} f(x_i) = \prod_{i=1}^{n} \frac{1}{\sqrt{2\pi\sigma^2}} e^{-(x_i-\mu)^2/(2\sigma^2)},$$

or

$$L(\mu, \sigma^2) = \left(2\pi\sigma^2\right)^{-n/2} \exp\left(-\frac{1}{2\sigma^2}\sum_{i=1}^{n}(x_i-\mu)^2\right),$$

with respect to μ and σ^2. It is easier to maximize the log likelihood function, which is

$$\ln L(\mu, \sigma^2) = -\frac{n}{2}\ln\left(2\pi\sigma^2\right) - \frac{1}{2\sigma^2}\sum_{i=1}^{n}(x_i-\mu)^2.$$

Find the maximum likelihood estimates by differentiating the log likelihood function with respect to μ and σ^2. Treating σ^2 as a single quantity, the elements of the score vector are

$$\frac{\partial \ln L(\mu, \sigma^2)}{\partial \mu} = \frac{1}{\sigma^2}\sum_{i=1}^{n}(x_i-\mu)$$

and

$$\frac{\partial \ln L(\mu, \sigma^2)}{\partial \sigma^2} = -\frac{n}{2\sigma^2} + \frac{1}{2\sigma^4}\sum_{i=1}^{n}(x_i-\mu)^2.$$

When the elements of the score vector are equated to zero, the resulting set of two equations in two unknowns is

$$\frac{1}{\sigma^2}\sum_{i=1}^{n}(x_i-\mu) = 0$$

and

$$-\frac{n}{2\sigma^2} + \frac{1}{2\sigma^4}\sum_{i=1}^{n}(x_i-\mu)^2 = 0.$$

Solving these equations gives the maximum likelihood estimates

$$\hat{\mu} = \bar{x} = \frac{1}{n}\sum_{i=1}^{n} x_i$$

and

$$\hat{\sigma}^2 = \frac{1}{n}\sum_{i=1}^{n}(x_i-\bar{x})^2.$$

For this particular population distribution, the maximum likelihood estimates are identical to the method of moments estimates from Example 2.5. But equating first derivatives to zero and solving does not necessarily mean that $\hat{\mu}$ and $\hat{\sigma}^2$ *maximize* the log likelihood function—they might minimize the log likelihood function. Under what conditions does the log likelihood function achieve a local maximum at the maximum likelihood estimates? Consider the previous examples in which there was just a single parameter θ to estimate that maximizes the log likelihood function. If the log likelihood function was twice differentiable and had a stationary point at $\hat{\theta}$, meaning that $\frac{\partial}{\partial\theta}\ln L(\theta)\big|_{\theta=\hat{\theta}} = 0$, then $\hat{\theta}$ maximizes $\ln L(\theta)$ if $\frac{\partial^2}{\partial\theta^2}\ln L(\theta)\big|_{\theta=\hat{\theta}} < 0$. The situation is analogous with two parameters, but slightly more complicated. Again assume that the log likelihood function is twice differentiable. Returning to the current example of the maximum likelihood estimates for a random sample from the normal distribution, to show that $\hat{\mu}$ and $\hat{\sigma}^2$ achieve a local maximum for the log likelihood function, it is sufficient to show that the two first-order partial derivatives are equal to zero at $\hat{\mu}$ and $\hat{\sigma}^2$ and that the 2×2 Hessian matrix of second partial derivatives evaluated at $\hat{\mu}$ and $\hat{\sigma}^2$

$$\begin{bmatrix} \frac{\partial^2 \ln L(\mu,\sigma^2)}{\partial\mu^2} & \frac{\partial^2 \ln L(\mu,\sigma^2)}{\partial\mu\partial\sigma^2} \\ \frac{\partial^2 \ln L(\mu,\sigma^2)}{\partial\sigma^2\partial\mu} & \frac{\partial^2 \ln L(\mu,\sigma^2)}{\partial(\sigma^2)^2} \end{bmatrix}_{\mu=\hat{\mu},\,\sigma^2=\hat{\sigma}^2}$$

is negative definite. This matrix is symmetric because the mixed partial derivatives are equal. There are several ways to show that a matrix is negative definite (for example, showing that both eigenvalues are negative). The simplest approach for sampling from a normal population is to show that the upper-left-hand element of the Hessian matrix is negative and the determinant is positive. The Hessian matrix evaluated at $\hat{\mu}$ and $\hat{\sigma}^2$ is

$$\begin{bmatrix} -\frac{n}{\sigma^2} & -\frac{1}{\sigma^4}\sum_{i=1}^n (x_i-\mu) \\ -\frac{1}{\sigma^4}\sum_{i=1}^n (x_i-\mu) & \frac{n}{2\sigma^4} - \frac{1}{\sigma^6}\sum_{i=1}^n (x_i-\mu)^2 \end{bmatrix}_{\mu=\hat{\mu},\,\sigma^2=\hat{\sigma}^2} = \begin{bmatrix} -\frac{n}{\hat{\sigma}^2} & 0 \\ 0 & -\frac{n}{2\hat{\sigma}^4} \end{bmatrix}$$

because $\sum_{i=1}^n (x_i - \bar{x}) = 0$. The upper-left-hand entry is negative as long as $\hat{\sigma}^2 > 0$ and the determinant of the Hessian matrix at $\hat{\mu}$ and $\hat{\sigma}^2$ is

$$\begin{vmatrix} -\frac{n}{\hat{\sigma}^2} & 0 \\ 0 & -\frac{n}{2\hat{\sigma}^4} \end{vmatrix} = \frac{n^2}{2\hat{\sigma}^6}.$$

The determinant is positive as long as $\hat{\sigma}^6 > 0$. To summarize, we can conclude that the Hessian matrix is negative definite as long as the maximum likelihood estimate of the population variance σ^2 is positive. The only data set that results in a value of zero for $\hat{\sigma}^2$ is one in which

$$x_1 = x_2 = \cdots = x_n.$$

So we can conclude that the maximum likelihood estimators $\hat{\mu}$ and $\hat{\sigma}^2$ maximize the log likelihood function as long as there are at least two distinct data values.

This derivation has produced maximum likelihood estimates of the population mean μ and the population variance σ^2. But what if you decide that you also want the maximum likelihood estimate of the population standard deviation σ? Do you need to begin the derivation from scratch, differentiating the log likelihood function with respect to σ rather than σ^2? This is a case in which the *invariance property* of maximum likelihood estimates comes to the rescue. In a nutshell, it states that "functions of maximum likelihood estimates are themselves maximum likelihood estimates." This means that all that is necessary to find the maximum likelihood estimate of σ is to take the square root of both sides of the expression for $\hat{\sigma}^2$. So in this case, the (biased) maximum likelihood estimate of σ is

$$\hat{\sigma} = \sqrt{\frac{1}{n}\sum_{i=1}^{n}(x_i - \bar{x})^2}.$$

We now formally state and prove the invariance property for maximum likelihood estimators. The proof considers just the case of a one-to-one function of the maximum likelihood estimator. In an advanced course in mathematical statistics, you will see the proof for more general functions of maximum likelihood estimators and the result for more than just one parameter.

Theorem 2.1 Let $X_1, X_2, \ldots, X_n$ be a random sample from a population distribution with unknown but fixed parameter θ, which is defined on the sample space Ω. Let $\hat{\theta}$ be the maximum likelihood estimator of θ. If g is a function on Ω, then $\hat{\beta} = g(\hat{\theta})$ is the maximum likelihood estimator of $\beta = g(\theta)$.

Proof (This proof considers only the case of a one-to-one function g.) The one-to-one function g maps the parameter space for θ, which is Ω, to the parameter space for β, which will be denoted here by Ω'. Since g is a one-to-one function on Ω, an inverse function $\theta = g^{-1}(\beta)$ exists, which maps Ω' to Ω. Let $L(\theta)$ be the likelihood function associated with the data values $X_1, X_2, \ldots, X_n$. Consider the likelihood function as a function of $g(\theta)$ rather than θ, which is $L\big(g(\theta)\big)$. This is the function that we want to maximize in the context of finding the maximum likelihood estimator of $\beta = g(\theta)$. The same maximum value is achieved by the following three functions.

$$\max_{\Omega} L(\theta) = \max_{\Omega} L\big(g(\theta)\big) = \max_{\Omega'} L\big(g^{-1}(\beta)\big).$$

The maximum of the likelihood function occurs when $g^{-1}\big(\hat{\beta}\big) = \hat{\theta}$, so the maximum likelihood estimator of β is

$$\hat{\beta} = g\big(\hat{\theta}\big).$$

□

Invariance is a useful property of maximum likelihood estimators. Since there are many characteristics of a distribution (for example, moments or percentiles) that we might be interested in estimating which are functions of θ, invariance allows us to easily estimate these characteristics.

All of the examples encountered so far have had closed-form expressions for the maximum likelihood estimators. There are many parametric distributions in which this is not the case. The Weibull distribution is one such case.

Example 2.14 Let $X_1, X_2, \ldots, X_n$ be a random sample from a Weibull population with probability density function

$$f(x) = \kappa\lambda^{\kappa}x^{\kappa-1}e^{-(\lambda x)^{\kappa}} \qquad x > 0$$

for positive unknown parameters λ and κ. Find the maximum likelihood estimators $\hat{\lambda}$ and $\hat{\kappa}$ and fit this distribution to the ball bearing failure times from Example 2.3.

As before, we assume that an analysis of the shape of the histogram reveals that a Weibull(λ, κ) model is warranted, so we proceed to determine the maximum likelihood estimates. The likelihood function is

$$\begin{aligned} L(\lambda, \kappa) &= \prod_{i=1}^{n} f(x_i) \\ &= \prod_{i=1}^{n} \kappa \lambda^{\kappa} x_i^{\kappa-1} e^{-(\lambda x_i)^{\kappa}} \\ &= \kappa^n \lambda^{n\kappa} \left(\prod_{i=1}^{n} x_i \right)^{\kappa-1} e^{-\lambda^{\kappa} \sum_{i=1}^{n} x_i^{\kappa}}. \end{aligned}$$

The log likelihood function is

$$\ln L(\lambda, \kappa) = n \ln \kappa + n\kappa \ln \lambda + (\kappa - 1) \sum_{i=1}^{n} \ln x_i - \lambda^{\kappa} \sum_{i=1}^{n} x_i^{\kappa}$$

and the 2×1 score vector has elements

$$\frac{\partial \ln L(\lambda, \kappa)}{\partial \lambda} = \frac{n\kappa}{\lambda} - \kappa \lambda^{\kappa-1} \sum_{i=1}^{n} x_i^{\kappa}$$

$$\frac{\partial \ln L(\lambda, \kappa)}{\partial \kappa} = \frac{n}{\kappa} + n \ln \lambda + \sum_{i=1}^{n} \ln x_i - \sum_{i=1}^{n} (\lambda x_i)^{\kappa} \ln(\lambda x_i).$$

When these equations are set equal to zero, the simultaneous equations have no closed-form solution for $\hat{\lambda}$ and $\hat{\kappa}$:

$$\frac{n\kappa}{\lambda} - \kappa \lambda^{\kappa-1} \sum_{i=1}^{n} x_i^{\kappa} = 0$$

$$\frac{n}{\kappa} + n \ln \lambda + \sum_{i=1}^{n} \ln x_i - \sum_{i=1}^{n} (\lambda x_i)^{\kappa} \ln(\lambda x_i) = 0.$$

One piece of good fortune, however, to avoid solving a 2×2 set of nonlinear equations numerically, is that this first equation can be solved for λ in terms of κ:

$$\lambda = \left(\frac{n}{\sum_{i=1}^{n} x_i^{\kappa}} \right)^{1/\kappa}.$$

Using this expression for λ in terms of κ in the second element of the score vector yields a single, albeit more complicated, expression with κ as the only unknown. After applying some algebra, this equation reduces to

$$g(\kappa) = \frac{n}{\kappa} + \sum_{i=1}^{n} \ln x_i - \frac{n \sum_{i=1}^{n} x_i^{\kappa} \ln x_i}{\sum_{i=1}^{n} x_i^{\kappa}} = 0,$$

which must be solved iteratively. One technique that can be used here is the Newton–Raphson procedure, which uses the iterative step

$$\kappa_{j+1} = \kappa_j - \frac{g(\kappa_j)}{g'(\kappa_j)},$$

where κ_0 is an initial estimate and the derivative of $g(\kappa)$ reduces to

$$g'(\kappa) = -\frac{n}{\kappa^2} - \frac{n}{\left(\sum_{i=1}^{n} x_i^\kappa\right)^2}\left[\left(\sum_{i=1}^{n} x_i^\kappa\right)\left(\sum_{i=1}^{n} (\ln x_i)^2 x_i^\kappa\right) - \left(\sum_{i=1}^{n} x_i^\kappa \ln x_i\right)^2\right].$$

The iterative procedure can be repeated until the desired accuracy for κ is achieved, that is, $|\kappa_{j+1} - \kappa_j| < \varepsilon$ for some small, positive real number ε. The only remaining issue associated with implementing the Newton–Raphson procedure is finding an initial estimate κ_0 for the procedure. Menon's estimate of κ can serve as a starting point for the procedure:

$$\kappa_0 = \left\{\frac{6}{(n-1)\pi^2}\left[\sum_{i=1}^{n} (\ln x_i)^2 - \frac{\left(\sum_{i=1}^{n} \ln x_i\right)^2}{n}\right]\right\}^{-1/2}.$$

An R function named `menon` for computing the initial estimate $\hat{\kappa}$ from a data set that is stored in the vector `data` is given below.

```
menon = function(data) {
  n = length(data)
  sum1 = sum(log(data) ^ 2)
  sum2 = sum(log(data)) ^ 2
  1 / sqrt(6 * (sum1 - sum2 / n) / ((n - 1) * pi ^ 2))
}
```

This function is called in the R code below that places the ball bearing failure times in the vector `bb`, then implements the Newton–Raphson procedure with $\varepsilon = 0.00001$ to determine the maximum likelihood estimates.

```
bb   = c(17.88, 28.92, 33.00, 41.52, 42.12, 45.60, 48.48, 51.84,
         51.96, 54.12, 55.56, 67.80, 68.64, 68.64, 68.88, 84.12,
         93.12, 98.64, 105.12, 105.84, 127.92, 128.04, 173.40)
n    = length(bb)
eps  = 0.00001
k    = 0
knew = menon(bb)
while(abs(knew - k) > eps) {
  k = knew
  sum1 = sum(log(bb))
  sum2 = sum(bb ^ k)
  sum3 = sum(bb ^ k * log(bb))
  sum4 = sum(bb ^ k * (log(bb)) ^ 2)
  top  = n / k + sum1 - n * sum3 / sum2
  bot  = -n / k ^ 2 - (n / sum2 ^ 2) * ((sum2 * sum4) - sum3 ^ 2)
  knew = k - top / bot
}
kappa.hat = knew
lambda.hat = (n / sum(bb ^ kappa.hat)) ^ (1 / kappa.hat)
print(c(lambda.hat, kappa.hat))
```

When this code is executed, Menon's initial estimate of κ is $\kappa_0 = 2.405$. The algorithm converges in just four iterations of the `while` loop to the maximum likelihood estimates

$$\hat{\lambda} = 0.01221 \qquad \text{and} \qquad \hat{\kappa} = 2.102,$$

which have been rounded to significant digits. A reliability engineer might consider this fitted distribution superior to the gamma distribution fitted in Example 2.4 because the Weibull distribution avoids the exponential right-hand tail of the gamma distribution and might be a more accurate lifetime model for ball bearings that are wearing out. A plot of the fitted cumulative distribution function and the empirical cumulative distribution function is given in Figure 2.12, which is quite similar (but not identical) to Figure 2.3, which involved fitting the gamma distribution to the ball bearing failure times via the method of moments.

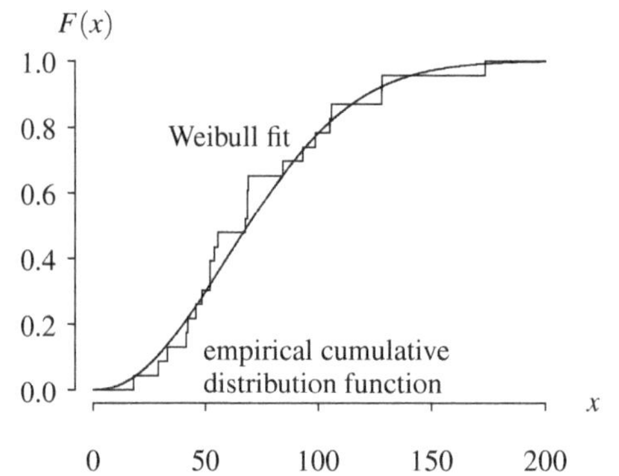

Figure 2.12: Empirical and fitted cumulative distribution functions for the ball bearing data.

The work done so far in this example (deriving the equations to be solved for $\hat{\lambda}$ and $\hat{\kappa}$, the development of the numerical method for determining $\hat{\lambda}$ and $\hat{\kappa}$, and the implementation of that algorithm) could have been avoided. An R function named `fitdist` in a package named `fitdistrplus` computes maximum likelihood estimates for several popular probability models. The first step is to install the package with the R command

```
install.packages("fitdistrplus")
```

This is a one-time investment on any one particular machine. The next step is to load the package with the R command

```
library(fitdistrplus)
```

This must be done at the beginning of an R session. Finally, the R statements

```
bb   = c(17.88, 28.92, 33.00, 41.52, 42.12, 45.60, 48.48, 51.84,
         51.96, 54.12, 55.56, 67.80, 68.64, 68.64, 68.88, 84.12,
         93.12, 98.64, 105.12, 105.84, 127.92, 128.04, 173.40)
fitdist(bb, distr = "weibull")
```

give the estimate of the scale parameter as 81.868 and the shape parameter as 2.102. The shape parameter estimate matches our earlier work, but the scale parameter is not even close. What went wrong? The problem here is that not everyone parameterizes the Weibull distribution in the same fashion. The `fitdist` designers used $1/\eta$ instead of λ when they parameterized the Weibull distribution. The reciprocal of their scale parameter estimate, $1/81.868 = 0.01221$, equals $\hat{\lambda}$. The `fitdist` function generalizes to include other distributions than the Weibull (for example, the beta distribution), estimation methods other than maximum likelihood (for example, the method of moments), and censored data in which only an upper or lower bound is known on a data value.

The Newton–Raphson method, which was used to find the maximum likelihood estimators for the Weibull distribution in the previous example, can fail to converge for certain data sets. There is an algorithm that uses fixed-point iteration that is much more reliable.

One question that has not been addressed by any of the previous examples is whether maximum likelihood estimators are unique. The next example shows that they are not unique for all population probability distributions.

Example 2.15 Let $X_1, X_2, \ldots, X_n$ be a random sample from a $U(\theta, \theta+1)$ population. Find the maximum likelihood estimator of θ.

We assume that a preliminary analysis of the data leads us to believe that a $U(\theta, \theta+1)$ distribution is an appropriate probability model for the random variable of interest. The probability density function of $X \sim U(\theta, \theta+1)$ is

$$f(x) = 1 \qquad \theta < x < \theta + 1.$$

So the likelihood function is

$$L(\theta) = \prod_{i=1}^{n} f(x_i) = 1^n = 1$$

for all x_i values that lie between θ and $\theta+1$, and 0 otherwise. The usual approach to finding the maximum likelihood estimator $\hat{\theta}$ using calculus does not look promising. Consider instead the thought experiment associated with the following tiny data set with $n = 3$ observations:

$$6.42, 6.93, 6.85.$$

For any θ satisfying $5.93 < \theta < 6.42$, the likelihood function is 1; otherwise, the likelihood function is 0, as illustrated in Figure 2.13. Thus, any $\hat{\theta}$ satisfying $5.93 < \hat{\theta} < 6.42$ could serve as a maximum likelihood estimate. This example shows that the maximum likelihood estimate is not necessarily a unique value as it was in all of the previous examples. To generalize, for the data values $x_1, x_2, \ldots, x_n$ and a $U(\theta, \theta+1)$ population, as long as $x_{(n)} - x_{(1)} < 1$, any θ value in the range

$$x_{(n)} - 1 < \theta < x_{(1)}$$

maximizes the likelihood function, and hence is a maximum likelihood estimate $\hat{\theta}$.

The series of examples presented here has revealed some weaknesses associated with maximum likelihood estimates. First, as just illustrated, they are not necessarily unique. Second, they are not necessarily unbiased. Third, they do not necessarily even exist. Fourth, oftentimes numerical methods are required to determine their values. Fifth, the mathematics associated with the second derivative test for maximization can be complicated. Despite these weaknesses, their statistical properties make them the estimators of choice in many statistical settings.

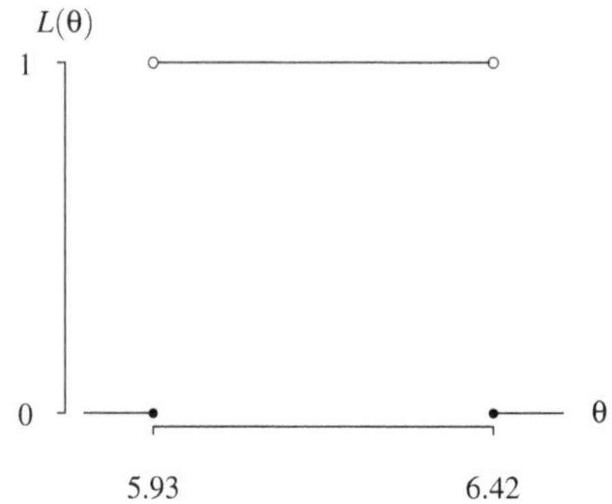

Figure 2.13: The likelihood function for $n = 3$ observations from a $U(\theta, \theta + 1)$ population.

2.4 Properties of Point Estimators

The two previous sections have introduced two techniques for finding a point estimator of an unknown population parameter θ from a random sample $X_1, X_2, \ldots, X_n$: the method of moments and maximum likelihood estimation. But the techniques only define the point estimator—they do not give us any information about the quality of the estimator. This section introduces several properties that reflect the quality of a point estimator.

Data which was not previously collected (due to the resources required to do so) is now being collected automatically, which has resulted in huge data sets. The notion of "big data" has arisen in the past few years as the ability to capture and store large amounts of data automatically and cheaply has increased. Most of the data captured throughout history has been collected in just the past few years. So how can these massive data sets be summarized? One way is with a statistical graphic. A second way is with some carefully selected statistics that capture the essence of the data values. These statistics should be selected so that they reduce the data set in a manner that keeps the information of interest and discards the information that is not of interest.

It is too much to expect that a point estimator $\hat{\theta}$ would always match the population parameter θ exactly. Since point estimates are statistics, they are also random variables. This means that point estimates will vary from one sample to the next. But some evidence that $\hat{\theta}$ will be reasonably close to θ would certainly be reassuring. This section presents some of the thinking and analysis along these lines.

The subsections presented here outline properties of point estimators that can be useful when comparing two or more point estimators. These properties help a statistician decide which is the best to use in a particular statistical setting. The properties outlined here are unbiasedness, efficiency, sufficiency, and consistency.

Unbiased estimators

An important property of point estimators is whether or not their expected value equals the unknown parameter that they are estimating. If θ is considered the target parameter, then we want $\hat{\theta}$ to be *on target*, as defined formally next.

Definition 2.1 Let $\hat{\theta}$ denote a statistic that is calculated from the sample $X_1, X_2, \ldots, X_n$. Then $\hat{\theta}$ is an *unbiased estimator* of the unknown parameter θ defined on the parameter space Ω if and only if, for all $\theta \in \Omega$,

$$E\left[\hat{\theta}\right] = \theta.$$

If $\hat{\theta}$ is not unbiased, that is, $E\left[\hat{\theta}\right] \neq \theta$ for some $\theta \in \Omega$, then $\hat{\theta}$ is a *biased estimator* of θ.

Determining whether a point estimator $\hat{\theta}$ is an unbiased estimator of θ typically arises in the following setting. The random sample $X_1, X_2, \ldots, X_n$ is drawn from the probability distribution described by $f(x)$, which defines a parametric distribution with a single unknown parameter θ. Knowing that $E\left[\hat{\theta}\right] = \theta$ is valuable information in that the sampling distribution of $\hat{\theta}$ is centered on the target value. Although the specific value of $\hat{\theta}$ for a specific data set $x_1, x_2, \ldots, x_n$ might fall above θ or below θ, knowing that the mean value of $\hat{\theta}$ is θ assures us that our (metaphorical) arrow, the point estimator $\hat{\theta}$, is pointing at the center of the target (the true value of the unknown parameter θ).

We have now effectively partitioned the set of all point estimators $\hat{\theta}$ for the unknown parameter θ into two sets: unbiased estimators and biased estimators. The Venn diagram in Figure 2.14 illustrates this partition. All other factors being the same, one would always prefer an unbiased estimate over a biased estimate. But as subsequent examples will show, the decision is not always that clear-cut. We begin with an example of determining whether or not a point estimate is unbiased.

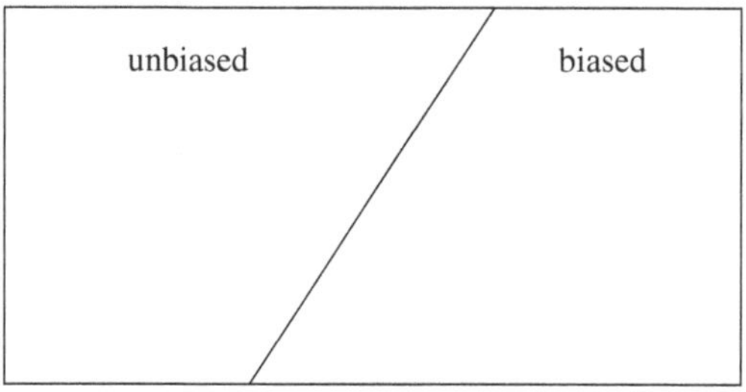

Figure 2.14: Venn diagram of unbiased and biased point estimators.

Example 2.16 Let $X_1, X_2, \ldots, X_n$ denote a random sample from a Bernoulli(p) population, where p is an unknown parameter satisfying $0 < p < 1$. Is the maximum likelihood estimator of p unbiased?

A statistician might encounter a data set of this kind in a political poll involving two candidates for elected office. Since the data is drawn from a Bernoulli population, each data value is a 0 or 1 corresponding to the respondent's preference for one of two candidates for a particular office. The Bernoulli distribution has probability mass function

$$f(x) = p^x (1-p)^{1-x} \qquad x = 0, 1.$$

So the likelihood function is

$$L(p) = \prod_{i=1}^{n} f(x_i) = \prod_{i=1}^{n} p^{x_i} (1-p)^{1-x_i} = p^{\sum_{i=1}^{n} x_i} (1-p)^{n - \sum_{i=1}^{n} x_i}.$$

The log likelihood function is

$$\ln L(p) = \left(\sum_{i=1}^{n} x_i\right) \ln p + \left(n - \sum_{i=1}^{n} x_i\right) \ln(1-p).$$

The score is the partial derivative of the log likelihood function with respect to p, which is

$$\frac{\partial \ln L(p)}{\partial p} = \frac{1}{p}\sum_{i=1}^{n} x_i - \frac{1}{1-p}\left(n - \sum_{i=1}^{n} x_i\right).$$

When the score is equated to zero,

$$\frac{1}{p}\sum_{i=1}^{n} x_i - \frac{1}{1-p}\left(n - \sum_{i=1}^{n} x_i\right) = 0,$$

and solved for p, the resulting maximum likelihood estimate is

$$\hat{p} = \frac{1}{n}\sum_{i=1}^{n} x_i,$$

which is the sample mean. In the context of sampling from a Bernoulli population, the maximum likelihood estimate also has the interpretation as the *fraction of successes*. This is an intuitively appealing estimator because the true probability of success is being estimated by the observed fraction of successes. The second derivative of the log likelihood function is negative at the maximum likelihood estimator for all possible data sets, except for the extreme cases of all zeros or all ones, so we are assured that $\hat{p}$ maximizes the log likelihood function. We now return to the original question concerning whether $\hat{p}$ is an unbiased estimator of p. The expected value of the maximum likelihood estimator is

$$E[\hat{p}] = E\left[\frac{1}{n}\sum_{i=1}^{n} X_i\right] = \frac{1}{n}E\left[\sum_{i=1}^{n} X_i\right] = \frac{1}{n}\cdot(np) = p$$

because $\sum_{i=1}^{n} X_i$ has the binomial(n, p) distribution with expected value np. So $\hat{p}$ is an unbiased estimator of p.

This example allows some insight into the sampling distribution of $\hat{p}$ because $\hat{p}$ is the ratio of a binomial(n, p) random variable and n. This is a "scaled binomial" random variable. Using the random variable Y to denote $\hat{p}$ for notational clarity, the probability mass function of the maximum likelihood estimator $Y = \hat{p}$ is

$$f_Y(y) = \binom{n}{ny} p^{ny}(1-p)^{n-ny} \qquad y = 0, \frac{1}{n}, \frac{2}{n}, \ldots, 1.$$

This probability mass function is plotted in Figure 2.15 for $n = 50$ and $p = 3/5$. This scenario corresponds to a political poll of a random sample of 50 voters from a large population for a candidate having the support of 60% of the electorate. The sample of 50 voters is large enough so that the central limit theorem has kicked in and the probability mass function is roughly bell shaped. Could such a poll give a point estimate which incorrectly concludes that the candidate with support of 60% of the electorate will not

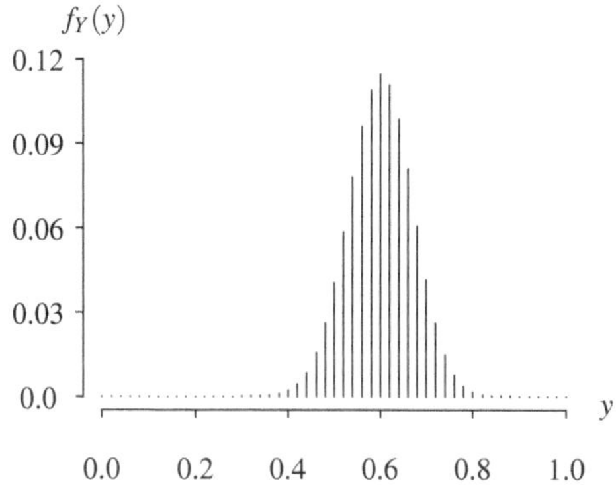

Figure 2.15: Maximum likelihood estimator probability mass function for $n = 50$ and $p = 3/5$.

win the election? This probability is $P(Y \le 0.5) \cong 0.09781$, which is calculated with the R statement `pbinom(25, 50, 3 / 5)`. If this value is unacceptably large, then a poll with a larger sample size n must be taken. The population variance of the maximum likelihood estimator $\hat{p}$ can also be easily calculated:

$$V[\hat{p}] = V\left[\frac{1}{n}\sum_{i=1}^{n} X_i\right] = \frac{1}{n^2} V\left[\sum_{i=1}^{n} X_i\right] = \frac{1}{n^2} \cdot np(1-p) = \frac{p(1-p)}{n}.$$

The precision of $\hat{p}$ depends only on the sample size and the true value of p. Notice that the population variance of the maximum likelihood estimator $\hat{p}$ goes to zero as the sample size goes to infinity. For $n = 50$ and $p = 0.6$,

$$V[\hat{p}] = \frac{0.6(1-0.6)}{50} = 0.0048.$$

So the standard error of the estimate of p is

$$\sigma_{\hat{p}} = \sqrt{0.0048} \cong 0.06928.$$

A standard error of $\hat{p}$ of 7% in a political poll might be unacceptably high. This is why sample sizes in political polls are substantially higher than $n = 50$. A sample size of $n = 1000$, for example, reduces the standard error of the estimate of p when $p = 0.6$ to $\sigma_{\hat{p}} \cong 0.01549$.

In the previous example, we considered a specific parametric population distribution—the Bernoulli distribution—with a single unknown parameter p. But the notion of an unbiased estimator is more general. The next result applies to any population distribution, with only the rather mild restriction that the first two population moments must be finite. Recall from Theorem 1.1 that

$$E\left[\bar{X}\right] = \mu,$$

where $\bar{X}$ is the sample mean, μ is the finite population mean, and $X_1, X_2, \ldots, X_n$ are the values from a random sample from *any* population distribution. This result indicated that $\bar{X}$ is an unbiased estimator of μ. The next result indicates that the sample variance S^2 is an unbiased estimator of the population variance σ^2.

Theorem 2.2 Let $X_1, X_2, \ldots, X_n$ be a random sample from a population with finite population mean μ and finite population variance σ^2. The expected value of the sample variance is

$$E\left[S^2\right] = \sigma^2,$$

which means S^2 is an unbiased estimate of σ^2.

Proof The expected value of the sample variance is

$$\begin{aligned}
E\left[S^2\right] &= E\left[\frac{1}{n-1}\sum_{i=1}^{n}\left(X_i - \bar{X}\right)^2\right] \\
&= \frac{1}{n-1}E\left[\sum_{i=1}^{n}\left(X_i^2 - 2X_i\bar{X} + \bar{X}^2\right)\right] \\
&= \frac{1}{n-1}E\left[\sum_{i=1}^{n}X_i^2 - 2\bar{X}\sum_{i=1}^{n}X_i + n\bar{X}^2\right] \\
&= \frac{1}{n-1}E\left[\sum_{i=1}^{n}X_i^2 - n\bar{X}^2\right] \\
&= \frac{1}{n-1}\left(\sum_{i=1}^{n}E\left[X_i^2\right] - nE\left[\bar{X}^2\right]\right) \\
&= \frac{1}{n-1}\left[\sum_{i=1}^{n}\left(V\left[X_i\right] + E\left[X_i\right]^2\right) - n\left(V\left[\bar{X}\right] + E\left[\bar{X}\right]^2\right)\right] \\
&= \frac{1}{n-1}\left[n\sigma^2 + n\mu^2 - n\left(\frac{\sigma^2}{n} + \mu^2\right)\right] \\
&= \sigma^2,
\end{aligned}$$

which, using Theorem 1.1, proves that S^2 is an unbiased estimator of σ^2. □

The fact that S^2 is an unbiased estimator of σ^2 for any population distribution is one of the most compelling reasons to use the $n-1$ in the denominator of S^2 in Definition 1.10. This result does not imply, however, that $E[S] = \sigma$. Even so, statisticians often use S to estimate σ, sometimes including an unbiasing constant for small values of the sample size n.

So far we have seen that $E[\hat{p}] = p$ for random sampling from a Bernoulli(p) population and $E\left[S^2\right] = \sigma^2$ for random sampling from any population with finite population mean and variance. It is helpful to define the bias explicitly in order to have a measure of the expected distance between a point estimator and its target value.

Definition 2.2 Let $\hat{\theta}$ denote a statistic that is calculated from the sample $X_1, X_2, \ldots, X_n$. The *bias* associated with using $\hat{\theta}$ as an estimator of θ is

$$B(\hat{\theta}, \theta) = E\left[\hat{\theta}\right] - \theta.$$

There is a subset of the biased estimators that is of interest. The classification is a bit of a consolation prize for biased estimators. Their redeeming feature is that although they are biased estimators for finite sample sizes n, they are unbiased in the limit as $n \to \infty$. These estimators are known as asymptotically unbiased estimators and are defined formally below.

Definition 2.3 Let $\hat{\theta}$ denote a statistic that is calculated from the sample $X_1, X_2, \ldots, X_n$. If

$$\lim_{n \to \infty} B(\hat{\theta}, \theta) = 0,$$

then $\hat{\theta}$ is an *asymptotically unbiased estimator* of θ.

All unbiased estimators are necessarily asymptotically unbiased. But only some of the biased estimators are asymptotically unbiased. To this end, we subdivide the biased portion of the Venn diagram from Figure 2.14 to include asymptotically unbiased estimators in Figure 2.16.

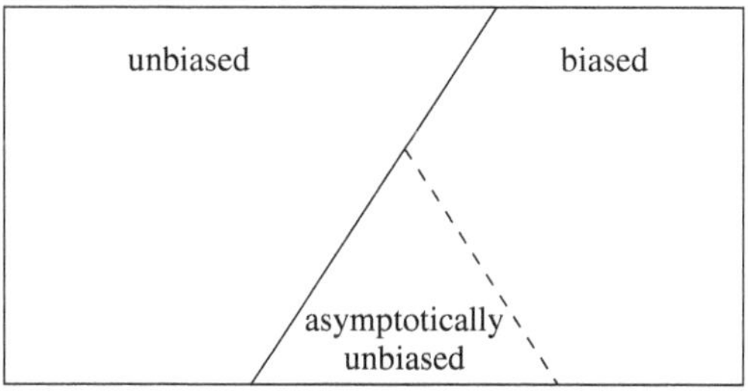

Figure 2.16: Venn diagram of unbiased, biased, and asymptotically unbiased point estimators.

Example 2.17 Let $X_1, X_2, \ldots, X_n$ denote a random sample from a $U(0, \theta)$ population, where θ is a positive unknown parameter. Classify the following point estimators of θ into the categories given in Figure 2.16 and select the best estimator:

- $2\bar{X}$,
- $3\bar{X}$,
- $X_{(n)}$,
- $(n+1)X_{(n)}/n$,
- $(n+1)X_{(1)}$,
- 17,

where $X_{(1)} = \min\{X_1, X_2, \ldots, X_n\}$ and $X_{(n)} = \max\{X_1, X_2, \ldots, X_n\}$.

When faced with a real data set, we oftentimes have to choose a point estimator from a set of potential point estimators such as this. The purpose of this example is to investigate the properties of these six point estimators.

- The first point estimator, $2\bar{X}$, is the method of moments estimator. The derivation was given in Example 2.2. Since the population mean of the $U(0, \theta)$ distribution is $\theta/2$ and

$$E\left[2\bar{X}\right] = 2E\left[\bar{X}\right] = 2 \cdot \frac{\theta}{2} = \theta,$$

via Theorem 1.1, the method of moments estimator is classified as an unbiased estimator.

- The point estimator $3\bar{X}$ is classified as a biased estimator because

$$E\left[3\bar{X}\right] = 3E\left[\bar{X}\right] = 3 \cdot \frac{\theta}{2} = \frac{3}{2}\theta.$$

This estimator overestimates the population parameter θ on average. The positive bias is

$$B(\hat{\theta}, \theta) = B(3\bar{X}, \theta) = E\left[3\bar{X}\right] - \theta = \frac{3}{2}\theta - \theta = \frac{\theta}{2}.$$

- The point estimator $X_{(n)}$ is the maximum likelihood estimator. The derivation, after some minor manipulation of the objective function or the support of the population distribution, was given in Example 2.10. Using an order statistic result, the expected value of $X_{(n)}$ is

$$E\left[X_{(n)}\right] = \frac{n\theta}{n+1}.$$

The maximum likelihood estimator underestimates the population parameter θ, on average, because $E\left[X_{(n)}\right]$ is less than θ. Since the expected value is not equal to θ for finite values of n, this estimator is biased. The bias is

$$B(\hat{\theta}, \theta) = B(X_{(n)}, \theta) = E\left[X_{(n)}\right] - \theta = \frac{n\theta}{n+1} - \theta = -\frac{\theta}{n+1}.$$

This estimator should be classified as asymptotically unbiased, however, because

$$\lim_{n\to\infty} B(\hat{\theta}, \theta) = \lim_{n\to\infty}\left(-\frac{\theta}{n+1}\right) = 0.$$

- The point estimator $(n+1)X_{(n)}/n$ was presented in Example 2.10 as a modification of the maximum likelihood estimator that included an unbiasing constant. The expected value of $(n+1)X_{(n)}/n$ is

$$E\left[\frac{n+1}{n}X_{(n)}\right] = \frac{n+1}{n} \cdot \frac{n\theta}{n+1} = \theta,$$

so this point estimator is classified as an unbiased estimator.

- The point estimator $(n+1)X_{(1)}$ is also an unbiased estimator of θ, that is,

$$E\left[(n+1)X_{(1)}\right] = \theta.$$

This can be seen by invoking an order statistic result and computing the appropriate expected value.

- The point estimator $\hat{\theta} = 17$ is quite bizarre. The statistician simply ignores the data values $X_1, X_2, \ldots, X_n$ and pulls 17 out of thin air as the estimate of θ. The expected value of $\hat{\theta}$ is $E\left[\hat{\theta}\right] = E[17] = 17$, which is not θ (unless θ just happens to be 17), so this estimator is classified as a biased estimator. Recall from Definition 2.1 that an estimator $\hat{\theta}$ is an unbiased estimator of θ if $E\left[\hat{\theta}\right] = \theta$ for *all* $\theta \in \Omega$.

We now know that three of the six suggested point estimators are unbiased. The results of our analysis are summarized in the Venn diagram in Figure 2.17.

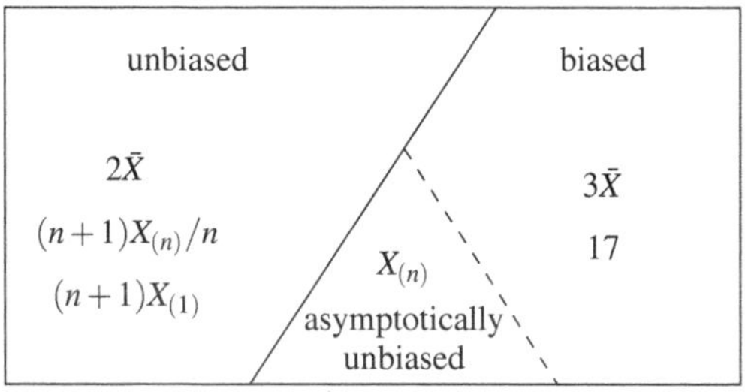

Figure 2.17: Venn diagram of several point estimates of θ for a $U(0, \theta)$ population.

Now to the more difficult question: which is the best of the six point estimators? This is a purposefully vague question at this point, so the question will be addressed from several different angles. The choice between the point estimators boils down to which point estimator will perform best for a higher fraction of data sets drawn from a $U(0, \theta)$ population than the other point estimators. This does not imply, of course, that the point estimator selected will be the best for every data set. We begin by plotting the sampling distributions of the three unbiased estimators to gain some additional insight. This can only be done for specific values of n and θ, so let's arbitrarily choose $n = 5$ and $\theta = 10$. For this choice, the probability density functions of $2\bar{X}$, $6X_{(5)}/5$, and $6X_{(1)}$ are plotted in Figure 2.18. APPL was used to calculate the probability density functions. The sampling distributions of $2\bar{X}$, $6X_{(5)}/5$, and $6X_{(1)}$ reveal vastly different shapes even though all three have expected value $\theta = 10$. The probability density function of $2\bar{X}$ is bell shaped (via the central limit theorem) and symmetric about $\theta = 10$; the probability

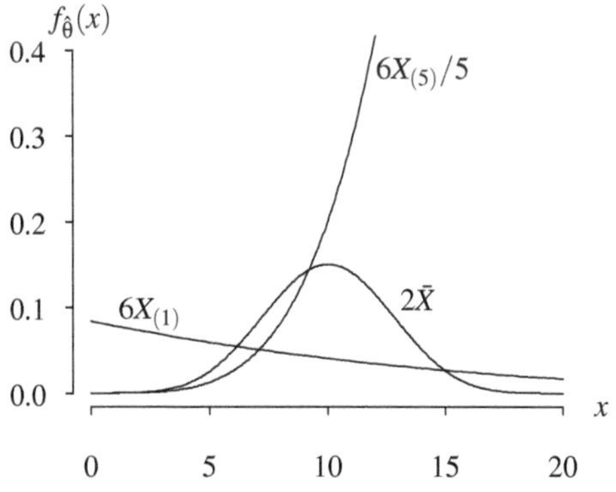

Figure 2.18: Sampling distributions of $2\bar{X}$, $6X_{(5)}/5$, and $6X_{(1)}$ when $n = 5$ and $\theta = 10$.

density functions of $6X_{(5)}/5$ and $6X_{(1)}$ are skewed distributions. Since the support of the population is $(0, 10)$, the support of $2\bar{X}$ is $(0, 20)$, the support of $6X_{(5)}/5$ is $(0, 12)$, and the support of $6X_{(1)}$ is $(0, 60)$. Figure 2.18 reveals that $6X_{(1)}$ has a significantly larger population variance than the other two unbiased estimators, so it is probably the weakest candidate of the three unbiased estimators.

Since the population variance of the point estimators plays a critical role in analyzing the sampling distributions of the three unbiased estimators, it is worthwhile calculating the population means and population variances of all six of the estimators. The values are summarized in Table 2.5. Notice that the three unbiased estimators, $2\bar{X}$, $(n+1)X_{(n)}/n$, and $(n+1)X_{(1)}$ all collapse to the same estimator when $n = 1$; the point estimator is just double the single observation. Choosing the point estimator with the smallest population variance is not appropriate here because this would result in choosing the strange point estimate $\hat{\theta} = 17$. Instead, it is advantageous to choose the unbiased estimator with the smallest variance. Using this criteria, $(n+1)X_{(n)}/n$ has the smallest population variance of the three unbiased estimators for samples of $n = 2$ or more observations.

Point estimate $\hat{\theta}$	$E\left[\hat{\theta}\right]$	$V\left[\hat{\theta}\right]$	Categorization
$2\bar{X}$	θ	$\dfrac{\theta^2}{3n}$	unbiased
$3\bar{X}$	$\dfrac{3\theta}{2}$	$\dfrac{3\theta^2}{4n}$	biased
$X_{(n)}$	$\dfrac{n\theta}{n+1}$	$\dfrac{n\theta^2}{(n+2)(n+1)^2}$	asymptotically unbiased
$\dfrac{(n+1)X_{(n)}}{n}$	θ	$\dfrac{\theta^2}{n(n+2)}$	unbiased
$(n+1)X_{(1)}$	θ	$\dfrac{n\theta^2}{n+2}$	unbiased
17	17	0	biased

Table 2.5: Population means and variances of the six point estimators of θ.

But the unbiased estimator with the smallest population variance is not the only criteria that can be used to select the preferred estimator. The R code below conducts a Monte Carlo simulation with 10,000 random samples of size $n = 5$ from a $U(0, \theta)$ population when $\theta = 10$. All six point estimators are calculated for each sample, and the point estimator that lies closest to θ is identified and tabulated. Finally, the fraction of times that each estimator is closest to $\theta = 10$ is printed.

```
set.seed(8)
n = 5
theta = 10
nrep = 10000
theta.hat = numeric(6)
count = numeric(6)
```

```
for (i in 1:nrep) {
  x = runif(n, 0, theta)
  theta.hat[1] = 2 * mean(x)
  theta.hat[2] = 3 * mean(x)
  theta.hat[3] = max(x)
  theta.hat[4] = (n + 1) * max(x) / n
  theta.hat[5] = (n + 1) * min(x)
  theta.hat[6] = 17
  index = which.min(abs(theta.hat - theta))
  count[index] = count[index] + 1
}
print(count / nrep)
```

The results of the Monte Carlo simulation are given in Table 2.6 for sample sizes $n = 5$, $n = 50$, and $n = 500$. The entries give the fractions of the simulations giving the closest estimator to the true parameter value $\theta = 10$. As expected, the column sums of the entries in the table equal 1. When $n = 5$, even the maligned $\hat{\theta} = 17$ is the closest to $\theta = 10$ for two of the 10,000 random samples. The reader is encouraged to imagine what type of data set would lead to this awful estimator outdoing the other estimators. Table 2.6 shows that, by a somewhat narrow margin, the unbiased estimator $(n+1)X_{(n)}/n$ dominates the other estimators for the sample sizes considered here.

Point estimate	$n = 5$	$n = 50$	$n = 500$
$2\bar{X}$	0.1765	0.0912	0.0328
$3\bar{X}$	0.1323	0.0000	0.0000
$X_{(n)}$	0.3178	0.3749	0.3905
$(n+1)X_{(n)}/n$	0.3262	0.5275	0.5762
$(n+1)X_{(1)}$	0.0470	0.0064	0.0005
17	0.0002	0.0000	0.0000

Table 2.6: Monte Carlo simulation results for a $U(0, 10)$ population.

In summary, based on $\hat{\theta} = (n+1)X_{(n)}/n$ being (a) an unbiased estimate, (b) the unbiased estimate with the smallest population variance, and (c) the estimate that is most likely to be the closest to the population value of θ for several sample sizes in a Monte Carlo experiment, we conclude that $\hat{\theta} = (n+1)X_{(n)}/n$ is the best of the six point estimators. It carries the additional bonus that all of the data values are necessarily less than $\hat{\theta}$, which is a desirable property for this particular population distribution.

Keep in mind that although the Monte Carlo simulation used 10,000 replications, the statistician only sees a single data set of n observations and must choose among the six candidate point estimators. We have chosen the unbiased estimator $(n+1)X_{(n)}/n$ as the best of the six because it performs the best on average.

This example has brought up three issues concerning point estimators that will be addressed in the paragraphs that follow.

The first issue is motivated by the Monte Carlo simulation experiment. The objective of the experiment was to find the point estimator that was most likely to be closest to the true parameter value. The distance between the estimator $\hat{\theta}$ and the true parameter value θ is an important quantity known as the error of estimation, which is formally defined next.

Definition 2.4 Let $\hat{\theta}$ denote a statistic that is calculated from the sample $X_1, X_2, \ldots, X_n$ that is used to estimate the population parameter θ. The *error of estimation* is

$$R(\hat{\theta}, \theta) = |\hat{\theta} - \theta|.$$

The second issue concerns the comparison of the three unbiased estimators and the three biased estimators. Could there ever be circumstances in which one would choose a biased estimator over an unbiased estimator? Consider the generic and idealized presentation of the sampling distributions of two point estimators $\hat{\theta}_1$ and $\hat{\theta}_2$ in Figure 2.19 for a fixed sample size n. The sampling distribution of $\hat{\theta}_1$ is centered over the true parameter value θ, so $\hat{\theta}_1$ is an unbiased estimator of θ, that is, $E\left[\hat{\theta}_1\right] = \theta$. The sampling distribution of $\hat{\theta}_2$, however, is not centered over the true parameter value θ, so $\hat{\theta}_2$ is a biased estimator of θ, that is, $E\left[\hat{\theta}_2\right] \neq \theta$. But the decision between the two is complicated by the fact that the population variance of the second estimator is much smaller than the population variance of the first estimator, that is, $V\left[\hat{\theta}_2\right] < V\left[\hat{\theta}_1\right]$. The choice between the unbiased estimator with the larger population variance and the biased estimator with the smaller population variance is a difficult one.

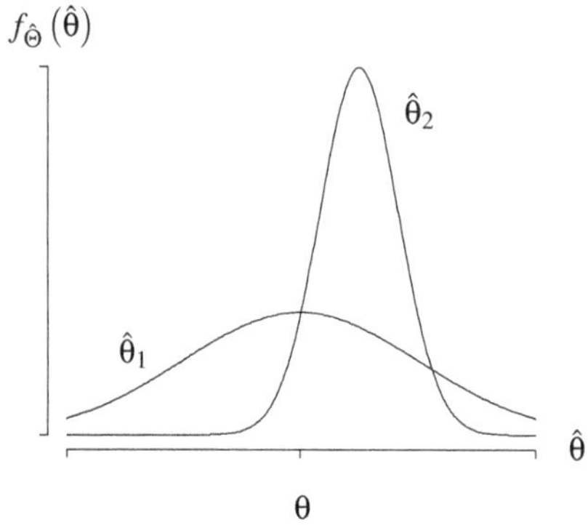

Figure 2.19: Two sampling distributions.

Statisticians have concocted a quantity that considers bias and variance simultaneously. The mean square error, often abbreviated MSE, is the expected squared error of $\hat{\theta}$ in estimating θ.

Definition 2.5 For a point estimator $\hat{\theta}$ which estimates a population parameter θ, the *mean square error* is

$$\text{MSE}(\hat{\theta}, \theta) = E\left[(\hat{\theta} - \theta)^2\right].$$

The defining formula for the mean square error given in Definition 2.5 does not make it clear how bias and variability are partitioned. The following theorem pins down the details.

Theorem 2.3 For a point estimator $\hat{\theta}$ which estimates a population parameter θ, the mean square error can be written as

$$\text{MSE}(\hat{\theta}, \theta) = V[\hat{\theta}] + (B(\hat{\theta}, \theta))^2.$$

Proof The mean square error can be broken into the variability and bias components by squaring the quantity $(\hat{\theta} - \theta)^2$ and simplifying:

$$\begin{aligned}
\text{MSE}(\hat{\theta}, \theta) &= E\left[(\hat{\theta} - \theta)^2\right] \\
&= E\left[\hat{\theta}^2 - 2\theta\hat{\theta} + \theta^2\right] \\
&= E\left[\hat{\theta}^2\right] - E\left[2\theta\hat{\theta}\right] + E\left[\theta^2\right] \\
&= E\left[\hat{\theta}^2\right] - 2\theta E\left[\hat{\theta}\right] + \theta^2 \\
&= E\left[\hat{\theta}^2\right] - E\left[\hat{\theta}\right]^2 + E\left[\hat{\theta}\right]^2 - 2\theta E\left[\hat{\theta}\right] + \theta^2 \\
&= V\left[\hat{\theta}\right] + \left(E\left[\hat{\theta}\right] - \theta\right)^2,
\end{aligned}$$

where the first term is the variability of the point estimator and the second term is the squared bias. □

Theorem 2.3 has established that an unbiased estimator has a mean square error equal to its population variance. We can expand Table 2.5 from the previous example, which contained population means and variances of the six point estimators, adding columns for the bias and mean square error. Table 2.7 reveals that our point estimator of choice, $(n+1)X_{(n)}/n$, also has the smallest mean square error of the six point estimates for sample sizes $n > 1$, unless, of course, θ just so happens to be very close to 17.

The mean square error is an expected value that is a function of the bias and variability of a point estimator. Bias and variability are two quantities that are extensions of the population mean and the population variance from probability theory, but now are applied to point estimators. Statisticians

Point estimate $\hat{\theta}$	$E[\hat{\theta}]$	$V[\hat{\theta}]$	$B(\hat{\theta}, \theta)$	$\text{MSE}(\hat{\theta}, \theta)$
$2\bar{X}$	θ	$\frac{\theta^2}{3n}$	0	$\frac{\theta^2}{3n}$
$3\bar{X}$	$\frac{3\theta}{2}$	$\frac{3\theta^2}{4n}$	$\frac{\theta}{2}$	$\frac{(3+n)\theta^2}{4n}$
$X_{(n)}$	$\frac{n\theta}{n+1}$	$\frac{n\theta^2}{(n+2)(n+1)^2}$	$-\frac{\theta}{n+1}$	$\frac{2\theta^2}{(n+1)(n+2)}$
$\frac{(n+1)X_{(n)}}{n}$	θ	$\frac{\theta^2}{n(n+2)}$	0	$\frac{\theta^2}{n(n+2)}$
$(n+1)X_{(1)}$	θ	$\frac{n\theta^2}{n+2}$	0	$\frac{n\theta^2}{n+2}$
17	17	0	$17-\theta$	$(17-\theta)^2$

Table 2.7: Population means, variances, biases, and MSEs of the six point estimators of θ.

often use the words "accuracy" and "precision" to describe the bias and variability of a point estimator. Figure 2.20 illustrates the notion of accuracy and precision for a point estimator $\hat{\theta}$ for a single unknown parameter θ with a 2×2 display of axes. Five data sets produce the five point estimators that are displayed as points along each of the four axes. The upper-left-hand axis is the ideal case, in which the five point estimators are tightly clustered about the true value of θ. As is apparent from the other three axes, an accurate point estimator is on target, in other words, is unbiased. A precise point estimator, on the other hand, has low variability.

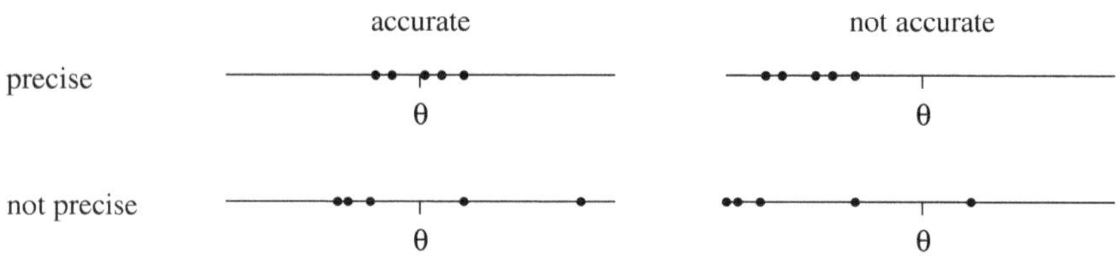

Figure 2.20: Accuracy vs. precision for a single parameter θ.

Figure 2.21 illustrates the notion of accuracy and precision using a 2×2 array of targets for two unknown parameters θ_1 and θ_2. The center of the bull's-eye of each target is at the coordinates of the true values of θ_1 and θ_2. In this case, there were again five samples of fixed sample size n, and the points plotted are the five point estimators $(\hat{\theta}_1, \hat{\theta}_2)$ for the two parameters.

The third issue raised by the previous example concerns a lingering doubt about our choice of the best point estimator. We have selected, with considerable effort, the best of the six candidate point estimators considered, and that estimator seems to have dominated the other five. But how do we know that there isn't yet another point estimator that dominates the best one we selected? That issue is tackled in the next subsection, which concerns a property of point estimators known as efficiency.

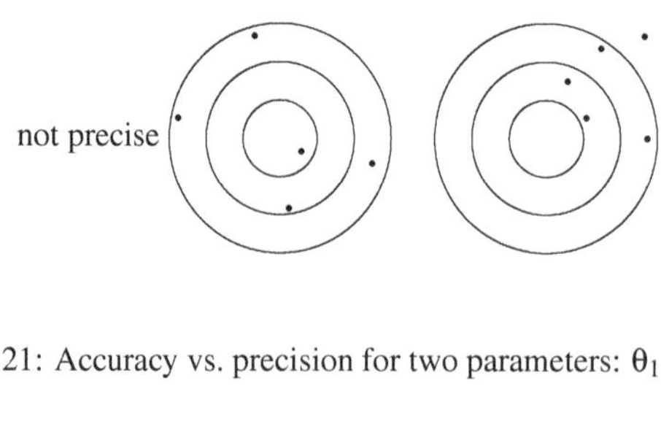

Figure 2.21: Accuracy vs. precision for two parameters: θ_1 and θ_2.

Efficient estimators

In our search for the best possible point estimator of an unknown parameter, we have seen that obtaining an unbiased estimator is a positive first step. But there are oftentimes several unbiased estimators from which to choose. It makes sense to choose the unbiased estimator with the smallest population variance of its sampling distribution. Once we have selected this point estimator, can we guarantee that there is no other point estimator that might provide a still lower population variance? The surprising answer to this question is yes, due to a remarkable result known as the *Cramér–Rao inequality*.

Before stating the Cramér–Rao inequality, it is important to first introduce the concept of the *information* contained in a single random variable X. This is often called the Fisher information, named after Sir Ronald Fisher, who was one of the pioneers in the development of modern statistics.

Definition 2.6 For a random variable X whose probability distribution is a parametric distribution with a single unknown parameter θ, the *Fisher information* associated with X in estimating θ is

$$I(\theta) = E\left[\left(\frac{\partial \ln f(X)}{\partial \theta}\right)^2\right]$$

under certain regularity conditions when the expectation exists.

This is the first case in which "regularity conditions" are a part of a definition or theorem. These conditions are critical in proving certain results in a more advanced course in mathematical statistics. For now, it is adequate to know that most of the common probability distributions satisfy the regularity conditions.

This expected value is almost certainly not the first one that would come to mind when quantifying the amount of information that a single instance of a random variable X contains in estimating θ. This is perhaps the first time that you have seen the probability density function or probability mass function $f(x)$ with a random variable as an argument. The following example illustrates how to interpret the Fisher information.

Example 2.18 Let X be a single observation from a Bernoulli(p) population. Calculate and interpret the Fisher information associated with X.

The probability mass function for the Bernoulli(p) distribution is

$$f(x) = p^x(1-p)^{1-x} \qquad x = 0, 1.$$

The logarithm of the probability mass function is

$$\ln f(x) = x \ln p + (1-x)\ln(1-p) \qquad x = 0, 1.$$

The derivative of the logarithm of the probability mass function with respect to p is

$$\frac{\partial \ln f(x)}{\partial p} = \frac{x}{p} - \frac{1-x}{1-p} \qquad x = 0, 1.$$

Recalling that the population mean and variance of a Bernoulli(p) random variable X

are $E[X] = p$ and $V[X] = p(1-p)$, the Fisher information associated with X is

$$\begin{aligned} I(p) &= E\left[\left(\frac{\partial \ln f(X)}{\partial p}\right)^2\right] \\ &= E\left[\left(\frac{X}{p} - \frac{1-X}{1-p}\right)^2\right] \\ &= E\left[\left(\frac{X-p}{p(1-p)}\right)^2\right] \\ &= \frac{E\left[(X-p)^2\right]}{p^2(1-p)^2} \\ &= \frac{1}{p(1-p)}. \end{aligned}$$

A plot of the Fisher information $I(p)$ as a function of p is given in Figure 2.22. The Fisher information is lowest at $p = 1/2$ and higher at the more extreme values of p. The interpretation is a bit more subtle. Let's say a Martian is visiting earth and asks a human a single question with a binary response. If the question is "Are you male?" and the response from the woman being surveyed is "no," then the Fisher information in the $X = 0$ response is low because $p \cong 0.5$ for that particular survey question. If instead the Martian's single question is "Are you an Olympic gold medal winner?" and again the response is "no," then in a sense more information has been conveyed in the $X = 0$ response because p is nearly zero. The variability of the Bernoulli(p) distribution when $p \cong 0$ is much smaller than when $p \cong 0.5$, so $X = 0$ conveys more information when $p \cong 0$.

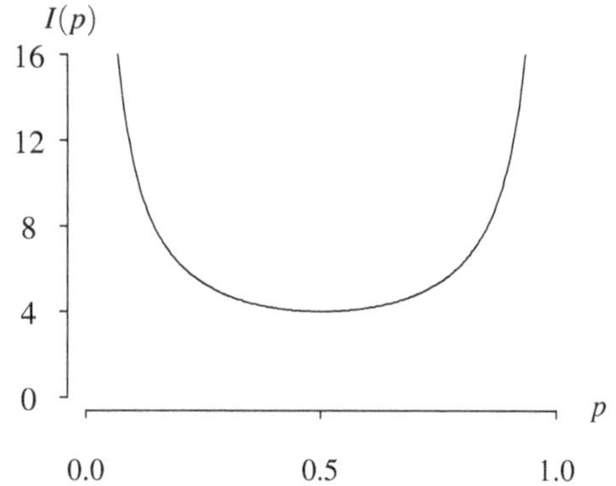

Figure 2.22: Fisher information for $X \sim$ Bernoulli(p).

The alert reader will have noticed that the Fisher information is the reciprocal of the population variance in the previous example. This is the case for many parametric distributions. The interpretation is as follows: if the population variance of X is high, then the Fisher information associated with X is low because X is not capable of accurately pinning down θ due to the high variability. Conversely, if the population variance of X is low, then the Fisher information associated with X is high because X is capable of accurately pinning down θ.

The Fisher information can be linked to the likelihood function. The expression

$$\frac{\partial \ln f(X)}{\partial \theta}$$

was encountered in the previous section on maximum likelihood estimation. It is the score in the case of a single ($n = 1$) observation. This insight leads to the following result.

Theorem 2.4 For a random variable X whose probability distribution is a parametric distribution with a single unknown parameter θ, the Fisher information $I(\theta)$ is the population variance of the score under certain regularity conditions.

Proof Consider only the case when X is a continuous random variable. The proof for a discrete random variable X is similar. For any continuous random variable X, the probability density function $f(x)$ must satisfy the normalizing equation

$$\int_{-\infty}^{\infty} f(x)\,dx = 1.$$

The partial derivative with respect to θ of both sides of the normalizing equation is

$$\frac{\partial}{\partial \theta}\int_{-\infty}^{\infty} f(x)\,dx = \int_{-\infty}^{\infty} \frac{\partial}{\partial \theta} f(x)\,dx = \int_{-\infty}^{\infty} \frac{\partial \ln f(x)}{\partial \theta} f(x)\,dx = E\left[\frac{\partial \ln f(X)}{\partial \theta}\right] = 0.$$

Using the shortcut formula for the population variance, the Fisher information is

$$I(\theta) = E\left[\left(\frac{\partial \ln f(X)}{\partial \theta}\right)^2\right] = V\left[\frac{\partial \ln f(X)}{\partial \theta}\right],$$

which proves the result. □

This proof of the result helps identify some of the regularity conditions for this result and the definition of the Fisher information: (a) the probability density function $f(x)$ must allow differentiation and integration to be interchanged as in the proof, (b) the expectations in the proof exist, and (c) the random variable X must have support that does not depend on θ. There is a second technique for computing the Fisher information when the second partial derivative of $\ln f(x)$ exists. This second technique can save considerable time when used to calculate the Fisher information for certain distributions.

Theorem 2.5 For a random variable X whose probability distribution is a parametric distribution with a single unknown parameter θ, the Fisher information $I(\theta)$ can be calculated by

$$I(\theta) = E\left[\frac{-\partial^2 \ln f(X)}{\partial \theta^2}\right]$$

under certain regularity conditions when the expectations exist.

Proof Consider only the case when X is a continuous random variable. The proof for a discrete random variable X is similar. For any continuous random variable X, the probability density function $f(x)$ must satisfy the normalizing equation

$$\int_{-\infty}^{\infty} f(x)\,dx = 1.$$

The second partial derivative with respect to θ of the left-hand side of the normalizing equation is

$$\begin{aligned}
\frac{\partial^2}{\partial\theta^2}\int_{-\infty}^{\infty} f(x)\,dx &= \int_{-\infty}^{\infty} \frac{\partial^2}{\partial\theta^2} f(x)\,dx \\
&= \int_{-\infty}^{\infty} \frac{\partial}{\partial\theta}\left[\frac{\partial \ln f(x)}{\partial\theta} f(x)\right] dx \\
&= \int_{-\infty}^{\infty} \left[\frac{\partial \ln f(x)}{\partial\theta}\cdot\frac{\partial f(x)}{\partial\theta} + \frac{\partial^2 \ln f(x)}{\partial\theta^2} f(x)\right] dx \\
&= \int_{-\infty}^{\infty} \left[\left(\frac{\partial \ln f(x)}{\partial\theta}\right)^2 f(x) + \frac{\partial^2 \ln f(x)}{\partial\theta^2} f(x)\right] dx
\end{aligned}$$

by the product rule for differentiation. Since the second partial derivative must be zero,

$$I(\theta) = E\left[\left(\frac{\partial \ln f(X)}{\partial\theta}\right)^2\right] = E\left[\frac{-\partial^2 \ln f(X)}{\partial\theta^2}\right],$$

which proves the result. □

The next example uses Theorem 2.5 to calculate the Fisher information for a Bernoulli(p) random variable using the alternative formula.

Example 2.19 Let X be a single observation from a Bernoulli(p) population. Calculate and interpret the Fisher information associated with X using Theorem 2.5.

Picking up the narrative from Example 2.18, the negative of the second partial derivative of the logarithm of the probability mass function is

$$\frac{-\partial^2 \ln f(x)}{\partial p^2} = \frac{x}{p^2} + \frac{1-x}{(1-p)^2} \qquad x = 0, 1.$$

Recalling that the population mean of $X \sim \text{Bernoulli}(p)$ is $E[X] = p$ and applying Theorem 2.5, the Fisher information is

$$E\left[\frac{-\partial^2 \ln f(X)}{\partial p^2}\right] = E\left[\frac{X}{p^2} + \frac{1-X}{(1-p)^2}\right] = \frac{p}{p^2} + \frac{1-p}{(1-p)^2} = \frac{1}{p} + \frac{1}{1-p} = \frac{1}{p(1-p)},$$

which matches the Fisher information calculated using the defining formula in Example 2.18.

In a subsequent chapter, the Fisher information will be generalized to more than just a single observation and more than just a single parameter in what is known as the *Fisher information matrix*. The work concerning the Fisher information now allows us to state the Cramér–Rao inequality.

Theorem 2.6 (Cramér–Rao inequality) Let $X_1, X_2, \ldots, X_n$ be a random sample drawn from a population probability distribution described by $f(x)$ whose support does not depend on any unknown parameters. Let θ be an unknown parameter associated with the population. Assume that the first-order partial derivative of $f(x)$ with respect to θ exists and is continuous. Let $\hat{\theta}$ be an unbiased estimator of θ. A lower bound on the population variance of $\hat{\theta}$ is given by

$$V\left[\hat{\theta}\right] \ge \frac{1}{nI(\theta)}$$

under certain regularity conditions when the expectation exists.

Proof Assume that the population distribution is continuous for this proof. The proof for a discrete population distribution is similar. Let $L(\theta)$ be the likelihood function and $U(\theta)$ be the score. The likelihood function satisfies the normalizing equation

$$\int_{-\infty}^{\infty}\int_{-\infty}^{\infty}\cdots\int_{-\infty}^{\infty} L(\theta)dx_1\,dx_2\ldots dx_n = 1$$

because $L(\theta) = \prod_{i=1}^{n} f(x_i)$. The partial derivative of this equation with respect to θ is

$$\begin{aligned}
\frac{\partial}{\partial\theta}\int_{-\infty}^{\infty}\int_{-\infty}^{\infty}\cdots\int_{-\infty}^{\infty} L(\theta)\,dx_1\,dx_2\ldots dx_n &= \int_{-\infty}^{\infty}\int_{-\infty}^{\infty}\cdots\int_{-\infty}^{\infty}\frac{\partial}{\partial\theta}L(\theta)\,dx_1\,dx_2\ldots dx_n\\
&= \int_{-\infty}^{\infty}\int_{-\infty}^{\infty}\cdots\int_{-\infty}^{\infty}\frac{\partial\ln L(\theta)}{\partial\theta}\cdot L(\theta)\,dx_1\,dx_2\ldots dx_n\\
&= E\left[\frac{\partial\ln L(\theta)}{\partial\theta}\right]\\
&= E\left[U(\theta)\right]\\
&= 0.
\end{aligned}$$

So $E\left[U(\theta)\right] = 0$. Now compute the population covariance between the random variables $\hat{\theta}$ and $U(\theta)$ using the shortcut formula for computing covariance:

$$\begin{aligned}
\text{Cov}\left(\hat{\theta}, U(\theta)\right) &= E\left[\hat{\theta}\cdot U(\theta)\right] - E\left[\hat{\theta}\right]E\left[U(\theta)\right]\\
&= E\left[\hat{\theta}\cdot U(\theta)\right]\\
&= \int_{-\infty}^{\infty}\int_{-\infty}^{\infty}\cdots\int_{-\infty}^{\infty}\hat{\theta}\cdot U(\theta)\cdot L(\theta)dx_1\,dx_2\ldots dx_n\\
&= \int_{-\infty}^{\infty}\int_{-\infty}^{\infty}\cdots\int_{-\infty}^{\infty}\hat{\theta}\left[\frac{\partial}{\partial\theta}\ln\left(\prod_{i=1}^{n} f(x_i)\right)\right]\left[\prod_{i=1}^{n} f(x_i)\right]dx_1\,dx_2\ldots dx_n\\
&= \int_{-\infty}^{\infty}\int_{-\infty}^{\infty}\cdots\int_{-\infty}^{\infty}\hat{\theta}\left[\frac{\partial}{\partial\theta}\prod_{i=1}^{n} f(x_i)\right]dx_1\,dx_2\ldots dx_n\\
&= \frac{\partial}{\partial\theta}\int_{-\infty}^{\infty}\int_{-\infty}^{\infty}\cdots\int_{-\infty}^{\infty}\hat{\theta}\left[\prod_{i=1}^{n} f(x_i)\right]dx_1\,dx_2\ldots dx_n\\
&= \frac{\partial}{\partial\theta}E\left[\hat{\theta}\right]\\
&= \frac{\partial}{\partial\theta}\theta\\
&= 1
\end{aligned}$$

because $\hat{\theta}$ is an unbiased estimator of θ by assumption. Since the correlation coefficient ρ must satisfy $|\rho| \leq 1$,

$$\left| \frac{\text{Cov}\left(\hat{\theta}, U(\theta)\right)}{V\left[\hat{\theta}\right] V\left[U(\theta)\right]} \right| \leq 1$$

or

$$\left| \frac{1}{V\left[\hat{\theta}\right] nI(\theta)} \right| \leq 1$$

or

$$V\left[\hat{\theta}\right] nI(\theta) \geq 1,$$

which proves the result. □

The Cramér–Rao inequality gives a smallest possible population variance for an unbiased estimator. When $V\left[\hat{\theta}\right]$ is equal to the right-hand side of the inequality, the unbiased estimator is known as an *efficient estimator*, which is defined next.

Definition 2.7 An unbiased estimator $\hat{\theta}$ of an unknown parameter θ is *efficient* if its population variance achieves the Cramér–Rao lower bound.

No other unbiased estimator can have a smaller population variance than an efficient estimator. Other terms used to describe an efficient estimator are *best unbiased estimator* and *minimum variance unbiased estimator* (MVUE). The existence of efficient point estimators updates our Venn diagram for point estimators to include this exclusive classification within unbiased estimators as shown in Figure 2.23. Efficient estimators are a subset of unbiased estimators that represent the very best of these estimators in terms of their variability.

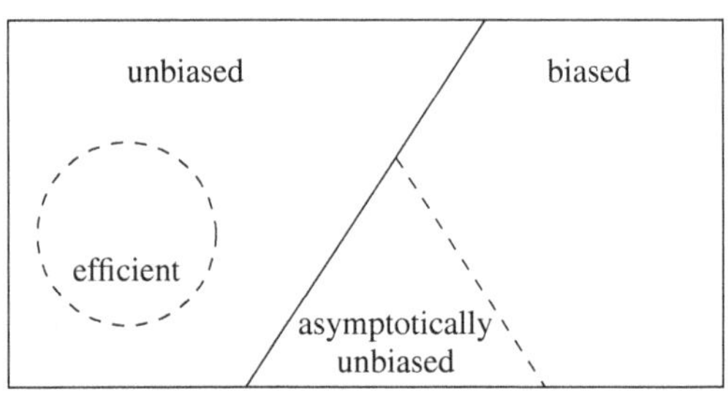

Figure 2.23: Venn diagram classifying point estimates.

As with many results in statistics, the assumptions deserve as much attention as the result itself. One important assumption is that the support of the unknown population does not depend on any unknown parameters. This implies that the Cramér–Rao inequality, for example,

- can be applied to a $N(\theta, 1)$ population because the support is $\mathcal{A} = \{x \mid -\infty < x < \infty\}$,
- can be applied to an exponential(θ) population because the support is $\mathcal{A} = \{x \mid x > 0\}$,
- cannot be applied to a $U(0, \theta)$ population because the support is $\mathcal{A} = \{x \mid 0 < x < \theta\}$.

The first example of the use of the Cramér–Rao inequality shows that the sample mean is an efficient estimator of the population mean when a random sample of size n is drawn from an exponential population.

Example 2.20 Show that $\hat{\theta} = \frac{1}{n}\sum_{i=1}^{n} X_i$ is an efficient estimator of θ for a random sample $X_1, X_2, \ldots, X_n$ drawn from an exponential population with population mean θ.

Recall that the sample mean is both the method of moments estimator from Example 2.3 and the maximum likelihood estimator from Example 2.8. It was shown in Example 2.8 that $\hat{\theta}$ is an unbiased estimator of θ and the support of the population distribution $\mathcal{A} = \{x \,|\, x > 0\}$ does not depend on the unknown parameter θ, so the Cramér–Rao inequality can be applied. Since the population probability density function is

$$f(x) = \frac{1}{\theta} e^{-x/\theta} \qquad\qquad x > 0$$

and the natural logarithm of the population probability density function is

$$\ln f(x) = -\ln\theta - \frac{x}{\theta} \qquad\qquad x > 0,$$

the partial derivative of the natural logarithm of the population probability density function with respect to θ is

$$\frac{\partial \ln f(x)}{\partial\theta} = -\frac{1}{\theta} + \frac{x}{\theta^2} \qquad\qquad x > 0.$$

Since $E[X] = \theta$ and $E\left[X^2\right] = V[X] + \left(E[X]\right)^2 = \theta^2 + \theta^2 = 2\theta^2$ for an exponential random variable with population mean θ, the Fisher information is

$$\begin{aligned} I(\theta) &= E\left[\left(\frac{\partial \ln f(X)}{\partial\theta}\right)^2\right] \\ &= E\left[\left(-\frac{1}{\theta} + \frac{X}{\theta^2}\right)^2\right] \\ &= E\left[\frac{1}{\theta^2} - \frac{2X}{\theta^3} + \frac{X^2}{\theta^4}\right] \\ &= \frac{1}{\theta^2} - \frac{2}{\theta^2} + \frac{2}{\theta^2} \\ &= \frac{1}{\theta^2} \end{aligned}$$

via Definition 2.6. So the right-hand side of the Cramér–Rao inequality is

$$\frac{1}{nI(\theta)} = \frac{1}{n/\theta^2} = \frac{\theta^2}{n},$$

which is the population variance of $\hat{\theta}$ because

$$V\big[\hat{\theta}\big] = V\left[\frac{1}{n}\sum_{i=1}^{n} X_i\right] = \frac{1}{n^2} V\left[\sum_{i=1}^{n} X_i\right] = \frac{1}{n^2}\sum_{i=1}^{n} V\left[X_i\right] = \frac{1}{n^2}\cdot n\theta^2 = \frac{\theta^2}{n}.$$

Therefore, the maximum likelihood estimator $\hat{\theta} = \frac{1}{n}\sum_{i=1}^{n} X_i$ is an efficient estimator of θ. For a random sample drawn from an exponential population, the sample mean falls in the efficient subset in Figure 2.23. The Cramér–Rao inequality tells us that there is no unbiased estimator of θ with a smaller population variance than $\hat{\theta}$.

The next example will show that the Cramér–Rao inequality is implemented in the same fashion for a random sample drawn from a discrete population as it was for a continuous population.

Example 2.21 Show that the maximum likelihood estimator of λ for a random sample $X_1, X_2, \ldots, X_n$ drawn from a Poisson(λ) population is efficient.

Random sampling from a Poisson(λ) population was encountered in Example 2.9, where it was determined that the maximum likelihood estimator

$$\hat{\lambda} = \frac{1}{n}\sum_{i=1}^{n} X_i$$

was an unbiased estimator of λ. In addition, the support of the population distribution $\mathcal{A} = \{x \,|\, x = 0, 1, 2, \ldots\}$ does not depend on the unknown parameter λ, so the Cramér–Rao inequality can be applied. Since the population probability mass function is

$$f(x) = \frac{\lambda^x e^{-\lambda}}{x!} \qquad x = 0, 1, 2, \ldots,$$

and the natural logarithm of the population probability mass function is

$$\ln f(x) = x \ln \lambda - \lambda - \ln x! \qquad x = 0, 1, 2, \ldots,$$

the partial derivative of the natural logarithm of the population probability mass function with respect to λ is

$$\frac{\partial \ln f(x)}{\partial \lambda} = \frac{x}{\lambda} - 1 \qquad x = 0, 1, 2, \ldots.$$

The negative of the second partial derivative of the natural logarithm of the population probability mass function with respect to λ is

$$\frac{-\partial^2 \ln f(x)}{\partial \lambda^2} = \frac{x}{\lambda^2} \qquad x = 0, 1, 2, \ldots.$$

Since $E[X] = \lambda$ for a Poisson random variable, the Fisher information is

$$I(\lambda) = E\left[\frac{-\partial^2 \ln f(X)}{\partial \lambda^2}\right] = E\left[\frac{X}{\lambda^2}\right] = \frac{1}{\lambda}$$

via Theorem 2.5. So the right-hand side of the Cramér–Rao inequality is

$$\frac{1}{nI(\lambda)} = \frac{1}{n/\lambda} = \frac{\lambda}{n},$$

which is the population variance of $\hat{\lambda}$ because

$$V[\hat{\lambda}] = \frac{V[X]}{n} = \frac{\lambda}{n}$$

because $V[X] = \lambda$. Therefore, the maximum likelihood estimator $\hat{\lambda} = \frac{1}{n}\sum_{i=1}^{n} X_i$ is an efficient estimator of λ. There is no unbiased estimator with a smaller population variance than $\hat{\lambda}$.

If there are two unbiased estimators of a parameter θ, a comparison between the two can be made with respect to the population variance of the estimators using the relative efficiency, which is defined next. The intuition associated with one point estimator being more efficient than another point estimator is as follows: the more efficient point estimator uses the data more efficiently in the sense that it results in a smaller population variance of the point estimator.

Definition 2.8 If $\hat{\theta}_1$ and $\hat{\theta}_2$ are two unbiased point estimators of the parameter θ, then

$$\frac{V[\hat{\theta}_2]}{V[\hat{\theta}_1]}$$

is the *relative efficiency* of $\hat{\theta}_1$ to $\hat{\theta}_2$.

If the relative efficiency is less than 1, then $\hat{\theta}_2$ is preferred; if the relative efficiency is greater than 1, then $\hat{\theta}_1$ is preferred. The calculation of the relative efficiency is illustrated for a random sample drawn from a generic population distribution in the next example.

Example 2.22 Let X_1, X_2, and X_3 be a random sample from a population with unknown finite population mean θ and known finite population variance $\sigma^2 > 0$. Consider the following point estimators of θ:

- the sample mean $\hat{\theta}_1 = \dfrac{X_1+X_2+X_3}{3}$,
- the weighted average $\hat{\theta}_2 = \dfrac{X_1+4X_2+X_3}{6}$.

Calculate the relative efficiency of $\hat{\theta}_1$ to $\hat{\theta}_2$.

The first point estimator $\hat{\theta}_1$ gives equal weight $(1/3)$ to each of the three observations. The second point estimator $\hat{\theta}_2$ gives more weight $(4/6 = 2/3)$ to the second observation and less weight $(1/6)$ to the first and third observations. The question is asking which of the point estimators is better in terms of the population variance of the estimator. First, show that both estimators are unbiased:

$$E[\hat{\theta}_1] = E\left[\frac{X_1+X_2+X_3}{3}\right] = \frac{1}{3}\left(E[X_1]+E[X_2]+E[X_3]\right) = \theta$$

and

$$E[\hat{\theta}_2] = E\left[\frac{X_1+4X_2+X_3}{6}\right] = \frac{1}{6}E[X_1]+\frac{2}{3}E[X_2]+\frac{1}{6}E[X_3] = \theta.$$

Next, the population variances of the two point estimators are

$$V[\hat{\theta}_1] = V\left[\frac{X_1+X_2+X_3}{3}\right] = \frac{1}{9}\left(V[X_1]+V[X_2]+V[X_3]\right) = \frac{\sigma^2}{3}$$

and

$$V[\hat{\theta}_2] = V\left[\frac{X_1+4X_2+X_3}{6}\right] = \frac{1}{36}\left(V[X_1]+16V[X_2]+V[X_3]\right) = \frac{\sigma^2}{2}.$$

Therefore, the relative efficiency is

$$\frac{V[\hat{\theta}_2]}{V[\hat{\theta}_1]} = \frac{3}{2}.$$

The sample mean $\hat{\theta}_1$ is the preferred unbiased estimator because it has a population variance that is two-thirds the population variance of $\hat{\theta}_2$. The increased weight placed on X_2 by $\hat{\theta}_2$ increases $V\left[\hat{\theta}_2\right]$. The intuition behind this conclusion can be seen by taking the argument to the extreme case. Let's say *all* of the weight (that is, weight 1) is allocated to x_2, leaving none (that is, weight 0) for x_1 and x_3. In this case, we have ignored the data values x_1 and x_3 and therefore effectively reduced the sample size from $n = 3$ to $n = 1$. It is no surprise that this is an inferior estimator. The take-away from this example is that the best approach for the point estimator of the population mean is to place equal weight on each of the three observations in the random sample.

We now revisit two of the point estimators from Example 2.17 associated with sampling n observations from a $U(0, \theta)$ population.

Example 2.23 Let $X_1, X_2, \ldots, X_n$ be a random sample from a $U(0, \theta)$ population. Find the relative efficiency of the following two estimators of θ:

$$\hat{\theta}_1 = 2\bar{X} \qquad \text{and} \qquad \hat{\theta}_2 = \frac{(n+1)X_{(n)}}{n}.$$

Recall from Example 2.2 that $\hat{\theta}_1$ is the method of moments estimator. Recall from Example 2.10 that $\hat{\theta}_2$ is the maximum likelihood estimator modified by an unbiasing constant. In Example 2.17, we concluded that $\hat{\theta}_2$ was the superior point estimate. That should also be reflected in the analysis with relative efficiency here. From Table 2.7, both point estimators are unbiased with population variances

$$V\left[\hat{\theta}_1\right] = \frac{\theta^2}{3n} \qquad \text{and} \qquad V\left[\hat{\theta}_2\right] = \frac{\theta^2}{n(n+2)}.$$

Therefore, the relative efficiency is

$$\frac{V\left[\hat{\theta}_2\right]}{V\left[\hat{\theta}_1\right]} = \frac{3}{n+2}.$$

Since this quantity is less than 1 for sample sizes $n > 1$, $\hat{\theta}_2$ is the preferred estimator. This is consistent with our previous conclusion.

As a third example illustrating relative efficiency, consider the choice between using the sample median and the sample mean for a population probability distribution having a symmetric probability density function.

Example 2.24 Let $X_1, X_2, \ldots, X_n$ be a random sample from a population having probability density function

$$f(x) = \frac{1}{2}e^{-|x-\theta|} \qquad -\infty < x < \infty,$$

where θ is an unknown parameter. Find the relative efficiency of the following two estimators of θ:

$$\hat{\theta}_1 = \bar{X} \qquad \text{and} \qquad \hat{\theta}_2 = M,$$

where M is the sample median. Assume that the sample size n is odd in order to ease the calculations related to the sample median.

The population probability distribution, which is a special case of the error, Laplace, or double exponential distribution, has a symmetric probability density function centered on θ, so the population mean and the population median are both equal to θ. Furthermore, the population variance is $V[X] = 2$, regardless of the value of θ. Using geometric arguments involving the symmetry of their sampling distributions, both $\hat{\theta}_1$ and $\hat{\theta}_2$ are unbiased estimates of θ. The population variance of the first point estimator is

$$V\left[\hat{\theta}_1\right] = V\left[\bar{X}\right] = \frac{V[X]}{n} = \frac{2}{n}$$

via Theorem 1.1. Computing the population variance of the second point estimator is more difficult. We leave the computations to APPL using the following code.

```
X := ErrorRV(1, 1, theta);
for n from 1 to 99 by 2 do
  Variance(OrderStat(X, n, (n + 1) / 2));
od;
```

Using the population variances calculated in the loop for odd values of n between 1 and 99, the relative efficiency is calculated and plotted in Figure 2.24. This figure reveals that the sample median $\hat{\theta}_2 = M$ has a smaller population variance than the sample mean $\hat{\theta}_1 = \bar{X}$ for $n > 1$. (The sample median and sample mean are identical for $n = 1$.) In addition, the relative efficiency is approaching a limit, known as the *asymptotic relative efficiency* (often abbreviated ARE), of

$$\lim_{n\to\infty} \frac{V\left[\hat{\theta}_2\right]}{V\left[\hat{\theta}_1\right]} = \lim_{n\to\infty} \frac{V[M]}{V\left[\bar{X}\right]} = 0.5,$$

which appears as a horizontal line in Figure 2.24. The sample median is asymptotically twice as efficient as the sample mean for estimating θ.

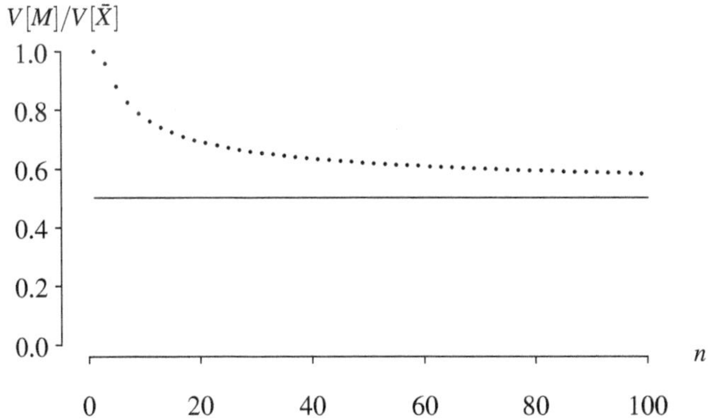

Figure 2.24: Relative efficiency as a function of the sample size n.

The previous example should not lead one to conclude that the sample median is always superior to the sample mean. The conclusion will vary depending upon the population distribution. For example, if $X_1, X_2, \ldots, X_n$ is a random sample from a $N(\theta, 1)$ population, then the asymptotic relative efficiency of the sample mean and sample median as estimates of θ can be found based on the following (see the preface for a pointer to the proofs) results:

- $\bar{X} \xrightarrow{D} N\left(\theta, \frac{1}{n}\right)$,
- $M \xrightarrow{D} N\left(\theta, \frac{\pi}{2n}\right)$.

The asymptotic relative efficiency is

$$\lim_{n\to\infty} \frac{V\left[\hat{\theta}_2\right]}{V\left[\hat{\theta}_1\right]} = \lim_{n\to\infty} \frac{V[M]}{V\left[\bar{X}\right]} = \lim_{n\to\infty} \frac{\pi/(2n)}{1/n} = \frac{\pi}{2} \cong 1.5708,$$

which implies that the sample mean is preferred to the sample median for estimating θ for a large random sample drawn from a normal population. The take-away here is that the relative efficiency of two unbiased estimators must be worked out for each population distribution.

In summary, point estimators that are efficient are by definition unbiased and have the smallest possible population variance, as determined by the Cramér–Rao inequality. We now consider a third property concerning point estimates: sufficiency.

Sufficient statistics

The two previous subsections have classified point estimators as unbiased (that is, on target) and efficient (that is, unbiased with the smallest possible population variance). This subsection introduces the notion of a sufficient statistic $Y = g(X_1, X_2, \ldots, X_n)$, which is a statistic that captures *all* of the information in a random sample $X_1, X_2, \ldots, X_n$ with respect to estimating θ.

To illustrate, let's say you want to estimate p, the probability that I make a free throw on a single shot. (Full disclosure: p is low.) In order to estimate p, you record the outcomes of $n = 3$ free throw shots X_1, X_2, and X_3, where $X_i = 0$ corresponds to a failure and $X_i = 1$ corresponds to a success, for $i = 1, 2, 3$. Further, you can assume that the three outcomes are mutually independent Bernoulli(p) random variables, so they constitute a random sample. Now consider the following four statistics as candidates for sufficient statistics to be used in estimating p:

- $g_1(X_1, X_2, X_3) = X_3$,
- $g_2(X_1, X_2, X_3) = X_{(3)}$,
- $g_3(X_1, X_2, X_3) = X_1X_2X_3$,
- $g_4(X_1, X_2, X_3) = X_1 + X_2 + X_3$.

These statistics are not point estimators of p. We are only interested in whether they capture all of the information in X_1, X_2, X_3 concerning p. If the three free throw shot outcomes are $x_1 = 0$, $x_2 = 0$ and $x_3 = 1$, for instance, then the statistics assume the values

- $g_1(0, 0, 1) = 1$,
- $g_2(0, 0, 1) = 1$,
- $g_3(0, 0, 1) = 0$,

- $g_4(0, 0, 1) = 1$.

Let's determine which of these are sufficient statistics for p using intuition alone. The first statistic, $g_1(X_1, X_2, X_3) = X_3$, is not a sufficient statistic for estimating p because the outcomes of the first two free throws are ignored. The first two data values contain pertinent information with respect to estimating p, which is lost. The second statistic, $g_2(X_1, X_2, X_3) = X_{(3)}$, is not a sufficient statistic for estimating p because the outcomes of all shots except for the outcome of the best shot are ignored, and these other shots might contain information that is useful in estimating p. The third statistic, $g_3(X_1, X_2, X_3) = X_1X_2X_3$, is not a sufficient statistic for estimating p because just a single missed shot sends this statistic to zero, wiping out the information associated with the other two shots. Finally, the fourth statistic, $g_4(X_1, X_2, X_3) = X_1 + X_2 + X_3$, is a sufficient statistic for estimating p. It is a count of the number of shots made. It encapsulates all of the information in the data set necessary to estimate p; no relevant information is lost. One could not reconstruct the data set from $g_4(X_1, X_2, X_3) = X_1 + X_2 + X_3$, but it turns out the trial number where I made my one free throw is not pertinent in estimating p.

Notice that the term "sufficient statistic" was used rather than "sufficient estimator." In the previous scenario, $g_4(X_1, X_2, X_3) = X_1 + X_2 + X_3$ is certainly not an estimator of p because p is a probability and $g_4(X_1, X_2, X_3)$ is an integer that can assume the values 0, 1, 2, or 3. In this setting, the sufficient statistic needs to be divided by $n = 3$ to result in an unbiased estimator of p, as was seen in Example 2.16.

So a sufficient statistic for a generic unknown parameter θ reduces a data set of n values to a single statistic in a manner that captures all of the information contained in the data values concerning θ. No information about θ leaks away in the transformation from the data to the sufficient statistic. One intuitive way to think about a sufficient statistic is that you would be neutral between having the raw data and a sufficient statistic with respect to estimating an unknown parameter θ. Once you know the value of the sufficient statistic for θ, any further information concerning the sample with respect to estimating θ is irrelevant. The sufficient statistic exhausts all of the information concerning θ from a data set. The formal definition is given next.

Definition 2.9 Let $X_1, X_2, \ldots, X_n$ be a random sample from a population with an unknown parameter θ. A statistic $Y = g(X_1, X_2, \ldots, X_n)$ is a *sufficient statistic* for θ if the conditional distribution of $X_1, X_2, \ldots, X_n$ given $Y = g(X_1, X_2, \ldots, X_n) = y$ does not depend on θ for any value of y.

One important aspect of sufficient statistics to keep in mind is that they are always oriented around an unknown parameter θ. A statistic cannot be known as just a "sufficient statistic." It must be stated that it is a sufficient statistic for θ. Likewise, there is no such thing as an "unbiased estimator" or an "efficient estimator." These must be stated as $\hat{\theta}$ is an unbiased estimator of the population parameter θ or $\hat{\theta}$ is an efficient estimator of the population parameter θ.

Definition 2.9 concerns a conditional distribution of the data values given a particular value of the sufficient statistic Y. For a random sample drawn from a discrete population, for instance, this conditional distribution is computed as

$$P(X_1 = x_1, X_2 = x_2, \ldots, X_n = x_n \,|\, Y = y) = \frac{P(X_1 = x_1, X_2 = x_2, \ldots, X_n = x_n, Y = y)}{P(Y = y)},$$

for each y where $P(Y = y) \neq 0$. For the particular population distribution from which the random sample was drawn (for example, a Poisson(λ) or Bernoulli(p) population) and for a particular candidate sufficient statistic, this conditional distribution must be computed and shown to not depend on the unknown parameter. The next example uses the definition of a sufficient statistic to extend the free throw illustration from three data values to n data values.

Example 2.25 Let $X_1, X_2, \ldots, X_n$ be a random sample drawn from a Bernoulli(p) population, where p is an unknown parameter satisfying $0 < p < 1$. Is

$$Y = \sum_{i=1}^{n} X_i$$

a sufficient statistic for p?

The population probability mass function is

$$f_X(x) = p^x(1-p)^{1-x} \qquad x = 0, 1.$$

Since $X_1, X_2, \ldots, X_n$ are mutually independent and identically distributed Bernoulli(p) random variables, the probability mass function of $Y = \sum_{i=1}^{n} X_i$ is

$$f_Y(y) = \binom{n}{y} p^y(1-p)^{n-y} \qquad y = 0, 1, 2, \ldots, n,$$

which can be recognized as the binomial(n, p) distribution. The conditional distribution of $X_1, X_2, \ldots, X_n$ given $Y = g(X_1, X_2, \ldots, X_n) = y$ is the ratio of the probability of one particular sequence of y ones and $n - y$ zeros in n trials to the probability of exactly y ones in n trials without regard to order:

$$\frac{p^y(1-p)^{n-y}}{\binom{n}{y} p^y(1-p)^{n-y}} = \frac{1}{\binom{n}{y}},$$

which does not depend on p. So $Y = \sum_{i=1}^{n} X_i$ is a sufficient statistic for p. Again, this does not imply that $Y = \sum_{i=1}^{n} X_i$ is necessarily a good *point estimator* of p (in this case, it is not), it simply implies that all of the information concerning the estimation of p is contained in the sufficient statistic $Y = \sum_{i=1}^{n} X_i$.

The intuition behind this conditional probability distribution is that once the value of the sufficient statistic Y is known, the conditional distribution of the data does not involve the unknown parameter. Hence, knowing the individual data values adds no information over the sufficient statistic with respect to the estimation of the unknown parameter.

For a random sample $X_1, X_2, \ldots, X_n$ from a continuous population described by the probability density function $f_X(x)$ with an unknown parameter θ, the statistic $Y = g(X_1, X_2, \ldots, X_n)$ is a sufficient statistic for the unknown parameter θ if for each y where $f_Y(y) \neq 0$ and each n-tuple $(x_1, x_2, \ldots, x_n)$ with $g(x_1, x_2, \ldots, x_n) = y$, the conditional probability distribution described by

$$\frac{f_X(x_1) f_X(x_2) \ldots f_X(x_n)}{f_Y(y)}$$

does not depend on θ. Notice that the numerator of this expression is the likelihood function $L(\theta)$.

Example 2.26 Let $X_1, X_2, \ldots, X_n$ be a random sample drawn from a $U(0, \theta)$ population, where θ is a positive unknown parameter. Is

$$Y = X_{(n)} = \max\{X_1, X_2, \ldots, X_n\}$$

a sufficient statistic for θ?

The population probability density function is

$$f_X(x) = \frac{1}{\theta} \qquad 0 < x < \theta.$$

The population cumulative distribution function on the support of X is

$$F_X(x) = \int_0^x \frac{1}{\theta}\, dw = \frac{x}{\theta} \qquad 0 < x < \theta.$$

Using this probability density function, cumulative distribution function, and an order statistic result, the probability density function of $Y = X_{(n)} = \max\{X_1, X_2, \ldots, X_n\}$ is

$$f_Y(y) = n \cdot \frac{1}{\theta}\left(\frac{y}{\theta}\right)^{n-1} = \frac{ny^{n-1}}{\theta^n} \qquad 0 < y < \theta.$$

The conditional distribution of $X_1, X_2, \ldots, X_n$ given $Y = g(X_1, X_2, \ldots, X_n) = y$ is

$$\frac{f_X(x_1) f_X(x_2) \ldots f_X(x_n)}{f_Y(y)} = \frac{\frac{1}{\theta} \cdot \frac{1}{\theta} \cdot \ldots \cdot \frac{1}{\theta}}{\frac{ny^{n-1}}{\theta^n}} = \frac{1}{ny^{n-1}},$$

which does not depend on θ. So $Y = \max\{X_1, X_2, \ldots, X_n\}$ is a sufficient statistic for θ. In this example, the sufficient statistic also happens to be a point estimator. Recall from Example 2.10 that the maximum likelihood estimator of θ for random sampling from this population is $Y = \hat{\theta} = \max\{X_1, X_2, \ldots, X_n\}$, which is an asymptotically unbiased estimator of θ.

The previous two examples have considered sufficient statistics for random sampling from the Bernoulli(p) and $U(0, \theta)$ distributions. The next example considers a less mainstream population distribution.

Example 2.27 Let $X_1, X_2, \ldots, X_n$ be a random sample drawn from a continuous population described by the probability density function

$$f_X(x) = \theta x^{\theta - 1} \qquad 0 < x < 1,$$

where θ is a positive unknown parameter. Is

$$Y = \prod_{i=1}^{n} X_i$$

a sufficient statistic for θ?

The population probability density function is

$$f_X(x) = \theta x^{\theta - 1} \qquad 0 < x < 1.$$

Determining the distribution of $Y = \prod_{i=1}^{n} X_i$ is not as easy as it was in the previous two examples. The APPL statements given below compute the probability density function of Y for $n = 2$, $n = 3$, and $n = 4$.

```
assume(theta > 0);
X := [[x -> theta * x ^ (theta - 1)], [0, 1], ["Continuous", "PDF"]];
Y := ProductIID(X, 2);
Y := ProductIID(X, 3);
Y := ProductIID(X, 4);
```

Generalizing the pattern, the probability density function of $Y = \prod_{i=1}^{n} X_i$ is

$$f_Y(y) = \frac{(-\ln y)^{n-1}}{(n-1)!}\,\theta^n y^{\theta-1} \qquad 0 < y < 1.$$

The negative sign in front of the logarithm is due to the fact that the population support is the unit interval. The conditional distribution of $X_1, X_2, \ldots, X_n$ given that $Y = g(X_1, X_2, \ldots, X_n) = y$ is

$$\frac{\theta x_1^{\theta-1}\theta x_2^{\theta-1}\ldots\theta x_n^{\theta-1}}{(-\ln x_1x_2\ldots x_n)^{n-1}\theta^n(x_1x_2\ldots x_n)^{\theta-1}/(n-1)!} = \frac{(n-1)!}{(-\ln x_1x_2\ldots x_n)^{n-1}},$$

which does not depend on θ. So $Y = \prod_{i=1}^{n} X_i$ is a sufficient statistic for θ.

A sufficient statistic has now been defined and illustrated by three examples. The examples of determining whether a statistic is a sufficient statistic by using the definition of a sufficient statistic have revealed two major weaknesses. First, Definition 2.9 is often inconvenient to work with because it requires that we know the distribution of Y. The distribution of $Y = g(X_1, X_2, \ldots, X_n)$ might be quite difficult or even impossible to determine. Second, Definition 2.9 gives no guidance as to what statistic might be a sufficient statistic. You are left to guess at potential sufficient statistics, then test each statistic to see if it is indeed a sufficient statistic for θ.

Fortunately, alternative ways for identifying sufficient statistics have been developed that overcome these two weaknesses. These are generally known as "factorization criterion," and they are typically easier to work with than Definition 2.9 for most population probability distributions. Although several of these results exist, we will state, prove, and illustrate just one of them, known as the Fisher–Neyman factorization criterion, named after Sir Ronald Fisher and Jerzy Neyman.

Theorem 2.7 (Fisher–Neyman factorization criterion) Let $X_1, X_2, \ldots, X_n$ be a random sample from a population described by $f_X(x)$ with unknown parameter θ. The statistic $Y = g(X_1, X_2, \ldots, X_n)$ is a sufficient statistic for the parameter θ if and only if there exist nonnegative functions h_1 and h_2 such that

$$f_X(x_1)f_X(x_2)\ldots f_X(x_n) = h_1\left(g(x_1, x_2, \ldots, x_n), \theta\right) \cdot h_2(x_1, x_2, \ldots, x_n).$$

Proof An outline of the proof for random sampling from a discrete population follows. First assume that the statistic $Y = g(X_1, X_2, \ldots, X_n)$ is a sufficient statistic for the parameter θ. By Definition 2.9, the conditional distribution of $X_1, X_2, \ldots, X_n$ given $Y = y$ does not depend on θ. Conditioning on $Y = y$,

$$\begin{aligned} f_X(x_1)f_X(x_2)\ldots f_X(x_n) &= P(X_1 = x_1, X_2 = x_2, \ldots, X_n = x_n \,|\, Y = y) \cdot P(Y = y) \\ &= h_2(x_1, x_2, \ldots, x_n) \cdot h_1(g(x_1, x_2, \ldots, x_n), \theta). \end{aligned}$$

This completes the forward direction of the proof. Conversely, assume that

$$f_X(x_1)f_X(x_2)\ldots f_X(x_n) = h_1(g(x_1, x_2, \ldots, x_n), \theta) \cdot h_2(x_1, x_2, \ldots, x_n).$$

The conditional distribution of $X_1, X_2, \ldots, X_n$ given $Y = y$ is

$$\begin{aligned} P(X_1 = x_1, X_2 = x_2, \ldots, X_n = x_n \,|\, Y = y) &= \frac{P(X_1 = x_1, X_2 = x_2, \ldots, X_n = x_n)}{P(Y = y)} \\ &= \frac{P(X_1 = x_1, X_2 = x_2, \ldots, X_n = x_n)}{\sum_{Y=y} P(X_1 = x_1, X_2 = x_2, \ldots, X_n = x_n)} \\ &= \frac{h_2(x_1, x_2, \ldots, x_n)}{\sum_{Y=y} h_2(x_1, x_2, \ldots, x_n)}, \end{aligned}$$

which does not depend on θ. So $Y = g(X_1, X_2, \ldots, X_n)$ is a sufficient statistic for θ by Definition 2.9. □

The key to using the Fisher–Neyman factorization criterion is to factor the likelihood function $f_X(x_1)f_X(x_2)\ldots f_X(x_n)$ into the product of a function h_1 that depends on the sample and the sufficient statistic and a function h_2 that depends on the sample but does not involve θ. We now rework the previous three examples using the time-saving Fisher–Neyman factorization criterion.

Example 2.28 Let $X_1, X_2, \ldots, X_n$ be a random sample drawn from a Bernoulli(p) population, where p is an unknown parameter satisfying $0 < p < 1$. Use the Fisher–Neyman factorization criterion to determine whether

$$Y = \sum_{i=1}^{n} X_i$$

is a sufficient statistic for p.

Since the population probability mass function is

$$f_X(x) = p^x(1-p)^{1-x} \qquad x = 0, 1,$$

the likelihood can be factored as

$$p^{\sum_{i=1}^n x_i}(1-p)^{n-\sum_{i=1}^n x_i} = \underbrace{\left(\frac{p}{1-p}\right)^{\sum_{i=1}^n x_i}(1-p)^n}_{h_1} \cdot \underbrace{1}_{h_2}$$

for $x_i = 0, 1$ and $i = 1, 2, \ldots, n$. Since h_1 is a function of p and $\sum_{i=1}^n x_i$ and h_2 is a function of the data values only, $Y = \sum_{i=1}^n X_i$ is a sufficient statistic for p by the Fisher–Neyman factorization criterion. Three key observations follow concerning this application of the factorization criterion. First, there was no need to determine the probability distribution of the candidate sufficient statistic $Y = g(X_1, X_2, \ldots, X_n)$. Second, the factorization criterion actually *suggested* the sufficient statistic $Y = \sum_{i=1}^n X_i$ for p by examining the functional form of h_1. Third, $h_2(x_1, x_2, \ldots, x_n)$ can arbitrarily be set to 1; the h_2 function does not need to overtly include the data values.

Applying the Fisher–Neyman factorization criterion to random sampling from a Bernoulli(p) population was straightforward. Applying it to a $U(0, \theta)$ population requires some finesse.

Example 2.29 Let $X_1, X_2, \ldots, X_n$ be a random sample drawn from a $U(0, \theta)$ population, where θ is a positive unknown parameter. Use the Fisher–Neyman factorization criterion to determine whether

$$Y = X_{(n)} = \max\{X_1, X_2, \ldots, X_n\}$$

is a sufficient statistic for θ.

The population probability density function is

$$f_X(x) = \frac{1}{\theta} \qquad\qquad 0 < x < \theta.$$

The joint probability density function, or likelihood function, of $X_1, X_2, \ldots, X_n$ can be written as

$$f(x_1, x_2, \ldots, x_n) = \frac{1}{\theta^n} \qquad\qquad 0 < x_i < \theta;\ i = 1, 2, \ldots, n.$$

The indicator function will be used to define the support of the likelihood function. The indicator function $I_R(x)$ for some set R equals 1 if $x \in R$ and 0 otherwise. Defining the sets $R_0 = \{x \,|\, x > 0\}$ and $R_\theta = \{x \,|\, 0 < x < \theta\}$, this joint probability density function can also be written as

$$f(x_1, x_2, \ldots, x_n) = \frac{1}{\theta^n} \prod_{i=1}^{n} I_{R_\theta}(x_i).$$

So the likelihood can be factored as

$$f(x_1, x_2, \ldots, x_n) = \underbrace{\frac{1}{\theta^n} I_{R_\theta}(\max\{x_1, x_2, \ldots, x_n\})}_{h_1} \cdot \underbrace{\prod_{i=1}^{n} I_{R_0}(x_i)}_{h_2}$$

because the n-tuple $(x_1, x_2, \ldots, x_n)$ falling in the hypercube is equivalent to having $x_1, x_2, \ldots, x_n$ all positive and $\max\{x_1, x_2, \ldots, x_n\}$ less than θ. Since h_1 is a function of θ and $\max\{x_1, x_2, \ldots, x_n\}$, and h_2 is a function of the data values only, the Fisher–Neyman factorization criterion can be used to conclude that $Y = X_{(n)} = \max\{X_1, X_2, \ldots, X_n\}$ is a sufficient statistic for θ.

This third and final example illustrates how it is much easier to work with the Fisher–Neyman factorization criterion than the definition because it is not necessary to find the sampling distribution of the sufficient statistic $Y = g(X_1, X_2, \ldots, X_n)$.

Example 2.30 Let $X_1, X_2, \ldots, X_n$ be a random sample drawn from a continuous population described by the probability density function

$$f_X(x) = \theta x^{\theta - 1} \qquad\qquad 0 < x < 1,$$

where θ is a positive unknown parameter. Use the Fisher–Neyman factorization criterion to determine whether

$$Y = \prod_{i=1}^{n} X_i$$

is a sufficient statistic for θ.

The population probability density function is

$$f_X(x) = \theta x^{\theta-1} \qquad\qquad 0 < x < 1.$$

Using the Fisher–Neyman factorization criterion, the joint probability density function of $X_1, X_2, \ldots, X_n$ can be factored as:

$$\theta^n (x_1 x_2 \ldots x_n)^{\theta-1} = \underbrace{\left[\theta^n (x_1 x_2 \ldots x_n)^{\theta}\right]}_{h_1} \cdot \underbrace{\left(\frac{1}{x_1 x_2 \ldots x_n}\right)}_{h_2}$$

where $0 < x_i < 1$ for $i = 1, 2, \ldots, n$. Since h_1 is a function of θ and $x_1 x_2 \ldots x_n$ and h_2 is a function of the data values only, $Y = X_1 X_2 \ldots X_n$ is a sufficient statistic for θ.

There is, however, a lurking issue concerning sufficient statistics that has not been addressed to this point. The factorization chosen for the previous three examples was not unique. In the previous example, for instance, the joint probability density function of $X_1, X_2, \ldots, X_n$ could also have been factored as

$$\underbrace{\left[\theta^n \left((x_1 x_2 \ldots x_n)^{1/n}\right)^{n\theta}\right]}_{h_1} \cdot \underbrace{\left(\frac{1}{x_1 x_2 \ldots x_n}\right)}_{h_2}$$

or instead it could have been factored as

$$\underbrace{\left[\theta^n n^{\theta} (x_1 x_2 \ldots x_n / n)^{\theta}\right]}_{h_1} \cdot \underbrace{\left(\frac{1}{x_1 x_2 \ldots x_n}\right)}_{h_2}$$

where $0 < x_i < 1$, for $i = 1, 2, \ldots, n$. So the Fisher–Neyman factorization criterion indicates that the sample geometric mean and the ratio of the product of the data values to n are also sufficient statistics for θ:

$$(X_1 X_2 \ldots X_n)^{1/n} \qquad \text{and} \qquad X_1 X_2 \ldots X_n / n.$$

So there is a limitless source of sufficient statistics by simply taking various functions of one particular sufficient statistic for θ. Stated more generally, any one-to-one function of a sufficient statistic for θ *that does not involve* θ is also a sufficient statistic.

The three examples that have been considered thus far have all involved random samples from a probability distribution with a single unknown parameter, which has been generically named θ. There is a multidimensional version of the Fisher–Neyman factorization criterion which is illustrated next for a two-parameter probability distribution.

Example 2.31 Let $X_1, X_2, \ldots, X_n$ be a random sample from a $N(\mu, \sigma^2)$ population, where μ and σ^2 are unknown parameters. Find sufficient statistics for μ and σ^2.

In this statement of the problem, there are no suggested candidates for sufficient statistics as in the previous three examples. The purpose here is to let the Fisher–Neyman factorization criterion suggest the sufficient statistics. The population probability density function is

$$f_X(x) = \frac{1}{\sqrt{2\pi}\sigma} e^{-\frac{1}{2}\left(\frac{x-\mu}{\sigma}\right)^2} \qquad\qquad -\infty < x < \infty.$$

The joint probability density function of $X_1, X_2, \ldots, X_n$ (or, equivalently, the likelihood function) is

$$\begin{aligned} f(x_1, x_2, \ldots, x_n) &= \prod_{i=1}^{n} f_X(x_i) \\ &= \prod_{i=1}^{n} \frac{1}{\sqrt{2\pi}\sigma} e^{-\frac{1}{2}\left(\frac{x_i-\mu}{\sigma}\right)^2} \\ &= (2\pi)^{-n/2}\sigma^{-n} e^{-\frac{1}{2}\sum_{i=1}^{n}\left(\frac{x_i-\mu}{\sigma}\right)^2} \\ &= \underbrace{\sigma^{-n} e^{-\frac{1}{2\sigma^2}\left(\sum_{i=1}^{n} x_i^2 - 2\mu\sum_{i=1}^{n} x_i + n\mu^2\right)}}_{h_1} \cdot \underbrace{(2\pi)^{-n/2}}_{h_2}, \end{aligned}$$

where $-\infty < x_i < \infty$ for $i = 1, 2, \ldots, n$. Since h_1 is a function of μ, σ^2, $\sum_{i=1}^{n} x_i$, and $\sum_{i=1}^{n} x_i^2$, and h_2 is a function of the data values only,

$$Y_1 = \sum_{i=1}^{n} X_i \qquad \text{and} \qquad Y_2 = \sum_{i=1}^{n} X_i^2$$

are joint sufficient statistics for μ and σ^2 by the two-dimensional version of the Fisher–Neyman factorization criterion. The sufficient statistics are two-dimensional because there are two unknown parameters.

You might have noticed that the likelihood function has appeared in many of the results and examples concerning sufficient statistics. There is a relationship between the maximum likelihood estimator $\hat{\theta}$ and a sufficient statistic for an unknown parameter θ, which is given next.

Theorem 2.8 Let $X_1, X_2, \ldots, X_n$ be a random sample from a population described by $f_X(x)$ with unknown parameter θ. If $\hat{\theta}$ is a maximum likelihood estimator of θ that exists uniquely and $Y = g(X_1, X_2, \ldots, X_n)$ is a sufficient statistic for θ, then $\hat{\theta}$ is a function of $Y = g(X_1, X_2, \ldots, X_n)$.

Proof The sufficient statistic $Y = g(X_1, X_2, \ldots, X_n)$ must satisfy

$$L(\theta) = f_X(x_1)f_X(x_2)\ldots f_X(x_n) = h_1(g(x_1, x_2, \ldots, x_n), \theta) \cdot h_2(x_1, x_2, \ldots, x_n)$$

for nonnegative functions h_1 and h_2 by the Fisher–Neyman factorization criterion. Selecting h_1 to be the probability mass function or probability density function of the sufficient statistic, the likelihood function $L(\theta)$ and $h_1\left(g(x_1, x_2, \ldots, x_n), \theta\right)$ are maximized at the same θ value, namely $\hat{\theta}$. Furthermore, the maximum likelihood estimator $\hat{\theta}$ must be a function of the sufficient statistic $Y = g(X_1, X_2, \ldots, X_n)$. □

So maximum likelihood estimators are functions of sufficient statistics. This result will be illustrated next for a random sample of n observations drawn from a population that has the Rayleigh distribution, which is a special case of the Weibull distribution with shape parameter 2.

Example 2.32 Let $X_1, X_2, \ldots, X_n$ be a random sample from a population with probability density function

$$f(x) = 2\lambda^2 x e^{-(\lambda x)^2} \qquad x > 0,$$

where λ is a positive unknown parameter.

(a) Find the maximum likelihood estimator $\hat{\lambda}$.

(b) Find a sufficient statistic for λ.

(c) Show that the maximum likelihood estimator is a function of the sufficient statistic.

(a) The likelihood function is

$$L(\lambda) = \prod_{i=1}^{n} f(x_i) = 2^n \lambda^{2n} \left(\prod_{i=1}^{n} x_i \right) e^{-\lambda^2 \sum_{i=1}^{n} x_i^2}.$$

The log likelihood function is

$$\ln L(\lambda) = n \ln 2 + 2n \ln \lambda + \sum_{i=1}^{n} \ln x_i - \lambda^2 \sum_{i=1}^{n} x_i^2.$$

The score is

$$\frac{\partial \ln L(\lambda)}{\partial \lambda} = \frac{2n}{\lambda} - 2\lambda \sum_{i=1}^{n} x_i^2.$$

Equating the score to zero and solving for λ yields a closed-form maximum likelihood estimator

$$\hat{\lambda} = \sqrt{\frac{n}{\sum_{i=1}^{n} x_i^2}}.$$

The second partial derivative of the log likelihood function

$$\frac{\partial^2 \ln L(\lambda)}{\partial \lambda^2} = -\frac{2n}{\lambda^2} - 2 \sum_{i=1}^{n} x_i^2$$

evaluates to $-4\sum_{i=1}^{n} x_i^2 < 0$ at the maximum likelihood estimator $\hat{\lambda}$, so the maximum likelihood estimator $\hat{\lambda}$ is the unique value that maximizes the log likelihood function.

(b) The likelihood function can be factored as

$$L(\lambda) = \underbrace{\lambda^{2n} e^{-\lambda^2 \sum_{i=1}^{n} x_i^2}}_{h_1} \cdot \underbrace{2^n \left(\prod_{i=1}^{n} x_i \right)}_{h_2},$$

so $Y = \sum_{i=1}^{n} X_i^2$ is a sufficient statistic by the Fisher–Neyman factorization criterion.

(c) The maximum likelihood estimator

$$\hat{\lambda} = \sqrt{\frac{n}{\sum_{i=1}^{n} x_i^2}}$$

is a function of the sufficient statistic $Y = \sum_{i=1}^{n} X_i^2$.

Sufficient statistics may be helpful in our search for high-quality point estimates. The Rao–Blackwell theorem provides a mechanism for combining a sufficient statistic and an unbiased estimator to arrive at a second unbiased estimator with a smaller population variance.

Theorem 2.9 (Rao–Blackwell theorem) Let $X_1, X_2, \ldots, X_n$ be a random sample from a population described by $f(x)$, a probability distribution with a single unknown parameter θ. Let $Y_1 = g_1(X_1, X_2, \ldots, X_n)$ be a sufficient statistic for θ and let $Y_2 = g_2(X_1, X_2, \ldots, X_n)$ be an unbiased estimator of θ. Then $E[Y_2 \,|\, Y_1]$ is an unbiased estimator of θ whose population variance is less than or equal to the population variance of Y_2, when the expectations exist.

Proof Recall from probability theory that for random variables X and Y,

$$E[X] = E\big[E[X \,|\, Y]\big]$$

and

$$V[X] = E\big[V[X \,|\, Y]\big] + V\big[E[X \,|\, Y]\big]$$

when the expectations exist. Letting Y_2 assume the role of X and Y_1 assume the role of Y, the statistic $E[Y_2 \,|\, Y_1]$ must be unbiased because

$$E\big[E[Y_2 \,|\, Y_1]\big] = E\left[Y_2\right] = \theta$$

when the expectations exist. Furthermore,

$$V[Y_2] \ge V\big[E[Y_2 \,|\, Y_1]\big]$$

when the expectations exist because population variances are always nonnegative. □

The Rao–Blackwell theorem is more complicated than some of the earlier results because there are three characters in the picture:

- $Y_1 = g_1(X_1, X_2, \ldots, X_n)$: a sufficient statistic for θ,
- $Y_2 = g_2(X_1, X_2, \ldots, X_n)$: an unbiased estimator of θ, and
- $E[Y_2 \,|\, Y_1]$: an unbiased estimator of θ like Y_2, but with $V\big[E[Y_2 \,|\, Y_1]\big] \le V[Y_2]$.

This gives us a mechanism for determining an improved unbiased estimator of θ if we are able to identify a statistic $Y_1 = g_1(X_1, X_2, \ldots, X_n)$ which is a sufficient statistic for θ, and a statistic $Y_2 = g_2(X_1, X_2, \ldots, X_n)$ which is an unbiased estimator of θ. The Cramér–Rao inequality can then be applied to the improved unbiased estimator to see if it is efficient.

Example 2.33 Let $X_1, X_2, \ldots, X_7$ be a random sample of size $n = 7$ from a $U(0, \theta)$ population, where θ is a positive unknown parameter. From Example 2.26, the maximum likelihood estimator $X_{(7)}$ is a sufficient statistic for θ. Apply the Rao–Blackwell theorem to

(a) show that twice the sample median, that is, $2X_{(4)}$ is an unbiased estimator of θ,

(b) determine the joint probability density function of $X_{(4)}$ and $X_{(7)}$,

(c) find $E\big[2X_{(4)} \,|\, X_{(7)}\big]$,

(d) compare the population variances of $2X_{(4)}$ and $E\big[2X_{(4)} \,|\, X_{(7)}\big]$.

The population probability density function is

$$f_X(x) = \frac{1}{\theta} \qquad 0 < x < \theta,$$

where θ is a positive unknown parameter. The population cumulative distribution function on the support of X is

$$F_X(x) = \int_0^x \frac{1}{\theta}\, dw = \frac{x}{\theta} \qquad 0 < x < \theta.$$

In implementing the Rao–Blackwell theorem, the sufficient statistic for θ is the sample maximum

$$Y_1 = g_1(X_1, X_2, \ldots, X_7) = X_{(7)}$$

and the unbiased estimator of θ is twice the sample median

$$Y_2 = g_2(X_1, X_2, \ldots, X_7) = 2X_{(4)}.$$

(a) Twice the sample median is an unbiased estimate of θ based on the symmetry of the population probability density function. We will perform the mathematics to verify that this is indeed the case just for the practice. Using an order statistic result, the sample median $X_{(4)}$ has probability density function

$$\begin{aligned}
f_{X_{(4)}}(x_4) &= \frac{7!}{3!1!3!}\left[F_X(x_4)\right]^3 f_X(x_4)\left[1 - F_X(x_4)\right]^3 \\
&= \frac{7!}{3!1!3!}\left[\frac{x_4}{\theta}\right]^3 \frac{1}{\theta}\left[1 - \frac{x_4}{\theta}\right]^3 \\
&= 140\left[\frac{x_4}{\theta}\right]^3 \frac{1}{\theta}\left[1 - \frac{x_4}{\theta}\right]^3 \qquad 0 < x_4 < \theta.
\end{aligned}$$

The expected value of $2X_{(4)}$ is

$$E\left[2X_{(4)}\right] = \int_0^\theta 2x_4 \cdot 140\left[\frac{x_4}{\theta}\right]^3 \frac{1}{\theta}\left[1 - \frac{x_4}{\theta}\right]^3 dx_4 = \theta.$$

So the point estimator $2X_{(4)}$ is an unbiased estimator of θ.

(b) In order to find the joint probability density function of $X_{(4)}$ and $X_{(7)}$, use the order statistic result: the joint probability density function $f_{X_{(i)},X_{(j)}}(x_i, x_j)$ for the order statistics $X_{(i)}$ and $X_{(j)}$ drawn from a continuous population with probability density function $f(x)$ and cumulative distribution function $F(x)$ with support $a < x < b$ is

$$\frac{n!}{(i-1)!(j-i-1)!(n-j)!}[F(x_i)]^{i-1} f(x_i)[F(x_j) - F(x_i)]^{j-i-1} f(x_j)[1 - F(x_j)]^{n-j}$$

for $a < x_i < x_j < b$. Setting $i = 4$, $j = 7$, $a = 0$, and $b = \theta$, the joint probability density function of $X_{(4)}$ and $X_{(7)}$ is

$$\begin{aligned}
f_{X_{(4)},X_{(7)}}(x_4, x_7) &= \frac{7!}{3!2!0!}\left[\frac{x_4}{\theta}\right]^3 \frac{1}{\theta}\left[\frac{x_7}{\theta} - \frac{x_4}{\theta}\right]^2 \frac{1}{\theta}\left[1 - \frac{x_7}{\theta}\right]^0 \\
&= \frac{420x_4^3(x_7 - x_4)^2}{\theta^7} \qquad 0 < x_4 < x_7 < \theta.
\end{aligned}$$

(c) In order to find $E\left[2X_{(4)}\,|\,X_{(7)}\right]$, we first need to calculate the conditional expectation $E\left[2X_{(4)}\,|\,X_{(7)}=x_7\right]$, which in turn requires the conditional distribution of $X_{(4)}$ given $X_{(7)}=x_7$. The required conditional probability density function is

$$f_{X_{(4)}\,|\,X_{(7)}=x_7}(x_4\,|\,X_{(7)}=x_7)=\frac{f_{X_{(4)},X_{(7)}}(x_4,x_7)}{f_{X_{(7)}}(x_7)} \qquad 0<x_4<x_7<\theta,$$

which requires that we first find the marginal distribution of $X_{(7)}$. Using the order statistic result for maximums,

$$f_{X_{(7)}}(x_7)=\frac{7!}{6!1!0!}\left[\frac{x_7}{\theta}\right]^6\frac{1}{\theta}\left[1-\frac{x_7}{\theta}\right]^0=\frac{7x_7^6}{\theta^7} \qquad 0<x_7<\theta.$$

So the conditional probability density function of $X_{(4)}$ given $X_{(7)}$ is the ratio of the joint probability density function of $X_{(4)}$ and $X_{(7)}$ to the marginal distribution of $X_{(7)}$:

$$f_{X_{(4)}\,|\,X_{(7)}=x_7}(x_4\,|\,X_{(7)}=x_7)=\frac{420x_4^3(x_7-x_4)^2/\theta^7}{7x_7^6/\theta^7}=\frac{60x_4^3(x_7-x_4)^2}{x_7^6}$$

for $0<x_4<x_7<\theta$. The expected value of twice the sample median given the value of the sample maximum equals x_7 is

$$\begin{aligned}
E\left[2X_{(4)}\,|\,X_{(7)}=x_7\right] &= 2E\left[X_{(4)}\,|\,X_{(7)}=x_7\right]\\
&= 2\int_0^{x_7}x_4\cdot f_{X_{(4)}\,|\,X_{(7)}=x_7}(x_4\,|\,X_{(7)}=x_7)\,dx_4\\
&= 2\int_0^{x_7}x_4\cdot\frac{60x_4^3(x_7-x_4)^2}{x_7^6}\,dx_4\\
&= \frac{8}{7}x_7.
\end{aligned}$$

So the Rao–Blackwell theorem suggests that the statistic $E\left[2X_{(4)}\,|\,X_{(7)}\right]=8X_{(7)}/7$ is an unbiased estimator whose population variance is smaller than that of $2X_{(4)}$. This will be verified in the next part of the problem.

(d) The population variance of the original unbiased estimator $2X_{(4)}$ is

$$V\left[2X_{(4)}\right]=4V\left[X_{(4)}\right]=4\int_0^{\theta}\left(x_4-\frac{\theta}{2}\right)^2\cdot 140\left[\frac{x_4}{\theta}\right]^3\frac{1}{\theta}\left[1-\frac{x_4}{\theta}\right]^3dx_4=\frac{\theta^2}{9}.$$

The population variance of the improved unbiased estimator $V[8X_{(7)}/7]$ is

$$V\left[\frac{8}{7}X_{(7)}\right]=\frac{64}{49}V\left[X_{(7)}\right]=\frac{64}{49}\int_0^{\theta}\left(x_7-\frac{7\theta}{8}\right)^2\cdot\frac{7x_7^6}{\theta^7}\,dx_7=\frac{\theta^2}{63},$$

because $E[X_{(7)}]=7\theta/8$.

To summarize this example so far, we started with $X_{(7)}$, a sufficient statistic for θ, and $2X_{(4)}$, an unbiased estimator of θ. The Rao–Blackwell theorem performs as advertised—the improved unbiased estimator $8X_{(7)}/7$ has a smaller population variance than the original unbiased estimator. In this case, the improved unbiased estimator is vastly superior to the original unbiased estimator: its population variance is decreased by a factor of 7. A plot of the probability density functions of the sampling distributions of the two unbiased estimators when the unknown parameter θ is arbitrarily set to $\theta = 50$ is given in Figure 2.25. Both probability density functions correspond to an expected value of $\theta = 50$, but the population variance of $8X_{(7)}/7$ is clearly much smaller. Notice in passing that the superior estimate happens to be the same one selected as the superior point estimate in Example 2.17. Unfortunately, the Cramér–Rao inequality cannot be applied in this case to see if $8X_{(7)}/7$ is efficient because the support of the population distribution depends on the unknown parameter θ.

The Rao–Blackwell theorem has produced an improved unbiased estimator of θ. But in practice only one sample $X_1, X_2, \ldots, X_7$ will be collected and therefore only one point estimator will be computed. There is a chance that the original unbiased estimator will provide a point estimator that is closer to the unknown parameter. We end this example by computing the probability that the improved unbiased estimator will be closer to θ than the original unbiased estimator, that is,

$$P\big(R\big(2X_{(4)}, \theta\big) > R\big(8X_{(7)}/7, \theta\big)\big)$$

or

$$P\big(\left|2X_{(4)} - \theta\right| > \left|8X_{(7)}/7 - \theta\right|\big),$$

where R is the error of estimation from Definition 2.4. Figure 2.26 shows the support of $X_{(4)}$ and $X_{(7)}$ as the isosceles triangle. The shaded regions within the isosceles triangle are the regions where the improved unbiased estimator $8X_{(7)}/7$ is closer to θ than the original unbiased estimator $2X_{(4)}$. By integrating the joint probability density function of $X_{(4)}$ and $X_{(7)}$ over the shaded regions, the probability that the improved estimator is closer to θ is

$$P\left(\left|2X_{(4)} - \theta\right| > \left|8X_{(7)}/7 - \theta\right|\right) = \frac{201304062979}{251976997888} \cong 0.7989$$

for all values of θ.

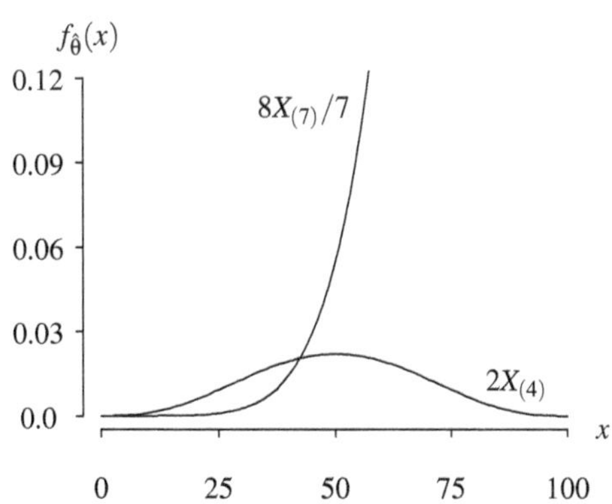

Figure 2.25: Sampling distributions of $2X_{(4)}$ and $8X_{(7)}/7$ when $\theta = 50$.

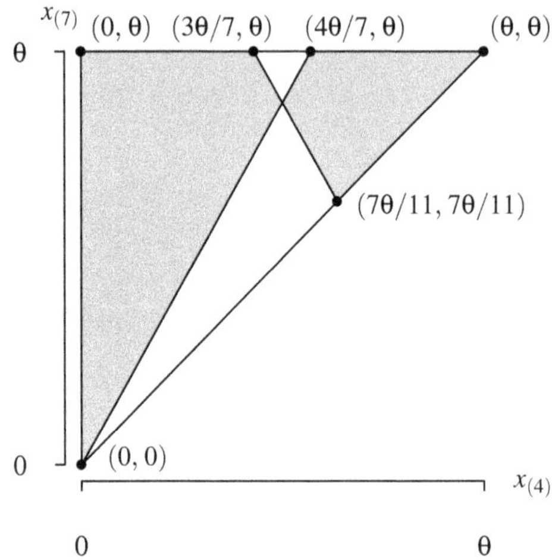

Figure 2.26: Regions (shaded) in which $8X_{(7)}/7$ is closer to θ than $2X_{(4)}$.

A Monte Carlo simulation experiment can be used to support this analytic result. Arbitrarily setting θ to one, the following code generates one million random samples of size $n = 7$ from a $U(0, 1)$ population distribution and prints the fraction of those random samples in which the improved unbiased estimator is closer to $\theta = 1$ than the original unbiased estimator.

```
nrep  = 1000000
count = 0
for (i in 1:nrep) {
  x      = runif(7, 0, 1)
  x      = sort(x)
  error1 = abs(2 * x[4] - 1)
  error2 = abs(8 * x[7] / 7 - 1)
  if (error1 > error2) count = count + 1
}
print(count / nrep)
```

After a call to `set.seed(3)` to initialize the random number stream, five runs of this simulation yield

0.7982 0.7990 0.7989 0.7991 0.7992.

Since these values hover around the analytic solution of 0.7989, the Monte Carlo results support the analytic solution. In conclusion, the Rao–Blackwell theorem has produced an improved unbiased estimator of θ whose population variance is seven times smaller than the original unbiased estimator of θ. In addition, the improved unbiased estimator is closer to θ than the original unbiased estimator approximately 80% of the time.

Before leaving sufficient statistics behind, we briefly survey three topics associated with sufficient statistics that would arise in an advanced course in mathematical statistics. First, a population distribution with a single unknown parameter θ that is described by $f_X(x)$ having the form

$$f_X(x) = e^{a(\theta)g(x)+b(\theta)+c(x)} \qquad x \in \mathcal{A},$$

for functions a, b, c, and g satisfying mild conditions over the support $\mathcal{A}$ that does not involve θ, belongs to what is known as an *exponential family* of distributions. Several well-known parametric distributions such as the Poisson, normal, and gamma distributions are members of this family. For any distribution belonging to the exponential family of distributions, a straightforward application of the Fisher–Neyman factorization criterion indicates that

$$Y = \sum_{i=1}^{n} g(X_i)$$

is a sufficient statistic. There is also a multi-parameter version of this result.

Second, some sufficient statistics are better than others. Recall that a sufficient statistic contains all of the information concerning an unknown parameter θ. But if one sufficient statistic achieves more data reduction than all other sufficient statistics, then it is known as a *minimal sufficient statistic*. More specifically, the sufficient statistic $Y_1 = g_1(X_1, X_2, \ldots, X_n)$ is a minimal sufficient statistic if for any other sufficient statistic $Y_2 = g_2(X_1, X_2, \ldots, X_n)$, Y_1 is a function of Y_2.

Third, for a random sample drawn from a population described by $f_X(x)$ with a single unknown parameter θ, if the condition $E[g(X)] = 0$ for all values of θ requires that $g(x)$ be identically zero, then $f_X(x)$ is referred to as a *complete family*. Once a distribution family has been identified as a complete family, then a result attributed to Erich Leo Lehmann and Henry Scheffé can be used to establish that an estimator is efficient. If $X_1, X_2, \ldots, X_n$ is a random sample from a complete family and $Y = g(X_1, X_2, \ldots, X_n)$ is a sufficient statistic, then any function of Y that results in an unbiased estimator of θ is efficient.

Sufficient statistics are one of the cornerstones of mathematical statistics. To summarize this subsection:

- sufficient statistics encapsulate all of the information regarding an unknown parameter,
- sufficient statistics can be identified using the definition of a sufficient statistic or by using the Fisher–Neyman factorization criterion,
- maximum likelihood estimators are functions of sufficient statistics, and
- the Rao–Blackwell theorem can be used to find an improved unbiased estimator given an unbiased estimator and a sufficient statistic.

Consistent estimators

The previous three subsections have introduced unbiased estimators, efficient estimators, and sufficient statistics. All three of these properties concern a finite sample size n. Consistency, on the other hand, is an asymptotic property whose focus is on the limiting distribution of a point estimator. A *consistent estimator* approaches the true parameter value as n increases.

Definition 2.10 Let $\hat{\theta}$ denote a statistic that is calculated from the sample $X_1, X_2, \ldots, X_n$ that is used to estimate the population parameter θ. The point estimator $\hat{\theta}$ is a *consistent estimator* of θ if it converges in probability to θ, that is, for any positive constant ε,

$$\lim_{n\to\infty} P\left(|\hat{\theta} - \theta| < \varepsilon\right) = 1.$$

Definition 2.10 establishes that for any arbitrarily small positive constant ε, a consistent estimator $\hat{\theta}$ will, with high probability, be arbitrarily close to the value of the parameter as the sample size increases.

Example 2.34 Show that $\hat{\theta} = \bar{X} = \frac{1}{n}\sum_{i=1}^{n} X_i$ is a consistent estimator of θ for a random sample $X_1, X_2, \ldots, X_n$ from a $N(\theta, 1)$ population, where θ is an unknown parameter.

Calculating the probability in Definition 2.10 and using the fact that the sum of independent normal random variables is itself normally distributed,

$$\begin{aligned}
P\big(|\hat{\theta}-\theta| < \varepsilon\big) &= P\big(|\bar{X}-\theta| < \varepsilon\big) \\
&= P\big(-\varepsilon < \bar{X}-\theta < \varepsilon\big) \\
&= P\big(\theta-\varepsilon < \bar{X} < \theta+\varepsilon\big) \\
&= P\left(n(\theta-\varepsilon) < \sum_{i=1}^{n} X_i < n(\theta+\varepsilon)\right) \\
&= P\big(n(\theta-\varepsilon) < N(n\theta, n) < n(\theta+\varepsilon)\big) \\
&= P\left(\frac{n\theta-n\varepsilon-n\theta}{\sqrt{n}} < \frac{N(n\theta, n)-n\theta}{\sqrt{n}} < \frac{n\theta+n\varepsilon-n\theta}{\sqrt{n}}\right) \\
&= P\big(-\varepsilon\sqrt{n} < Z < \varepsilon\sqrt{n}\big),
\end{aligned}$$

where Z is a standard normal random variable. Taking the limit as the sample size goes to infinity,

$$\lim_{n\to\infty} P\big(|\hat{\theta}-\theta| < \varepsilon\big) = 1,$$

which implies that the sample mean is a consistent estimator of the population mean for a random sample drawn from a $N(\theta, 1)$ population.

Although using the definition of consistency is tedious, one redeeming quality is that the calculations can be used for sample size determination. In the previous example, to determine the smallest sample size to be at least 95% certain that $\hat{\theta}$ will be within 0.2 units from θ, the last line of the derivation becomes

$$P\big(-0.2\sqrt{n} < Z < 0.2\sqrt{n}\big) = 0.95.$$

Using the R statement `qnorm(0.975)` to calculate the 0.975 percentile of the standard normal distribution, 1.96, this value is equated to the upper limit of the inequality:

$$0.2\sqrt{n} = 1.96,$$

or

$$n = \left\lceil (1.96/0.2)^2 \right\rceil = 97.$$

A Monte Carlo simulation experiment with one million replications to check whether this sample size of $n = 97$ observations is adequate for an arbitrary population mean $\theta = 37$ in R is given below.

```
nrep  = 1000000
count = 0
theta = 37
eps   = 0.2
n     = 97
```

```
for (i in 1:nrep) {
  xbar = mean(rnorm(n, theta, 1))
  if (xbar < theta + eps && xbar > theta - eps) count = count + 1
}
print(count / nrep)
```

After a call to `set.seed(3)`, five replications of this Monte Carlo simulation experiment yield

0.9508 0.9510 0.9510 0.9510 0.9513.

The simulation supports the analytic result that a sample size of $n = 97$ makes us at least 95% certain that $\bar{X}$ will be within 0.2 of the true value of θ. The output of each of the five replications just slightly exceeds 0.95.

The previous example considered an unbiased estimate of a population parameter; the next example considers a biased estimate of a population parameter.

Example 2.35 Show that the biased estimator $\hat{\theta} = X_{(n)}$ is a consistent estimator of θ for $X_1, X_2, \ldots, X_n$ sampled randomly from a $U(0, \theta)$ population.

Recall from Example 2.8 that $\hat{\theta} = X_{(n)}$ is the maximum likelihood estimator of θ and is also a biased estimator of θ. Using an order statistic result, the probability density function of the maximum likelihood estimator $X_{(n)}$ is

$$f_{X_{(n)}}(x) = \frac{nx^{n-1}}{\theta^n} \qquad 0 < x < \theta.$$

Calculating the probability in Definition 2.10 and adjusting for the fact that $X_{(n)}$ cannot exceed θ when sampling from a $U(0, \theta)$ population,

$$\begin{aligned}
P\big(|\hat{\theta} - \theta| < \varepsilon\big) &= P\big(|X_{(n)} - \theta| < \varepsilon\big) \\
&= P\big(-\varepsilon < X_{(n)} - \theta < \varepsilon\big) \\
&= P\big(\theta - \varepsilon < X_{(n)} < \theta\big) \\
&= \int_{\theta-\varepsilon}^{\theta} \frac{nx^{n-1}}{\theta^n}\, dx \\
&= \left[\left(\frac{x}{\theta}\right)^n\right]_{\theta-\varepsilon}^{\theta} \\
&= 1 - \left(\frac{\theta - \varepsilon}{\theta}\right)^n.
\end{aligned}$$

Taking the limit as the sample size goes to infinity,

$$\lim_{n \to \infty} P\big(|\hat{\theta} - \theta| < \varepsilon\big) = 1,$$

which implies that the largest data value $X_{(n)}$ is a consistent estimator of θ for a random sample drawn from a $U(0, \theta)$ population.

The two previous examples have determined that the point estimators are consistent by appealing to the definition of a consistent estimator. This can be tedious for some distributions. The result that follows allows us to show consistency by showing that the point estimator is asymptotically unbiased and that the population variance of the distribution of the point estimator goes to zero as n goes to infinity. Keep in mind when reading the following theorem that $\hat{\theta}$ is not just a single point estimator, but rather a sequence of point estimators indexed by the sample size n. The intuition here is that as n increases, if the bias is shrinking to zero and the population variance of the point estimator is shrinking to zero, then the point estimator is converging in probability to θ.

Theorem 2.10 Let $\hat{\theta}$ denote a statistic that is calculated from the sample $X_1, X_2, \ldots, X_n$ that is used to estimate the population parameter θ. The point estimator $\hat{\theta}$ is a *consistent estimator* of θ if

(a) $\lim_{n\to\infty} B(\hat{\theta}, \theta) = 0$, and

(b) $\lim_{n\to\infty} V\left[\hat{\theta}\right] = 0$.

Proof Chebyshev's inequality states that for a random variable X with a finite population mean μ and a finite population variance σ^2, and for every real, positive constant k,

$$P(|X-\mu| \ge k\sigma) \le \frac{1}{k^2}.$$

Now the point estimator $\hat{\theta}$ is itself a random variable. So applying Chebyshev's inequality to the point estimator $\hat{\theta}$ gives

$$P\left(\left|\hat{\theta} - E\left[\hat{\theta}\right]\right| \ge k\sigma_{\hat{\theta}}\right) \le \frac{1}{k^2},$$

where $\sigma_{\hat{\theta}}$ is the standard error (that is, the population standard deviation) of the sampling distribution of $\hat{\theta}$. Replacing k with $\varepsilon/\sigma_{\hat{\theta}}$, where ε is any real, positive constant, gives

$$P\left(\left|\hat{\theta} - E\left[\hat{\theta}\right]\right| \ge \varepsilon\right) \le \frac{V\left[\hat{\theta}\right]}{\varepsilon^2}.$$

Using the two limiting assumptions, as $n \to \infty$,

$$\lim_{n\to\infty} P\left(\left|\hat{\theta} - \theta\right| \ge \varepsilon\right) \le 0$$

which implies that $\hat{\theta}$ is a consistent estimator of θ according to Definition 2.10. □

Example 2.36 Let $X_1, X_2, \ldots, X_n$ denote a random sample from a $N\left(\mu, \sigma^2\right)$ population. Show that the sample variance S^2 is a consistent estimator of σ^2.

Instead of using the definition of a consistent estimator, we appeal to Theorem 2.10 and show that the two conditions are met for this point estimator of the population variance. Theorem 2.2 showed that

$$E\left[S^2\right] = \sigma^2,$$

so we know that S^2 is an unbiased estimator of σ^2 (and is therefore asymptotically unbiased). So all that remains is to show that the population variance of S^2 goes to zero as $n \to \infty$. Instead of calculating the population variance of S^2 directly, we can use Theorem 1.6, which states that

$$\frac{(n-1)S^2}{\sigma^2} \sim \chi^2(n-1).$$

Since the population variance of a chi-square random variable is just twice its degrees of freedom, we know that

$$V\left[\frac{(n-1)S^2}{\sigma^2}\right] = 2n-2.$$

Pulling the constant out of the quadratic variance operator gives

$$\frac{(n-1)^2}{\sigma^4} V\left[S^2\right] = 2n-2$$

or

$$V\left[S^2\right] = \frac{2\sigma^4}{n-1}.$$

This expression goes to zero in the limit as $n \to \infty$, so we can conclude that the conditions of Theorem 2.10 are met, which implies that S^2 is a consistent estimator of σ^2 for a random sample drawn from a $N\left(\mu, \sigma^2\right)$ population.

To summarize this chapter, the focus has been on methods for determining a point estimator $\hat{\theta}$ of an unknown population parameter θ (or k point estimators of unknown population parameters $\theta_1, \theta_2, \ldots, \theta_k$) from a data set $X_1, X_2, \ldots, X_n$, and investigating properties of point estimator(s). Two methods for calculating $\hat{\theta}$ are the method of moments and maximum likelihood estimation.

- **The method of moments** forces the first k sample moments to match the first k population moments, where k is the number of unknown parameters in the population distribution.

- **Maximum likelihood estimation** produces population parameter estimates that are the most likely ones to have generated the observed data values. Maximum likelihood estimators maximize the likelihood function $L(\theta)$, or, equivalently, they maximize the log likelihood function $\ln L(\theta)$.

The final section considered properties of point estimators. Three important properties are unbiasedness, efficiency, and consistency. In addition, a sufficient statistic captures all of the information in a data set concerning an unknown parameter θ.

- **Unbiased estimators** satisfy $E\left[\hat{\theta}\right] = \theta$ for all $\theta \in \Omega$, so, in a sense, they are pointed at the correct target. Point estimators that are not unbiased estimators are biased estimators. Point estimators satisfying $\lim_{n\to\infty} E\left[\hat{\theta}\right] = \theta$ are known as asymptotically unbiased point estimators.

- **Efficient estimators** are unbiased estimators that achieve the lower bound of the Cramér–Rao inequality. Efficient estimators have the smallest possible population variance among all unbiased estimators of a population parameter.

- **Sufficient statistics** encapsulate all of the information in a random sample $X_1, X_2, \ldots, X_n$ concerning an unknown parameter θ in a statistic $Y = g(X_1, X_2, \ldots, X_n)$. Sufficient statistics can be identified by appealing to their definition or using a factorization criterion, such as the Fisher–Neyman factorization criterion. One-to-one functions of sufficient statistics are sufficient statistics. Maximum likelihood estimators are functions of sufficient statistics. The Rao–Blackwell theorem can be used to find an improved unbiased estimator with a smaller population variance given an unbiased estimator and a sufficient statistic.

- **Consistent estimators** come arbitrarily close to the associated population parameter in the limit as the sample size n goes to infinity. Consistent estimators can be identified by appealing to their definition or by confirming that they are asymptotically unbiased and their population variance goes to zero in the limit as the sample size $n \to \infty$.

2.5 Exercises

2.1 Let $X_1, X_2, \ldots, X_n$ be a random sample from a population with probability mass function

$$f(x) = \begin{cases} \theta & x = -1 \\ 2\theta & x = 0 \\ 1-3\theta & x = 1, \end{cases}$$

where θ is an unknown parameter satisfying $0 \le \theta \le 1/3$.

(a) Find the method of moments estimator of θ.

(b) Is the method of moments estimator an unbiased estimator of θ?

2.2 A beta distribution is often used as a probability model for random variables that are defined on the interval $(0, 1)$. A beta random variable X has probability density function

$$f(x) = \frac{\Gamma(\alpha+\beta)}{\Gamma(\alpha)\Gamma(\beta)} x^{\alpha-1}(1-x)^{\beta-1} \qquad 0 < x < 1,$$

where α and β are positive unknown parameters.

(a) Show that the population mean of the beta distribution is $\alpha/(\alpha+\beta)$. *Hint*: The fact that the probability density function for a beta distribution integrates to one may be used to help prove this result.

(b) Find general formulas for the method of moments estimators of the parameters α and β when $X_1, X_2, \ldots, X_n$ denote a random sample from a beta population.

(c) The ancient Greeks referred to a rectangle with a height h and a length l satisfying

$$\frac{h}{l} = \frac{l}{h+l}$$

as a *golden rectangle*. Rectangles which adhere to the *golden ratio* appear on the Parthenon, as well as the dimensions of drivers' licenses. The height that satisfies this equation must be $(\sqrt{5}-1)/2 \cong 0.6180$ times the length. They found this particular rectangle the most pleasing aesthetically. The Shoshone Indians decorated $n = 20$ of their leather goods with beaded rectangles that have the following ratios of their heights to their lengths (the data is from page 15 of Larsen, R.J., and Marx, M.L. (2006), *An Introduction to Mathematical Statistics and Its Applications*, Fourth Edition, Pearson Prentice–Hall):

0.693	0.662	0.690	0.606	0.570
0.749	0.672	0.628	0.609	0.844
0.654	0.615	0.668	0.601	0.576
0.670	0.606	0.611	0.553	0.933.

Fit this data set to the beta distribution using the method of moments. Also, calculate the population mean of the fitted distribution. *Hint*: Comment on whether $\hat{\alpha}$ and $\hat{\beta}$ are in the parameter space.

2.3 Let $X_1, X_2, \ldots, X_n$ be a random sample from a population with probability density function

$$f(x) = \theta x^{\theta-1} \qquad 0 < x < 1,$$

where θ is a positive unknown parameter. Find the method of moments estimator of θ.

2.4 Let $X_1, X_2, \ldots, X_n$ be a random sample from a population with probability density function

$$f(x) = \theta(1-x)^{\theta-1} \qquad 0 < x < 1,$$

where θ is a positive unknown parameter. Find the method of moments estimator of θ.

2.5 Let $X_1, X_2, \ldots, X_n$ be a random sample from a population with probability density function

$$f(x) = \theta\left(x - \frac{1}{2}\right) + 1 \qquad 0 < x < 1,$$

with unknown parameter θ and parameter space $\Omega = \{\theta \mid -2 \le \theta \le 2\}$.

(a) What is the method of moments estimator of θ?

(b) Is the method of moments estimator an unbiased estimator of θ?

2.6 Let $X_1, X_2, \ldots, X_n$ be a random sample from a population having the *power distribution* with probability density function

$$f(x) = \frac{\beta x^{\beta-1}}{\alpha^\beta} \qquad 0 < x < \alpha,$$

where α and β are positive unknown parameters. Find the method of moments estimators of α and β for the $n = 3$ times between nuclear power plant accidents (in operating years): 1728, 1986, 10746. Plot the empirical and fitted cumulative distribution functions.

2.7 Let $X_1, X_2, \ldots, X_n$ be a random sample from a Poisson(λ) population, where λ is a positive unknown parameter. Find the method of moments estimator of λ.

2.8 Let $X_1, X_2, \ldots, X_n$ be a random sample from a population with probability mass function

$$f(x) = \begin{cases} p + (1-p)e^{-\lambda} & x = 0 \\ (1-p)\lambda^x e^{-\lambda}/x! & x = 1, 2, \ldots \end{cases}$$

with associated parameter space $\Omega = \{(p, \lambda) \mid 0 < p < 1, \lambda > 0\}$. This probability distribution is known as the "zero-inflated Poisson distribution," and is often abbreviated by ZIP. Applications include modeling annual property insurance claims on a policy or number of cigarettes smoked per day. Find the method of moments estimators of p and λ.

2.9 Let $X_1, X_2, \ldots, X_n$ be a random sample from a population with probability mass function

$$f(x) = \frac{p^x}{x \ln\left(1/(1-p)\right)} \qquad x = 1, 2, \ldots,$$

where p is an unknown parameter satisfying $0 < p < 1$. This distribution is known as the *logarithmic* distribution.

(a) Find the maximum likelihood estimator of p for the following $n = 25$ observations of the number of puppies in the litters of 25 Labrador Retrievers (note that $\bar{x} = 7$ exactly):

8, 4, 6, 3, 5, 7, 8, 10, 5, 14, 7, 9, 8, 5, 8, 9, 5, 7, 6, 9, 8, 6, 6, 4, 8.

(b) Find the method of moments estimator of p for the data set given above.

2.10 The "zero-truncated" Poisson distribution is appropriate for modeling certain discrete random variables in which zero is not a potential outcome (for example, the number of puppies in a litter).

(a) Show that if the probability mass function value at zero for a Poisson(λ) random variable is distributed proportionally to all other probability mass function values, the resulting probability mass function for such a random variable is

$$f(x) = \frac{\lambda^x e^{-\lambda}}{x!(1-e^{-\lambda})} \qquad x = 1, 2, \ldots,$$

where λ is a positive parameter.

(b) Find the method of moments estimator of λ to four digits for the $n = 25$ observed number of puppies in the litter data set from the previous exercise.

(c) Find the maximum likelihood estimator of λ to four digits for the puppy data set.

2.11 Let $X_1, X_2, \ldots, X_n$ be a random sample from an *inverse Gaussian* population with probability density function

$$f(x) = \left(\frac{\lambda}{2\pi x^3}\right)^{1/2} e^{-\lambda(x-\mu)^2/(2\mu^2 x)} \qquad x > 0,$$

where $\lambda > 0$ is a scale parameter and $\mu > 0$ is a location parameter (which is also the population mean).

(a) Find the maximum likelihood estimators of μ and λ and determine their values for the puppy data from the previous two problems.

(b) Under what conditions does the log likelihood function achieve a local maximum at the maximum likelihood estimators? *Hint*: To show that a real-valued function of *one variable* with continuous second derivatives achieves a local maximum at a particular value in question, it is sufficient to show that the first derivative is zero at the particular value in question and that the second derivative is negative at the particular value in question. Now consider the analogous problem for a real-valued function of *two variables* with continuous second-order partial derivatives. In this case, it is sufficient to show that the two first-order partial derivatives are equal to zero at the particular values in question and that the 2×2 Hessian matrix of second partial derivatives is negative definite at the particular values in question. This matrix is symmetric because the mixed partial derivatives are equal. Recall that there are three equivalent ways to show that a symmetric 2×2 matrix A is negative definite:

- $y'Ay < 0$ for all real $y \neq 0$ in $\Re^2$,
- both of the eigenvalues of A are negative,
- the determinant A is positive and the upper-left-hand element is negative.

For the inverse Gaussian distribution, it is sufficient to show that

$$\begin{bmatrix} \dfrac{\partial^2 \ln L(\lambda, \mu)}{\partial \lambda^2} & \dfrac{\partial^2 \ln L(\lambda, \mu)}{\partial \lambda \partial \mu} \\ \dfrac{\partial^2 \ln L(\lambda, \mu)}{\partial \mu \partial \lambda} & \dfrac{\partial^2 \ln L(\lambda, \mu)}{\partial \mu^2} \end{bmatrix}$$

is negative definite at its maximum likelihood estimators $\hat{\lambda}$ and $\hat{\mu}$.

2.12 On the previous three questions, you have fit three distributions to the puppy data: logarithmic, truncated Poisson, and the inverse Gaussian. Considering just the maximum likelihood estimates, which distribution provides the best fit? Why?

2.13 Let $X_1, X_2, \ldots, X_n$ be a random sample from a population with probability density function

$$f(x) = \frac{\theta x^{\theta-1}}{2^\theta} \qquad 0 < x < 2,$$

where θ is a positive unknown parameter. Find the maximum likelihood estimator of θ.

2.14 If $X_1 \sim \text{Poisson}(\lambda)$, $X_2 \sim \text{Poisson}(2\lambda)$, and $X_3 \sim \text{Poisson}(3\lambda)$ are mutually independent random variables, what is the maximum likelihood estimator of λ?

2.15 Let $X_1, X_2, \ldots, X_n$ be a random sample from a population with probability density function

$$f(x) = \theta(1-x)^{\theta-1} \qquad 0 < x < 1,$$

where θ is a positive unknown parameter.

(a) Find the maximum likelihood estimator of θ.

(b) Show that the log likelihood function is maximized at $\hat{\theta}$.

2.16 Let X be a single observation from a population with probability density function

$$f(x) = \frac{1}{2\theta} - \frac{x}{12\theta^2} \qquad \theta < x < 5\theta,$$

where θ is a positive unknown parameter. Find the maximum likelihood estimator $\hat{\theta}$.

2.17 Let $X_1, X_2, \ldots, X_n$ be a random sample from a population with probability density function

$$f(x) = 1 - \frac{2}{3}\theta + \theta\sqrt{x} \qquad 0 < x < 1,$$

where θ is an unknown parameter.

(a) Find the population parameter space Ω.

(b) Plot the population probability density function for several values of θ. Describe the shape of the population probability density function and describe the types of data that would be appropriate for this population distribution.

(c) Find the maximum likelihood estimator of θ for $n = 2$ data values x_1 and x_2. The expression for $\hat{\theta}$ should be expressed in closed form.

(d) Find the maximum likelihood estimate of θ for the $n = 9$ data values

$$x_1 = \frac{1}{10}, x_2 = \frac{2}{10}, x_3 = \frac{3}{10}, \ldots, x_9 = \frac{9}{10}.$$

Finding $\hat{\theta}$ will require numerical methods.

2.18 Let $X_1, X_2, \ldots, X_n$ be a random sample from a $N\left(8, \sigma^2\right)$ population, where $\sigma^2 > 0$ is an unknown parameter. Find the maximum likelihood estimator of σ^2.

2.19 Sir Ronald Fisher analyzed the counts of the number of ticks appearing on $n = 82$ sheep, which are tabulated in the table below.

number of ticks	frequency
0	4
1	5
2	11
3	10
4	9
5	11
6	3
7	5
8	3
9	2
10	2
11	5
12	0
13	2
14	2
15	1
16	1
17	0
18	0
19	1
20	0
21	1
22	1
23	1
24	0
25	2

Fit the negative binomial distribution with parameters r and p to this data set using maximum likelihood estimation.

2.20 Let $X_1, X_2, \ldots, X_n$ be a random sample drawn from a $U(0, \theta)$ population, where θ is a positive unknown parameter. Find

(a) the maximum likelihood estimator of the population variance of the population distribution (that is, the maximum likelihood estimator of $V[X]$), and

(b) the population variance of the maximum likelihood estimator $\hat{\theta}$.

2.21 Let $X_1, X_2, \ldots, X_n$ be a random sample from a population with probability density function

$$f(x) = \frac{\theta}{x^{\theta+1}} \qquad x > 1,$$

where θ is a positive unknown parameter. Find the maximum likelihood estimator of the *population median* of the distribution.

2.22 Let $X_1, X_2, \ldots, X_n$ be a random sample from a population with probability density function

$$f(x) = \theta x^{\theta-1} \qquad 0 < x < 1,$$

where θ is a positive unknown parameter. What is the maximum likelihood estimator of the population mean?

2.23 Let X be a single observation from a continuous probability distribution with probability density function $f(x)$ defined on the support $\mathcal{A} = \{x \,|\, 0 < x < 1\}$. One of the three probability density functions plotted below corresponds to the fitted distribution via maximum likelihood estimation for the single observation $x = 0.1$. Indicate which of the three is the fitted distribution and write a sentence or two explaining your choice.

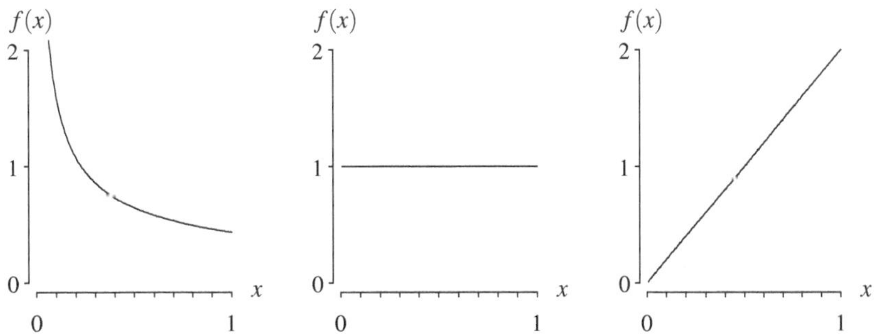

2.24 Consider the continuous random variable X with probability density function

$$f(x) = \theta x^{\theta-1} \qquad 0 < x < 1,$$

where θ is a positive unknown shape parameter.

(a) Find the cumulative distribution function for this distribution.

(b) Find the hazard function for this distribution, where $h(x) = f(x)/\big(1 - F(x)\big)$.

(c) Find the population mean of this distribution.

(d) Find the population variance of this distribution.

(e) Find the 99th percentile of this distribution.

(f) Consider the random variable Y which corresponds to the random variable X truncated on the left at a and truncated on the right at b, where $0 < a < b < 1$. Find $E[Y]$.

(g) If X_1 and X_2 are a random sample from this distribution, find the cumulative distribution function of their product.

(h) If X_1, X_2, and X_3 are a random sample from this distribution, find the expected value of the sample median of the three observations.

(i) If X_1, X_2, and X_3 are a random sample from this distribution, find the maximum likelihood estimator $\hat{\theta}$ of the unknown parameter θ.

(j) Give the code and output to solve all of the problems given above using APPL. If APPL fails, then write the Maple code to solve the problem.

2.25 Consider a population of items having exponential(λ) lifetimes, where λ is a positive unknown failure rate parameter. A random sample of n items is drawn from the population and placed on a *progressive Type II right-censored life test* which is conducted as follows. All n items are simultaneously placed on test at time 0. The first item fails at the random time $t_{(1)}$. Upon the first failure, r_1 items are selected at random from the surviving $n-1$ items and are *immediately* removed from the life test. The second failure occurs at the random time $t_{(2)}$. Upon the second failure, r_2 items are selected at random from the surviving $n-2-r_1$ items and are immediately removed from the life test. This pattern continues until the mth failure at random time $t_{(m)}$. Upon this failure, the remaining surviving $r_m = n-m-r_1-r_2-\cdots-r_{m-1}$ items are immediately removed from the life test, and the life test is terminated. The fixed nonnegative integers $r_1, r_2, \ldots, r_m$ are determined prior to conducting the life test. Derive an expression for the maximum likelihood estimator $\hat{\lambda}$.

2.26 The number of copies until a particular copy machine jams is well modeled by a geometric distribution with probability mass function

$$f(x) = p(1-p)^{x-1} \qquad x = 1, 2, \ldots,$$

where p is an unknown parameter satisfying $0 < p < 1$. What is the maximum likelihood estimator of p associated with the random sample $X_1, X_2, \ldots, X_n$?

2.27 Woody Hayes was the football coach at The Ohio State University from 1951 to 1979. He did not like pass plays because "there are three things that can happen when you throw a pass, and two of them are bad." Consider a football season for a team that runs n pass plays. Let x_1 be the number of completions, x_2 be the number of incompletions, and x_3 be the number of interceptions, where $x_1 + x_2 + x_3 = n$. Consider fitting the *trinomial distribution* to this data set. This is actually a bivariate distribution because $x_3 = n - x_1 - x_2$. The joint probability mass function of X_1 and X_2 is

$$f(x_1, x_2) = \frac{n!}{x_1! x_2! (n - x_1 - x_2)!} p_1^{x_1} p_2^{x_2} p_3^{n-x_1-x_2}$$

for (x_1, x_2) in the support set $\mathcal{A} = \{x_1 = 0, 1, 2, \ldots;\ x_2 = 0, 1, 2, \ldots;\ x_1 + x_2 \le n\}$. The parameter space is

$$\Omega = \{(p_1, p_2, p_3, n) \,|\, p_1 > 0;\ p_2 > 0;\ p_3 > 0;\ p_1 + p_2 + p_3 = 1;\ n = 1, 2, \ldots\}.$$

For a fixed positive integer n, find the maximum likelihood estimators of p_1, p_2, and p_3.

2.28 Let $X_1, X_2, \ldots, X_n$ be a random sample from a population with probability density function

$$f(x) = \frac{2x}{\theta^2} \qquad 0 < x < \theta,$$

where θ is a positive unknown parameter.

(a) Find the maximum likelihood estimator $\hat{\theta}$.

(b) Find the maximum likelihood estimator of the population mean.

(c) Find an unbiasing constant c_n that satisfies $E\left[c_n \hat{\theta}\right] = \theta$.

2.29 A chest contains two drawers. A fair coin is in the top drawer. A biased coin which results in a head with probability p when tossed, where $0 \le p \le 1$, is in the bottom drawer. The following random experiment is conducted three times. A drawer is selected at random and the coin from the selected drawer is tossed and returned to its drawer. The results of these three random experiments yield X_1, X_2, X_3, where $X_i = 0$ denotes a tail being tossed on experiment i and $X_i = 1$ denotes a head being tossed on experiment i, for $i = 1, 2, 3$. Find the maximum likelihood estimator of p for all possible outcomes of the three random experiments.

2.30 Let X be a single observation from a population with probability density function

$$f(x) = \frac{1}{8\theta}\left(5 - \frac{x}{\theta}\right) \qquad \theta < x < 5\theta,$$

where θ is a positive unknown parameter. Find the maximum likelihood estimator of θ and determine whether it is an unbiased estimator of θ.

2.31 Let $X_1, X_2, \ldots, X_n$ be a random sample from a triangular population with probability density function

$$f(x) = \begin{cases} \dfrac{2(x-a)}{(m-a)(b-a)} & a < x \le m \\ \dfrac{2(b-x)}{(b-a)(b-m)} & m < x < b, \end{cases}$$

where a is the minimum, m is the mode, and b is the maximum, which are real constants satisfying $a < b$ and $a \le m \le b$.

(a) Find the maximum likelihood estimates of a, m, and b for the $n = 4$ data values:

$$-3, -1, 1, 3.$$

(b) Find the maximum likelihood estimates of a, m, and b for the $n = 3$ data values:

$$2, 3, 5.$$

2.32 Let $X_1, X_2, \ldots, X_n$ be a random sample from a population with probability density function

$$f(x) = \frac{1}{2\theta} e^{-|x|/\theta} \qquad -\infty < x < \infty,$$

where θ is a positive unknown parameter. This is a special case of the Laplace or double exponential distribution.

(a) Find the maximum likelihood estimator $\hat{\theta}$.

(b) Find $E\left[\hat{\theta}\right]$.

(c) Find $V\left[\hat{\theta}\right]$.

(d) Conduct a Monte Carlo simulation experiment that gives convincing numerical evidence to support your values for $E\left[\hat{\theta}\right]$ and $V\left[\hat{\theta}\right]$ for arbitrary values of the population parameter θ and sample size n of your choosing.

2.33 Let $X_1, X_2, \ldots, X_n$ be a random sample from a population with probability density function

$$f(x) = \frac{e^{-1/(\beta x)}}{\beta x^2} \qquad x > 0,$$

where β is a positive unknown parameter. This is a special case of the inverse gamma distribution. Find the maximum likelihood estimator of $P(X > 4)$.

2.34 A class of 100 students take a five-question test. Each question is a true/false question. The five questions are of equal difficulty. The 100 students are of equal capability. Let p be the probability that a student gets one of the questions correct. The results of the test are given in the table below.

number of correct answers	0	1	2	3	4	5
frequency	1	3	9	38	40	9

(a) Find the maximum likelihood estimate of p.

(b) Find the maximum likelihood estimate of a student getting two or fewer correct answers on the five-question test.

2.35 Let $X_1, X_2, \ldots, X_n$ be a random sample from a $U(-\theta, \theta)$, where θ is a positive unknown parameter.

(a) Find the maximum likelihood estimator of θ.

(b) Find the population mean and population variance of $\hat{\theta}$.

(c) Show that $\hat{\theta}$ is asymptotically unbiased.

(d) Calculate the numeric values of $E\left[\hat{\theta}\right]$ and $V\left[\hat{\theta}\right]$ when $\theta = 2$ and $n = 3$.

(e) Conduct a Monte Carlo simulation experiment to support the values from part (d).

2.36 Let $X_1, X_2, \ldots, X_n$ be a random sample from a discrete uniform distribution with support values from $-\theta$ to θ with probability mass function

$$f(x) = \frac{1}{2\theta + 1} \qquad x = -\theta, -\theta + 1, \ldots, \theta,$$

where θ is a positive integer unknown parameter.

(a) Find the maximum likelihood estimator of θ.

(b) Find expressions for the population mean and population variance of $\hat{\theta}$.

(c) Show that $\hat{\theta}$ is asymptotically unbiased.

(d) Calculate the numeric values of $E\left[\hat{\theta}\right]$ and $V\left[\hat{\theta}\right]$ when $\theta = 2$ and $n = 3$.

(e) Conduct a Monte Carlo simulation experiment to support the values from part (d).

2.37 Let $X_1, X_2, \ldots, X_n$ be a random sample from a population with probability mass function

$$f(x) = -\frac{p^x}{x \ln(1 - p)} \qquad x = 1, 2, \ldots,$$

where p is an unknown parameter satisfying $0 < p < 1$. This distribution is known as the *logarithmic distribution*. Use the Fisher–Neyman factorization criterion to determine a sufficient statistic for p.

2.38 Vilfredo Pareto (1848–1923) was an Italian economist who devised the Pareto distribution, which can be used for modeling the probability distribution of wealth, income, etc. The Pareto principle, or 80–20 rule, for example, states that 80% of the wealth in a country is owned by 20% of the population. The probability density function of a Pareto random variable X is

$$f(x) = \frac{\theta_2 \theta_1^{\theta_2}}{x^{\theta_2+1}} \qquad x \ge \theta_1,$$

where θ_1 is a positive scale parameter and θ_2 is a positive shape parameter. In this problem, the Pareto distribution will be used to model hourly wages. In an economy with a minimum wage, θ_1 denotes the minimum wage, and θ_2 reflects the probability distribution of incomes exceeding the minimum wage. Assume that an economy has an hourly minimum wage of \$10 per hour, that is, $\theta_1 = 10$.

(a) Find the maximum likelihood estimator of θ_2 for a random sample of hourly wages $X_1, X_2, \ldots, X_n$.

(b) Find the maximum likelihood estimator of θ_2 for the $n = 3$ hourly wages: \$14, \$22, \$63.

(c) Plot the fitted cumulative distribution function and the empirical cumulative distribution function from part (b).

(d) What is the maximum vertical distance between the fitted cumulative distribution function and the empirical cumulative distribution function from part (c)?

(e) What do you think that this maximum vertical distance measures?

(f) For arbitrary data values of x_1, x_2, x_3, write a paragraph that describes how you would find the smallest and largest possible values of the maximum vertical difference between the empirical and fitted cumulative distribution functions.

2.39 Let $X_1, X_2, \ldots, X_n$ be a random sample from a population with probability density function

$$f(x) = \sqrt{\frac{\lambda}{2\pi x^3}}\, e^{-\lambda(x-1)^2/(2x)} \qquad x > 0,$$

where λ is a positive unknown parameter. This distribution is known as the *standard Wald distribution*, which is a special case of the inverse Gaussian distribution.

(a) Use the Fisher–Neyman factorization criterion to determine a sufficient statistic for λ.

(b) Find the maximum likelihood estimator of λ.

2.40 Alexis samples a single observation X from a $N\left(0, \sigma^2\right)$ population, where σ^2 is a positive unknown parameter. Determine whether $|X|^3$ is a sufficient statistic for σ^2.

2.41 Let $X_1, X_2, \ldots, X_n$ be a random sample from a population with probability density function

$$f(x) = \frac{\Gamma(2\theta)}{[\Gamma(\theta)]^2} x^{\theta-1}(1-x)^{\theta-1} \qquad 0 < x < 1,$$

where θ is a positive unknown parameter. This is a special case of the beta distribution in which the two shape parameters are equal. Use the Fisher–Neyman factorization criterion to determine a sufficient statistic for θ.

2.42 Let $X_1, X_2, \ldots, X_n$ be a random sample from a beta(α, β) population with probability density function

$$f(x) = \frac{\Gamma(\alpha+\beta)}{\Gamma(\alpha)\Gamma(\beta)} x^{\alpha-1}(1-x)^{\beta-1} \qquad 0 < x < 1,$$

where α and β are positive unknown shape parameters. Use the Fisher–Neyman factorization criterion to determine joint sufficient statistics for α and β.

2.43 Not all sufficient statistics are identical in nature. Some sufficient statistics have a special status known as *minimal sufficient statistics*. A sufficient statistic $g(X_1, X_2, \ldots, X_n)$ is a minimal sufficient statistic if it is a function of any other sufficient statistic. As a result of the Fisher–Neyman factorization criterion, $g(X_1, X_2, \ldots, X_n)$ is a minimal sufficient statistic for an unknown parameter θ if and only if

$$\frac{f(x_1, x_2, \ldots, x_n)}{f(y_1, y_2, \ldots, y_n)} \text{ is independent of } \theta \qquad \Longleftrightarrow \qquad g(x_1, x_2, \ldots, x_n) = g(y_1, y_2, \ldots, y_n).$$

(a) Let $X_1, X_2, \ldots, X_n$ be a random sample from a Poisson(θ) population, where θ is a positive unknown parameter. Show that $\sum_{i=1}^{n} X_i$ is a minimal sufficient statistic for θ.

(b) Let $X_1, X_2, \ldots, X_n$ be a random sample from an exponential population with population mean θ, where θ is a positive unknown parameter. Show that $\sum_{i=1}^{n} X_i$ is a minimal sufficient statistic for θ.

(c) Let $X_1, X_2, \ldots, X_n$ be a random sample from a beta(α, β) population, where α and β are positive unknown shape parameters. Show that $\prod_{i=1}^{n} X_i$ and $\prod_{i=1}^{n}(1-X_i)$ are minimal sufficient statistics for α and β.

2.44 Let $X_1, X_2, \ldots, X_n$ be a random sample from a $U(\theta-1, \theta+1)$ distribution, where θ is an unknown parameter. Let $X_{(1)}, X_{(2)}, \ldots, X_{(n)}$ be the associated order statistics. Find the maximum likelihood estimator of θ. *Hint*: This is a case of a non-unique maximum likelihood estimator.

2.45 Let $X_1, X_2, \ldots, X_n$ be a random sample from a population with probability density function

$$f(x) = \theta x^{\theta-1} \qquad 0 < x < 1,$$

where θ is a positive unknown parameter.

(a) Find the maximum likelihood estimator of θ.

(b) Find the probability density function of the maximum likelihood estimator $\hat{\theta}$.

(c) Show that $\hat{\theta}$ is a consistent estimator of θ.

2.46 Let $\hat{\theta}_1$ and $\hat{\theta}_2$ be two unbiased estimators of an unknown parameter θ.

(a) Show that $\hat{\theta}_3 = p\hat{\theta}_1 + (1-p)\hat{\theta}_2$ is an unbiased estimator of θ for any real constant p.

(b) Find the value of p that minimizes $V\left[\hat{\theta}_3\right]$ when $\hat{\theta}_1$ and $\hat{\theta}_2$ are independent.

(c) Find the value of p that minimizes $V\left[\hat{\theta}_3\right]$ when $\hat{\theta}_1$ and $\hat{\theta}_2$ are dependent.

The particular estimator derived here is known as the *best linear unbiased estimator*, often given the colorful abbreviation BLUE.

2.47 A *sufficient statistic* for θ contains all of the information concerning the unknown parameter θ contained in a data set. An *ancillary statistic* for θ, on the other hand, is the opposite of a sufficient statistic for θ in that it contains *no information* about the unknown parameter θ. The probability distribution of an ancillary statistic does not involve the unknown parameter θ. Between these two extremes are statistics that contain some, but not all of the information in a data set concerning the unknown parameter θ. For each of the population distributions given below, classify each statistic as a sufficient statistic for θ, an ancillary statistic for θ, or neither a sufficient statistic nor an ancillary statistic for θ.

(a) Let X_1, X_2, X_3 be a random sample from a $N(\theta, 1)$ population, where θ is an unknown parameter. Classify the statistics

$$X_2 \qquad X_2 - X_3 \qquad X_1 + X_2 + X_3 \qquad \bar{X} \qquad X_{(3)}.$$

(b) Let X_1, X_2, X_3 be a random sample from a $N\left(0, \theta^2\right)$ population, where θ is a positive unknown parameter. Classify the statistics

$$X_1 \qquad X_2 - X_3 \qquad X_1^2 + X_2^2 + X_3^2 \qquad \bar{X} \qquad X_{(3)} - X_{(1)}.$$

(c) Let X_1 and X_2 be a random sample from an exponential population with population mean θ, where θ is a positive unknown parameter. Classify the statistics

$$X_1/X_2 \qquad X_{(2)} - X_{(1)} \qquad X_1^2 + X_2^2 \qquad \bar{X} \qquad X_1.$$

2.48 Let $X_1, X_2, \ldots, X_n$ be a random sample from an exponential distribution with probability density function

$$f(x) = \lambda e^{-\lambda x} \qquad x > 0,$$

where λ is a positive unknown rate parameter. Find an unbiasing constant c_n so that $c_n X_{(1)}$ is an unbiased estimator of $1/\lambda$, where $X_{(1)} = \min\{X_1, X_2, \ldots, X_n\}$ is the first order statistic. *Hint*: The unbiasing constant c_n is a function of the sample size n.

2.49 Let X_1 and X_2 be a random sample from a population with finite population mean μ and finite population variance σ^2. Clarck would like to use $(X_1 + X_2)/2$ to estimate μ whereas Fifi wants to use $(2X_1 + X_2)/3$ to estimate μ. Choose the better of the two estimators. Base your choice on bias and variance. Explain your reasoning.

2.50 Let $X_1, X_2, \ldots, X_n$ be a random sample from an exponential population with population mean θ, where θ is a positive unknown parameter. Find an unbiasing constant c_n such that

$$c_n \left(\prod_{i=1}^{n} X_i \right)^{1/n}$$

is an unbiased estimator of θ. *Hint*: The unbiasing constant c_n is a function of the sample size n.

2.51 Let $X_1, X_2, \ldots, X_n$ be a random sample from a $N\left(\mu, \sigma^2\right)$ population, where μ and $\sigma^2 > 0$ are unknown parameters. Find the constant c_n that minimizes the MSE of $c_n S^2$.

2.52 Let $X_1, X_2, \ldots, X_n$ be a random sample from a Geometric(p) population parameterized with support beginning at 1 with probability mass function

$$f(x) = p(1-p)^{x-1} \qquad\qquad x = 1, 2, \ldots,$$

where p is an unknown parameter satisfying $0 < p < 1$.

(a) Find the method of moments estimator of p.

(b) Calculate the Cramér–Rao lower bound.

(c) Use the Fisher–Neyman factorization criterion to suggest a sufficient statistic for p.

2.53 Let $X_1, X_2, \ldots, X_n$ be a random sample from a geometric(p) population with probability mass function

$$f(x) = p(1-p)^{x} \qquad\qquad x = 0, 1, 2, \ldots,$$

where p is an unknown parameter satisfying $0 < p < 1$. Find the maximum likelihood estimator $\hat{p}$ and the Cramér–Rao lower bound.

2.54 Let X_1, X_2, and X_3 be a random sample from a population with finite population mean θ and finite population variance $\sigma^2 > 0$. Consider the following estimators of θ:

- the sample mean $\hat{\theta}_1 = \dfrac{X_1 + X_2 + X_3}{3}$,
- the weighted average $\hat{\theta}_2 = \dfrac{X_1 + 4X_2 + X_3}{6}$.

Calculate the relative efficiency of $\hat{\theta}_1$ to $\hat{\theta}_2$ and plot the sampling distributions of the two estimators for a random sample drawn from an exponential(1) population.

2.55 Sixty-watt light bulb lifetimes are known to be exponentially distributed with unknown population mean θ. The company that produces these light bulbs would like to estimate θ by testing n bulbs to failure at one facility and m bulbs to failure at a second facility. Let $X_1, X_2, \ldots, X_n$ be the mutually independent lifetimes of the bulbs tested at the first facility; let $Y_1, Y_2, \ldots, Y_m$ be the mutually independent lifetimes of the bulbs tested at the second facility. An unbiased estimate of θ is the convex combination

$$\hat{\theta} = p\hat{\theta}_X + (1-p)\hat{\theta}_Y,$$

where $0 < p < 1$, $\hat{\theta}_X = \bar{X}$ is the maximum likelihood estimator of θ for the data from the first facility, and $\hat{\theta}_Y = \bar{Y}$ is the maximum likelihood estimator of θ for the data from the second facility. Find the value of p that minimizes $V\left[\hat{\theta}\right]$.

2.56 Let X_1 and X_2 be independent random variables. Their population means and variances are

$$E[X_1] = 5 \qquad V[X_1] = \sigma^2 \qquad E[X_2] = 3 \qquad V[X_2] = \sigma^2,$$

where σ^2 is a positive unknown parameter. Find the value of the constant c so that

$$c\left(X_2^2 - X_1^2\right) + X_1^2$$

is an unbiased estimate of σ^2.

2.57 Let $X_1, X_2, \ldots, X_n$ be a random sample drawn from a population having probability density function

$$f(x) = 2x\theta^{-2}e^{-(x/\theta)^2} \qquad x > 0,$$

where θ is a positive unknown parameter.

(a) Use the Fisher–Neyman factorization criterion to find a sufficient statistic for θ.

(b) Find the maximum likelihood estimator $\hat{\theta}$.

(c) Find $E\left[\hat{\theta}\right]$.

(d) Show that $\hat{\theta}$ is an asymptotically unbiased estimator of θ (using the `limit` function in Maple on your answer to part (c) is acceptable).

2.58 If $X \sim \text{Poisson}(\lambda)$, find a function of X only that is an unbiased estimator of λ^2.

2.59 Let $X_1, X_2, \ldots, X_n$ be a random sample drawn from a Poisson(λ) population with probability mass function

$$f(x) = \frac{\lambda^x e^{-\lambda}}{x!} \qquad x = 0, 1, 2, \ldots,$$

where λ is a positive unknown parameter.

(a) What is the maximum likelihood estimator $\hat{\lambda}$?

(b) Is $\hat{\lambda}$ an unbiased estimator of λ?

(c) Is $\hat{\lambda}$ an efficient estimator of λ?

(d) Is $\hat{\lambda}$ a consistent estimator of λ?

(e) Does the quantity

$$\frac{\sqrt{n}\left(\hat{\lambda} - \lambda\right)}{\sqrt{\hat{\lambda}}}$$

converge in distribution as $n \to \infty$? If so, give the probability distribution or constant to which it converges.

2.60 Let $X_1, X_2, \ldots, X_n$ be a random sample drawn from a $U(0, \theta)$ distribution, where θ is a positive unknown parameter.

(a) Find an unbiasing constant a_n such that $E\left[a_n X_{(1)}\right] = \theta$.

(b) Find an unbiasing constant b_n such that $E\left[b_n X_{(n)}\right] = \theta$.

(c) Find an unbiasing constant c_n such that $E\left[c_n \bar{X}\right] = \theta$.

(d) Find the population variances of the three unbiased estimators of θ, namely $a_n X_{(1)}$, $b_n X_{(n)}$, $c_n \bar{X}$.

(e) Which of the three unbiased estimators is preferred and why?

2.61 Let $X_1, X_2, \ldots, X_8$ be a random sample from a population with finite unknown population mean θ. Five point estimators of θ are listed below. Indicate which point estimators are unbiased estimators of θ for all possible population distributions.

$$X_6 \qquad X_{(6)} \qquad \frac{X_{(1)} + X_{(8)}}{2} \qquad 2X_1 - 8X_3 + 10X_5 - 3X_7 \qquad 17.$$

2.62 Let $X_1, X_2, \ldots, X_n$ be a random sample from a $N(\mu_X, \theta)$ population and let $Y_1, Y_2, \ldots, Y_m$ be a random sample from a $N(\mu_Y, \theta)$ population, which is independent of the first sample. Assume that μ_X and μ_Y are known. Assume that θ is unknown.

(a) Find the maximum likelihood estimator $\hat{\theta}$ of the common unknown population variance θ.

(b) Determine whether the maximum likelihood estimator $\hat{\theta}$ is an unbiased estimator of θ.

2.63 Let $X_1, X_2, \ldots, X_n$ be mutually independent observations of a continuous random variable X. The *sample kurtosis*

$$\hat{\gamma}_4 = \frac{1}{n}\sum_{i=1}^{n}\left(\frac{X_i - \bar{X}}{S}\right)^4$$

is an estimate of the population kurtosis

$$\gamma_4 = E\left[\left(\frac{X-\mu}{\sigma}\right)^4\right].$$

(a) Find $E\left[\hat{\gamma}_4\right]$ when $n = 2$.

(b) Use Monte Carlo simulation with sample sizes $n = 10$, $n = 100$, $n = 1000$, $n = 10000$, and $n = 100000$, to provide convincing numerical evidence that $\hat{\gamma}_4$ is an asymptotically unbiased estimate of γ_4 when the observations are sampled from an exponential(1) population.

2.64 Let $X_{(1)} < X_{(2)} < X_{(3)}$ be the order statistics of a random sample of size $n = 3$ drawn from a $U(0, \theta)$ population, where θ is a positive unknown parameter.

(a) Show that $2X_{(2)}$ is an unbiased estimator of θ.

(b) Find a sufficient statistic for θ.

(c) Use the Rao–Blackwell theorem to find an improved estimator of θ using the results from parts (a) and (b).

(d) Calculate and compare the population variances of the two unbiased estimates.

2.65 Let $X_1, X_2, \ldots, X_n$ be a random sample from a Bernoulli(p) population, where p is an unknown parameter satisfying $0 < p < 1$. The maximum likelihood estimator

$$\hat{p} = \frac{1}{n}\sum_{i=1}^{n} X_i = \frac{X}{n}$$

is an unbiased estimator of p. Find an unbiased estimator of $p(1-p)$.

2.66 Use the Fisher–Neyman factorization criterion to determine a sufficient statistic for θ for a random sample $X_1, X_2, \ldots, X_n$ drawn from a population with probability density function

$$f(x) = \frac{\theta}{(1+x)^{\theta+1}} \qquad x > 0,$$

where θ is a positive unknown parameter.

2.67 Let $X_1, X_2, \ldots, X_n$ be a random sample drawn from a population with probability density function

$$f(x) = \frac{\theta}{(1+x)^{\theta+1}} \qquad x > 0,$$

where θ is a positive unknown parameter. Calculate the Cramér–Rao lower bound for the population variance of an unbiased estimator of θ.

2.68 The assumptions that are made on the population distribution in statistical results are critical. Consider the following population assumptions, listed in decreasing order of strength.

1. All population distributions.
2. All population distributions with finite first and second moments.
3. $N\left(\mu, \sigma^2\right)$ population distribution.

Let μ be the population mean when it exists; let σ^2 be the population variance when it exists; let n be the sample size; let $\bar{X}$ be the sample mean; let S^2 be the sample variance; let M be the sample median. For each of the five results given below, indicate the weakest assumptions about the population distribution from the list above that make the result true.

(a) $V\left[\bar{X}\right] = \sigma^2/n$,

(b) $E\left[M\right] = \mu$,

(c) $(n-1)S^2/\sigma^2 \sim \chi^2(n-1)$,

(d) $E\left[S^2\right] = \sigma^2$,

(e) $\dfrac{\bar{X}-\mu}{\sigma/\sqrt{n}} \xrightarrow{D} N(0, 1)$.

2.69 Let $X_1, X_2, \ldots, X_n$ be a random sample from a $U(0, \theta)$ population, where θ is a positive unknown parameter. Show that

$$E\left[\hat{\theta}\right] = \frac{n}{n+1}\theta,$$

where $\hat{\theta}$ is the maximum likelihood estimator of θ.

2.70 Let $X_1, X_2, \ldots, X_n$ be a random sample from a Laplace population with probability density function

$$f(x) = \frac{1}{2}e^{-|x-\theta|} \qquad -\infty < x < \infty,$$

where θ is an unknown parameter.

(a) Find the probability density function of the sample median. Assume that n is odd to simplify your calculations.

(b) Is the sample median for an odd sample size n an unbiased estimator of θ?

2.71 Let X_1, X_2, X_3 be a random sample from a population with probability density function

$$f(x) = \frac{3x^2}{\theta^3} \qquad 0 < x < \theta,$$

where θ is a positive unknown parameter. Find an unbiasing constant c such that the statistic $c(X_1 + 2X_2 + 3X_3)$ is an unbiased estimator of θ.

2.72 Let $X_1, X_2, \ldots, X_n$ be a random sample from a population with probability density function

$$f(x) = \frac{1}{2}\lambda^3 x^2 e^{-\lambda x} \qquad x > 0,$$

where λ is a positive unknown parameter. Use the Fisher–Neyman factorization criterion to find a sufficient statistic for λ.

2.73 Let $X_1, X_2, \ldots, X_n$ be a random sample from a $N\left(\mu_0, \sigma^2\right)$ population with *known* population mean μ_0 and *unknown* population variance σ^2.

(a) Find the maximum likelihood estimator of σ^2.

(b) Is the maximum likelihood estimator of σ^2 an unbiased estimator of σ^2?

2.74 Let X_1 and X_2 be a random sample from a population with probability mass function

$$f(x) = \begin{cases} \theta & x = -1 \\ 2\theta & x = 0 \\ 1 - 3\theta & x = 1, \end{cases}$$

where θ is an unknown parameter satisfying $0 \le \theta \le 1/3$.

(a) Find the maximum likelihood estimator of θ.

(b) Find the expected value of the maximum likelihood estimator $\hat{\theta}$.

2.75 In Example 2.12, ten motors were stored in each of five sheds. A shed was selected at random, and three motors were selected at random from that shed and tested. The maximum likelihood estimator of θ, the shed that was selected, given the number of defective motors in the sample x was determined to be

$$\hat{\theta} = \begin{cases} 1 & x = 0 \\ 4 & x = 1 \\ 5 & x = 2, 3. \end{cases}$$

(a) Find the probability mass function of the maximum likelihood estimator $\hat{\theta}$.

(b) Determine whether $\hat{\theta}$ is an unbiased estimator of θ.

2.76 Let $X_1, X_2, \ldots, X_n$ be a random sample from a population with probability density function

$$f(x) = 2\theta x e^{-\theta x^2} \qquad x > 0,$$

where θ is a positive unknown parameter. What is the Cramér–Rao lower bound on the population variance of unbiased estimators of θ?

2.77 Let X_1 and X_2 be a random sample from a $U(0, \theta)$ population, where θ is a positive unknown parameter. The three point estimators $2\bar{X}$, $3X_{(2)}/2$, and $3X_{(1)}$ are all unbiased estimators of θ. What are the probabilities, stated as exact fractions, that each of the three estimators is closest to the true value of θ when $\theta = 10$?

2.78 Let $X_1, X_2, \ldots, X_n$ be a random sample from a population with probability density function

$$f(x) = \frac{\Gamma(\theta+2)}{\Gamma(\theta)} x^{\theta-1}(1-x) \qquad 0 < x < 1.$$

This is a special case of the beta distribution. Use the Fisher–Neyman factorization criterion to find a sufficient statistic for θ.

2.79 A population distribution with a single unknown parameter θ that is described by $f(x)$ having the form

$$f(x) = e^{a(\theta)g(x)+b(\theta)+c(x)} \qquad x \in \mathcal{A},$$

for functions a, b, c, and g satisfying mild conditions over the support $\mathcal{A}$ that does not involve θ, belongs to what is known as an *exponential family of distributions*.

(a) Determine whether the exponential(λ) distribution, with positive rate parameter λ, belongs to the exponential family of distributions.

(b) Determine whether the Poisson(λ) distribution, with positive parameter λ, belongs to the exponential family of distributions.

(c) Determine whether the $N\left(\mu, \sigma^2\right)$ distribution, with fixed known parameter μ and positive parameter σ, belongs to the exponential family of distributions.

(d) Determine whether the binomial(n, p) distribution, with fixed known positive integer parameter n and parameter p satisfying $0 < p < 1$, belongs to the exponential family of distributions.

(e) Determine whether the geometric(p) distribution, with parameter p satisfying $0 < p < 1$, belongs to the exponential family of distributions.

(f) Determine whether the $U(0, \theta)$ distribution, with positive parameter θ, belongs to the exponential family of distributions.

2.80 Let $X_1, X_2, \ldots, X_n$ be a random sample from the exponential family of distributions having unknown parameter θ described by $f_X(x)$ having the form

$$f_X(x) = e^{a(\theta)g(x)+b(\theta)+c(x)} \qquad x \in \mathcal{A},$$

for functions a, b, c, and g satisfying mild conditions over the support $\mathcal{A}$ that does not involve θ, belongs to what is known as an *exponential family of distributions*. Show that

$$\sum_{i=1}^{n} g(X_i)$$

is a sufficient statistic for θ.

2.81 Let $X_1, X_2, \ldots, X_n$ be a random sample from a shifted unit exponential population with probability density function

$$f(x) = e^{-(x-\theta)} \qquad x \geq \theta,$$

where θ is an unknown parameter.

(a) Find the maximum likelihood estimator of θ.

(b) Show that the maximum likelihood estimator of θ is an asymptotically unbiased estimator of θ.

(c) Show that the maximum likelihood estimator of θ is a consistent estimator of θ.

2.82 Let $X_1, X_2, \ldots, X_n$ be a random sample from a $U(0, \theta)$ distribution, where θ is a positive unknown parameter. Determine whether $(n+1)X_{(1)}$ is a consistent estimator of θ.

2.83 Let $X_1, X_2, \ldots, X_n$ be a random sample from a population with finite population mean μ and finite population variance σ^2. Show that the sample mean $\bar{X}$ is a consistent estimator of the population mean μ.

2.84 Let $X_1, X_2, \ldots, X_n$ be a random sample from a $U(\theta, \theta+1)$ distribution, where θ is an unknown parameter.

(a) Find the method of moments estimator of θ.

(b) Show that the method of moments estimator of θ is unbiased.

(c) Find an unbiasing constant c_n such that $X_{(1)} - c_n$ is an unbiased estimator of θ.

(d) Compare the two unbiased estimators of θ.

Chapter 3

Interval Estimation

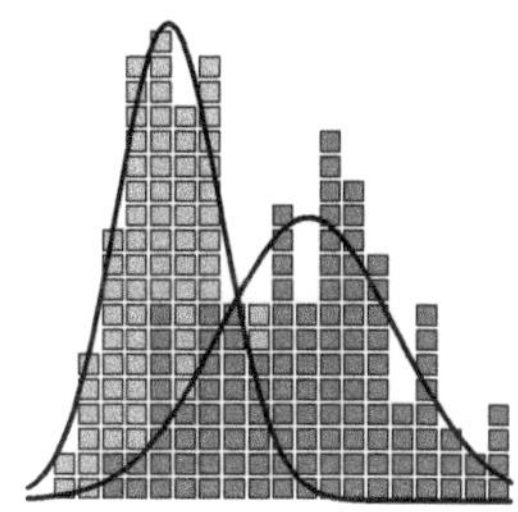

One inherent weakness associated with point estimators is that they do not quantify the precision of the estimator. As an illustration, assume that the annual salaries of n people are collected, yielding the sample mean annual salary

$$\bar{x} = \$35,281.$$

You would certainly have much more faith in this point estimate of the population mean annual salary μ if $n = 500$ salaries were collected than if only $n = 5$ salaries were collected. It would be beneficial to augment a point estimator $\hat{\mu}$ with supplemental information concerning its precision. Giving the value of the sample size n, or even better $V[\hat{\mu}]$, would be helpful, but statisticians tend to use an interval (sometimes known as a statistical interval), primarily a special type of interval known as a confidence interval, to quantify the uncertainty associated with a point estimator. The four sections of this chapter consider exact confidence intervals, approximate confidence intervals, asymptotically exact confidence intervals, and other types of intervals.

3.1 Exact Confidence Intervals

Edward Russo and Paul Schoemaker designed the following ten-question quiz.

1. Martin Luther King's age at death.
2. Length of the Nile River in miles.
3. Number of countries that are members of OPEC.
4. Number of books in the Old Testament.
5. Diameter of the moon in miles.
6. Weight of an empty Boeing 747 in pounds.
7. Year in which Wolfgang Amadeus Mozart was born.
8. Gestation period, in days, of an Asian elephant.
9. Air distance in miles from London to Tokyo.
10. Deepest known point in the oceans in feet.

If you were to simply guess at an answer to each of these ten questions, you would be providing ten *point estimates* of the true answers, which was the topic of the last chapter. But this was not what Russo and Schoemaker had in mind. Instead, the authors asked for a lower bound and upper bound for the answer to each question, chosen so that you are 90 percent confident that the correct answer will fall in the interval that you specify. Provide ten intervals that are wide enough so that the probability an individual interal contains the correct answer is 0.9. Go ahead and take the quiz now before reading on; the answers are given in the preface at the beginning of the Chapter 3 reference material. Count how many of your intervals contain the correct answer.

Russo and Schoemaker had more than 1000 American and European managers take the quiz. If the managers had accurate intervals that correctly expressed their uncertainty with their knowledge, you would expect that many of them would have exactly nine of the ten intervals contain the correct answers. But that was not the case. Less than 1 percent of the managers had nine or ten of the intervals contain the correct answers. Most managers had between three and six of their intervals contain the correct answers. What happened? The managers were *overconfident* in choosing their lower and upper bounds to their intervals, and chose intervals that were too narrow.

Russo and Schoemaker's quiz provides a good illustration of the use of an interval to express the *precision* associated with how much faith we have in our point estimator. They were assessing the ability to accurately measure how well you know what you don't know. To use some of the language that will be introduced subsequently in this chapter, although the *stated* coverage of the confidence intervals was 90%, the *actual* coverage turned out to be much lower than 90%, somewhere between 30% and 60%. The confidence intervals that the managers gave were *approximate* 90% confidence intervals rather than *exact* 90% confidence intervals.

A confidence interval gives an estimate of the precision associated with a point estimate. An exact confidence interval is the preferred type of confidence interval.

Definition 3.1 A random interval of the form

$$L < \theta < U$$

is an *exact two-sided* $100(1-\alpha)\%$ *confidence interval* for the unknown parameter θ provided

$$P(L < \theta < U) = 1 - \alpha$$

for all values of θ.

Some comments regarding exact confidence intervals are listed below.

- The confidence interval $L < \theta < U$ is a random interval because the endpoints of the interval, L and U, are random variables. It is for this reason that they are set in upper case.
- The random variable L is known as the *lower bound* of the confidence interval.
- The random variable U is known as the *upper bound* of the confidence interval.
- The probability $1-\alpha$ will be referred to as the *stated coverage*. Other terms to describe this quantity are the nominal coverage, the confidence level, and the confidence coefficient.
- Although α can assume any value between 0 and 1, the most common values used by statisticians are 0.1, 0.05, and 0.01. The confidence intervals associated with these values of α are referred to as 90%, 95%, and 99% confidence intervals, respectively. Because of the increased confidence, smaller values of α result in wider confidence intervals for identical data values.

- The notation in Definition 3.1 has been radically oversimplified. The random variables L and U are typically functions of the sample size n, the stated coverage $1-\alpha$, and the data values $X_1, X_2, \ldots, X_n$. Since the notation

$$L = g_1(n, \alpha, X_1, X_2, \ldots, X_n) \qquad \text{and} \qquad U = g_2(n, \alpha, X_1, X_2, \ldots, X_n)$$

is cumbersome, we let just L and U suffice.

- As was the case with point estimates, a confidence interval associated with a random sample drawn from a population is not necessarily unique. There might be several exact confidence intervals from which to choose. Criteria that can be used to select among them will be introduced in this chapter.

- The confidence interval defined in Definition 3.1 is exact because $P(L < \theta < U)$ is exactly $1-\alpha$. But not all confidence intervals are exact. Approximate confidence intervals and asymptotically exact confidence intervals will be introduced in the next two sections.

- There are three ways to report a confidence interval. The first follows Definition 3.1:

$$L < \theta < U.$$

This can also be reported as the open interval

$$(L, U).$$

Finally, when the confidence interval bounds are symmetric about the point estimator $\hat{\theta}$, the confidence interval can be reported as

$$\hat{\theta} \pm H,$$

where $H = (U-L)/2$ is the confidence interval *halfwidth*. The advantage of expressing a confidence interval in this format is that the point estimate $\hat{\theta}$ and the confidence interval halfwidth H are both immediately apparent.

- Reporting a confidence interval in the form $L < \theta < U$ does not imply that θ is between L and U. For this reason, it might be clearer writing a confidence interval as simply (L, U).

- Using the experimental values $x_1, x_2, \ldots, x_n$ of the random variables $X_1, X_2, \ldots, X_n$, the confidence intervals can be written in lower case:

$$l < \theta < u, \qquad (l, u), \qquad \text{or} \qquad \hat{\theta} \pm h.$$

- The confidence interval in Definition 3.1 is a two-sided confidence interval. Some applications require a *one-sided confidence interval* when the interest is only in an upper or lower bound on a parameter. A one-sided confidence interval of the form $(-\infty, U)$, which is typically reported as $\theta < U$, gives only an upper bound on the parameter θ. Conversely, a one-sided confidence interval of the form (L, ∞), which is typically reported as $\theta > L$, gives only a lower bound on the parameter θ. One-sided confidence intervals will be illustrated subsequently. The default assumption for confidence intervals is that they are two-sided.

- As was the case with point estimates, care must be taken to report the confidence interval bounds to the appropriate number of significant digits for a particular data set.

We begin with three examples of deriving an exact two-sided confidence interval from $n = 1$ observation drawn from a continuous population. Statisticians seldom encounter a data set with just a single observation, but if data are expensive (for example, if it involves crashing a car into a wall) or difficult to collect (for example, if it involves an experiment conducted by a rover on the moon), reporting a point and interval estimate from a single observation might be appropriate.

Example 3.1 Consider a single ($n = 1$) observation X drawn from a population with probability density function

$$f(x) = \theta x^{\theta-1} \qquad 0 < x < 1,$$

where θ is a positive unknown parameter.

(a) Find the method of moments and the maximum likelihood estimators of θ.

(b) Find an exact two-sided 90% confidence interval for θ.

(c) Calculate the point estimators and the 90% confidence interval for the data value $x = 0.63$.

The usual notation for a random sample is $X_1, X_2, \ldots, X_n$, but since $n = 1$, the subscripts are dropped entirely to keep the notation as simple as possible. This population distribution is a special case of the beta distribution with its second parameter equal to one. The population probability density function is plotted for several values of θ in Figure 3.1. It is clear from examining these curves that it would be preferable for a point estimate $\hat{\theta}$ to be monotonically increasing with the data value X, that is, small values of X should result in small values of $\hat{\theta}$ and large values of X should result in large values of $\hat{\theta}$. The population mean of this distribution is

$$E[X] = \int_0^1 \theta x^{\theta}\, dx = \left[\frac{\theta x^{\theta+1}}{\theta+1}\right]_0^1 = \frac{\theta}{\theta+1}.$$

The cumulative distribution function of the population distribution on the support of X is

$$F(x) = \int_0^x \theta w^{\theta-1}\, dw = \left[w^{\theta}\right]_0^x = x^{\theta} \qquad 0 < x < 1.$$

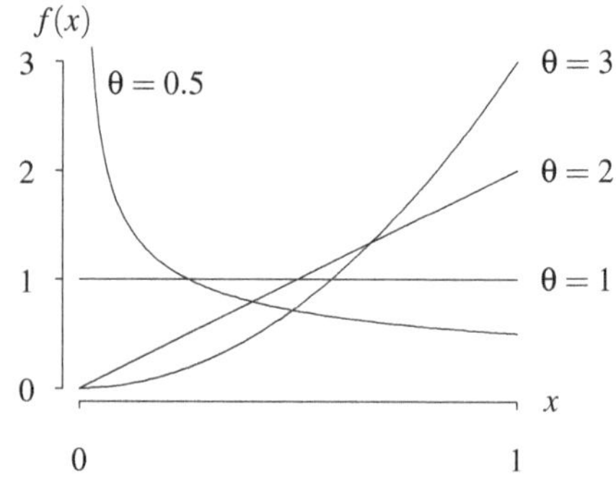

Figure 3.1: Population probability density functions.

(a) Since there is only a single positive unknown parameter θ, the population mean should be equated to the sample mean to determine the method of moments estimator. Since the sample mean for $n = 1$ is just the value of the single observation x, this equation is

$$\frac{\theta}{\theta+1} = x.$$

Solving for θ gives the method of moments estimate

$$\hat{\theta}_1 = \frac{x}{1-x}.$$

The subscript on $\hat{\theta}$ is needed to distinguish this estimate from the maximum likelihood estimate. To determine the maximum likelihood estimate, the likelihood function for the single observation is just the probability density function:

$$L(\theta) = f(x) = \theta x^{\theta-1}.$$

The log likelihood function is

$$\ln L(\theta) = \ln \theta + (\theta - 1) \ln x.$$

The score is

$$\frac{\partial \ln L(\theta)}{\partial \theta} = \frac{1}{\theta} + \ln x.$$

Equating the score to zero and solving for θ gives the maximum likelihood estimate

$$\hat{\theta}_2 = -\frac{1}{\ln x}.$$

Figure 3.2 shows these two point estimates as functions of the data value x. There is a vertical asymptote at $x = 1$. Both estimates are monotonically increasing in x as desired. The maximum likelihood estimate $\hat{\theta}_2$ is slightly larger than the method of moments estimator $\hat{\theta}_1$ for the same value of x.

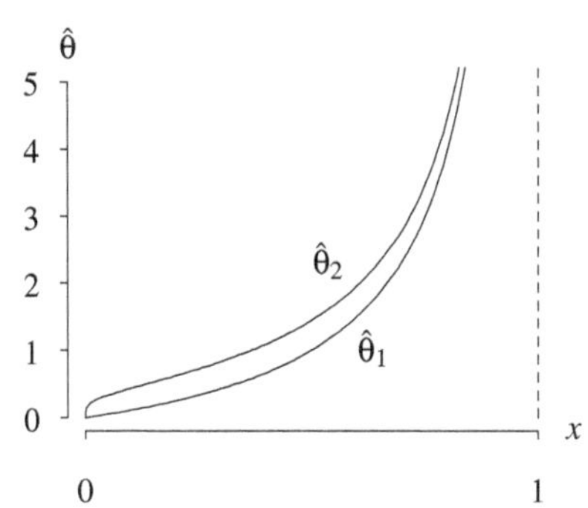

Figure 3.2: Method of moments and maximum likelihood estimators of θ.

(b) In our quest to find an exact two-sided 90% confidence interval for θ, we can begin with the fact that the probability that X lies between a and b can be calculated with

$$P(a < X < b) = \int_a^b f(x)\,dx,$$

where a and b are real constants satisfying $0 < a < b < 1$. Integration is used because the population probability distribution is continuous. Since this is to be an exact two-sided 90% confidence interval, this probability can be set to 0.9:

$$P(a < X < b) = \int_a^b f(x)\,dx = 0.9.$$

This probability statement is not yet of the form $P(L < \theta < U) = 0.9$ as in Definition 3.1, but this can be overcome. Unfortunately, there are an infinite number of pairs of a and b values satisfying this equation. One solution to the problem is to find a and b that minimize $b - a$, that is, the narrowest possible confidence interval. Another solution is to let a be the 5th percentile of the probability distribution of X and let b be the 95th percentile of the probability distribution of X. It is this second solution that is almost universally taken by statisticians. Using the cumulative distribution function of the population distribution, these percentiles can be found by solving

$$F(a) = a^{\theta} = 0.05$$

and

$$F(b) = b^{\theta} = 0.95$$

for a and b, which results in

$$a = 0.05^{1/\theta} \qquad \text{and} \qquad b = 0.95^{1/\theta}.$$

So the probability statement $P(a < X < b) = 0.9$ with this choice of a and b becomes

$$P\left(0.05^{1/\theta} < X < 0.95^{1/\theta}\right) = 0.9.$$

The next step is to perform algebraic operations on the inequality so as to place θ in the center of the inequality. Taking the natural logarithm of all parts of the inequality results in

$$P\left(\frac{1}{\theta}\ln 0.05 < \ln X < \frac{1}{\theta}\ln 0.95\right) = 0.9.$$

Finally, multiplying by θ (which is positive) and dividing by $\ln X$ (which is negative because the support of X is $0 < x < 1$),

$$P\left(\frac{\ln 0.95}{\ln X} < \theta < \frac{\ln 0.05}{\ln X}\right) = 0.9.$$

So an exact two-sided 90% confidence interval for the unknown parameter θ is

$$\frac{\ln 0.95}{\ln X} < \theta < \frac{\ln 0.05}{\ln X}.$$

This confidence interval is a random interval in that it is expressed in terms of the random variable X. For a specific data value x, it is written as

$$\frac{\ln 0.95}{\ln x} < \theta < \frac{\ln 0.05}{\ln x}.$$

(c) For the specific single observation $x = 0.63$, the two point estimators are

$$\hat{\theta}_1 = \frac{0.63}{1-0.63} = 1.70 \qquad \text{and} \qquad \hat{\theta}_2 = -\frac{1}{\ln 0.63} = 2.16.$$

The exact two-sided 90% confidence interval for θ is

$$\frac{\ln 0.95}{\ln 0.63} < \theta < \frac{\ln 0.05}{\ln 0.63}$$

or

$$0.11 < \theta < 6.48.$$

Since there is just one observation, the 90% confidence interval is quite wide, as one would expect. We simply cannot expect much precision on $\hat{\theta}$ from just a single observation. The point estimates and the confidence interval bounds were all reported to two digits to the right of the decimal point, which is consistent with the data value.

The initial example of an exact confidence interval considered a single observation X drawn from a population distribution with support $0 < x < 1$. The second example of an exact confidence interval considers a single observation X drawn from a population distribution with positive support $x > 0$. Population distributions of this type are used in a field known as *survival analysis*, where the interest is in a positive random variable, such as the remission time for a cancer, the duration of a strike by a labor union, or the lifetime of a coil spring.

Example 3.2 Consider a single ($n = 1$) observation X drawn from a population whose cumulative distribution function on its support is

$$F(x) = \frac{x^2}{\theta^2 + x^2} \qquad x > 0,$$

where θ is a positive unknown parameter.

(a) Find the method of moments and the maximum likelihood estimators of θ and determine which is the preferred point estimator.

(b) Find an exact two-sided 80% confidence interval for θ.

Writing the cumulative distribution function of the population distribution as

$$F(x) = \frac{(x/\theta)^2}{1+(x/\theta)^2} \qquad x > 0$$

emphasizes the fact that θ is a scale parameter of the population probability distribution. This population distribution is a special case of a *log logistic distribution*. The probability density function can be found by differentiating the cumulative distribution function with respect to x:

$$f(x) = \frac{2\theta^2 x}{\left(\theta^2 + x^2\right)^2} \qquad x > 0.$$

Leaving out the integration details, the population mean of the population distribution is

$$E[X] = \int_0^\infty x \cdot \frac{2\theta^2 x}{\left(\theta^2 + x^2\right)^2}\, dx = \frac{\pi\theta}{2}.$$

(a) Since there is only a single unknown parameter, the method of moments estimate of θ is found by equating the population mean to the sample mean, which is just x in this case:

$$\frac{\pi\theta}{2} = x.$$

Solving for θ gives the method of moments estimate

$$\hat{\theta}_1 = \frac{2x}{\pi}.$$

Switching the notation associated with this point estimator to upper case, the expected value of the method of moments estimator is

$$E\left[\hat{\theta}_1\right] = E\left[\frac{2X}{\pi}\right] = \frac{2}{\pi}E\left[X\right] = \frac{2}{\pi}\cdot\frac{\pi\theta}{2} = \theta,$$

which indicates that the method of moments estimator is an unbiased estimator of θ. For the maximum likelihood estimator, the likelihood function is again just the probability density function because there is only a single observation:

$$L(\theta) = f(x) = \frac{2\theta^2 x}{\left(\theta^2 + x^2\right)^2}.$$

The log likelihood function is

$$\ln L(\theta) = \ln 2 + 2\ln\theta + \ln x - 2\ln\left(\theta^2 + x^2\right).$$

The score is

$$\frac{\partial \ln L(\theta)}{\partial\theta} = \frac{2}{\theta} - \frac{4\theta}{\theta^2 + x^2}.$$

Equating the score to zero and solving for θ results in the maximum likelihood estimate

$$\hat{\theta}_2 = x.$$

Switching to upper case, the expected value of the maximum likelihood estimator is

$$E\left[\hat{\theta}_2\right] = E[X] = \frac{\pi\theta}{2},$$

so the maximum likelihood estimator is a biased estimator of θ. For this reason, the method of moments estimator is the preferred point estimator. Since the maximum likelihood estimator is a constant times the method of moments estimator, that is, $\hat{\theta}_2 = (\pi/2)\hat{\theta}_1$, the method of moments estimator is equivalent to the maximum likelihood estimator with an unbiasing constant. The subscript on the point estimator will be dropped and the method of moments estimate will be referred to as just $\hat{\theta} = 2x/\pi$.

A Monte Carlo simulation experiment can be used to assess the random sampling distribution of $\hat{\theta}$ for one particular value of the population parameter θ. Random variates from the population distribution are generated by inverting the cumulative distribution function of the population:

$$F^{-1}(u) = \theta\sqrt{\frac{u}{1-u}} \qquad 0 < u < 1,$$

where u is a random number. Consider 100 replications of the point estimator for a population with θ arbitrarily set to 5. The R code below generates and plots 100 point estimators.

```
theta = 5
nrep = 100
u = runif(nrep)
x = theta * sqrt(u / (1 - u))
thetahat = 2 * x / pi
plot(1:nrep, thetahat)
```

After a call to `set.seed(20)` to initialize the random number seed, the code is executed and the resulting graph is shown in Figure 3.3, which includes a spectacular $\hat{\theta} = 18.3$ in the 12th replication. Since the method of moments estimator $\hat{\theta}$ is unbiased, the 100 point estimators hover about $\theta = 5$, which is indicated by the horizontal line. The distribution of $\hat{\theta}$ is not symmetric; it has a long right-hand tail. The sample mean of the 100 simulated point estimates is 4.2, which is slightly lower than expected.

(b) The technique for determining an exact two-sided 80% confidence interval for θ will be similar to the technique used in the previous example. We begin by choosing constants a and b such that the probability that X lies between a and b is 0.8:

$$P(a < X < b) = \int_a^b f(x)\,dx = 0.8.$$

As before, there are an infinite number of pairs of a and b satisfying this equation, but common statistical practice is to choose a to be the 10th percentile of the population distribution and b to be the 90th percentile of the population distribution. These percentiles are found by solving

$$F(a) = \frac{a^2}{\theta^2 + a^2} = 0.1$$

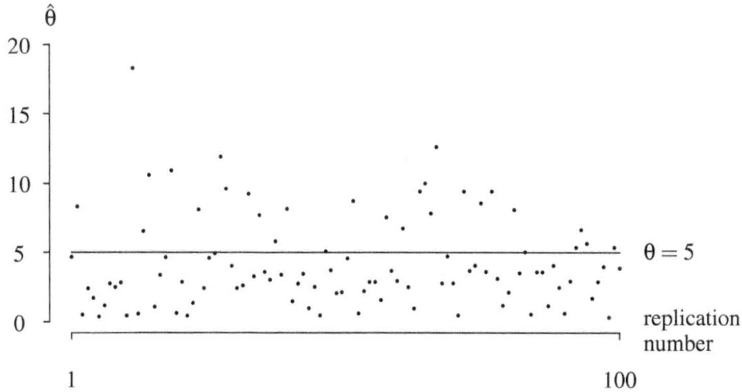

Figure 3.3: Monte Carlo simulation of 100 point estimates $\hat{\theta}$.

and

$$F(b) = \frac{b^2}{\theta^2 + b^2} = 0.9$$

for a and b, which results in

$$a = \sqrt{\frac{0.1}{0.9}}\,\theta = \frac{\theta}{3} \qquad \text{and} \qquad b = \sqrt{\frac{0.9}{0.1}}\,\theta = 3\theta.$$

So the probability statement $P(a < X < b) = 0.8$ becomes

$$P\left(\frac{\theta}{3} < X < 3\theta\right) = 0.8.$$

Dividing the inequality by θ, then dividing the inequality by X (both of which are positive), then inverting, θ now appears at the center of the inequality:

$$P\left(\frac{X}{3} < \theta < 3X\right) = 0.8.$$

So we have achieved a probability statement of the form $P(L < \theta < U)$. Thus, an exact two-sided 80% confidence interval for the unknown parameter θ is

$$\frac{X}{3} < \theta < 3X.$$

For a particular observation x, the 80% confidence interval is

$$\frac{x}{3} < \theta < 3x.$$

Monte Carlo simulation can again be used to evaluate the confidence interval procedure. Using the same random number stream as in Figure 3.3, 100 confidence intervals are generated from a population distribution with $\theta = 5$. Confidence intervals are drawn as vertical lines that intersect the point estimators in Figure 3.4. Several conclusions can immediately be drawn from Figure 3.4: (a) the confidence interval bounds are highly nonsymmetric about the point estimator, (b) the widths of the confidence intervals increase with $\hat{\theta}$, and (c) twice as many confidence intervals fall below $\theta = 5$ as those that fall above $\theta = 5$ for this particular simulation. Since 79 out of the 100 simulated confidence intervals contained the population parameter $\theta = 5$, the Monte Carlo simulation supports our analytic derivation of the exact 80% confidence interval. The number of confidence intervals that cover θ will vary from one simulation experiment to the next. The number of exact two-sided 80% confidence intervals that contain θ in the Monte Carlo experiment with 100 replications is a binomial random variable with parameters $n = 100$ and $p = 0.8$. The number of exact two-sided 80% confidence intervals that miss high and the number that miss low are binomial random variables with $n = 100$ and $p = 0.1$, although the three binomial random variables are correlated (the number of confidence intervals that miss low, cover $\theta = 5$, and miss high has the trinomial distribution). Care should be taken to not draw any meaningful conclusions from a Monte Carlo simulation experiment with only 100

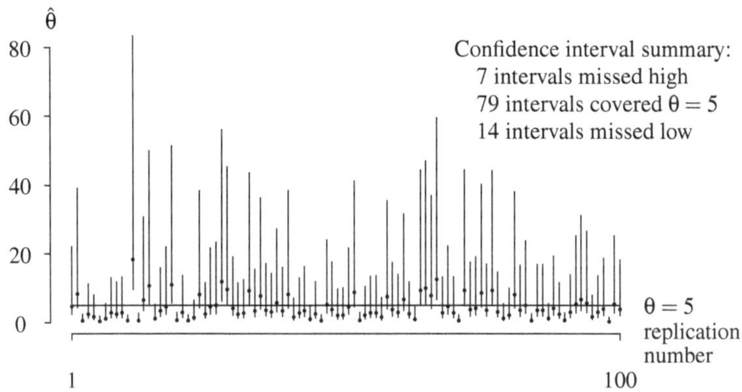

Figure 3.4: Monte Carlo simulation of 100 exact 80% confidence intervals.

replications. To that end, the number of replications was increased to one million, and the simulation yielded the following fractions of the confidence intervals generated that covered $\theta = 5$ for five runs:

0.7995 0.7997 0.8001 0.8003 0.8002.

These Monte Carlo simulation results provide strong evidence to support the exact coverage probability of the two-sided 80% confidence interval. In these simulation experiments, approximately 10% of the confidence intervals miss low and approximately 10% miss high. The fact that twice as many intervals missed low as those that missed high in Figure 3.4 is due to random sampling variability.

The Monte Carlo simulation in the previous example has highlighted the importance of the *width* of an exact confidence interval, which is defined next.

Definition 3.2 For the two-sided confidence interval $L < \theta < U$, the random confidence interval *width* is

$$W = U - L$$

and the random confidence interval *halfwidth* is

$$H = \frac{U - L}{2}.$$

Since the confidence interval width W and the confidence interval halfwidth H are random variables, their expected values and population variances are of interest. For any confidence interval, being exact, that is, $P(L < \theta < U) = 1 - \alpha$ for all values of θ and all sample sizes n, is a primary goal. For an exact confidence interval, one would prefer a small value of $E[W]$ or $E[H]$ as a secondary criterion because this is an indicator of increased precision. A tertiary criterion would be to have a small value of $V[W]$ or $V[H]$ because this indicates a stable width or halfwidth to the interval. The next example again considers the formulation of an exact two-sided confidence interval in the case of a single observation drawn from a continuous population, but this time the expected value and population variance of the confidence interval width will be calculated.

Example 3.3 Consider a single ($n = 1$) observation X drawn from a population with probability density function

$$f(x) = 2\theta^{-2}xe^{-(x/\theta)^2} \qquad x > 0,$$

for a positive unknown scale parameter θ.

(a) Find the method of moments and the maximum likelihood estimator of θ. Compute their expected values and population variances and decide which estimator is superior.

(b) Find an exact two-sided 82% confidence interval for θ.

(c) Find the expected value and population variance of the width of the confidence interval.

The population distribution is a special case of the Weibull distribution with shape parameter 2. This probability distribution is known as the *Rayleigh distribution*. Leaving out the integration details, the population mean is

$$E[X] = \int_0^{\infty} x \cdot 2\theta^{-2}xe^{-(x/\theta)^2}\,dx = \frac{\sqrt{\pi}}{2}\theta$$

and the population variance is

$$V[X] = \int_0^{\infty} \left(x - \frac{\sqrt{\pi}}{2}\theta\right)^2 \cdot 2\theta^{-2}xe^{-(x/\theta)^2}\,dx = \left(1 - \frac{\pi}{4}\right)\theta^2.$$

The cumulative distribution function of X on its support, which will be needed for determining an exact 82% confidence interval, is

$$F(x) = 1 - e^{-(x/\theta)^2} \qquad x > 0$$

for $\theta > 0$.

(a) With just a single random observation X, the sample mean is just the observed value x, so the method of moments estimate of θ can be found by equating the population mean and the sample mean:

$$\frac{\sqrt{\pi}}{2}\theta = x.$$

Solving for θ yields the method of moments point estimate

$$\hat{\theta}_1 = \frac{2}{\sqrt{\pi}}x.$$

The expected value and the population variance of the method of moments estimator are

$$E\left[\hat{\theta}_1\right] = E\left[\frac{2}{\sqrt{\pi}}X\right] = \frac{2}{\sqrt{\pi}}E[X] = \frac{2}{\sqrt{\pi}} \cdot \frac{\sqrt{\pi}}{2}\theta = \theta$$

and

$$V\left[\hat{\theta}_1\right] = V\left[\frac{2}{\sqrt{\pi}}X\right] = \frac{4}{\pi}V\left[X\right] = \frac{4}{\pi}\cdot\left(1-\frac{\pi}{4}\right)\theta^2 = \left(\frac{4}{\pi}-1\right)\theta^2.$$

So the method of moments estimator $\hat{\theta}_1$ is an unbiased estimator of θ with population variance $V\left[\hat{\theta}_1\right] = \left(\frac{4}{\pi}-1\right)\theta^2 \cong 0.2732\theta^2$. The likelihood function associated with the single observation is

$$L(\theta) = f(x) = 2\theta^{-2}xe^{-(x/\theta)^2}.$$

The log likelihood function is

$$\ln L(\theta) = \ln 2 - 2\ln\theta + \ln x - \left(\frac{x}{\theta}\right)^2.$$

The score is

$$\frac{\partial \ln L(\theta)}{\partial \theta} = -\frac{2}{\theta} + \frac{2x^2}{\theta^3}.$$

Equating the score to zero and solving for θ gives the maximum likelihood estimate

$$\hat{\theta}_2 = x.$$

The expected value and population variance of the maximum likelihood estimator are

$$E\left[\hat{\theta}_2\right] = E\left[X\right] = \frac{\sqrt{\pi}}{2}\theta$$

and

$$V\left[\hat{\theta}_2\right] = V\left[X\right] = \left(1-\frac{\pi}{4}\right)\theta^2.$$

So the maximum likelihood estimator is a biased estimator of θ with population variance $V\left[\hat{\theta}_2\right] = \left(1-\frac{\pi}{4}\right)\theta^2 \cong 0.2146\theta^2$. Since the method of moments estimator is unbiased, we will use it as the point estimator of choice for this problem, and refer to it as simply $\hat{\theta}$. The maximum likelihood estimator with an unbiasing constant will also result in the method of moments estimator. A separate analysis using the Cramér–Rao inequality shows that the method of moments estimator is not an efficient estimator, but comes very close to achieving the Cramér–Rao lower bound.

(b) To derive an exact two-sided 82% confidence interval for θ, we again begin with the observation from probability theory:

$$P(a < X < b) = \int_a^b f(x)\,dx = 0.82$$

for any two real constants a and b satisfying $0 < a < b$ because the population distribution is continuous. As before, there are an infinite number of choices for a and b, but we choose the 9th percentile for a and the 91st percentile for b. Using the cumulative distribution function for the population given earlier, these values can be found by solving

$$F(a) = 1 - e^{-(a/\theta)^2} = 0.09$$

and

$$F(b) = 1 - e^{-(b/\theta)^2} = 0.91$$

for a and b, which results in

$$a = \theta\sqrt{-\ln 0.91}$$

and

$$b = \theta\sqrt{-\ln 0.09}.$$

So the probability statement $P(a < X < b) = 0.82$ becomes

$$P\left(\theta\sqrt{-\ln 0.91} < X < \theta\sqrt{-\ln 0.09}\right) = 0.82.$$

The inequality in the probability operator can be manipulated to achieve an exact two-sided confidence interval of the form $P(L < \theta < U)$:

$$P\left(\frac{X}{\sqrt{-\ln 0.09}} < \theta < \frac{X}{\sqrt{-\ln 0.91}}\right) = 0.82.$$

So an 82% confidence interval for θ is

$$\frac{X}{\sqrt{-\ln 0.09}} < \theta < \frac{X}{\sqrt{-\ln 0.91}}.$$

If many confidence intervals were generated from a single observation from a Rayleigh population for a fixed value of θ, we would expect 9% of the intervals to miss θ low, 82% of the intervals to cover θ, and 9% of the intervals to miss θ high.

(c) The width of the confidence interval is

$$W = U - L = \frac{X}{\sqrt{-\ln 0.91}} - \frac{X}{\sqrt{-\ln 0.09}}.$$

The expected interval width is

$$\begin{aligned} E[W] &= E\left[\frac{X}{\sqrt{-\ln 0.91}} - \frac{X}{\sqrt{-\ln 0.09}}\right] \\ &= \left[\frac{1}{\sqrt{-\ln 0.91}} - \frac{1}{\sqrt{-\ln 0.09}}\right] E[X] \\ &= \left[\frac{1}{\sqrt{-\ln 0.91}} - \frac{1}{\sqrt{-\ln 0.09}}\right] \frac{\sqrt{\pi}}{2}\theta \\ &\cong 2.31470. \end{aligned}$$

The population variance of the interval width is

$$\begin{aligned} V[W] &= V\left[\frac{X}{\sqrt{-\ln 0.91}} - \frac{X}{\sqrt{-\ln 0.09}}\right] \\ &= \left[\frac{1}{\sqrt{-\ln 0.91}} - \frac{1}{\sqrt{-\ln 0.09}}\right]^2 V[X] \\ &= \left[\frac{1}{\sqrt{-\ln 0.91}} - \frac{1}{\sqrt{-\ln 0.09}}\right]^2 \left(1 - \frac{\pi}{4}\right)\theta^2 \\ &\cong 1.4639\theta^2. \end{aligned}$$

A Monte Carlo simulation experiment can again be used to produce a graphic which shows how 100 random confidence intervals with $n = 1$ would perform for one particular value of θ. The population cumulative distribution function should be inverted to generate a random variate which assumes the role of the single data value:

$$x = F^{-1}(u) = \theta\sqrt{-\ln(1-u)} \qquad 0 < u < 1.$$

The R code below generates 100 point estimators for θ, which are stored in the vector `thetahat`, and 100 exact two-sided 82% confidence intervals, whose lower and upper bounds are stored in the vectors `l` and `u`, for data values drawn from a population distribution with θ arbitrarily set to 2. The graph of the 100 exact two-sided 82% confidence intervals, after a call to `set.seed(20)` to initialize the random number stream, is shown in Figure 3.5.

```
theta    = 2
nrep     = 100
x        = theta * sqrt(-log(1 - runif(nrep)))
thetahat = 2 * x / sqrt(pi)
l        = x / sqrt(-log(0.09))
u        = x / sqrt(-log(0.91))
```

Since the method of moments estimator $\hat{\theta}$ is unbiased, the 100 point estimators hover about $\theta = 2$, which is indicated by the horizontal line in Figure 3.5. As before, the point estimates are given by points and the exact two-sided 82% confidence intervals are given by vertical lines. The confidence intervals are again highly nonsymmetric about the point estimator.

The Monte Carlo simulation experiments in the two previous examples each generated 100 confidence intervals for Figures 3.4 and 3.5. In a real-world statistical problem involving a real data set, however, you only calculate a single confidence interval. Monte Carlo simulation is used here to assess the performance of the confidence interval procedures.

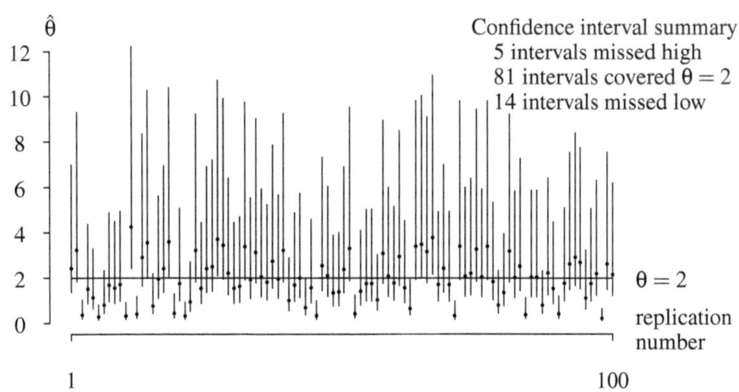

Figure 3.5: Monte Carlo simulation of 100 exact 82% confidence intervals.

The proper interpretation of a confidence interval is crucial. The following discussion concerns an exact 95% confidence interval for θ, that is, $\alpha = 0.05$. It is incorrect to say

> "The probability that my confidence interval contains the unknown parameter θ is 0.95"

because once the data has been collected and the interval is calculated, it either contains θ or it does not. A probability statement like this one does not make sense because there are no random variables after the data values are collected. The correct interpretation of an exact confidence interval for θ with nominal coverage 0.95 is as follows.

> "The confidence interval I have calculated might contain θ or it might not. However, if (a) all of the assumptions that I have made concerning the population probability distribution are correct, (b) many data sets consisting of random samples of size n are collected, and (c) the same procedure was used for calculating a confidence interval for each of the data sets, then 0.95 is the expected fraction of these confidence intervals that will contain the true parameter θ."

Obviously, one would not want to repeat this tedious explanation every time a confidence interval is calculated. So statisticians shorten this by simply saying:

> "I am 95% confident that my confidence interval contains the unknown parameter θ."

The brevity and avoidance of the use of "probability" in this statement aids the proper interpretation of the confidence interval.

With these definitions and this interpretation in place, we are now ready to move beyond $n = 1$. The next example considers $n = 2$ observations from a familiar distribution, and shows that a number of new issues arise.

Example 3.4 Let X_1 and X_2 be a random sample from a $U(0, \theta)$ population, where θ is a positive fixed but unknown parameter. Find an exact two-sided 84% confidence interval for θ.

The point estimate in this setting was investigated in Examples 2.17, 2.23, and 2.33, and it was concluded that the maximum likelihood estimator with an unbiasing constant, that is,

$$\hat{\theta} = \frac{3}{2} X_{(2)},$$

is the preferred point estimate. The method for deriving an exact confidence interval is not obvious. In the previous three examples, the random variable X was used in the inequality $P(a < X < b) = 1 - \alpha$ to arrive at $P(L < \theta < U) = 1 - \alpha$. But it is not clear what quantity belongs between a and b in the inequality for this particular example. There are an endless list of possibilities, for example:

$$X_1 + X_2 \qquad X_{(2)} \qquad X_{(1)} \qquad X_1 \cdot X_2 \qquad \sqrt{X_1/X_2}.$$

We will work with the first two listed above to derive two different exact two-sided 84% confidence intervals for θ. We will then describe how the two confidence intervals can be compared in order to select the superior confidence interval.

For a $U(0, \theta)$ population distribution, the sum of the two data values, $X_1 + X_2$, has the triangular distribution with minimum 0, mode θ, and maximum 2θ. Since this probability distribution involves the parameter θ, we can simplify by considering the quantity $(X_1 + X_2)/\theta$ which has the triangular distribution with minimum 0, mode 1, and maximum 2, which is a distribution that is free of θ. To determine an exact two-sided 84% confidence interval, we want a and b values that satisfy

$$P\left(a < \frac{X_1 + X_2}{\theta} < b\right) = 0.84.$$

More specifically, since $\alpha = 0.16$, we want to split the two tail probabilities $\alpha/2 = 0.08$ so that a and b are the 8th percentile and 92nd percentile of the triangular(0, 1, 2) distribution. We can avoid calculus because of the tractability of the triangular distribution. Using the one-half base times height formula for the area of a triangle, a must satisfy

$$\frac{1}{2}a^2 = 0.08.$$

Solving for a gives $a = 0.4$. By symmetry, $b = 1.6$. The geometry associated with finding the 8th percentile and 92nd percentile of a triangular(0, 1, 2) distribution is shown in Figure 3.6, where the areas associated with the various probabilities are labeled under the probability density function. So the probability statement becomes

$$P\left(0.4 < \frac{X_1 + X_2}{\theta} < 1.6\right) = 0.84.$$

Performing the algebra required to isolate θ in the center of the inequality results in

$$P\left(\frac{X_1 + X_2}{1.6} < \theta < \frac{X_1 + X_2}{0.4}\right) = 0.84,$$

so an exact two-sided 84% confidence interval for θ is

$$\frac{X_1 + X_2}{1.6} < \theta < \frac{X_1 + X_2}{0.4}.$$

This confidence interval, however, is but one among many.

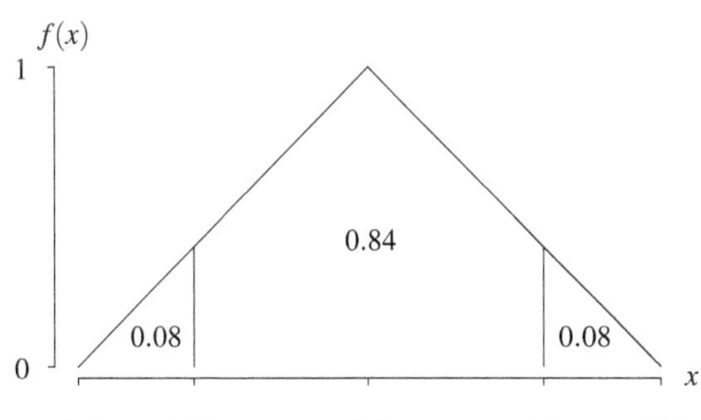

Figure 3.6: Triangular(0, 1, 2) distribution probability density function.

We now derive a second exact two-sided 84% confidence interval that is based on the distribution of $X_{(2)}$ rather than $X_1 + X_2$.

For the $U(0, \theta)$ population, the population probability density function is

$$f_X(x) = \frac{1}{\theta} \qquad 0 < x < \theta.$$

The associated cumulative distribution function on the support of X is

$$F_X(x) = \int_0^x \frac{1}{\theta}\, dw = \left[\frac{w}{\theta}\right]_0^x = \frac{x}{\theta} \qquad 0 < x < \theta.$$

Using an order statistic result, the maximum of two independent $U(0, \theta)$ random variables, $X_{(2)} = \max\{X_1, X_2\}$, has probability density function

$$f_{X_{(2)}}(x) = \frac{2!}{1!\,1!\,0!}\left(\frac{x}{\theta}\right)^1\left(\frac{1}{\theta}\right)^1\left(1 - \frac{x}{\theta}\right)^0 = \frac{2x}{\theta^2} \qquad 0 < x < \theta.$$

The associated cumulative distribution function of $X_{(2)}$ is

$$F_{X_{(2)}}(x) = \int_0^x \frac{2w}{\theta^2}\, dw = \left[\frac{w^2}{\theta^2}\right]_0^x = \frac{x^2}{\theta^2} \qquad 0 < x < \theta.$$

As with the first exact two-sided 84% confidence interval, this is a distribution that involves θ. Dividing $X_{(2)}$ by θ will result in a distribution that is free of θ. Let $Y = X_{(2)}/\theta$. Using the cumulative distribution function technique,

$$\begin{aligned}
F_Y(y) &= P(Y \le y) \\
&= P\left(X_{(2)}/\theta \le y\right) \\
&= P\left(X_{(2)} \le \theta y\right) \\
&= F_{X_{(2)}}(\theta y) \\
&= y^2 \qquad 0 < y < 1.
\end{aligned}$$

To find the exact two-sided 84% confidence interval for θ, we begin with a probability statement of the form

$$P\left(a < \frac{X_{(2)}}{\theta} < b\right) = 0.84.$$

Letting a be the 8th percentile of the probability distribution of $Y = X_{(2)}/\theta$ and b be the 92nd percentile of the probability distribution of $Y = X_{(2)}/\theta$, these constants can be found by solving

$$F_Y(a) = a^2 = 0.08 \qquad \text{and} \qquad F_Y(b) = b^2 = 0.92$$

for a and b, yielding

$$a = \sqrt{0.08} \qquad \text{and} \qquad b = \sqrt{0.92}.$$

The probability statement can be written as

$$P\left(\sqrt{0.08} < \frac{X_{(2)}}{\theta} < \sqrt{0.92}\right) = 0.84.$$

Isolating θ in the middle of the inequality leads to

$$P\left(\frac{X_{(2)}}{\sqrt{0.92}} < \theta < \frac{X_{(2)}}{\sqrt{0.08}}\right) = 0.84.$$

So an exact two-sided 84% confidence interval for θ based on $X_{(2)}$ is

$$\frac{X_{(2)}}{\sqrt{0.92}} < \theta < \frac{X_{(2)}}{\sqrt{0.08}}.$$

Now the two confidence intervals need to be compared.

For a particular data set consisting of observed values x_1 and x_2 of the independent random variables X_1 and X_2, the two exact two-sided 84% confidence intervals for θ are

$$\frac{x_1+x_2}{1.6} < \theta < \frac{x_1+x_2}{0.4} \qquad \text{and} \qquad \frac{x_{(2)}}{\sqrt{0.92}} < \theta < \frac{x_{(2)}}{\sqrt{0.08}}.$$

But which confidence interval is better? Since both intervals are exact, the tie-breaking criteria is to choose the narrower interval because this implies greater precision. Try plugging some data values into these intervals. You can easily concoct data sets in which the first confidence interval is narrower and other data sets in which the second confidence interval is narrower. So it is necessary to compute the expected value and population variance of the interval width for the two confidence intervals. The width of the first confidence interval is a triangular random variable because

$$W = U - L = \frac{X_1+X_2}{0.4} - \frac{X_1+X_2}{1.6} = \frac{15}{8}(X_1+X_2) \sim \text{triangular}\left(0,\ \frac{15}{8}\theta,\ \frac{15}{4}\theta\right).$$

Using the formulas for the population mean and population variance of a triangular random variable, the expected value and population variance of the width of the first confidence interval are

$$E[W] = \frac{15}{8}\theta = 1.875\theta \qquad \text{and} \qquad V[W] = \frac{75}{128}\theta^2 \cong 0.5859\theta^2.$$

The width of the second confidence interval is just a constant c times the second order statistic because

$$W = U - L = \frac{X_{(2)}}{\sqrt{0.08}} - \frac{X_{(2)}}{\sqrt{0.92}} = \frac{\sqrt{0.92}-\sqrt{0.08}}{\sqrt{0.08\cdot 0.92}}X_{(2)} = cX_{(2)}.$$

Since the expected value and population variance of $X_{(2)}$ are

$$E\left[X_{(2)}\right] = \int_0^\theta x \cdot \frac{2x}{\theta^2}\,dx = \left[\frac{2x^3}{3\theta^2}\right]_0^\theta = \frac{2\theta}{3}$$

and

$$V\left[X_{(2)}\right] = \int_0^\theta \left(x - \frac{2\theta}{3}\right)^2 \cdot \frac{2x}{\theta^2}\,dx = \frac{\theta^2}{18},$$

the expected value and population variance of the width of the second confidence interval are

$$E[W] = E\left[cX_{(2)}\right] = cE\left[X_{(2)}\right] = c\frac{2}{3}\theta \cong 1.6620\theta$$

and

$$V[W] = V\left[cX_{(2)}\right] = c^2 V\left[X_{(2)}\right] = c^2 \frac{1}{18}\theta^2 \cong 0.3453\theta^2.$$

The second confidence interval procedure produces narrower confidence intervals, on average, and their widths are also more stable. Based on this analysis, we prefer the second confidence interval.

There is a second way to compare the two confidence intervals. Since the lower confidence bound L and the upper confidence bound U are random variables, they have a joint probability distribution. For the first confidence interval, the values of L and U are

$$L = \frac{5}{8}(X_1 + X_2) \qquad \text{and} \qquad U = \frac{5}{2}(X_1 + X_2),$$

which fall on a line of slope $(5/2)/(5/8) = 4$ in the (L, U) plane. Likewise, for the second confidence interval, the values of L and U are

$$L = \frac{1}{\sqrt{0.92}} X_{(2)} \qquad \text{and} \qquad U = \frac{1}{\sqrt{0.08}} X_{(2)},$$

which fall on a line of slope $\sqrt{0.92/0.08} \cong 3.3912$ in the (L, U) plane. Since both sets of confidence interval bounds fall in lines with positive slope, the population correlation between L and U for both confidence intervals is 1. The Monte Carlo simulation experiment in R given below plots 100 random exact two-sided 84% confidence intervals of each type associated with a population with θ arbitrarily set to 2. The bounds for the first confidence interval are stored in the vectors `l1` and `u1`; the bounds for the second confidence interval are stored in the vectors `l2` and `u2`. These confidence interval bounds are plotted as points in the (l, u) plane in Figure 3.7.

```
theta = 2
n     = 2
nrep  = 100
l1    = u1 = l2 = u2 = rep(0, nrep)
for (i in 1:nrep) {
  x     = runif(n, 0, theta)
  l1[i] = sum(x) / 1.6
  u1[i] = sum(x) / 0.4
  l2[i] = max(x) / sqrt(0.92)
  u2[i] = max(x) / sqrt(0.08)
}
plot(c(l1, l2), c(u1, u2))
```

Notice that the scales on the two axes differ so that the line $l = u$ is not inclined at 45°. It is not possible for a confidence interval of the form $l < \theta < u$ to fall below the line $l = u$. The following three regions are associated with the three outcomes for a particular confidence interval.

- Any (l, u) pair that falls below $u = \theta = 2$ corresponds to a confidence interval that misses θ low.
- Any (l, u) pair that falls to the northwest of $(\theta, \theta) = (2, 2)$ corresponds to a confidence interval that covers θ.

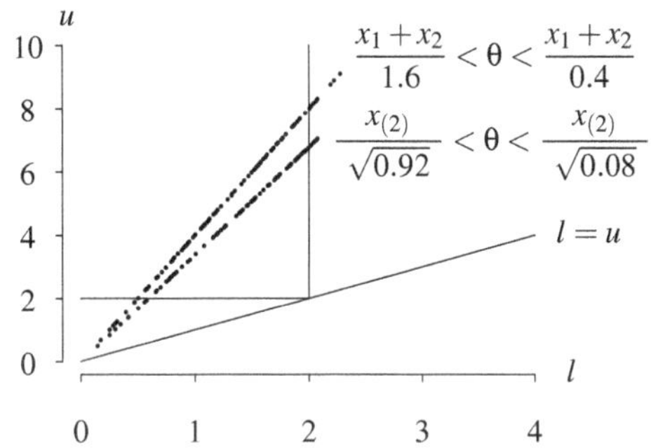

Figure 3.7: Monte Carlo simulation of 100 (l, u) pairs for $\theta = 2$.

- Any (l, u) pair that falls to the right of $l = \theta = 2$ corresponds to a confidence interval that misses θ high.

The confidence intervals associated with the second order statistic fall closer to the line $l = u$ which means that they are narrower, and therefore this is the preferred confidence interval procedure. This is consistent with our calculation of the expected confidence interval width $E[W]$.

So far, our development of confidence intervals has been somewhat ad hoc. That will soon change. Notice from the example concerning random sampling from a $U(0, \theta)$ population that we worked with two quantities,

$$\frac{X_1 + X_2}{\theta} \qquad \text{and} \qquad \frac{X_{(2)}}{\theta},$$

whose probability distributions did not depend on the unknown parameter θ. For both of these quantities, dividing by θ eliminated θ from the probability density function of the quantity. Quantities such as these are known as *pivotal quantities* and they are defined next.

Definition 3.3 Let $X_1, X_2, \ldots, X_n$ be a random sample from a population with unknown parameter θ. A random variable $g(X_1, X_2, \ldots, X_n, \theta)$, which is a function of the data and the unknown parameter, whose probability distribution does not involve θ, is known as a *pivotal quantity*.

A pivotal quantity is also known as a pivot or a pivotal statistic. The term *pivotal statistic* is avoided because while a statistic is a function of the data only, a pivotal quantity is a function of the data and the unknown parameter. The next two examples illustrate the identification of a pivotal quantity.

Example 3.5 Let X_1 and X_2 be a random sample from a $U(0, \theta)$ population, where θ is a positive unknown parameter.

- $X_1 \sim U(0, \theta)$ is not a pivotal quantity because its distribution depends on θ.
- $X_1 + X_2 \sim \text{triangular}(0, \theta, 2\theta)$ is not a pivotal quantity because its distribution depends on θ.

- $(X_1 + X_2)/\theta \sim$ triangular(0, 1, 2) is a pivotal quantity because its distribution does not depend on θ.
- $X_{(1)}/\theta \sim$ beta(1, 2) is a pivotal quantity because its distribution does not depend on θ.

Example 3.6 Let $X_1, X_2, \ldots, X_n$ be a random sample from a $N(\mu, 1)$ population, where μ is an unknown parameter.

- $X_2 \sim N(\mu, 1)$ is not a pivotal quantity because its distribution depends on μ.
- $X_1 + X_2 + \cdots + X_n \sim N(n\mu, n)$ is not a pivotal quantity because its distribution depends on μ.
- $X_2 - \mu \sim N(0, 1)$ is a pivotal quantity because its distribution does not depend on μ.
- $(\bar{X} - \mu) / (1/\sqrt{n}) \sim N(0, 1)$ is a pivotal quantity because its distribution does not depend on μ.

The "pivotal method" for constructing an exact two-sided $100(1-\alpha)\%$ confidence interval for the unknown parameter θ consists of the following three steps. We assume that the preliminary work has been done to identify the population distribution (a histogram is typically part of this preliminary work) and a high quality point estimator has been identified.

(a) Identify a pivotal quantity. As seen in the two previous examples, there can be several pivotal quantities. Is there a way to identify more promising pivotal quantities? *It is common practice in statistics to choose a pivotal quantity that is a function of a sufficient statistic.* The confidence interval in Example 3.4 that was associated with the pivotal quantity that was a function of the sufficient statistic $X_{(2)}$ outperformed the confidence interval associated with the pivotal quantity that was a function of $X_1 + X_2$, which is not a sufficient statistic. This choice for a pivotal quantity is related to the notion that a sufficient statistic captures all of the information with respect to estimating θ from a random sample.

(b) Letting $g(X_1, X_2, \ldots, X_n, \theta)$ denote the pivotal quantity that has been identified, find real constants a and b such that

$$P\big(a < g(X_1, X_2, \ldots, X_n, \theta) < b\big) = 1 - \alpha.$$

From the previous examples, it should now be clear that a and b are traditionally taken to be the $\alpha/2$ and $1-\alpha/2$ percentiles of the distribution of the pivotal quantity. Sometimes these percentiles can be expressed in closed form, as was the case in the first four examples in this chapter. In other cases, the pivotal quantity has a well-known distribution (for example, the normal or chi-square distribution), so the percentiles can be looked up in a table or computed in a statistical programming language. As a worst-case scenario, the percentiles might not be available in closed form and might not correspond to a well-known distribution, so numerical methods might be required to compute the percentiles.

(c) Once the values of a and b have been established, all that remains is to perform algebraic manipulations to the inequality in the probability statement that results in the unknown parameter alone in the center of the inequality, that is, of the form $L < \theta < U$. In the first four examples in this chapter, this could be done in closed form. Situations might arise in which numerical methods are required on this step.

These steps will now be applied to a random sample drawn from an exponential population. Unlike the previous examples in which n and α were specified constants, an exact two-sided confidence interval is derived for general values of n and α.

Example 3.7 Let $X_1, X_2, \ldots, X_n$ be a random sample from a population with probability density function

$$f_X(x) = \frac{1}{\theta} e^{-x/\theta} \qquad\qquad x > 0,$$

where θ is a positive unknown parameter. Find an exact two-sided $100(1-\alpha)\%$ confidence interval for θ.

This population distribution is recognized as the exponential distribution with population mean θ. From Example 2.8 the maximum likelihood estimator of the population mean θ is the sample mean

$$\hat{\theta} = \frac{1}{n}\sum_{i=1}^{n} X_i,$$

which is an unbiased estimator of θ. Furthermore, Example 2.20 showed that the maximum likelihood estimator is efficient. The search for a pivotal quantity begins with identifying a sufficient statistic. The Fisher–Neyman factorization criterion in Theorem 2.7 indicates that the statistic $Y = g(X_1, X_2, \ldots, X_n)$ is a sufficient statistic for the parameter θ if there exist nonnegative functions h_1 and h_2 such that

$$f_X(x_1)f_X(x_2)\ldots f_X(x_n) = h_1\left(g(x_1, x_2, \ldots, x_n), \theta\right) \cdot h_2(x_1, x_2, \ldots, x_n).$$

For a random sample drawn from an exponential population, the joint probability density function of $X_1, X_2, \ldots, X_n$ can be written as

$$f_X(x_1)f_X(x_2)\ldots f_X(x_n) = \frac{1}{\theta}e^{-x_1/\theta}\frac{1}{\theta}e^{-x_2/\theta}\cdots\frac{1}{\theta}e^{-x_n/\theta} = \underbrace{\frac{1}{\theta^n}e^{-\sum_{i=1}^{n} x_i/\theta}}_{h_1} \cdot \underbrace{1}_{h_2},$$

so $\sum_{i=1}^{n} X_i$ is a sufficient statistic for θ. The sum of n mutually independent and identically distributed exponential random variables has the Erlang distribution, that is,

$$\sum_{i=1}^{n} X_i \sim \text{Erlang}(1/\theta, n).$$

The sufficient statistic is not a pivotal quantity because its probability distribution depends on θ. We would like to transform the sufficient statistic in a manner so that the distribution of the transformed random variable (the pivotal quantity) does not depend on θ. As a bonus, it would be helpful if the pivotal quantity has tractable percentiles for establishing confidence intervals. The transformation that will take the sufficient statistic to a pivotal quantity is not immediately obvious, so the details are given below.

Since the sufficient statistic is the sum of the data values, we will work with moment generating functions. The moment generating function of the population distribution is

$$M_X(t) = E\left[e^{tX}\right] = \int_0^\infty e^{tx} \cdot \frac{1}{\theta}e^{-x/\theta}\,dx = \cdots = \frac{1}{1-\theta t} \qquad\qquad t < \frac{1}{\theta}.$$

Since $X_1, X_2, \ldots, X_n$ are mutually independent and identically distributed random variables, the moment generating function of their sum is

$$M_{\sum_{i=1}^n X_i}(t) = \prod_{i=1}^{n} M_{X_i}(t) = \left(\frac{1}{1-\theta t}\right)^n \qquad t < \frac{1}{\theta},$$

which is the moment generating function of an Erlang$(1/\theta, n)$ random variable. At this point we would like to transform the sufficient statistic so that it does not involve θ to arrive at a pivotal quantity. After some experimentation, the sum can be multiplied by $2/\theta$, which has moment generating function

$$M_{2\sum_{i=1}^n X_i/\theta}(t) = M_{\sum_{i=1}^n X_i}(2t/\theta) = \left(\frac{1}{1-2t}\right)^n \qquad t < \frac{1}{2},$$

which does not depend on θ. This moment generating function is recognized as that of a chi-square random variable with $2n$ degrees of freedom. So our pivotal quantity has been established:

$$\frac{2}{\theta}\sum_{i=1}^{n} X_i \sim \chi^2(2n),$$

which is a probability distribution that does not depend on θ. The second step of the pivotal method is to find real constants a and b such that

$$P\left(a < \frac{2}{\theta}\sum_{i=1}^{n} X_i < b\right) = 1-\alpha.$$

As in the previous cases, a is set to the $100(\alpha/2)$th percentile of the $\chi^2(2n)$ distribution and b is set to the $100(1-\alpha/2)$th percentile of the $\chi^2(2n)$ distribution. Denoting these quantities with a subscript that includes the degrees of freedom and the right-hand tail probabilities (a common practice in statistics) for the chi-square percentiles,

$$P\left(\chi^2_{2n,1-\alpha/2} < \frac{2}{\theta}\sum_{i=1}^{n} X_i < \chi^2_{2n,\alpha/2}\right) = 1-\alpha.$$

These two percentiles, along with the probability density function of a chi-square random variable with $2n$ degrees of freedom, and the associated areas are shown in Figure 3.8. The vertical axis label, $f(x)$, refers to the probability density function of a chi-square random variable with $2n$ degrees of freedom rather than the parent population distribution.

The third and final step in the pivotal method is to use algebra to manipulate the inequality in the probability statement so that θ appears alone in the middle of the inequality. Dividing by $2\sum_{i=1}^n X_i$ and inverting yields

$$P\left(\frac{2\sum_{i=1}^n X_i}{\chi^2_{2n,\alpha/2}} < \theta < \frac{2\sum_{i=1}^n X_i}{\chi^2_{2n,1-\alpha/2}}\right) = 1-\alpha.$$

So an exact two-sided $100(1-\alpha)\%$ confidence interval for θ for a random sample of size n drawn from an exponential population with unknown population mean θ is

$$\frac{2\sum_{i=1}^n X_i}{\chi^2_{2n,\alpha/2}} < \theta < \frac{2\sum_{i=1}^n X_i}{\chi^2_{2n,1-\alpha/2}}.$$

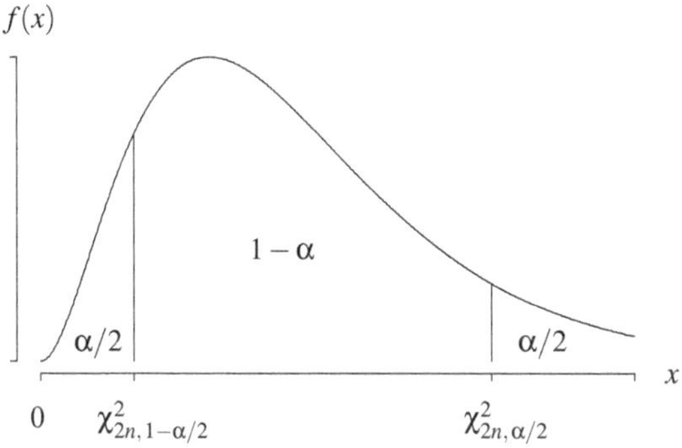

Figure 3.8: Probability density function of a $\chi^2(2n)$ random variable with percentiles.

The expected width of this confidence interval is

$$\begin{aligned}
E[W] &= E[U-L] \\
&= E\left[\frac{2\sum_{i=1}^{n} X_i}{\chi^2_{2n,1-\alpha/2}} - \frac{2\sum_{i=1}^{n} X_i}{\chi^2_{2n,\alpha/2}}\right] \\
&= \left[\frac{2}{\chi^2_{2n,1-\alpha/2}} - \frac{2}{\chi^2_{2n,\alpha/2}}\right] E\left[\sum_{i=1}^{n} X_i\right] \\
&= 2n\theta \left[\frac{1}{\chi^2_{2n,1-\alpha/2}} - \frac{1}{\chi^2_{2n,\alpha/2}}\right].
\end{aligned}$$

A Monte Carlo simulation can be used to visualize 100 exact two-sided confidence intervals for observations drawn from an exponential population, with n, θ, and α set to arbitrary values. Letting $n = 5$, $\theta = 1$, and $\alpha = 0.1$, the R code below generates 100 point estimates and 100 exact two-sided 90% confidence intervals. Notice that the first parameter in the `qchisq` function corresponds to a left-hand tail probability, contrary to the mathematical notation in this example.

```
nrep     = 100
n        = 5
alpha    = 0.1
thetahat = l = u = numeric(nrep)
for (i in 1:nrep) {
  x           = rexp(n, 1)
  thetahat[i] = mean(x)
  l[i]        = 2 * sum(x) / qchisq(1 - alpha / 2, 2 * n)
  u[i]        = 2 * sum(x) / qchisq(alpha / 2, 2 * n)
}
```

The plot of these point and interval estimators in Figure 3.9, after a call to `set.seed(7)`, again reveals nonsymmetric confidence intervals about the associated point estimators.

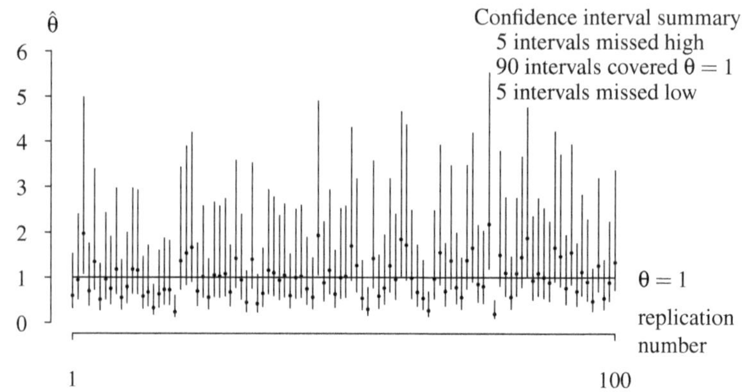

Figure 3.9: Monte Carlo simulation of 100 exact 90% confidence intervals ($n = 5$).

This is due to the nonsymmetry of the chi-square distribution. The width of the confidence intervals increases with the point estimators. The fact that five of the confidence intervals missed θ low, 90 of the confidence intervals covered θ, and five of the confidence intervals missed θ high should not be taken as strong evidence that the confidence interval formula is necessarily correct. Many more than 100 replications of the simulation experiment should be conducted to be sure that the proportions converge to 5%, 90%, and 5%.

What would these confidence intervals look like if the sample size were increased from $n = 5$ to $n = 25$? This question is addressed with another Monte Carlo simulation, whose results are displayed in Figure 3.10, using the same vertical scale as in Figure 3.9.

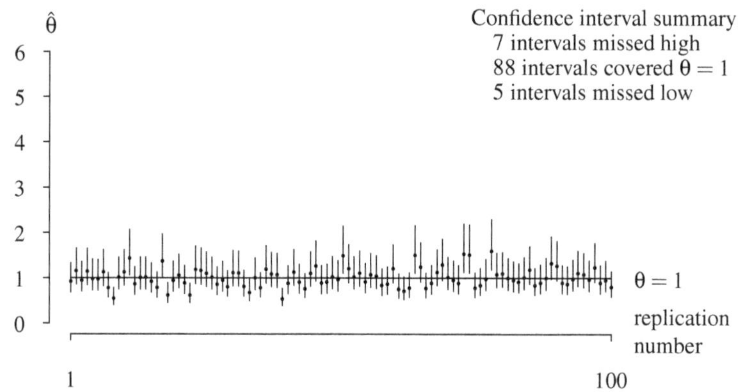

Figure 3.10: Monte Carlo simulation of 100 exact 90% confidence intervals ($n = 25$).

Two important differences are immediately apparent. First, the confidence intervals are much narrower, as is expected with a much larger sample size. Second, the confidence intervals are more symmetric than with the smaller sample size, as is expected because the chi-square distribution is more symmetric for a larger number of degrees of freedom.

Although the Monte Carlo experiment is helpful for visualizing the behavior of many confidence intervals, one only constructs a single confidence interval for a particular data set. To illustrate, consider the $n = 27$ times between failures (in hours) for an air conditioning system collected by Frank Proschan on a Boeing 720 jet airplane:

$$97, 51, 11, 4, 141, 18, 142, 68, 77, 80, 1, 16, 106, 206,$$

$$82, 54, 31, 216, 46, 111, 39, 63, 18, 191, 18, 163, 24.$$

The histogram of this data in Figure 3.11 indicates that the exponential distribution might provide a reasonable fit because its shape is similar to that of the probability density function of an exponential random variable.

The fitting process begins by calculating the point estimate of the population mean θ, which is just the sample mean:

$$\hat{\theta} = 77 \text{ hours.}$$

The next step is to calculate a confidence interval for θ which gives an indication of the precision of the point estimate. Arbitrarily setting $\alpha = 0.1$, an exact two-sided 90% confidence interval for θ is calculated with

$$\frac{2\sum_{i=1}^{n} x_i}{\chi^2_{2n,\alpha/2}} < \theta < \frac{2\sum_{i=1}^{n} x_i}{\chi^2_{2n,1-\alpha/2}}$$

or

$$\frac{2 \cdot 2074}{72.15} < \theta < \frac{2 \cdot 2074}{38.12}$$

or

$$57 < \theta < 109.$$

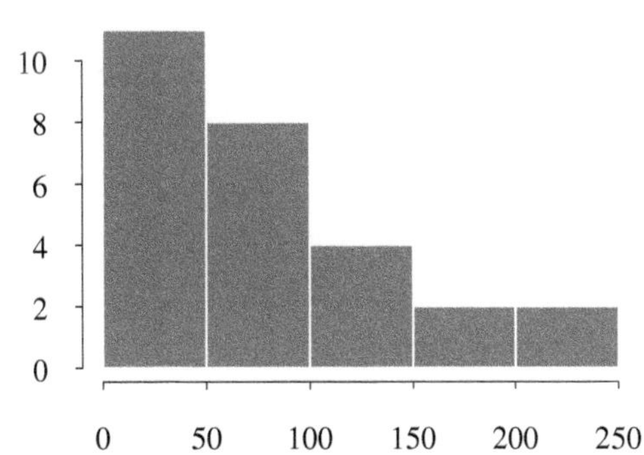

Figure 3.11: Histogram of the times between failures for an air conditioning system.

We are 90% confident that the population mean time between failure for the air conditioning system lies between 57 and 109 hours. Figure 3.12 contains a plot of the empirical and the fitted cumulative distribution functions. The exponential distribution appears to provide a reasonable fit to the data.

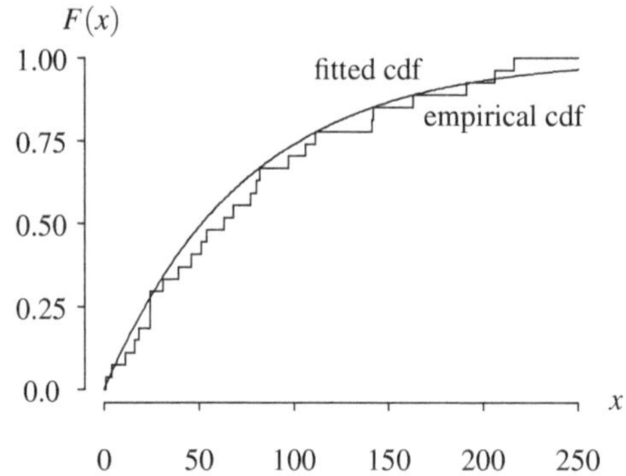

Figure 3.12: Cumulative distribution functions for the air conditioning failure data.

The fitting of the exponential distribution to the air conditioning system failure data provides the opportunity to bring up three important topics that have not yet been addressed: one-sided confidence intervals, confidence intervals for quantities other than θ, and sample size determination. The following three paragraphs discuss these three important topics.

All of the confidence intervals discussed thus far have been two-sided confidence intervals, but some applications favor the use of a one-sided confidence interval. In the case of the air conditioning system failure data, having a particularly *long* time between failure is, from Boeing's perspective, a good thing. It is quite likely that they are only concerned with *short* times between failure. So instead of a two-sided confidence interval, they might only want a lower bound on the population mean time between failure θ. Since the 90% confidence interval calculated in Example 3.7, which was

$$57 < \theta < 109,$$

used the 5th and 95th percentiles of the chi-square distribution with $2n$ degrees of freedom, we can conclude that an exact one-sided 95% confidence interval for θ is

$$\theta > 57.$$

We are 95% confident that the population mean time between failure exceeds 57 hours. Generalizing this approach, the formula for an exact one-sided confidence interval that satisfies $P(\theta > L) = 1 - \alpha$ for a random sample from an exponential population which includes only a lower bound on θ is

$$\theta > \frac{2\sum_{i=1}^{n} x_i}{\chi^2_{2n,\alpha}}.$$

Likewise, the formula for an exact one-sided confidence interval that includes only an upper bound on θ is

$$\theta < \frac{2\sum_{i=1}^{n} x_i}{\chi^2_{2n,1-\alpha}}.$$

The derivation of these formulas is analogous to their two-sided counterparts, except that there is no longer a need to split the probability α between the two tails of the distribution. When should a one-sided confidence interval be used in place of the usual two-sided confidence interval? The context of the problem will typically dictate which type of interval is appropriate. If the confidence interval concerns the population mean diameter of ball bearings, for example, and a large diameter and a small diameter are detrimental to the product, then a two-sided confidence interval is appropriate. On the other hand, if the unknown parameter is the population mean time between air conditioning failures, and an unusually large population mean time between failure is not a concern (but an unusually small population mean time between failures is a concern), then a one-sided confidence interval is appropriate.

The second issue is determining a confidence interval for a quantity other than θ. In the air conditioning system failure data scenario, the quantity of interest might be the probability that the air conditioning system survives to 200 hours. How would we develop a point and interval estimate for this probability? Let X be an exponential random variable with population mean θ. The probability that an air conditioner lasts 200 hours without failure is

$$P(X > 200) = \int_{200}^{\infty} \frac{1}{\theta} e^{-x/\theta}\, dx = \left[-e^{-x/\theta}\right]_{200}^{\infty} = e^{-200/\theta}.$$

Using the invariance property and the fact that the maximum likelihood estimator of θ is $\hat{\theta} = \bar{x} = 77$, the maximum likelihood estimator of $P(X > 200)$ is

$$\hat{P}(X > 200) = e^{-200/\hat{\theta}} = e^{-200/77} = 0.074.$$

(Keep in mind that all of the accuracy from the original data set has been used to compute this probability. Roundoff from reporting intermediate calculations, such as $\hat{\theta} = 77$, to significant digits, should not be allowed to propagate.) This point estimate is certainly reasonable, because two out of the $n = 27$ air conditioning systems had a time between failures larger than 200 hours, and $2/27 = 0.074$, which matches to two digits. Geometrically, these values are the vertical differences between the fitted and empirical cumulative distribution functions in Figure 3.12 and 1 at $x = 200$. Finding an exact two-sided 90% confidence interval for $P(X > 200)$ involves algebraic manipulation of the exact two-sided 90% confidence interval for θ from Example 3.7:

$$57 < \theta < 109.$$

The goal is to transform the inequality so that $e^{-200/\theta}$ is in the center. Begin by taking reciprocals:

$$\frac{1}{109} < \frac{1}{\theta} < \frac{1}{57}.$$

Multiplying by 200, negating, then exponentiating gives the exact two-sided 90% confidence interval

$$e^{-200/57} < e^{-200/\theta} < e^{-200/109}$$

or

$$0.031 < P(X > 200) < 0.16.$$

We are 90% confident that the probability an air conditioning system operates without failure for 200 hours is between 0.031 and 0.16.

The third issue raised by the previous example concerns sample size determination. The exact two-sided 90% confidence interval for θ,

$$57 < \theta < 109,$$

has width $w = 109 - 57 = 52$ hours. This might be too wide a range for management to make decisions with the precision that they would prefer. What sample size would be necessary to achieve a narrower confidence interval with a width of, say, 30 hours?

The formula derived in Example 3.7 for the expected interval width is

$$E[W] = 2n\theta \left[\frac{1}{\chi^2_{2n,1-\alpha/2}} - \frac{1}{\chi^2_{2n,\alpha/2}} \right].$$

One could simply increase n until the right-hand side of this equation dips below 30. The only problem here is that θ is unknown. If one were willing to use $\hat{\theta}$ in its place, then the following R code would compute a sample size that will decrease the expected confidence interval width to 30.

```
alpha = 0.1
s     = 2074
n     = 27
while (2 * s / qchisq(alpha / 2, 2 * n) - 2 * s / qchisq(1 - alpha / 2, 2 * n) > 30)
  n = n + 1
print(n)
```

The code returns $n = 39$. An additional 12 observations should be collected to shrink the width of the confidence interval to 30 hours. There is no guarantee that the interval width will actually be less than 30 hours if a total of $n = 37$ observations are collected because the interval width is a random variable and we used $\hat{\theta}$ to approximate θ.

In Example 3.4, we used trial and error to develop a confidence interval for θ for a $U(0, \theta)$ population with $n = 2$ and $\alpha = 0.16$. We now determine a confidence interval for general n and α using the pivotal method.

Example 3.8 Let $X_1, X_2, \ldots, X_n$ be a random sample from a $U(0, \theta)$ population, where θ is a positive unknown parameter. Find an exact two-sided $100(1-\alpha)\%$ confidence interval for θ.

As shown in Example 2.17, the maximum likelihood estimator modified by an unbiasing constant is

$$\hat{\theta} = \frac{n+1}{n} X_{(n)}.$$

Also, Example 2.26 established that $X_{(n)}$ is a sufficient statistic for θ. So we begin by finding the probability density function of the sufficient statistic. For the $U(0, \theta)$ population, the population probability density function is

$$f_X(x) = \frac{1}{\theta} \qquad 0 < x < \theta.$$

The associated cumulative distribution function on its support is

$$F_X(x) = \int_0^x \frac{1}{\theta}\, dw = \left[\frac{w}{\theta} \right]_0^x = \frac{x}{\theta} \qquad 0 < x < \theta.$$

Using an order statistic result, the maximum of n mutually independent $U(0, \theta)$ random variables, $X_{(n)} = \max\{X_1, X_2, \ldots, X_n\}$, has probability density function

$$f_{X_{(n)}}(x) = \frac{n!}{(n-1)!\,1!\,0!} \left(\frac{x}{\theta} \right)^{n-1} \left(\frac{1}{\theta} \right)^1 \left(1 - \frac{x}{\theta} \right)^0 = \frac{nx^{n-1}}{\theta^n} \qquad 0 < x < \theta.$$

So the sufficient statistic $X_{(n)}$ is not a pivotal quantity because its distribution depends on θ. We seek a function of $X_{(n)}$ that does not depend on θ, which will be the pivotal quantity. After some trial and error, we find that simply dividing $X_{(n)}$ by θ results in a pivotal quantity. The transformation technique is used to find the probability density function of the pivotal quantity $Y = g\left(X_{(n)}\right) = X_{(n)}/\theta$. The function $y = g\left(x_{(n)}\right) = x_{(n)}/\theta$ is a one-to-one transformation from the support of $X_{(n)}$, which is

$$\mathcal{A} = \{x \,|\, 0 < x < \theta\}$$

to the support of Y, which is

$$\mathcal{B} = \{y \,|\, 0 < x < 1\},$$

with inverse

$$x_{(n)} = g^{-1}(y) = \theta y$$

and

$$\frac{dx_{(n)}}{dy} = \theta.$$

Using the transformation technique, the probability density function of Y on its support is

$$f_Y(y) = \frac{n}{\theta^n}(\theta y)^{n-1} \cdot |\theta| = ny^{n-1} \qquad\qquad 0 < y < 1.$$

So $Y = g\left(X_{(n)}\right) = X_{(n)}/\theta$ has a distribution that does not depend on θ, which makes it a pivotal quantity. The associated cumulative distribution function on the support of Y, which will be helpful for finding percentiles, is

$$F_Y(y) = \int_0^y nw^{n-1}\,dw = \left[w^n\right]_0^y = y^n \qquad\qquad 0 < y < 1.$$

The next step in the pivotal method is to find constants a and b such that

$$P\left(a < \frac{X_{(n)}}{\theta} < b\right) = 1 - \alpha.$$

Let a be the $100(\alpha/2)$th percentile of the probability distribution of $Y = X_{(n)}/\theta$ and b be the $100(1-\alpha/2)$th percentile of the probability distribution of $Y = X_{(n)}/\theta$. These constants can be found by solving

$$F_Y(a) = a^n = \alpha/2 \qquad \text{and} \qquad F_Y(b) = b^n = 1 - \alpha/2$$

for a and b, yielding

$$a = (\alpha/2)^{1/n} \qquad \text{and} \qquad b = (1-\alpha/2)^{1/n}.$$

The probability statement can be written as

$$P\left((\alpha/2)^{1/n} < \frac{X_{(n)}}{\theta} < (1-\alpha/2)^{1/n}\right) = 1 - \alpha.$$

Isolating θ in the middle of the inequality results in

$$P\left(\frac{X_{(n)}}{(1-\alpha/2)^{1/n}} < \theta < \frac{X_{(n)}}{(\alpha/2)^{1/n}}\right) = 1 - \alpha.$$

So an exact two-sided $100(1-\alpha)\%$ confidence interval for θ is

$$\frac{X_{(n)}}{(1-\alpha/2)^{1/n}} < \theta < \frac{X_{(n)}}{(\alpha/2)^{1/n}}.$$

This is a rare and particularly desirable type of exact confidence interval because it can be written in closed form and does not require any table look-up or numerical methods to compute the lower and upper bounds for general values of n and α.

The previous two examples have constructed exact two-sided $100(1-\alpha)\%$ confidence intervals for an unknown population parameter θ for a random sample $X_1, X_2, \ldots, X_n$ drawn from an exponential and a uniform population. In both examples, the pivotal method was used to identify a pivotal quantity, which led to the exact confidence interval.

Random sampling from a normal population can also lead to exact confidence intervals for various population parameters. Section 1.5 contained eight results related to random sampling from normal populations. We derive and illustrate an exact confidence interval for the population mean μ for a random sample drawn from a normal population next.

Example 3.9 Let $X_1, X_2, \ldots, X_n$ be a random sample from a $N\left(\mu, \sigma^2\right)$ population, where μ and $\sigma^2 > 0$ are unknown parameters. Find an exact two-sided $100(1-\alpha)\%$ confidence interval for μ.

Example 2.13 showed that the maximum likelihood estimator of the population mean μ is the sample mean $\bar{X}$. The first step in the pivotal method to find an exact two-sided $100(1-\alpha)\%$ confidence interval for μ is to find a pivotal quantity whose probability distribution does not depend on the unknown parameter μ. Theorem 1.7 states that

$$\frac{\bar{X}-\mu}{S/\sqrt{n}} \sim t(n-1),$$

which is a probability distribution that does not depend on μ, so it will be adopted as the pivotal quantity. The second step is to find constants a and b such that

$$P\left(a < \frac{\bar{X}-\mu}{S/\sqrt{n}} < b\right) = 1-\alpha$$

for any $0 < \alpha < 1$. Again adopting the practice that the two constants a and b are the $100(\alpha/2)$th and the $100(1-\alpha/2)$th percentiles of the probability distribution of the pivotal quantity, these two percentiles could be written as $a = t_{n-1,1-\alpha/2}$ and $b = t_{n-1,\alpha/2}$. For the first time in all of the confidence intervals we have encountered thus far, however, the two percentiles are simply negatives of each other, as shown in Figure 3.13, so they are typically written as $a = -t_{n-1,\alpha/2}$ and $b = t_{n-1,\alpha/2}$. The vertical axis label $f(x)$ in the figure refers to the probability density function of the t distribution with $n-1$ degrees of freedom rather than the population distribution from which the random sample was drawn. So the probability statement can be written as

$$P\left(-t_{n-1,\alpha/2} < \frac{\bar{X}-\mu}{S/\sqrt{n}} < t_{n-1,\alpha/2}\right) = 1-\alpha.$$

The final step in the pivotal method is to perform algebra on the inequality in order to isolate μ in the center, which results in

$$P\left(\bar{X} - t_{n-1,\alpha/2}\frac{S}{\sqrt{n}} < \mu < \bar{X} + t_{n-1,\alpha/2}\frac{S}{\sqrt{n}}\right) = 1-\alpha.$$

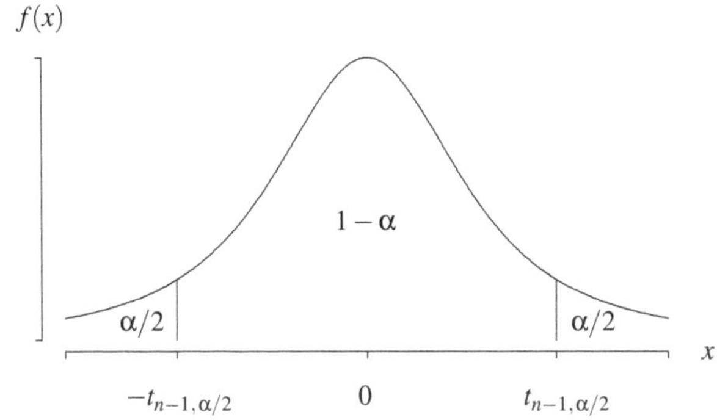

Figure 3.13: Probability density function of a $t(n-1)$ random variable with percentiles.

So an exact two-sided $100(1-\alpha)\%$ confidence interval for μ is

$$\bar{X} - t_{n-1,\alpha/2}\frac{S}{\sqrt{n}} < \mu < \bar{X} + t_{n-1,\alpha/2}\frac{S}{\sqrt{n}}.$$

This is the first confidence interval that we have encountered that is symmetric about the point estimator, so it can also be written as the point estimator plus or minus the halfwidth, that is,

$$\bar{X} \pm t_{n-1,\alpha/2}\frac{S}{\sqrt{n}}.$$

To illustrate the use of this confidence interval, consider the $n = 78$ IQ scores from seventh-grade students at a rural school in the midwest of the United States given below. IQ scores are often characterized by bell-shaped probability distributions.

111	107	100	107	114	115	111	97	100	112	104	89	104
102	91	114	114	103	106	105	113	109	108	113	130	128
128	118	113	120	132	111	124	127	128	136	106	118	119
123	124	126	116	127	119	97	86	102	110	120	103	115
93	72	111	103	123	79	119	110	110	107	74	105	112
105	110	107	103	77	98	90	96	112	112	114	93	106

The sample mean and sample standard deviation are

$$\bar{x} = 109 \qquad \text{and} \qquad s = 13.$$

An exact two-sided 99% confidence interval for the population mean IQ score μ is

$$\bar{x} - t_{n-1,\alpha/2}\frac{s}{\sqrt{n}} < \mu < \bar{x} + t_{n-1,\alpha/2}\frac{s}{\sqrt{n}}$$

or

$$109 - 2.6412\frac{13}{\sqrt{78}} < \mu < 109 + 2.6412\frac{13}{\sqrt{78}}$$

or

$$105 < \mu < 113.$$

As always, all intermediate quantities are stored to machine precision, then rounded to significant digits when reported.

A Monte Carlo simulation can add some insight into this particular confidence interval formula. Arbitrarily setting $\mu = 1$, $\sigma = 1$, $\alpha = 0.1$, and $n = 16$, the following R code generates and plots one hundred exact two-sided 90% confidence intervals in the (l, u) plane.

```
mu    = 1
sigma = 1
alpha = 0.1
n     = 16
nrep  = 100
l     = u = numeric(nrep)
for (i in 1:nrep) {
  x    = rnorm(n, mu, sigma)
  xbar = mean(x)
  sdev = sd(x)
  crit = qt(1 - alpha / 2, n - 1)
  l[i] = xbar - crit * sdev / sqrt(n)
  u[i] = xbar + crit * sdev / sqrt(n)
}
plot(l, u, type = "p")
```

The resulting figure, with guide lines added to show which confidence intervals miss low, cover μ, and miss high, is shown in Figure 3.14. Three of the confidence intervals miss μ low and three of the confidence intervals miss μ high.

Rotating Figure 3.14 clockwise by 45° gives another view of the simulated confidence intervals, which is shown in Figure 3.15. After this rotation, the horizontal axis becomes the midpoint of the confidence interval, which is $\bar{x}$ in this case. The vertical

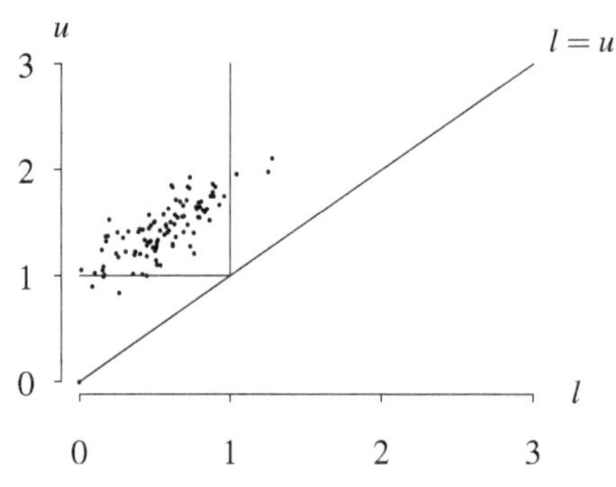

Figure 3.14: Monte Carlo simulation of 100 (l, u) pairs for a $N(1, 1)$ population.

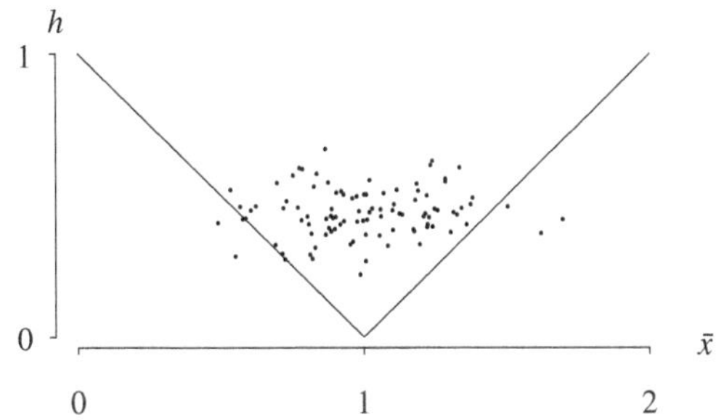

Figure 3.15: Monte Carlo simulation of 100 $(\bar{x}, h)$ pairs for a $N(1, 1)$ population.

axis becomes the confidence interval halfwidth $h = t_{n-1,\alpha/2}\, s/\sqrt{n}$. This view of the simulated confidence intervals highlights the fact that the midpoint $\bar{X}$ and the halfwidth $H = t_{n-1,\alpha/2}\, S/\sqrt{n}$ are independent random variables. This follows from Theorem 1.5, which stated that the sample mean and the sample variance are independent random variables for a random sample drawn from a $N\left(\mu, \sigma^2\right)$ population. This view of the simulated confidence intervals gives a visual estimate of $E[H]$ based on the average height of the points. It also gives a visual estimate of $V[H]$ based on the vertical spread of the points. The effect of decreasing α is to move the points vertically upward because $t_{n-1,\alpha/2}$ increases.

The previous example investigated an exact two-sided confidence interval concerning a random sample drawn from a normally distributed population. But there are many exact confidence intervals that can be constructed from such a random sample. Table 3.1 lists the pivotal quantities from Section 1.5 and the associated exact two-sided $100(1-\alpha)\%$ confidence intervals for the parameters of interest. The following items relate to Table 3.1.

- The assumptions associated with the random sampling are given in Section 1.5.
- Fractiles of distributions in Table 3.1, such as $t_{n-1,\alpha/2}$, have an $\alpha/2$ in their subscript, which denotes a right-hand tail probability.
- The confidence intervals associated with Theorems 1.3 and 1.4 for μ (when σ^2 is known) and σ^2 (when μ is known) are seldom used in practice because typically neither of the parameters are known.
- The confidence intervals associated with Theorems 1.6 and 1.7 for σ^2 (when μ is estimated from the data values) and μ (when σ^2 is estimated from the data values) cover the one-sample cases that are typically encountered in practice.
- The confidence intervals below the double line in the middle of the table correspond to the case in which two independent random samples are drawn from two normal populations.
- The confidence interval associated with Theorem 1.8 should only be applied when both σ_X^2 and σ_Y^2 are known, which is rarely the case in practice.
- The confidence intervals associated with Theorems 1.8 and 1.9 for $\mu_X - \mu_Y$ were too wide for the page, so they were written in the form $\hat{\theta} \pm H$, where H is the confidence interval halfwidth.

Theorem	Pivotal quantity	Exact two-sided $100(1-\alpha)\%$ confidence interval
1.3	$\frac{\bar{X}-\mu}{\sigma/\sqrt{n}} \sim N(0,1)$	$\bar{X} - z_{\alpha/2}\frac{\sigma}{\sqrt{n}} < \mu < \bar{X} + z_{\alpha/2}\frac{\sigma}{\sqrt{n}}$
1.4	$\sum_{i=1}^{n}\left(\frac{X_i-\mu}{\sigma}\right)^2 \sim \chi^2(n)$	$\frac{\sum_{i=1}^{n}(X_i-\mu)^2}{\chi^2_{n,\alpha/2}} < \sigma^2 < \frac{\sum_{i=1}^{n}(X_i-\mu)^2}{\chi^2_{n,1-\alpha/2}}$
1.6	$\frac{(n-1)S^2}{\sigma^2} \sim \chi^2(n-1)$	$\frac{(n-1)S^2}{\chi^2_{n-1,\alpha/2}} < \sigma^2 < \frac{(n-1)S^2}{\chi^2_{n-1,1-\alpha/2}}$
1.7	$\frac{\bar{X}-\mu}{S/\sqrt{n}} \sim t(n-1)$	$\bar{X} - t_{n-1,\alpha/2}\frac{S}{\sqrt{n}} < \mu < \bar{X} + t_{n-1,\alpha/2}\frac{S}{\sqrt{n}}$
1.8	$\frac{\bar{X}-\bar{Y}-(\mu_X-\mu_Y)}{\sqrt{\frac{\sigma_X^2}{n}+\frac{\sigma_Y^2}{m}}} \sim N(0,1)$	$\bar{X}-\bar{Y} \pm z_{\alpha/2}\sqrt{\frac{\sigma_X^2}{n}+\frac{\sigma_Y^2}{m}}$
1.9	$\frac{\bar{X}-\bar{Y}-(\mu_X-\mu_Y)}{S_P\sqrt{\frac{1}{n}+\frac{1}{m}}} \sim t(n+m-2)$	$\bar{X}-\bar{Y} \pm t_{n+m-2,\alpha/2}S_P\sqrt{\frac{1}{n}+\frac{1}{m}}$
1.10	$\frac{S_X^2/\sigma_X^2}{S_Y^2/\sigma_Y^2} \sim F(n-1,m-1)$	$\frac{S_X^2}{S_Y^2 F_{n-1,m-1,\alpha/2}} < \frac{\sigma_X^2}{\sigma_Y^2} < \frac{S_X^2 F_{m-1,n-1,\alpha/2}}{S_Y^2}$

Table 3.1: Pivotal quantities and exact $100(1-\alpha)\%$ confidence intervals for normal sampling.

- The confidence interval associated with Theorem 1.9 should only be applied when there is evidence that the two population variances are equal. This can be checked by using the confidence interval associated with Theorem 1.10.

The next two examples illustrate the application of some of the confidence intervals listed in Table 3.1.

Example 3.10 Find an exact two-sided 99% confidence interval for σ^2 for the $n = 78$ IQ scores from Example 3.9.

The implicit assumption that is being made in this example is that the population distribution is normal. This assumption would be verified prior to finding point and interval estimates. Per Theorem 2.2, the unbiased point estimate of σ^2 is just the sample variance:

$$s^2 = \frac{1}{n-1}\sum_{i=1}^{n}(x_i-\bar{x})^2 = 173.$$

An exact two-sided 99% confidence interval for σ^2 is

$$\frac{(n-1)s^2}{\chi^2_{n-1,\alpha/2}} < \sigma^2 < \frac{(n-1)s^2}{\chi^2_{n-1,1-\alpha/2}}$$

or

$$\frac{77 \cdot 173}{112.7038} < \sigma^2 < \frac{77 \cdot 173}{48.7884}$$

or

$$118 < \sigma^2 < 273.$$

If the population standard deviation is of interest, then a point estimate of σ is $s = 13.2$ and an exact two-sided 99% confidence interval for σ is found by taking the square roots of the three elements in the confidence interval for σ^2:

$$10.9 < \sigma < 16.5.$$

The confidence interval is somewhat narrow because of the large ($n = 78$) sample size.

The next example has random sampling from two normal populations, and the interest is in knowing whether there is a statistically significant difference between the population means of the two normal populations.

Example 3.11 The tomato yields (in pounds) for Fertilizer X and Fertilizer Y are given in Table 3.2. Find an exact two-sided 90% confidence interval for the difference between the population mean yields for the two fertilizers.

Fertilizer X yields	29.9	11.4	25.3	16.5	21.2	
Fertilizer Y yields	26.6	23.7	28.5	14.2	17.9	24.3

Table 3.2: Tomato yields for Fertilizer X and Fertilizer Y.

Fertilizer Y is a proposed fertilizer and the interest is whether it differs from Fertilizer X in terms of expected yield. Some sample statistics from this data set are

$$\bar{x} = 20.9 \qquad \bar{y} = 22.5 \qquad s_X = 7.25 \qquad s_Y = 5.43.$$

There is an increased yield of 1.6 pounds by using Fertilizer Y instead of Fertilizer X. But there is also significant spread to the data values as evidenced by the two sample standard deviations. So is the 1.6 pound increase in the yield a matter of random sampling variability or is it due to superior performance by Fertilizer Y?

There are a host of problems associated with constructing the 90% confidence interval for $\mu_X - \mu_Y$. There are only $n = 5$ observed yields from Fertilizer X and $m = 6$ observed yields from Fertilizer Y. These sample sizes are too small to plot a histogram to determine whether the normality assumption is justified. On the other hand, previous experience with tomato yields might allow one to conclude that the assumption of normality is reasonable for tomato yields. For now, let's assume that the normality assumption is justified based on previous experiments. The appropriate formula for a confidence interval for the difference between the population means from Table 3.1 is the confidence interval associated with Theorem 1.9. This confidence interval not only requires that the random samples come from normally distributed populations—it also requires that the population variances are equal. Although the sample standard deviations, $s_X = 7.25$ and $s_Y = 5.43$, are roughly equal, we would like some objective measure to indicate that they could indeed be considered equal for constructing the confidence interval. One way to proceed is to calculate a confidence interval for the ratio of the population variances using the entry associated with Theorem 1.10 from Table 3.1. So an exact two-sided 90% confidence interval for the ratio of the population variances is

$$\frac{s_X^2}{s_Y^2 F_{n-1,m-1,\alpha/2}} < \frac{\sigma_X^2}{\sigma_Y^2} < \frac{s_X^2 F_{m-1,n-1,\alpha/2}}{s_Y^2}$$

or

$$\frac{7.25^2}{5.43^2 F_{4,5,0.05}} < \frac{\sigma_X^2}{\sigma_Y^2} < \frac{7.25^2 F_{5,4,0.05}}{5.43^2}$$

or

$$0.343 < \frac{\sigma_X^2}{\sigma_Y^2} < 11.1.$$

This confidence interval can be calculated with the following four R statements.

```
x = c(29.9, 11.4, 25.3, 16.5, 21.2)
y = c(26.6, 23.7, 28.5, 14.2, 17.9, 24.3)
l = var(x) / (var(y) * qf(0.95, 4, 5))
u = var(x) * qf(0.95, 5, 4) / var(y)
```

Since this confidence interval covers 1 by a wide margin, it is reasonable to conclude that the differences between the two sample standard deviations can be attributed to random sampling variability rather than a statistically significant difference in the population variances. We can now proceed with the calculation of the confidence interval for the population mean difference in the yields under the assumption that the population variances are equal. The pooled sample variance is

$$s_p^2 = \frac{(n-1)s_X^2 + (m-1)s_Y^2}{n+m-2} = \frac{4 \cdot 7.25^2 + 5 \cdot 5.43^2}{5+6-2} = 39.7.$$

The associated pooled sample standard deviation is $s_p = \sqrt{39.7} = 6.30$. So assuming normality and equal population variances in the two populations, the exact two-sided 90% confidence interval for $\mu_X - \mu_Y$ is

$$\bar{x} - \bar{y} - t_{n+m-2,\alpha/2}\, s_p \sqrt{\frac{1}{n} + \frac{1}{m}} < \mu_X - \mu_Y < \bar{x} - \bar{y} + t_{n+m-2,\alpha/2}\, s_p \sqrt{\frac{1}{n} + \frac{1}{m}}$$

or

$$20.9 - 22.5 - t_{9,0.05}\, 6.30 \sqrt{\frac{1}{5} + \frac{1}{6}} < \mu_X - \mu_Y < 20.9 - 22.5 + t_{9,0.05}\, 6.30 \sqrt{\frac{1}{5} + \frac{1}{6}}$$

or

$$-8.67 < \mu_X - \mu_Y < 5.32.$$

This confidence interval can be calculated with the following R code.

```
x = c(29.9, 11.4, 25.3, 16.5, 21.2)
y = c(26.6, 23.7, 28.5, 14.2, 17.9, 24.3)
n = length(x)
m = length(y)
s = sqrt(((n - 1) * var(x) + (m - 1) * var(y)) / (n + m - 2))
l = mean(x) - mean(y) - qt(0.95, n + m - 2) * s * sqrt(1 / n + 1 / m)
u = mean(x) - mean(y) + qt(0.95, n + m - 2) * s * sqrt(1 / n + 1 / m)
```

Since confidence intervals of this type are calculated routinely by statisticians, some keystrokes can be saved by using the built-in function `t.test` with the specified stated coverage included in the `conf.level` argument, and the `var.equal` argument set to `TRUE`, as illustrated in the following R code.

```
x = c(29.9, 11.4, 25.3, 16.5, 21.2)
y = c(26.6, 23.7, 28.5, 14.2, 17.9, 24.3)
t.test(x, y, conf.level = 0.90, var.equal = TRUE)
```

Since this confidence interval includes 0, it can be concluded that there is not enough statistical evidence here to conclude that Fertilizer Y differs from Fertilizer X in terms of expected yield. Larger sample sizes for the two populations might be needed to conclude that the difference between the population means is nonzero.

This concludes the discussion of exact confidence intervals. The next section considers confidence intervals in which $P(L < \theta < U)$ does not equal $1 - \alpha$. These confidence intervals are known as *approximate* confidence intervals.

3.2 Approximate Confidence Intervals

All of the examples from the previous section considered exact confidence intervals, which is the ideal case. But exact confidence intervals are oftentimes impossible to construct, so *approximate confidence intervals* are the only alternative.

Definition 3.4 A random interval of the form

$$L < \theta < U$$

is an *approximate two-sided* $100(1-\alpha)\%$ *confidence interval* for the unknown parameter θ provided

$$P(L < \theta < U) \neq 1 - \alpha$$

for some value of θ.

So now the universe of all confidence intervals has been broken into two sets: exact confidence intervals and approximate confidence intervals, as illustrated in the Venn diagram in Figure 3.16.

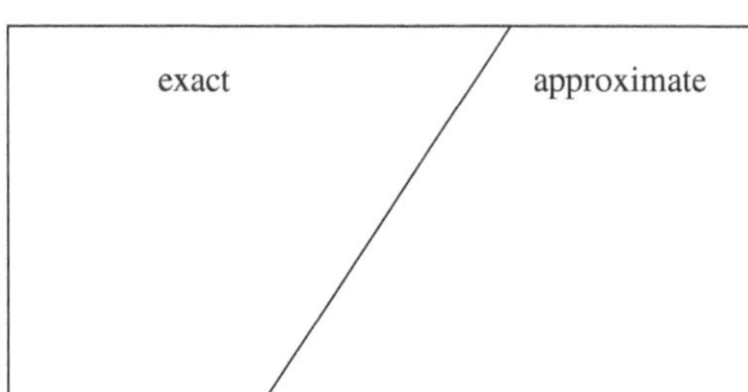

Figure 3.16: Venn diagram for exact and approximate confidence intervals.

There are several types of circumstances that can lead to an approximate confidence interval. First, it might be difficult or impossible to determine a pivotal quantity when using the pivotal method. Second, the nature of a discrete population distribution might make the derivation of an exact confidence interval impossible. (You might have noticed that all of the examples from the previous section involved random samples drawn from a continuous population.) Third, some of the population assumptions might not be perfectly satisfied, making an exact confidence interval procedure effectively an approximate confidence interval procedure. Finally, the population distribution might not be apparent due to a small sample size or an unusual population distribution.

There are four techniques for constructing an approximate confidence interval introduced in this section: discrete populations, heuristic methods, bootstrapping, and confidence intervals for percentiles.

Discrete populations

Random sampling from discrete populations poses a challenge for calculating confidence intervals because their discrete nature often makes achieving an exact confidence interval which satisfies $P(L < \theta < U) = 1 - \alpha$ for all values of an unknown parameter θ impossible. One approximate method that can be used for a random sample $X_1, X_2, \ldots, X_n$ from discrete population begins by identifying a quantity $Y = g(X_1, X_2, \ldots, X_n, \theta)$ that has cumulative distribution function $F_Y(y \mid \theta)$. As before, denote the confidence interval for the unknown parameter θ as $L < \theta < U$, where L is the lower bound and U is the upper bound. If $F_Y(y \mid \theta)$ is a decreasing function of θ, then choose the lower confidence bound l and the upper confidence bound u such that

$$P(Y \ge y \mid \theta = l) = \alpha/2 \qquad \text{and} \qquad P(Y \le y \mid \theta = u) = \alpha/2.$$

The intuition associated with the first of these equations is that for an observed value of y, the confidence interval lower bound l should be chosen so that decreasing the parameter θ to l results in a right-hand tail probability of $\alpha/2$ in the probability distribution of Y. The intuition associated with the second of these equations is that for an observed value of y, the confidence interval upper bound u should be chosen so that increasing the parameter θ to u results in a left-hand tail probability of $\alpha/2$ in the probability distribution of Y. Any parameter value smaller than l or greater than u is incompatible with the observed value of Y at the stated coverage $1 - \alpha$. If $F_Y(y \mid \theta)$ is an increasing function of θ, then choose l and u such that

$$P(Y \le y \mid \theta = l) = \alpha/2 \qquad \text{and} \qquad P(Y \ge y \mid \theta = u) = \alpha/2.$$

The first example uses this procedure to develop an approximate confidence interval for data values drawn from a Poisson(λ) population.

Example 3.12 Let $X_1, X_2, \ldots, X_n$ be a random sample from a Poisson(λ) population, where λ is a positive unknown parameter. Find an approximate two-sided $100(1-\alpha)\%$ confidence interval for λ and apply this confidence interval procedure to the horse kick data set from Example 2.9 to give a 95% confidence interval for λ.

Example 2.9 showed that the maximum likelihood estimator of a random sample from a Poisson population is

$$\hat{\lambda} = \frac{1}{n} \sum_{i=1}^{n} X_i,$$

which is the sample mean. There is no guidance as to what should be used for Y in order to construct the approximate confidence interval, so we will find a sufficient statistic to serve that purpose. The Fisher–Neyman factorization criterion in Theorem 2.7 indicates that the statistic $Y = g(X_1, X_2, \ldots, X_n)$ is a sufficient statistic for the parameter θ if there exist nonnegative functions h_1 and h_2 such that

$$f_X(x_1) f_X(x_2) \ldots f_X(x_n) = h_1\left(g(x_1, x_2, \ldots, x_n), \theta\right) \cdot h_2(x_1, x_2, \ldots, x_n).$$

For a random sample drawn from a Poisson(λ) population, the joint probability density function of $X_1, X_2, \ldots, X_n$ can be written as

$$f_X(x_1)f_X(x_2)\ldots f_X(x_n) = \frac{\lambda^{x_1}e^{-\lambda}}{x_1!}\cdot\frac{\lambda^{x_2}e^{-\lambda}}{x_2!}\cdot\ldots\cdot\frac{\lambda^{x_n}e^{-\lambda}}{x_n!} = \underbrace{\lambda^{\sum_{i=1}^{n}x_i}e^{-n\lambda}}_{h_1}\cdot\underbrace{\frac{1}{\prod_{i=1}^{n}x_i!}}_{h_2},$$

so $Y = \sum_{i=1}^{n} X_i$ is a sufficient statistic. From probability theory, we know that

$$Y = \sum_{i=1}^{n} X_i \sim \text{Poisson}(n\lambda)$$

because $X_1, X_2, \ldots, X_n$ are mutually independent and identically distributed random variables. The cumulative distribution function of the random variable Y on its support is

$$F_Y(y\,|\,\lambda) = \sum_{w=0}^{y}\frac{(n\lambda)^w e^{-n\lambda}}{w!} \qquad y = 0, 1, 2, \ldots.$$

Since $F_Y(y\,|\,\lambda)$ is a decreasing function of λ for a fixed value of y, the lower confidence bound l and the upper confidence bound u of the approximate two-sided $100(1-\alpha)\%$ confidence interval must satisfy

$$\sum_{w=y}^{\infty}\frac{(nl)^w e^{-nl}}{w!} = \alpha/2 \qquad \text{and} \qquad \sum_{w=0}^{y}\frac{(nu)^w e^{-nu}}{w!} = \alpha/2.$$

At this point, we could employ numerical methods to solve for l and u to arrive at a confidence interval for a particular data set. But there is some additional mathematics that can be performed which allow us to express l and u in a more compact manner. Let's begin with the first of these equations, which can also be written as

$$1 - \sum_{w=0}^{y-1}\frac{(nl)^w e^{-nl}}{w!} = \alpha/2.$$

The left-hand side of this equation can be recognized as the cumulative distribution function of an Erlang(nl, y) random variable evaluated at 1. Consequently, this equation can be written as

$$P\big(\text{Erlang}(nl, y) \le 1\big) = \alpha/2.$$

Since $2nl$ multiplied by an Erlang(nl, y) random variable is a chi-square random variable with $2y$ degrees of freedom (checking this via the transformation technique is left to the reader), this probability statement can be written as

$$P\big(\chi^2(2y) \le 2nl\big) = \alpha/2$$

or

$$l = \frac{1}{2n}\chi^2_{2y,1-\alpha/2}.$$

Using a similar line of reasoning, the upper limit for the approximate confidence interval is

$$u = \frac{1}{2n}\chi^2_{2(y+1),\alpha/2}.$$

Hence, an approximate two-sided $100(1-\alpha)\%$ confidence interval for λ of the form $L < \lambda < U$ is

$$\frac{1}{2n}\chi^2_{2Y,1-\alpha/2} < \lambda < \frac{1}{2n}\chi^2_{2(Y+1),\alpha/2}.$$

The confidence bounds have been expressed as functions of the percentiles of a chi-square distribution, which is much better than giving a pair of equations that need to be solved numerically for a particular data set.

Recall that the horse kick data from Example 2.9 consisted of $n = 200$ corps-years of data, which is listed below

number of deaths per corps per year	0	1	2	3	4
number of observed values	109	65	22	3	1

The $n = 200$ data values consist of 109 zeros, 65 ones, 22 twos, 3 threes, and 1 four. For this data set, there are

$$y = \sum_{i=1}^{n} x_i = 109\cdot 0 + 65\cdot 1 + 22\cdot 2 + 3\cdot 3 + 1\cdot 4 = 122$$

total deaths from horse kicks. The maximum likelihood estimator of λ is

$$\hat{\lambda} = \frac{1}{n}\sum_{i=1}^{n} x_i = \frac{122}{200} = 0.61$$

fatalities per corps-year. The approximate two-sided 95% confidence interval for λ is

$$\frac{1}{2n}\chi^2_{2y,1-\alpha/2} < \lambda < \frac{1}{2n}\chi^2_{2(y+1),\alpha/2}$$

or

$$\frac{1}{400}\chi^2_{244,0.975} < \lambda < \frac{1}{400}\chi^2_{246,0.025}$$

or

$$0.51 < \lambda < 0.73.$$

The 95% confidence interval can be computed with the R statements that follow.

```
x = 0:4
h = c(109, 65, 22, 3, 1)
y = sum(x * h)
n = sum(h)
a = 0.05
l = qchisq(a / 2, 2 * y) / (2 * n)
u = qchisq(1 - a / 2, 2 * (y + 1)) / (2 * n)
```

We are approximately 95% confident that the rate of fatalities from horse kicks per corps-year lies between 0.51 and 0.73 based on the data. The fairly narrow confidence interval is a result of the large ($n = 200$) sample size. Notice that the confidence interval is not symmetric about the maximum likelihood estimate $\hat{p} = 0.61$.

The second example of applying this confidence interval procedure considers a random sample from a Bernoulli(p) population. Normal and Bernoulli populations are arguably the two most common sampling distributions in applied statistics. Sampling from a normal population is common because so many random variables (for example, crop yields, dimensions on manufactured items, adult heights) have bell-shaped probability distributions. Sampling from a Bernoulli population is common because so many random variables have just two outcomes. Four examples illustrating data values associated with a binary response are given below.

- Did the patient survive five years after the surgery?
- Did the package get delivered by 10:30 AM?
- Do you support a particular political candidate?
- Do you like a particular breakfast cereal?

Although the Bernoulli distribution is one of the simplest probability distributions mathematically, no exact confidence interval for p exists because of the discrete nature of the population distribution. The following example uses the same procedure as in the previous example to develop an approximate confidence interval for p.

Example 3.13 Let $X_1, X_2, \ldots, X_n$ be a random sample from a Bernoulli(p) population, where $0 < p < 1$ is an unknown parameter. Find an approximate two-sided $100(1-\alpha)\%$ confidence interval for p.

As shown in Example 2.16, the maximum likelihood estimator of p for a random sample from a Bernoulli population,

$$\hat{p} = \frac{1}{n}\sum_{i=1}^{n} X_i,$$

is an unbiased estimate of p. Furthermore, as shown in Example 2.25,

$$Y = \sum_{i=1}^{n} X_i$$

is a sufficient statistic for p. Since the observations in the random sample are mutually independent and identically distributed Bernoulli(p) random variables,

$$Y = \sum_{i=1}^{n} X_i \sim \text{binomial}(n, p).$$

The cumulative distribution function of the random variable Y on its support is

$$F_Y(y \mid p) = \sum_{w=0}^{y} \binom{n}{w} p^w (1-p)^{n-w} \qquad y = 0, 1, 2, \ldots, n.$$

Since $F_Y(y \mid p)$ is a decreasing function of p for a fixed value of y, the lower confidence bound l and the upper confidence bound u of the approximate two-sided $100(1-\alpha)\%$ confidence interval must satisfy

$$\sum_{w=y}^{n} \binom{n}{w} l^w (1-l)^{n-w} = \alpha/2 \qquad \text{and} \qquad \sum_{w=0}^{y} \binom{n}{w} u^w (1-u)^{n-w} = \alpha/2.$$

As was the case in the previous example, numerical methods could be employed at this point to solve these two equations, which are polynomial equations, for l and u for a particular data set. As an alternative, it is possible to write l and u as percentiles of the F distribution, which makes the calculation of the confidence interval easier. Leaving the details as an exercise at the end of the chapter, the approximate two-sided $100(1-\alpha)\%$ confidence interval for p of the form $L < p < U$ is.

$$\frac{1}{1+\frac{n-Y+1}{YF_{2Y,2(n-y+1),1-\alpha/2}}} < p < \frac{1}{1+\frac{n-Y}{(Y+1)F_{2(Y+1),2(n-Y),\alpha/2}}},$$

where the first two values in the subscripts are the degrees of freedom for the F distribution and the third value in the subscript is a right-hand tail probability. The bounds of the confidence interval have been expressed as functions of percentiles of the F distribution, which is better than giving a pair of equations to solve numerically for the lower and upper confidence bounds. This confidence interval is known as the *Clopper–Pearson* confidence interval after C.J. Clopper and E.S. Pearson who devised it in 1934. In practice this interval works fine for $y = 1, 2, \ldots, n-1$, but a minor adjustment needs to be made at the extreme values of $y = 0$ and $y = n$ because of the problems associated with having zero for one of the degrees of freedom parameters for the F distribution. When $y = 0$, use $l = 0$; when $y = n$, use $u = 1$. The Clopper–Pearson confidence interval for p can be stated even more succinctly in terms of percentiles of the beta distribution:

$$B_{y,n-y+1,1-\alpha/2} < p < B_{y+1,n-y,\alpha/2},$$

for $y = 0, 1, 2, \ldots, n$, where the first two values in the subscripts are the parameters of the beta distribution and the third value in the subscript is a right-hand tail probability. As with the F distribution, when $y = 0$ and $y = n$, this results in zero for one of the parameters in the beta distribution. Fortunately, the `qbeta` function in R gives the correct bounds in these extreme cases. The rest of this example will be restricted to the case of $n = 5$ and $\alpha = 0.1$ as the Clopper–Pearson confidence interval is illustrated on a small data set.

A lady believes that tea tastes different depending on whether the tea is poured into the milk or the milk is poured into the tea. Ronald is skeptical. So an experiment with $n = 5$ cups was conducted. Letting 1 denote the correct identification and 0 being an error, the observed values of the five Bernoulli random variables are:

$$1, 0, 1, 1, 1.$$

She did pretty well. She identified four of the five correctly. Is this enough evidence to conclude that she can taste the difference? Perhaps she cannot tell the difference and just happened to get lucky. We will construct a confidence interval for p with an interest in seeing whether the confidence interval covers $p = 1/2$, which corresponds to a random guess on each Bernoulli trial.

For this data set,

$$y = \sum_{i=1}^{n} x_i = 4$$

correct identifications. The maximum likelihood estimator of p is

$$\hat{p} = \frac{1}{n}\sum_{i=1}^{n} x_i = \frac{4}{5} = 0.8.$$

An approximate two-sided 90% confidence interval for p has a lower bound l and an upper bound u that satisfy

$$\sum_{w=4}^{5}\binom{5}{w}l^w(1-l)^{5-w}=0.05 \qquad \text{and} \qquad \sum_{w=0}^{4}\binom{5}{w}u^w(1-u)^{5-w}=0.05.$$

The left-hand sides of these equations are polynomials, so the equations can be expressed as

$$-4l^5+5l^4=0.05 \qquad \text{and} \qquad 1-u^5=0.05.$$

(For a general sample size n, these will always be nth order polynomial equations.) The second equation has solution $u=0.95^{1/5}\cong 0.9898$, but the first equation does not have an analytic solution. Rather than using numerical methods to solve the first polynomial equation, we can instead use the percentiles of the F distribution. The approximate two-sided 90% confidence interval for p is

$$\frac{1}{1+\frac{n-y+1}{yF_{2y,2(n-y+1),1-\alpha/2}}}<p<\frac{1}{1+\frac{n-y}{(y+1)F_{2(y+1),2(n-y),\alpha/2}}}$$

or

$$\frac{1}{1+\frac{2}{4F_{8,4,0.95}}}<p<\frac{1}{1+\frac{1}{5F_{10,2,0.05}}}$$

or

$$0.34<p<0.99.$$

This confidence interval is highly nonsymmetric about the point estimate $\hat{p}=0.8$. This confidence interval can also be computed directly using the appropriate percentiles of the beta distribution:

$$B_{y,n-y+1,1-\alpha/2}<p<B_{y+1,n-y,\alpha/2}$$

or

$$B_{4,2,0.95}<p<B_{5,1,0.05}$$

or

$$0.34<p<0.99.$$

These confidence intervals can be computed with the R statements that follow using the `qbeta` function to find the appropriate percentiles of the beta distribution.

```
x = c(1, 0, 1, 1, 1)
n = length(x)
y = sum(x)
l = qbeta(0.05, y, n - y + 1)
u = qbeta(0.95, y + 1, n - y)
```

This calculation is performed so often by statisticians that a built-in function in R named `binom.test` can be used to save a few keystrokes. This confidence interval can also be computed in R with the single command

```
binom.test(4, 5, conf.level = 0.9)
```

We are 90% confident that the probability of correct identification p lies between 0.34 and 0.99 based on the data. The wide confidence interval is a result of the small ($n = 5$) sample size. Since $p = 1/2$ is included in the interval, we are not able to conclude from this data set that the lady can correctly identify the order that the milk and tea are poured. More data will need to be collected in order to draw that conclusion. Try running `binom.test` with $n = 100$ and $y = 80$ (which yields the same $\hat{p}$ as the test with $n = 5$ and $y = 4$) and see if the confidence interval shrinks so as to not include $p = 1/2$.

The Clopper–Pearson confidence interval for p has been deemed "approximate," but how well does it perform? To answer this, we calculate the actual coverage, which is a function of p. We would like this actual coverage to come as close as possible to the stated coverage. Begin by fixing the values of n and α at $n = 5$ and $\alpha = 0.1$ (a 90% confidence interval). Table 3.3 gives confidence intervals associated with all of the possible values of y, that is, $y = 0, 1, 2, \ldots, 5$.

y	l solves	u solves	approximate 90% confidence interval
0	$l = 0$	$\sum_{w=0}^{0} \binom{5}{w} u^w (1-u)^{5-w} = 0.05$	$0 < p < 0.45$
1	$\sum_{w=1}^{5} \binom{5}{w} l^w (1-l)^{5-w} = 0.05$	$\sum_{w=0}^{1} \binom{5}{w} u^w (1-u)^{5-w} = 0.05$	$0.01 < p < 0.66$
2	$\sum_{w=2}^{5} \binom{5}{w} l^w (1-l)^{5-w} = 0.05$	$\sum_{w=0}^{2} \binom{5}{w} u^w (1-u)^{5-w} = 0.05$	$0.08 < p < 0.81$
3	$\sum_{w=3}^{5} \binom{5}{w} l^w (1-l)^{5-w} = 0.05$	$\sum_{w=0}^{3} \binom{5}{w} u^w (1-u)^{5-w} = 0.05$	$0.19 < p < 0.92$
4	$\sum_{w=4}^{5} \binom{5}{w} l^w (1-l)^{5-w} = 0.05$	$\sum_{w=0}^{4} \binom{5}{w} u^w (1-u)^{5-w} = 0.05$	$0.34 < p < 0.99$
5	$\sum_{w=5}^{5} \binom{5}{w} l^w (1-l)^{5-w} = 0.05$	$u = 1$	$0.55 < p < 1$

Table 3.3: Confidence intervals ($\alpha = 0.1$) for p when $n = 5$.

The stated coverage of the interval is 0.90. We would like to calculate the actual coverage

$$c(p) = P(L < p < U) = P\left(B_{Y,n-Y+1,1-\alpha/2} < p < B_{Y+1,n-Y,\alpha/2}\right)$$

for $0 < p < 1$. Consider one particular value of p, say $p = 0.3$. When the probability of success on each Bernoulli trial is $p = 0.3$, all six values of Y are possible, and their probabilities of occurrence can be calculated by using the binomial distribution with $n = 5$ and $p = 0.3$. Since only the first four of the confidence intervals from Table 3.3 cover the true probability of success $p = 0.3$,

$$c(0.3) = \sum_{x=0}^{3} \binom{5}{x} (0.3)^x (0.7)^{5-x} = 0.9692.$$

This actual coverage at $p = 0.3$ is much higher than the stated coverage of $1 - \alpha = 0.9$. The R code that follows computes and plots the actual coverage for *all* values of p between 0 and 1 using this same line of reasoning. The vectors l and u contain the lower and upper confidence interval bounds given in the last column of Table 3.3. There are discontinuities in $c(p)$ at each of these confidence interval bounds. These discontinuities correspond to one of the confidence intervals in Table 3.3 changing its status with respect to covering p. The vector b holds all of the confidence interval bounds in sorted order. The loop over i plots the various segments of $c(p)$. Each piecewise segment that is plotted consists of 50 points connected by lines, which appear to be smooth curves. The built-in R function dbinom is used to calculate the actual coverage values by accumulating the probabilities associated with the confidence intervals that cover p.

```
n = 5
a = 0.1
y = 0:n
l = qbeta(a / 2, y, n - y + 1)
u = qbeta(1 - a / 2, y + 1, n - y)
b = sort(c(l, u))
plot(NA, xlim = c(0, 1), ylim = c(0.8, 1), xlab = "p", ylab = "c(p)")
abline(h = 1 - a)
for (i in 1:(length(b) - 1)) {
  p = seq(b[i], b[i + 1], length = 50)
  actual = rep(0, 50)
  mid = (b[i] + b[i + 1]) / 2
  for (y in 0:n)
    if (l[y + 1] < mid && mid < u[y + 1])
      actual = actual + dbinom(y, n, p)
  lines(p, actual)
}
```

Figure 3.17 shows the graph of $c(p)$ on $0 < p < 1$. Not surprisingly, the graph is symmetric about $p = 1/2$. This must be the case because arbitrarily changing "success" to "failure" in the experiment should result in a complementary confidence interval.

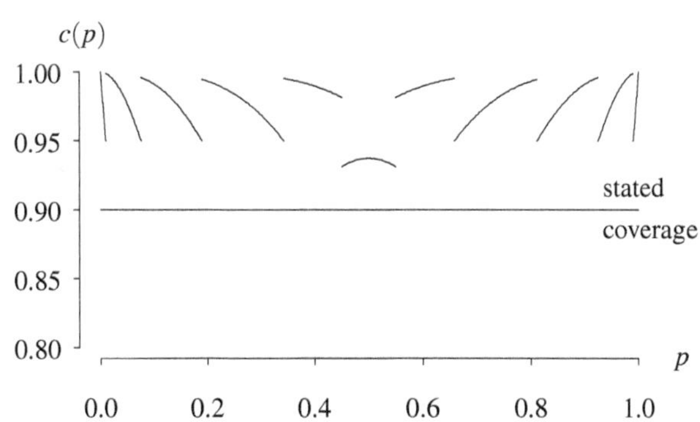

Figure 3.17: Clopper–Pearson actual coverage function for $\alpha = 0.1$ and $n = 5$.

The stated coverage is indicated by a solid horizontal line at $1-\alpha = 0.9$. If the actual coverage equaled the stated coverage $1-\alpha = 0.9$ for all values of p, then this would be an exact confidence interval. The actual coverage function, however, stays above the stated coverage for all values of p, so this is an approximate confidence interval. The Clopper–Pearson confidence interval is known as a "conservative" approximate confidence interval because $c(p) \geq 1-\alpha$ for all values of p. In this sense the confidence interval is wider than necessary, which is certainly the better direction to err. The approximate two-sided 90% confidence interval always gives you *less precision* than stated.

The two previous examples have illustrated a technique for constructing an approximate confidence interval from a random sample drawn from two discrete population distributions—the Poisson distribution with unknown parameter λ and the Bernoulli distribution with unknown parameter p. The procedure outlined at the beginning of the section can, in principle, be applied to any random sample drawn from any discrete distribution.

The example concerning random sampling from a Bernoulli(p) population brought up two new concepts, the actual coverage function and a conservative confidence interval, both of which are defined formally next.

Definition 3.5 For a confidence interval of the form

$$L < \theta < U$$

for the unknown parameter θ in the parameter space Ω, the *actual coverage function* $c(\theta)$ is

$$c(\theta) = P(L < \theta < U)$$

for all $\theta \in \Omega$.

If $c(\theta) = 1-\alpha$ for all values of $\theta \in \Omega$, the confidence interval is an exact confidence interval. If $c(\theta) \neq 1-\alpha$ for one or more values of $\theta \in \Omega$, the confidence interval is an approximate confidence interval. In any practical problem involving a confidence interval, we would like the actual coverage function to fall as close to the stated coverage $1-\alpha$ as possible. In Figure 3.17, we found that the actual coverage fell above the stated coverage, resulting in what was called a conservative confidence interval, which is defined next.

Definition 3.6 The confidence interval interval of the form

$$L < \theta < U$$

for the unknown parameter θ in the parameter space Ω with stated coverage $1-\alpha$ is a *conservative confidence interval* if

$$c(\theta) \geq 1-\alpha$$

for all $\theta \in \Omega$.

Conservative confidence intervals have an actual coverage function that is greater than or equal to the stated coverage. What does it mean to have an actual coverage that exceeds the stated coverage for all values of an unknown parameter? It means that the confidence interval is wider than necessary for the particular choice of the stated coverage $1-\alpha$. Conservative confidence intervals never over-promise on their assessment of the precision of $\hat{\theta}$. These confidence intervals are careful in the sense that their width is an upper bound on the width associated with an exact confidence interval. This means that the Venn diagram that classifies confidence intervals can be modified slightly now to include conservative confidence intervals. Although all exact confidence intervals are also conservative by Definition 3.6, we do not include them in the Venn diagram because it is only in the case of approximate confidence intervals in which a conservative confidence interval is of interest. The updated Venn diagram which includes conservative confidence intervals is shown in Figure 3.18.

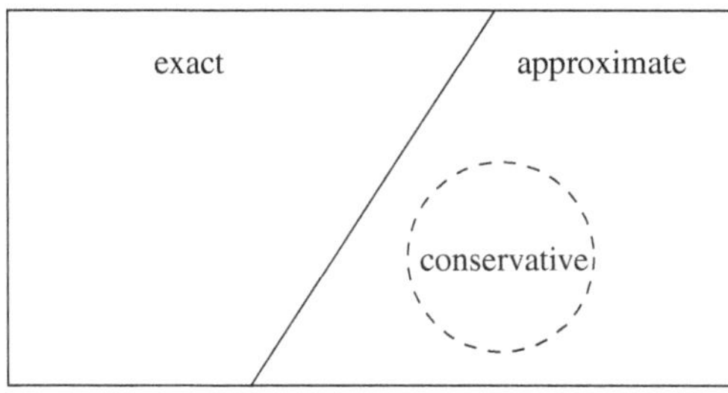

Figure 3.18: Venn diagram for exact, approximate, and conservative confidence intervals.

Since the approximate confidence interval for the binomial distribution is used so widely by statisticians in practical problems, we will spend a few pages analyzing this confidence interval and comparing its performance to some of its competitors. We begin by computing the expected confidence interval width W for a prescribed sample size n and stated coverage $1-\alpha$:

$$E[W] = \sum_{y=0}^{n} (u-l)\binom{n}{y} p^y(1-p)^{n-y},$$

for $0 < p < 1$. A plot of the expected confidence interval width for 90% Clopper–Pearson confidence intervals for various sample sizes n is given in Figure 3.19. Not surprisingly, the confidence intervals narrow as n increases. Also, the confidence intervals narrow around the extreme values of p.

Using the same reasoning as in the previous example, the actual coverage $c(p)$ of any confidence interval for the binomial proportion is

$$c(p) = P(L < p < U) = \sum_{y=0}^{n} I(y, p)\binom{n}{y} p^y(1-p)^{n-y} \qquad 0 < p < 1,$$

where $I(y, p)$ is an indicator function that is 0 when a confidence interval does not contain the binomial proportion p and is 1 when a confidence interval contains the binomial proportion p, when the number of successes is $Y = y$. The fact that the lower bounds and upper bounds on *any* confidence interval procedure for the binomial proportion p are nondecreasing functions of y means that the actual coverage $c(p)$ must lie on one of the *acceptance curves* defined as

$$b(p, y_0, y_1) = \sum_{y=y_0}^{y_1} \binom{n}{y} p^y(1-p)^{n-y}$$

for any value of p satisfying $0 < p < 1$, and for any integers y_0 and y_1 satisfying $0 \le y_0 \le y_1 \le n$. This is because the actual coverage function is the sum of an uninterrupted sequence of probability

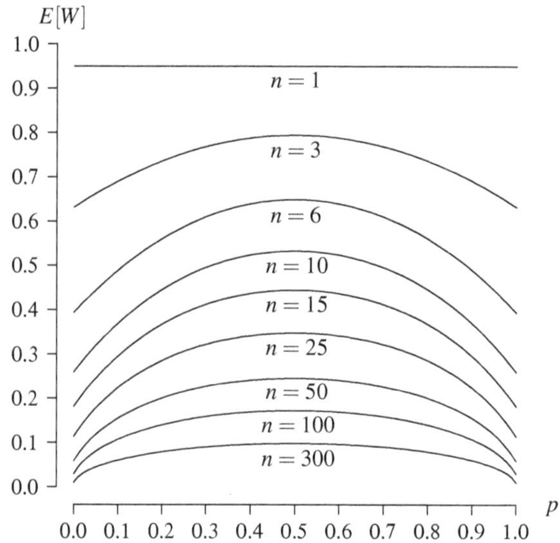

Figure 3.19: Expected Clopper–Pearson 90% confidence interval widths for various n.

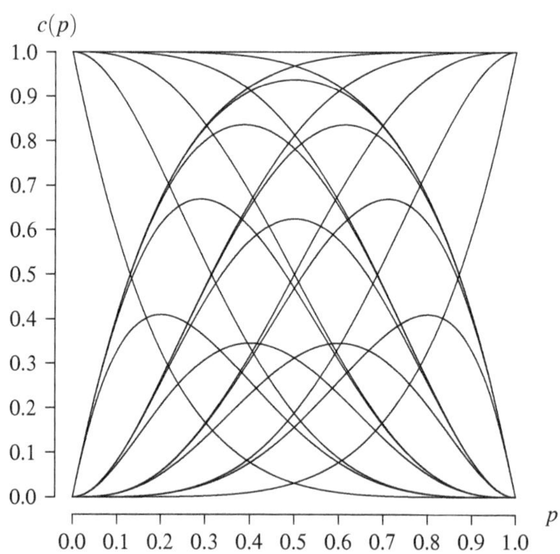

Figure 3.20: Actual coverage acceptance curves for $n = 5$.

mass function values of the binomial distribution. These acceptance curves are graphed in Figure 3.20 for $n = 5$ for all possible values of y_0 and y_1. Since the actual coverage function for all confidence interval procedures must lie on one of these curves, there will never be an exact confidence interval for p.

Figure 3.21 shows the 11 pieces associated with the Clopper–Pearson 90% confidence interval for p actual coverage function falling on the acceptance curves for $n = 5$. The scale on the vertical axis in Figure 3.21 is altered to range from 0.8 to 1. The acceptance curves from Figure 3.20 for $n = 5$ are plotted in gray. The actual coverage function $c(p)$ for a 90% Clopper–Pearson confidence interval is given by black curves. A solid horizontal line at 0.9 marks the stated coverage of a 90% confidence interval. As indicated in the previous paragraph, the actual coverage function $c(p)$ for the unknown binomial proportion p must lie on one of these acceptance curves for one particular value of p. The Clopper–Pearson confidence interval is conservative because $c(p) \geq 1 - \alpha$ for all values of p. The take-away from Figure 3.21 is that all actual coverage functions for a confidence interval procedure, either the Clopper–Pearson procedure or some other procedure, will jump from one of the acceptance curves to another. The values of p associated with the jumps are the confidence interval limits.

We now consider some competitors to the Clopper–Pearson confidence interval. One legitimate critique of the Clopper–Pearson confidence interval for p is that it is wider than necessary, and there are applications in which one would like to have a confidence interval whose actual coverage is close to the stated coverage, as opposed to exceeding the stated coverage. The existing approximate confidence interval procedures for p that will be presented here typically have two significant shortcomings: they do not perform well in terms of coverage (a) for small sample sizes, and (b) near the extremes, that is, near $p = 0$ and $p = 1$. We consider alternative two-sided confidence intervals for p with stated coverage $1 - \alpha$. The first of these competitors is known as the Wald confidence interval,

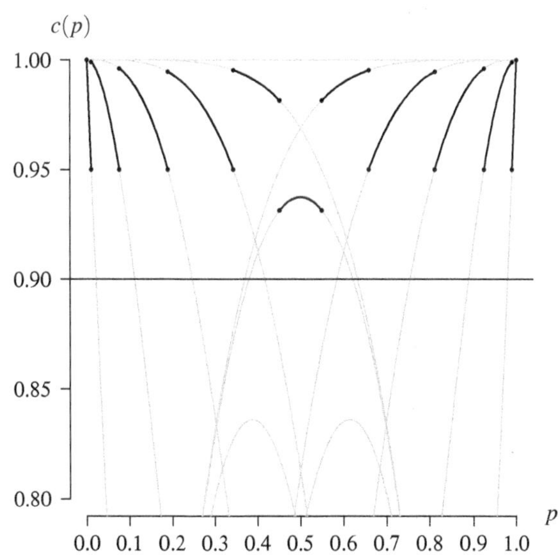

Figure 3.21: Clopper–Pearson actual coverage and acceptance curves for $n = 5$ and $\alpha = 0.1$.

named after Abraham Wald. The maximum likelihood estimator of p remains

$$\hat{p} = \frac{Y}{n} = \frac{1}{n}\sum_{i=1}^{n} X_i.$$

The expected value of the maximum likelihood estimator is

$$E[\hat{p}] = E\left[\frac{1}{n}\sum_{i=1}^{n} X_i\right] = \frac{1}{n}\sum_{i=1}^{n} E[X_i] = \frac{1}{n}(np) = p,$$

which indicates that $\hat{p}$ is an unbiased estimator of p. The population variance of the maximum likelihood estimator is

$$V[\hat{p}] = V\left[\frac{1}{n}\sum_{i=1}^{n} X_i\right] = \frac{1}{n^2}\sum_{i=1}^{n} V[X_i] = \frac{1}{n^2}\left(np(1-p)\right) = \frac{p(1-p)}{n}.$$

This expected value of $\hat{p}$ and population variance of $\hat{p}$ indicates, via Theorem 2.10, that $\hat{p}$ is also a consistent estimator of p. Since $\hat{p}$ is a sample mean, the central limit theorem indicates that

$$\hat{p} \xrightarrow{D} N\left(p, \frac{p(1-p)}{n}\right).$$

A problem that arises in constructing a confidence interval for p is that p is unknown. However, since $\hat{p}$ is a consistent estimator of p, an approximation to the population variance can be made by replacing p with $\hat{p}$. So

$$\frac{\hat{p} - p}{\sqrt{\hat{p}(1-\hat{p})/n}} \xrightarrow{D} N(0, 1).$$

Choosing the endpoints of an interval so that the probability of missing low and the probability of missing high both equal $\alpha/2$,

$$\lim_{n\to\infty} P\left(-z_{\alpha/2} < \frac{\hat{p} - p}{\sqrt{\hat{p}(1-\hat{p})/n}} < z_{\alpha/2}\right) = 1-\alpha.$$

Rearranging the terms in the inequality so that p appears in the center of the inequality yields the symmetric $100(1-\alpha)\%$ Wald confidence interval

$$\hat{p} \pm z_{\alpha/2}\sqrt{\frac{\hat{p}(1-\hat{p})}{n}},$$

where $z_{\alpha/2}$ is the $1-\alpha/2$ percentile of the standard normal distribution. When $x = 0$ ($x = n$), the point estimate of p is $\hat{p} = 0$ ($\hat{p} = 1$), so the Wald confidence interval degenerates to a zero-width confidence interval at the extremes. This is not ideal. When a confidence interval bound falls outside of $(0, 1)$, the bound is typically set to 0 or 1. The next four paragraphs briefly describe four other confidence intervals for p.

The bounds on the Wilson–score $100(1-\alpha)\%$ confidence interval for p are

$$\frac{1}{1+z_{\alpha/2}^2/n}\left[\hat{p} + \frac{z_{\alpha/2}^2}{2n} \pm z_{\alpha/2}\sqrt{\frac{\hat{p}(1-\hat{p})}{n} + \frac{z_{\alpha/2}^2}{4n^2}}\right],$$

where $z_{\alpha/2}$ is the $1-\alpha/2$ percentile of the standard normal distribution. The center of the Wilson–score confidence interval is

$$\frac{\hat{p}+z_{\alpha/2}^2/(2n)}{1+z_{\alpha/2}^2/n},$$

which is a weighted average of the point estimator $\hat{p}=y/n$ and $1/2$, with more weight on $\hat{p}$ as n increases.

The Jeffreys $100(1-\alpha)\%$ confidence interval for p is a Bayesian credible interval that uses a Jeffreys non-informative prior distribution for p. (Bayesian credible intervals will be discussed later in the chapter.) As was the case with the Clopper–Pearson confidence interval, the bounds of the Jeffreys confidence interval for p are percentiles of a beta random variable:

$$B_{y+1/2,n-y+1/2,1-\alpha/2} < p < B_{y+1/2,n-y+1/2,\alpha/2}$$

for $y=1,2,\ldots,n-1$. When $y=0$, the lower bound is set to zero and the upper bound is calculated using the formula above; when $y=n$, the upper bound is set to one and the lower bound is calculated using the formula above.

The bounds of the Agresti–Coull $100(1-\alpha)\%$ confidence interval, which was originally developed to approximate the Wilson–score confidence interval, are

$$\tilde{p}\pm z_{\alpha/2}\sqrt{\frac{\tilde{p}(1-\tilde{p})}{\tilde{n}}},$$

where $\tilde{n}=n+z_{\alpha/2}^2$ and $\tilde{p}=(y+z_{\alpha/2}^2/2)/\tilde{n}$. In the special case of $\alpha=0.05$, if one is willing to round $z_{\alpha/2}=1.96$ to 2, this interval can be interpreted as "add two successes and add two failures and use the Wald confidence interval formula."

The arcsine transformation uses a variance-stabilizing transformation when constructing a confidence interval for p. The bounds on a $100(1-\alpha)\%$ confidence interval for p are

$$\sin^2\left(\arcsin\left(\sqrt{\tilde{p}}\right)\pm\frac{z_{\alpha/2}}{2\sqrt{n}}\right),$$

where $\tilde{p}=(y+3/8)/(n+3/4)$. In the rare cases in which a confidence interval does not include the point estimator, one of the bounds is adjusted to include the point estimator.

These are not the only confidence intervals for p. Some of these confidence intervals have variations that include a continuity correction. In addition, there are other intervals, such as the logit interval, which perform well.

Example 3.14 When $n=5$, $y=4$ and $\alpha=0.1$, the point estimate of p is $\hat{p}=0.8$. The Wald, Clopper–Pearson, Wilson–score, Jeffreys, Agresti–Coull, and arcsine transformation 90% confidence intervals can be calculated with the following R commands after installing and loading the `conf` package.

```
binomTest(5, 4, alpha = 0.1, intervalType = "Wald")
binomTest(5, 4, alpha = 0.1, intervalType = "Clopper-Pearson")
binomTest(5, 4, alpha = 0.1, intervalType = "Wilson-Score")
binomTest(5, 4, alpha = 0.1, intervalType = "Jeffreys")
binomTest(5, 4, alpha = 0.1, intervalType = "Agresti-Coull")
binomTest(5, 4, alpha = 0.1, intervalType = "Arcsine")
```

The 90% confidence intervals for p are given in Table 3.4. The confidence interval bounds vary significantly between confidence interval procedures. The Clopper–Pearson confidence interval for p is the widest of the six; the Wald confidence interval for p is the narrowest of the six. Since the upper limit of the Wald confidence interval exceeds 1, it is reported as 1.

Interval name	90% confidence interval
Wald	$0.51 < p < 1.00$
Clopper–Pearson	$0.34 < p < 0.99$
Wilson–score	$0.44 < p < 0.95$
Jeffreys	$0.44 < p < 0.96$
Agresti–Coull	$0.42 < p < 0.97$
Arcsine	$0.41 < p < 0.98$

Table 3.4: Approximate 90% confidence intervals for p for $n = 5$ and $y = 4$.

With this array of strikingly different confidence intervals from Table 3.4, how does one choose the best one for a particular data set? One way to proceed is to inspect the actual coverage functions for each of the procedures. These actual coverage functions for the six confidence interval procedures are shown in Figure 3.22 for $n = 5$ and $\alpha = 0.1$. Each actual coverage function consists of $2(n+1) - 1 = (2)(6) - 1 = 11$ piecewise segments. The Wald confidence interval performs very poorly. If you are fortunate and $p = 0.4$, your 90% confidence interval will achieve an actual coverage of $c(p) = 0.84$, or 84% coverage. On the other hand, if $p = 0.1$, your 90% confidence interval will achieve an actual coverage of $c(p) = 0.40$, or 40% coverage, which is off of the graph in Figure 3.22. It is for this reason that the Wald interval, even though it is easy to compute and appears in many elementary statistics texts, should never be used, and this is particularly the case for small values of n. It gives a very narrow confidence interval that claims much more precision concerning $\hat{p}$ than it should. Unlike the Clopper–Pearson confidence interval, the other confidence intervals have an actual coverage that falls below $1 - \alpha$ for some values of p, which means that they could potentially give narrow confidence intervals that claim more precision than they should.

Some experimentation indicates that the Wilson–score confidence interval for p performs slightly better than the others among the non-conservative confidence intervals for moderate to large values of n. The actual coverage function for $n = 30$ is shown in Figure 3.23 and can be plotted in R with just a single command

```
binomTestCoveragePlot(n = 30, alpha = 0.1, intervalType = "Wilson-Score")
```

after installing and loading the `conf` package.

Heuristic methods

There are other practical situations in which it is not possible to find a pivotal quantity, and we are left to resort to what are best referred to as "heuristic methods" for determining approximate confidence intervals. One such heuristic method involves sampling from two normal populations. Let $X_1, X_2, \ldots, X_n$ be a random sample from a $N\left(\mu_X, \sigma_X^2\right)$ population. Let $Y_1, Y_2, \ldots, Y_m$ be a random sample from a $N\left(\mu_Y, \sigma_Y^2\right)$ population. The interest is to compute a $100(1-\alpha)\%$ two-sided confidence interval for $\mu_X - \mu_Y$. When σ_X^2 and σ_Y^2 are known, the fact that

$$\frac{\bar{X} - \bar{Y} - (\mu_X - \mu_Y)}{\sqrt{\frac{\sigma_X^2}{n} + \frac{\sigma_Y^2}{m}}} \sim N(0, 1)$$

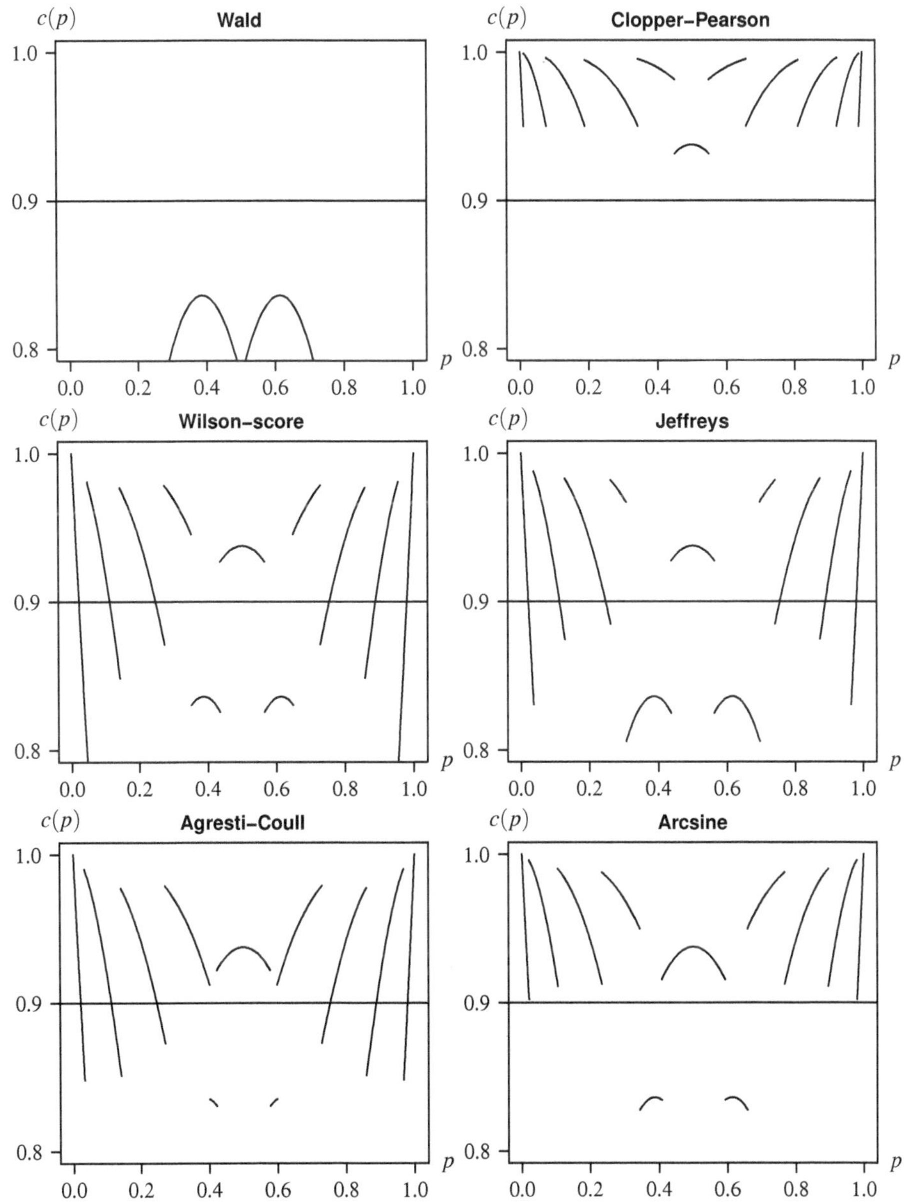

Figure 3.22: Actual coverage $c(p)$ for sampling from a Bernoulli population for $n = 5$.

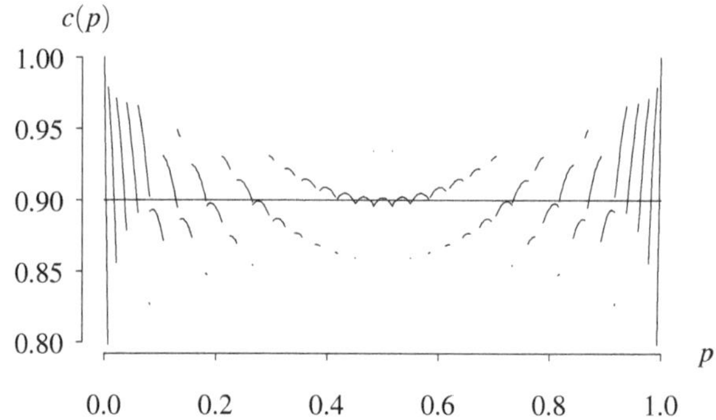

Figure 3.23: Wilson–score actual coverage function for $\alpha = 0.1$ and $n = 30$.

indicates that an exact two-sided $100(1-\alpha)\%$ confidence interval for $\mu_X - \mu_Y$ is

$$\bar{X} - \bar{Y} - z_{\alpha/2}\sqrt{\frac{\sigma_X^2}{n} + \frac{\sigma_Y^2}{m}} < \mu_X - \mu_Y < \bar{X} - \bar{Y} + z_{\alpha/2}\sqrt{\frac{\sigma_X^2}{n} + \frac{\sigma_Y^2}{m}},$$

as listed in Table 3.1. Furthermore, when σ_X^2 and σ_Y^2 are unknown, but can be presumed to be equal, the fact that

$$\frac{\bar{X} - \bar{Y} - (\mu_X - \mu_Y)}{S_p\sqrt{\frac{1}{n} + \frac{1}{m}}} \sim t(n+m-2)$$

indicates that an exact two-sided $100(1-\alpha)\%$ confidence interval for $\mu_X - \mu_Y$ is

$$\bar{X} - \bar{Y} - t_{n+m-2,\alpha/2}S_p\sqrt{\frac{1}{n} + \frac{1}{m}} < \mu_X - \mu_Y < \bar{X} - \bar{Y} + t_{n+m-2,\alpha/2}S_p\sqrt{\frac{1}{n} + \frac{1}{m}},$$

as listed in Table 3.1, where

$$S_p^2 = \frac{(n-1)S_X^2 + (m-1)S_Y^2}{n+m-2}$$

is the pooled sample variance. But what of the case when σ_X^2 and σ_Y^2 are unknown, but cannot be assumed equal? It does not make sense to compute a single pooled estimate of the population variance in this case. If n and m are large enough so that S_X^2 and S_Y^2 are reasonably accurate estimates of σ_X^2 and σ_Y^2, then it is approximately true that

$$\frac{\bar{X} - \bar{Y} - (\mu_X - \mu_Y)}{\sqrt{\frac{S_X^2}{n} + \frac{S_Y^2}{m}}} \sim N(0, 1),$$

resulting in an approximate two-sided $100(1-\alpha)\%$ confidence interval for $\mu_X - \mu_Y$ of

$$\bar{X} - \bar{Y} - z_{\alpha/2}\sqrt{\frac{S_X^2}{n} + \frac{S_Y^2}{m}} < \mu_X - \mu_Y < \bar{X} - \bar{Y} + z_{\alpha/2}\sqrt{\frac{S_X^2}{n} + \frac{S_Y^2}{m}}.$$

But what of the case when n and m are small? A heuristic method devised by Bernard Welch and modified by Alice Aspin gives surprisingly accurate approximate confidence intervals. They assumed that a t distribution is a close approximation to

$$\frac{\bar{X}-\bar{Y}-(\mu_X-\mu_Y)}{\sqrt{\frac{S_X^2}{n}+\frac{S_Y^2}{m}}} \sim t(r),$$

where the degrees of freedom r is

$$r = \frac{\left(\frac{s_X^2}{n}+\frac{s_Y^2}{m}\right)^2}{\frac{1}{n-1}\left(\frac{s_X^2}{n}\right)^2+\frac{1}{m-1}\left(\frac{s_Y^2}{m}\right)^2},$$

resulting in an approximate two-sided $100(1-\alpha)\%$ confidence interval for $\mu_X-\mu_Y$ of

$$\bar{X}-\bar{Y}-t_{r,\alpha/2}\sqrt{\frac{S_X^2}{n}+\frac{S_Y^2}{m}} < \mu_X-\mu_Y < \bar{X}-\bar{Y}+t_{r,\alpha/2}\sqrt{\frac{S_X^2}{n}+\frac{S_Y^2}{m}}.$$

Some authors apply the floor function to r in order to force the degrees of freedom to be an integer, but we do not here because the `qt` function in R accepts any positive value for the degrees of freedom.

Example 3.15 Consider again the tomato yield data from Example 3.11 shown below.

Fertilizer X yields ($n=5$)	29.9	11.4	25.3	16.5	21.2	
Fertilizer Y yields ($m=6$)	26.6	23.7	28.5	14.2	17.9	24.3

Use Welch's approximation to give an approximate two-sided 90% confidence interval for $\mu_X-\mu_Y$.

The sample statistics that are required for computing the confidence interval are

$$\bar{x}=20.9 \qquad \bar{y}=22.5 \qquad s_X^2=52.5 \qquad s_Y^2=29.5.$$

Using Welch's approximation, an approximate two-sided 90% confidence interval for $\mu_X-\mu_Y$ is

$$\bar{x}-\bar{y}-t_{r,\alpha/2}\sqrt{\frac{s_X^2}{n}+\frac{s_Y^2}{m}} < \mu_X-\mu_Y < \bar{x}-\bar{y}+t_{r,\alpha/2}\sqrt{\frac{s_X^2}{n}+\frac{s_Y^2}{m}},$$

where

$$r = \frac{\left(\frac{s_X^2}{n}+\frac{s_Y^2}{m}\right)^2}{\frac{1}{n-1}\left(\frac{s_X^2}{n}\right)^2+\frac{1}{m-1}\left(\frac{s_Y^2}{m}\right)^2} = \frac{\left(\frac{52.5}{5}+\frac{29.5}{6}\right)^2}{\frac{1}{4}\left(\frac{52.5}{5}\right)^2+\frac{1}{5}\left(\frac{29.5}{6}\right)^2} = 7.3362,$$

or

$$20.9-22.5-t_{7.3,0.05}\sqrt{\frac{52.5}{5}+\frac{29.5}{6}} < \mu_X-\mu_Y < 20.9-22.5+t_{7.3,0.05}\sqrt{\frac{52.5}{5}+\frac{29.5}{6}}$$

or

$$-9.06 < \mu_X-\mu_Y < 5.72.$$

This approximate 90% confidence interval is slightly wider than the exact confidence interval from Example 3.11. This confidence interval can be calculated with the following R code.

```
x = c(29.9, 11.4, 25.3, 16.5, 21.2)
y = c(26.6, 23.7, 28.5, 14.2, 17.9, 24.3)
n = length(x)
m = length(y)
r = (var(x) / n + var(y) / m) ^ 2 / ((var(x) / n) ^ 2 / (n - 1) +
    (var(y) / m) ^ 2 / (m - 1))
l = mean(x) - mean(y) - qt(0.95, r) * sqrt(var(x) / n + var(y) / m)
u = mean(x) - mean(y) + qt(0.95, r) * sqrt(var(x) / n + var(y) / m)
```

Once again, the R function `t.test` can be used to calculate the interval, this time with the `var.equal` argument set to `FALSE`, as shown below.

```
x = c(29.9, 11.4, 25.3, 16.5, 21.2)
y = c(26.6, 23.7, 28.5, 14.2, 17.9, 24.3)
t.test(x, y, conf.level = 0.90, var.equal = FALSE)
```

Since this confidence interval includes 0, it can be concluded that there is not enough statistical evidence in the data values to conclude that the population mean tomato yield using Fertilizer Y differs from the population mean tomato yield using Fertilizer X.

Since the confidence interval based on Welch's approximation is an approximate confidence interval, the next question concerns how well the confidence interval performs. Unfortunately, the mathematics associated with determining the actual coverage for this confidence interval is complicated. We instead settle for a Monte Carlo simulation experiment in the next example.

Example 3.16 Estimate the actual coverage of Welch's approximate confidence interval for the difference between the two population means $\mu_X - \mu_Y$ for the following set of parameters:

$$\alpha = 0.1, \qquad n = 3, \qquad m = 4, \qquad \mu_X = 7, \qquad \mu_Y = 5, \qquad \sigma_X = 2, \qquad \sigma_Y = 8.$$

The analytic derivation of the coverage of the interval is too difficult in this case, so we resort to Monte Carlo simulation. The following R code prints the fraction of approximate two-sided 90% confidence intervals that cover the true difference between the population means $\mu_X - \mu_Y = 7 - 5 = 2$ in one million replications of the experiment. The large number of replications is an effort to get several digits of accuracy in our estimate of the actual coverage. The `conf.int` extension to the call to `t.test` is used to extract the lower and upper bounds of the approximate confidence interval.

```
nrep = 1000000
count = 0
for (i in 1:nrep) {
  x = rnorm(3, 7, 2)
  y = rnorm(4, 5, 8)
  w = t.test(x, y, conf.level = 0.90, var.equal = FALSE)
  l = w$conf.int[1]
  u = w$conf.int[2]
  if (l < 2 && u > 2) count = count + 1
}
print(count / nrep)
```

After a call to `set.seed(3)` to establish the random number generator seed, five runs of this simulation yield the following actual coverages:

0.8996 0.8998 0.9001 0.8999 0.8998.

Based on these results, all of which round to 0.900, we conclude that the actual coverage is very close to the stated coverage $1 - \alpha = 0.9$. For this particular set of parameters and tiny sample sizes, Welch's heuristic approximation works surprisingly well.

Bootstrapping

You might have noticed that all of the examples presented in this chapter thus far have assumed that the data values are drawn from an identifiable population distribution involving an unknown parameter, which is generically referred to as θ. This assumes that the population distribution is identifiable, which is oftentimes not the case due to: (*a*) a small sample size, (*b*) lack of previous experience with the measurement of interest, and (*c*) a probability distribution that is not among the usual list of suspects. Statistical procedures in which there are no distribution assumptions placed on the population are known as *nonparametric methods*. One such nonparametric method, known as *bootstrapping*, can be used to construct approximate confidence intervals. The mathematical assumptions associated with bootstrapping are minimal, which makes it particularly appealing. Bradley Efron is a statistician who developed the bootstrapping methodology in the 1970s. Bootstrapping is also known as "resampling" because observations are drawn with replacement from the data set. Bootstrapping is a computer-intensive statistical technique that can be used to construct confidence intervals as an alternative to the parametric techniques discussed so far that rely on algebra and calculus to construct confidence intervals.

Consider the following setting. You have n data values drawn from an unknown distribution and you would like a point and interval estimate for some quantity associated with the population distribution; use the population median for the sake of discussion. Assuming that you have absolutely no knowledge of the population distribution, your best guess for the population distribution is the empirical distribution from Definition 1.4, which assigns a probability of $1/n$ to each data value. This discrete distribution has n support values when there are no ties in the data set. A reasonable point estimate of the population median is the sample median, but how do you construct a confidence interval? One way is to "resample" from your data set. That is, choose B samples of n observations drawn with replacement from the data set. (The number of bootstrap replications B is a large fixed positive integer; a lower-case b is more appropriate, but I decided to be consistent with the bootstrappers.) By the multiplication rule, there are n^n such equally-likely samples when the order of the resampled values is taken into account, again assuming that there are no ties in the data set. For each of the B samples generated, the measure of interest, in this case the sample median, is calculated. So we now have B sample medians. What to do with these B sample medians can take several different paths at this point. The series of four examples that follow illustrate four paths that can be taken with bootstrapping. These examples will all be applied to a single data set that is described next. The data set is from an application area known to biostatisticians as survival analysis. The data set is given in Efron's book (with co-author Robert Tibshirani) on bootstrapping.

An experiment is conducted to determine the effectiveness of an experimental treatment designed to prolong survival time after surgery. Seven out of sixteen rats are selected at random for the experimental treatment. The other nine rats do not receive the treatment and are classified as the control group. This setting is common in biostatistical research. The sorted survival times, in days, are shown in Table 3.5. The question of interest is whether the experimental treatment increases survival time. More specifically, does the experimental treatment increase the population median

Group	Data	Sample size	Sample median
Treatment	16, 23, 38, 94, 99, 141, 197	$n = 7$	94
Control	10, 27, 30, 40, 46, 51, 52, 104, 146	$m = 9$	46

Table 3.5: Rat survival times in days.

survival time? A cursory analysis of the data indicates that this might indeed be the case. The sample median for the treatment group is 94 days and the sample median for the control group is only 46 days. But because the sample sizes are small, the difference of $94 - 46 = 48$ days might not be statistically significant. The sample sizes are so small that it is not possible to assume a parametric distribution with any degree of certainty, so using the bootstrap method is appropriate.

Bootstrapping will be used in four different manners in the four examples that follow, all of which concern the rat survival times, in order to find

- bootstrap confidence intervals for the population median survival time for each group,
- bootstrap estimates of the standard error of the sample median survival time for each group,
- parametric bootstrap confidence intervals for the population median survival time for each group,
- a bootstrap confidence interval for the difference between the population median survival times for each group.

So we begin by calculating bootstrap confidence intervals for the population median survival time for the treatment and control groups. In order to do so, consider the treatment group separately, and execute the following algorithm, which is stated generically.

1. Collect n data values on the random variable of interest.
2. Calculate a point estimator $\hat{\theta}$ of the parameter of interest θ.
3. Repeat the following steps B times, where B is a large integer.
   a. Generate a *bootstrap sample* of n observations drawn *with replacement* from the data set.
   b. Calculate and store a bootstrap point estimator $\hat{\theta}$ on the bootstrap sample.
4. Sort the B bootstrap point estimators.
5. Report the $100(\alpha/2)$th and $100(1-\alpha/2)$th percentiles of the sorted bootstrap point estimators in order to create an approximate two-sided $100(1-\alpha)\%$ bootstrap confidence interval for θ.

We now apply these steps on the rat survival data to construct bootstrap confidence intervals for the population median survival times of the two groups.

Example 3.17 For the rat survival data from Table 3.5, compute an approximate two-sided 90% bootstrap confidence intervals for the population median survival times of the two groups.

Begin the process by focusing solely on the treatment group. Since the sample size $n = 7$ is so small and there is no previous experience with rat survival times with the experimental treatment, it is not possible to identify a parametric population distribution. This makes bootstrapping a reasonable approach for calculating confidence intervals for the population median survival time. Since the only information that we have on the survival times for the treatment group are the seven survival times, the best estimate of the survival time probability distribution is the empirical probability distribution. The empirical probability distribution is simply a discrete random variable X with support $\mathcal{A} = \{x \mid x = 16, 23, 38, 94, 99, 141, 197\}$, with each element associated with a mass value of $1/7$, that is, a probability distribution with empirical probability mass function

$$f_X(x) = \frac{1}{7} \qquad\qquad x = 16, 23, 38, 94, 99, 141, 197.$$

A graph of this probability mass function is shown in Figure 3.24. This is the probability distribution from which the B bootstrap samples of size $n = 7$ will be drawn.

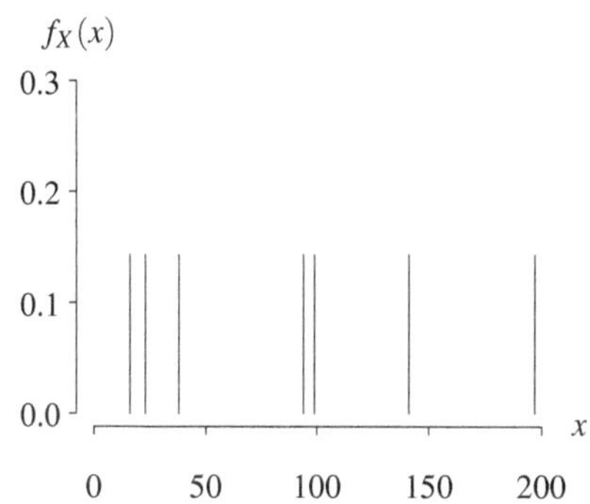

Figure 3.24: Probability mass function of the empirical distribution (treatment case).

Table 3.6 contains four of the $7^7 = 823{,}543$ possible equally-likely bootstrap random samples (assuming that the order that the values are sampled is significant), along with the associated bootstrap sample medians. Notice that the bootstrap sample medians must always equal one of the data values. There will be B random samples like these resampled from the seven rat survival times from the treatment group in the process of bootstrapping.

Although one could, in theory, perform a bootstrap experiment by hand by randomly

Bootstrap sample	Bootstrap sample median
16, 16, 16, 16, 16, 16, 16	16
16, 99, 16, 99, 16, 99, 16	16
197, 141, 99, 94, 38, 23, 16	94
197, 197, 197, 197, 197, 197, 197	197

Table 3.6: Four bootstrap samples.

pulling $n = 7$ balls associated with the survival times from an urn with replacement, bootstrapping is generally performed on a computer. The following R statements perform bootstrapping on the survival times from the treatment group in order to construct an approximate two-sided 90% confidence interval for the population median survival time for the treatment group m_X. The variable `B` holds the number of bootstrap replications. The vector `d` holds the data values, that is, the rat survival times in the treatment group. The `sample` function with the `replace` argument set to `TRUE` creates a bootstrap sample. The vector `x` holds the resampled values. The vector `y` holds the $B = 1000$ sample medians calculated from each bootstrap replication. This vector is sorted, and the 50th and 950th elements are printed as the bootstrap confidence interval bounds.

```
B = 1000
y = numeric(B)
d = c(16, 23, 38, 94, 99, 141, 197)
for(i in 1:B) {
  x    = sample(d, replace = TRUE)
  y[i] = median(x)
}
y = sort(y)
print(c(y[0.05 * B], y[0.95 * B]))
```

The code prints the 90% bootstrap confidence interval for the population median m_X:

$$23 < m_X < 141.$$

We are approximately 90% confident that the population median survival time for the treatment group is between 23 and 141 days. It is not surprising that the lower bound and upper bound for this confidence interval also happen to be data values. The resampling only samples from the data values, and the sample median must be one of the data values. Hence, the vector `y` in the R code consists of only data values. To save some keystrokes, this can also be accomplished using the built-in R function `boot`, which is contained in the R package named `boot`. This package is capable of performing many variations of the bootstrapping algorithm described here.

The bootstrap experiment included $B = 1000$ bootstrap samples drawn from the data set. One should set a B value that is large enough so that the resampling variability associated with bootstrapping is minimized. (The resampling variability associated with bootstrapping is separate from the sampling variability associated with the data set. The resampling variability is an artifact of performing bootstrapping and is something that we would like to drive to zero by increasing B, which is just a matter of increased CPU cycles. The sampling variability associated with the data set can only be decreased by increasing n.) We would always prefer a B value of ∞ because we do not want to add any additional error to the process from not including enough bootstrap samples. Since the sample size $n = 7$ is small and the support of the bootstrap median values includes only the seven values, APPL can be used to determine the distribution of the bootstrap medians as $B \to \infty$. The APPL code below calculates the probability mass function of the bootstrap sample medians.

```
d := [16, 23, 38, 94, 99, 141, 197];
X := BootstrapRV(d);
Y := OrderStat(X, 7, 4);
```

The `BootstrapRV` procedure creates a discrete random variable X with the probability mass function shown in Figure 3.24, that is, the empirical distribution associated with the data set. The `OrderStat` procedure determines the probability distribution of the sample median associated with the random sample of size $n = 7$ drawn from the empirical distribution. This APPL code computes the sample median associated with all $7^7 = 823{,}543$ different samples, and accumulates their probabilities. The probability mass function of the bootstrap sample median is

$$f_Y(y) = \begin{cases} 8359/823543 & y = 16 \\ 80809/823543 & y = 23 \\ 196519/823543 & y = 38 \\ 252169/823543 & y = 94 \\ 196519/823543 & y = 99 \\ 80809/823543 & y = 141 \\ 8359/823543 & y = 197, \end{cases}$$

which is plotted in Figure 3.25. The 5th percentile and the 95th percentile of this distribution are 23 and 141, which are the confidence interval bounds calculated by bootstrapping. So by good fortune the approximate 90% confidence interval computed with $B = 1000$ bootstrapping samples is identical to the approximate 90% confidence interval in the limit as $B \to \infty$. When this same line of reasoning is applied to the survival times of the control group of $m = 9$ rats, the approximate two-sided 90% confidence interval for the population median survival time m_Y is

$$30 < m_Y < 52.$$

We are approximately 90% confident that the population median survival time for the control group is between 30 and 52 days. The confidence interval for the population median survival time for the control group is narrower than that for the treatment group because the sample size is slightly larger and the survival times are less variable for the control group. The fact that the confidence intervals for m_X and m_Y overlap gives us some preliminary evidence that the 48 day difference between the two sample medians is not statistically significant. A subsequent example will compute an approximate two-sided confidence interval for $m_X - m_Y$ to address this question more directly.

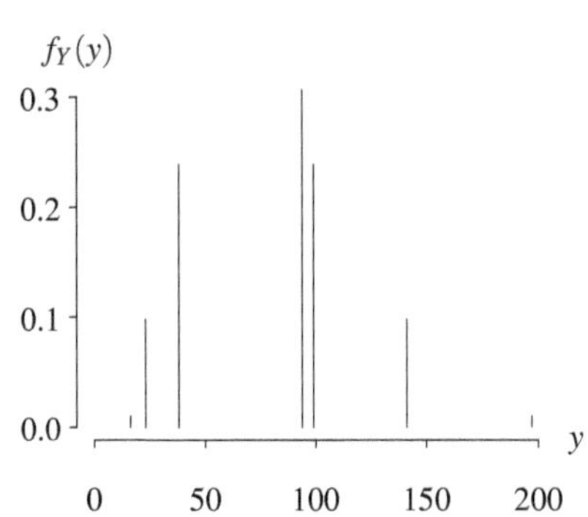

Figure 3.25: Probability mass function of the bootstrap sample medians.

There is always an interest in the standard error, or standard deviation, of a point estimate $\hat{\theta}$. This standard error is a measure of the precision of the point estimate $\hat{\theta}$. In many applications, the interest is in the estimation of the population mean μ_X. When using the sample mean $\bar{X}$ to estimate the population mean, the formula for the population variance of $\bar{X}$ from Theorem 1.1 is

$$V\left[\bar{X}\right] = \frac{\sigma_X^2}{n},$$

so the standard error of $\bar{X}$ is

$$\sigma_{\bar{X}} = \frac{\sigma_X}{\sqrt{n}}.$$

This formula works for random sampling from any population distribution that has a finite population mean μ_X and finite population variance σ_X^2.

But for point estimates $\hat{\theta}$ of quantities other than the population mean, there are oftentimes no general formulas for their standard error $\sigma_{\hat{\theta}}$. Bootstrapping can fill this gap by providing an estimate of the standard error of $\hat{\theta}$ by again using resampling. As in the previous cases, the original data set is resampled B times and $\hat{\theta}$ is computed for each bootstrap sample. But this time rather than sorting the data and printing two percentiles, the sample standard deviation of the B values of $\hat{\theta}$ is computed as an estimate of the standard error of $\hat{\theta}$. Estimating the standard error of the sample medians for the treatment and control groups for the rat data will be illustrated next.

Example 3.18 Find the bootstrap estimate of the standard error of the sample medians for the treatment and control groups of the rat data.

We again focus initially on the $n = 7$ rat survival times in the treatment group. The point estimate of the population median is the sample median: 94 days. Bootstrapping is used to estimate the standard error of this point estimate. The following R code (*a*) stores the data values in the vector `d`, (*b*) generates $B = 50$ bootstrap samples, (*c*) calculates the $B = 50$ sample medians of the bootstrap samples, and (*d*) estimates the standard error of the sample median by computing the sample standard deviation of the $B = 50$ bootstrap sample medians.

```
B = 50
y = numeric(B)
d = c(16, 23, 38, 94, 99, 141, 197)
for (i in 1:B) y[i] = median(sample(d, replace = TRUE))
sd(y)
```

This code returns the estimated standard error of the sample median of the treatment group as 41.18. This is a much smaller value of B than was used in the previous example. The estimates of the standard error of the sample median can vary wildly when smaller values of B are used. Table 3.7 shows the estimates of the standard error of the sample median when this code is run for several values of B for both the treatment and the control cases. Considering the fact that all of the bootstrapping error is artificial and computers are fast, it seems reasonable to use a minimum of $B = 10{,}000$ resamplings when using bootstrapping. There are two ways to determine what values should go in the $B = \infty$ column of Table 3.7. The first way is to increase B to a huge value and run the two lines of R code given above. The second way is a little more sophisticated. Since Y from the previous example contains the probability mass function of the bootstrap sample medians, all that is necessary is to compute the population standard deviation of that

	$B = 50$	$B = 100$	$B = 250$	$B = 1000$	$B = \infty$
Treatment	41.18	37.63	36.88	38.98	37.83
Control	20.30	12.68	9.538	13.82	13.08

Table 3.7: Bootstrap estimates of the standard error of the sample median.

probability distribution and this will be the limiting value of the standard error of the sample median as $B \to \infty$. The standard error for the sample median for the treatment group (σ_Y from Example 3.17) is

$$\frac{2}{823543}\sqrt{242712738519382} \cong 37.8347.$$

The standard error for the sample median for the control group is

$$\frac{1}{387420489}\sqrt{25662937134123797402} \cong 13.0759.$$

Finally, we return to the original question of whether there is a statistically significant difference between the two population medians of the two groups based on the difference between the two sample medians of $94 - 46 = 48$ days. Since the two data sets are considered to be independent random samples from two populations (the treatment and control populations), one can compute

$$\frac{48}{\sqrt{37.8347^2 + 13.0759^2}} \cong 1.19,$$

which indicates that the seemingly large difference between the two sample medians amounts to only 1.19 standard-deviation units. The conclusion that can be drawn from this analysis is that there is no statistically significant difference between the population medians based on these small sample sizes.

The beauty of bootstrapping is that no assumptions about the population distribution are necessary. But there is a type of bootstrapping known as *parametric bootstrapping* that can be performed when the analyst has knowledge of the population distribution. Instead of drawing the B bootstrap samples of size n directly from the data, n random variates are generated from a distribution that is fitted to the data set for each bootstrap replication. A parametric bootstrap will be applied to the survival times for the rats in the treatment group in order to calculate a confidence interval on the population median survival time.

Example 3.19 Find an approximate two-sided 90% confidence interval for the population median rat survival time in the treatment group using parametric bootstrapping. Assume an exponential population for the rat survival times in the treatment group.

Perhaps in light of experience with previous data sets, we are now comfortable assuming that the survival data in the treatment case represents a random sample of $n = 7$ observations from an exponential population. Example 2.8 showed that the maximum likelihood estimator of the population mean θ is the sample mean, so

$$\hat{\theta} = \frac{16 + 23 + 38 + 94 + 99 + 141 + 197}{7} = \frac{608}{7} = 86.9$$

days, which defines the fitted exponential distribution. In the code that follows, the R function `rexp` is used to generate the random exponential variates for the bootstrap samples. This code should be compared to the code used in Example 3.17.

```
B = 50
y = numeric(B)
d = c(16, 23, 38, 94, 99, 141, 197)
n = length(d)
m = mean(d)
for(i in 1:B) {
  x = rexp(n, 1 / m)
  y[i] = median(x)
}
y = sort(y)
print(c(y[0.05 * B], y[0.95 * B]))
```

The results of executing this bootstrapping code for several values of B, rounded to significant digits, are shown in Table 3.8. The confidence interval bounds are still quite sensitive to the value of B. One difference between the parametric bootstrap and the original bootstrap is that the bootstrap values are no longer limited to the data values, which results in more variability in the confidence interval bounds.

$B = 50$	$B = 100$	$B = 250$	$B = 1000$	$B = \infty$
$18 < m_X < 138$	$18 < m_X < 118$	$22 < m_X < 134$	$21 < m_X < 134$	$22 < m_X < 129$

Table 3.8: Parametric bootstrap approximate 90% confidence intervals for the population median.

All three of the previous examples concerning bootstrapping have considered the treatment group and the control group separately concerning the rat survival times. This does not need to be the case. The final example of this subsection concerns computing a confidence interval for the difference between the population median survival times for the two groups.

Example 3.20 Find an approximate two-sided 90% confidence interval for the difference between the population median survival times for the treatment and control groups via bootstrapping.

The quantity that will reside in the center of the approximate 90% confidence interval is $m_X - m_Y$. The bootstrap algorithm must be adjusted slightly in order to account for the two populations, as indicated below.

1. Collect $n = 7$ survival times from the treatment group and $m = 9$ survival times from the control group.
2. Calculate the point estimator $94 - 46 = 48$ days of the population parameter of interest $m_X - m_Y$.
3. Repeat the following steps B times, where B is a large integer.
   a. Generate a *bootstrap sample* of $n = 7$ observations drawn *with replacement* from the treatment group survival times.

   b. Generate a *bootstrap sample* of $m = 9$ observations drawn *with replacement* from the control group survival times.
   c. Calculate and store a bootstrap point estimator that is the difference between the sample medians of the two bootstrap samples.

4. Sort the B bootstrap point estimators.
5. Report the $100(\alpha/2)$th and $100(1-\alpha/2)$th percentiles of the sorted bootstrap point estimators in order to create an approximate two-sided $100(1-\alpha)\%$ bootstrap confidence interval for $m_X - m_Y$.

The following R code implements this algorithm for $B = 10,000$ bootstrap replicates.

```
B  = 10000
y  = rep(0, B)
d1 = c(16, 23, 38, 94, 99, 141, 197)
d2 = c(10, 27, 30, 40, 46, 51, 52, 104, 146)
for(i in 1:B) {
  x1   = sample(d1, replace = TRUE)
  x2   = sample(d2, replace = TRUE)
  y[i] = median(x1) - median(x2)
}
y = sort(y)
print(c(y[0.05 * B], y[0.95 * B]))
```

This code returns the approximate two-sided 90% confidence interval for $m_X - m_Y$ via bootstrapping as

$$-28 < m_X - m_Y < 95.$$

We are approximately 90% confident that the difference in the population medians lies between -28 and 95. Since this approximate confidence interval contains 0, there is not enough statistical evidence in the data to conclude that the experimental treatment increases the population median survival time over the control group.

The previous four examples have illustrated the use of bootstrapping, or resampling, to construct approximate confidence intervals. These methods are nonparametric, or distribution-free in the sense that they do not require a parametric distribution to be assumed for the population. Another nonparametric method can be used to construct approximate confidence intervals for percentiles which does not require resampling, which is introduced next.

Confidence intervals for percentiles

Let $X_1, X_2, \ldots, X_n$ be a random sample drawn from a continuous population with probability density function $f_X(x)$ and associated cumulative distribution function $F_X(x)$. The associated order statistics are denoted by $X_{(1)}, X_{(2)}, \ldots, X_{(n)}$. The procedure for constructing approximate confidence intervals for percentiles described next is nonparametric in the sense that it does not require any distributional assumptions about the population beyond the assumption that the population distribution is continuous. We want to find a point and interval estimate for the $100p$th percentile of the population distribution, which will be denoted by x_p, where the subscript denotes a left-hand tail probability. The use of the subscript here differs from their use elsewhere, where oftentimes the subscript denotes a right-hand tail probability. For example,

- the 10th percentile of the population is $x_{0.1}$,
- the population median, or the 50th percentile of the population, is $x_{0.5}$,
- the 95th percentile of the population is $x_{0.95}$.

The pth percentile of the population satisfies

$$P(X \le x_p) = F_X(x_p) = p$$

for $0 < p < 1$.

First, consider finding a point estimate of x_p. If $p = 1/2$, for example, then we seek a point estimate of the population median. Consider the following thought experiment for estimating the population median. If n is odd, then the middle sorted data value, the sample median, seems like a reasonable point estimator. Now we seek a formula for general p for the point estimator that will return this middle value when given a data set with odd sample size n. Since $np+1$ always falls in the interval $1 < np+1 < n+1$ for a p value on the interval $0 < p < 1$, we know that the floor of $np+1$ will always be an integer between 1 and n (inclusive). Hence a point estimator of x_p that behaves properly when n is odd and $p = 1/2$, and always returns one of the order statistics is

$$\hat{x}_p = X_{(\lfloor np+1 \rfloor)}.$$

When $p = 1/2$, this estimator correctly selects the middle data value, that is, the sample median, as the point estimator of the population median as desired for odd n. But there is trouble brewing for even n. When n is even, averaging the two middle sorted values seems like a reasonable point estimator of the population median, particularly if the population distribution is symmetric. But the previous point estimator does not do so. If there are $n = 4$ values, for example, it uses $X_{(\lfloor 4(1/2)+1 \rfloor)} = X_{(3)}$ to estimate the population median; this is a biased estimate. This makes the original point estimator a poor one for small values of n. Nearly a dozen solutions have cropped up for overcoming this deficiency, and one such solution will be presented here. Perhaps the simplest approach conceptually is to compute a convex combination of two adjacent order statistics in order to estimate x_p. Although the formula for the point estimate is complicated, x_p can be estimated with the convex combination

$$\hat{x}_p = X_{(\lfloor m \rfloor)} + (m - \lfloor m \rfloor)\left(X_{(\lfloor m \rfloor+1)} - X_{(\lfloor m \rfloor)}\right),$$

where $m = np + 1/2$. This point estimator is a convex combination of $X_{(\lfloor m \rfloor)}$ and $X_{(\lfloor m+1 \rfloor)}$, with mixing parameter $(m - \lfloor m \rfloor)$. When $p = 1/2$, this formula correctly selects the middle sorted data value for odd values of n, and correctly averages the two middle sorted data values for even values of n. For values of p other than $1/2$, the formula uses an appropriate convex combination of two order statistics as a point estimator. As long as p is not too close to 0 or 1, that is, values of p on the interval $1/(2n) < p \le 1 - 1/(2n)$, this formula gives an adequate point estimate of x_p.

The next step is to construct an approximate confidence interval for x_p. For the time being, consider just a single observation X. The observation is either less than x_p or greater than x_p. The probability that the observation is less than x_p is p. Now consider n such observations. Since the observations are a random sample, each observation being less than x_p is a mutually independent Bernoulli trial. Therefore, the number of observations that are less than x_p has the binomial distribution with parameters n and p.

One possibility for constructing an approximate confidence interval for x_p is to choose indices i and j such that

$$P\left(X_{(i)} < x_p < X_{(j)}\right)$$

is close to $1-\alpha$. The discrete nature of the sampling makes achieving this probability exactly for prescribed values of n and α virtually impossible. Returning to the relationship with the binomial distribution, the probability that $X_{(i)} < x_p$ is found by computing the probability that i or more order statistics fall to the left of x_p, that is,

$$P\left(X_{(i)} < x_p\right) = \sum_{x=i}^{n} \binom{n}{x} p^x (1-p)^{n-x}.$$

Likewise, the probability that $X_{(j)} > x_p$ is found by computing the probability that $j-1$ or fewer order statistics fall to the left of x_p, that is,

$$P\left(X_{(j)} > x_p\right) = \sum_{x=0}^{j-1} \binom{n}{x} p^x (1-p)^{n-x}.$$

Combining these two events,

$$P\left(X_{(i)} < x_p < X_{(j)}\right) = \sum_{x=i}^{j-1} \binom{n}{x} p^x (1-p)^{n-x}.$$

In constructing a confidence interval for x_p, the appropriate binomial probabilities are summed so as to determine the values of i and j such that the probability is as close as possible to the stated coverage $1-\alpha$.

Example 3.21 The $n=7$ rat survival times in the treatment group from Table 3.5, in days, are

$$16, 23, 38, 94, 99, 141, 197.$$

Find a point estimator and an approximate two-sided 90% nonparametric confidence interval for the population median survival time.

The point estimator of the population median is the sample median: $\hat{x}_{0.5} = 94$ days. Since $n=7$, $\alpha = 0.1$ and $p = 1/2$, we want a left-hand tail probability associated with a binomial(7, 1/2) distribution of approximately $\alpha/2 = 0.05$. Since

$$P\big(\text{binomial}(7, 1/2) \le 0\big) = \left(\frac{1}{2}\right)^7 = \frac{1}{128} = 0.0078125$$

and

$$P\big(\text{binomial}(7, 1/2) \le 1\big) = \left(\frac{1}{2}\right)^7 + 7\left(\frac{1}{2}\right)^7 = \frac{8}{128} = \frac{1}{16} = 0.0625,$$

we choose the second and sixth order statistics as the approximate 90% confidence interval based on the symmetry of the binomial distribution with $p = 1/2$. So an approximate two-sided 90% confidence interval for the population median is

$$X_{(2)} < x_{0.5} < X_{(6)}$$

or

$$23 < x_{0.5} < 141.$$

We are approximately 90% confident that the population median survival time for the rats given the treatment is between 23 and 141 days. This matches the 90% confidence interval from Example 3.17. The actual coverage for this interval is $1 - 2 \cdot 0.0625 = 0.875$, which is close to the stated coverage $1-\alpha = 0.9$.

The confidence interval for a percentile of a distribution given here is approximate. Determining the actual coverage is nontrivial because one would need to consider

- all continuous population distributions,
- all parameter settings for the population distribution,
- all sample sizes n,
- all stated coverages $1-\alpha$,
- all p values for the percentile x_p.

We will not be so ambitious. Rather, the next example gives a Monte Carlo simulation for one setting of these five factors.

Example 3.22 Find the actual coverage for an approximate two-sided 90% confidence interval for the population median for a random sample $X_1, X_2, \ldots, X_{21}$ drawn from an exponential population with population mean $\theta = 2$.

The cumulative distribution function of the population is

$$F_X(x) = 1 - e^{-x/\theta} \qquad x > 0,$$

so the true value of the population 100pth percentile x_p can be found by solving

$$F_X(x_p) = 1 - e^{-x_p/\theta} = p$$

for x_p yielding

$$x_p = -\theta \ln(1-p).$$

So the true value of the population median for an exponential population with population mean $\theta = 2$ is

$$x_{0.5} = -2\ln(1/2) \cong 1.3863.$$

The $\alpha/2 = 0.05$ percentile of a binomial(21, 1/2) random variable is 6, so the approximate confidence interval is

$$X_{(7)} < x_{0.5} < X_{(15)}.$$

The actual coverage is $P\left(X_{(7)} < x_{0.5} < X_{(15)}\right)$. Using an order statistic result, the joint probability density function of $X_{(7)}$ and $X_{(15)}$ is

$$f_{X_{(7)},X_{(15)}}(x_7, x_{15}) = \frac{21!}{6!7!6!}\left(1-e^{-x_7/2}\right)^6 \frac{1}{2}e^{-x_7/2}\left(e^{-x_7/2}-e^{-x_{15}/2}\right)^7 \frac{1}{2}e^{-x_{15}/2}\left(e^{-x_{15}/2}\right)^6$$

for $0 < x_7 < x_{15}$. Leaving the integration to a computer algebra system, the actual coverage is

$$P\left(X_{(7)} < x_{0.5} < X_{(15)}\right) = \int_0^{x_{1/2}} \int_{x_{1/2}}^{\infty} f_{X_{(7)},X_{(15)}}(x_7, x_{15})\, dx_{15}\, dx_7 = \frac{60401}{65536} \cong 0.9216,$$

which is near the stated coverage 0.9. The analytic calculation of the actual coverage can be assessed using the Monte Carlo simulation experiment in R given next. The population median is stored in `m`. The R function `rexp` generates the random exponential data values, which are stored in the vector `x`. This code prints the fraction of the approximate confidence intervals that cover the population median.

```
n = 21
m = -2 * log(1 / 2)
nrep  = 1000000
count = 0
for (i in 1:nrep) {
  x = sort(rexp(n, 1 / 2))
  if (x[7] < m && m < x[15]) count = count + 1
}
print(count / nrep)
```

After a call to `set.seed(3)` to initialize the random number stream, five calls to this program yield

$$0.9214 \qquad 0.9214 \qquad 0.9213 \qquad 0.9217 \qquad 0.9219.$$

Since these values hover about the analytic value for the actual coverage 0.9216, the Monte Carlo simulation supports the analytic result. The actual coverage of 0.9216 exceeds the stated coverage $1-\alpha = 0.9$.

This concludes the discussion of approximate confidence intervals. An exact confidence interval is always preferred to an approximate confidence interval when possible, but it is often impossible to construct one for a particular sampling distribution. An approximate confidence interval for an unknown parameter θ whose actual coverage exceeds the stated coverage for all values of θ is known as a conservative confidence interval. Some approximate confidence intervals have the appealing property that their actual coverage converges to the exact coverage in the limit as $n \to \infty$. Such a confidence interval is known as an asymptotically exact confidence interval, which is the topic of the next section.

3.3 Asymptotically Exact Confidence Intervals

The previous two sections in this chapter have divided the universe of confidence intervals into two subsets: exact confidence intervals in which $P(L < \theta < U) = 1-\alpha$ for all values of θ, and approximate confidence intervals in which $P(L < \theta < U) \neq 1-\alpha$ for one or more values of θ. An exact confidence interval gives an accurate assessment of the precision of a point estimator. An approximate confidence interval claims either too much or too little precision of a point estimator.

There is a consolation prize for some approximate confidence intervals. When the actual coverage of an approximate confidence interval approaches the stated coverage in the limit as the sample size n goes to infinity, the confidence interval is known as an *asymptotically exact confidence interval*, as defined next.

Definition 3.7 A confidence interval of the form

$$L < \theta < U$$

is an *asymptotically exact two-sided* $100(1-\alpha)\%$ *confidence interval* for the unknown parameter θ provided

$$\lim_{n\to\infty} P(L < \theta < U) = 1-\alpha$$

for all values of θ.

All exact confidence intervals are asymptotically exact. But only some of the approximate confidence intervals are asymptotically exact. So the Venn diagram in Figure 3.18 can be refined to include asymptotically exact confidence intervals as a subset of the approximate confidence intervals in Figure 3.26. An exact confidence interval remains the gold standard, but if we must settle for an approximate confidence interval, we would typically prefer it to be asymptotically exact.

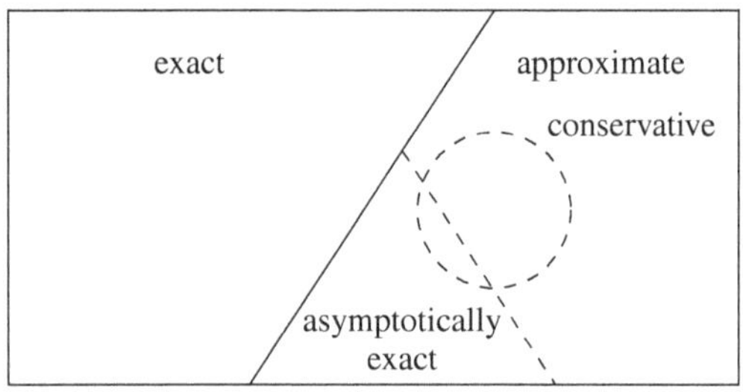

Figure 3.26: Venn diagram of exact, approximate, and asymptotically exact confidence intervals.

To enhance the Venn diagram a bit further, consider the case of two exact confidence interval procedures for an unknown parameter θ. This was the case for sampling from the $U(0, \theta)$ population in Example 3.4. In that case we chose the exact confidence interval that had the smaller expected width. So in general, among all exact confidence intervals we prefer the confidence interval with the minimum expected width. This subset of the exact confidence intervals, shown in Figure 3.27, is labeled MEWECI for "Minimum Expected Width Exact Confidence Interval." These confidence intervals assume an analogous position to the efficient estimators (also known as minimum variance unbiased estimators) in Figure 2.23. Unfortunately, no one has formulated a result that is analogous to the Cramér–Rao inequality that can identify whether a confidence interval belongs in the MEWECI set. Perhaps you will.

Asymptotically exact confidence intervals can be a good option when an exact confidence interval cannot be found or when the coverage of an approximate confidence interval is not satisfactory. This section will survey two methods for constructing asymptotically exact confidence intervals: using the asymptotic normality of maximum likelihood estimators and using the likelihood ratio statistic. Both methods are surprisingly general.

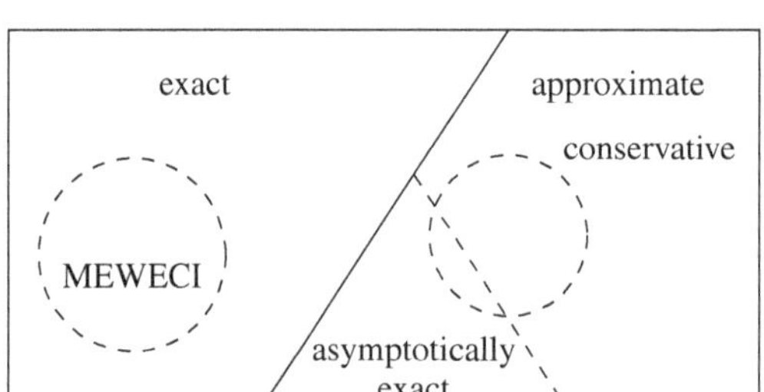

Figure 3.27: Venn diagram classifying confidence intervals.

Asymptotic normality of maximum likelihood estimators

You might have noticed that $\sqrt{n}$, where n is the sample size, has shown up quite frequently in the equations used in statistics. When dealing with the sample mean $\bar{X}$ associated with random sampling from a population distribution with finite population mean μ_X and finite population variance σ_X^2, for instance, we know from Theorem 1.1 that

$$E[\bar{X}] = \mu_X \qquad \text{and} \qquad V[\bar{X}] = \frac{\sigma_X^2}{n},$$

which can also be written as

$$\mu_{\bar{X}} = \mu_X \qquad \text{and} \qquad \sigma_{\bar{X}} = \frac{\sigma_X}{\sqrt{n}}.$$

So the sample mean $\bar{X}$ is an unbiased estimate of the population mean μ_X and the standard error of $\bar{X}$ is inversely related to $\sqrt{n}$. But this $\sqrt{n}$ is a scourge to statisticians because:

- in order to double the precision (that is, halve the standard error), you need four times as many observations,
- in order to triple the precision, you need nine times as many observations,
- in order to obtain an additional digit of accuracy in the point estimate (that is, decrease the standard error by a factor of ten), you need 100 times as many observations.

When obtaining data values is expensive, difficult, or time-consuming, getting additional precision can be a real burden.

To investigate the influence of $\sqrt{n}$ further, we will conduct a Monte Carlo simulation experiment. Assume that $X_1, X_2, \ldots, X_n$ is a random sample from an exponential population with population mean θ. The probability density function of the population is

$$f_X(x) = \frac{1}{\theta} e^{-x/\theta} \qquad x > 0.$$

Examples 2.3 and 2.8 showed that both the method of moments and the maximum likelihood estimators of θ are the sample mean

$$\hat{\theta} = \frac{1}{n} \sum_{i=1}^{n} X_i,$$

which is an unbiased estimator of θ. Furthermore, Example 2.20 showed that $\hat{\theta}$ is an efficient estimator with standard error

$$\sigma_{\hat{\theta}} = \frac{\theta}{\sqrt{n}}.$$

For the Monte Carlo simulation experiment, random variates representing data values from an exponential population with $\theta = 1$ (chosen arbitrarily) are generated for sample sizes $n = 1$, $n = 10$, and $n = 100$. The value of $\hat{\theta} - \theta$ is plotted on the vertical axis in Figure 3.28. There are 25 replications of the experiment conducted at each sample size n. The results are not at all surprising. The sample mean $\hat{\theta}$ is closing in on the population mean $\theta = 1$ as the sample size increases, so the points plotted approach $E\left[\hat{\theta} - \theta\right] = \theta - \theta = 0$ as n increases. Since the support of $\hat{\theta}$ is all positive real numbers, the support of $\hat{\theta} - \theta$ is all real numbers that are greater than -1. This Monte Carlo simulation does not provide the basis for an asymptotic confidence interval for θ. But a small tweak to the Monte Carlo simulation will provide that basis.

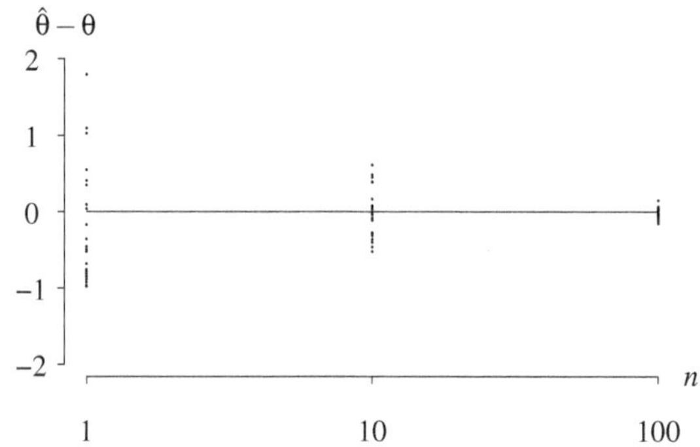

Figure 3.28: Monte Carlo experiment plotting the error $\hat{\theta}-\theta$ for $n=1$, $n=10$, and $n=100$.

Instead of plotting $\hat{\theta}-\theta$, consider plotting $\sqrt{n}\left(\hat{\theta}-\theta\right)$. This random variable involves the product of $\sqrt{n}$, a quantity that is headed off to infinity as $n \to \infty$, and $\hat{\theta}-\theta$, a quantity that is headed to zero as $n \to \infty$. The results of the Monte Carlo simulation are shown in Figure 3.29. It appears that the factor $\sqrt{n}$ is just the right multiplier so that the observations do not collapse to zero or explode to $\pm\infty$. To see why this is the case, let's calculate the expected value and population variance of $\sqrt{n}\left(\hat{\theta}-\theta\right)$. For a finite value of n, the expected value is

$$E\left[\sqrt{n}\left(\hat{\theta}-\theta\right)\right] = \sqrt{n}\,(\theta-\theta) = 0$$

and the population variance is

$$V\left[\sqrt{n}\left(\hat{\theta}-\theta\right)\right] = nV\left[\hat{\theta}\right] = n \cdot \frac{\theta^2}{n} = \theta^2.$$

The $\sqrt{n}$ multiplier provides just enough expansion as n grows so that $\sqrt{n}\left(\hat{\theta}-\theta\right)$ has an expected value and population variance that do not depend on n.

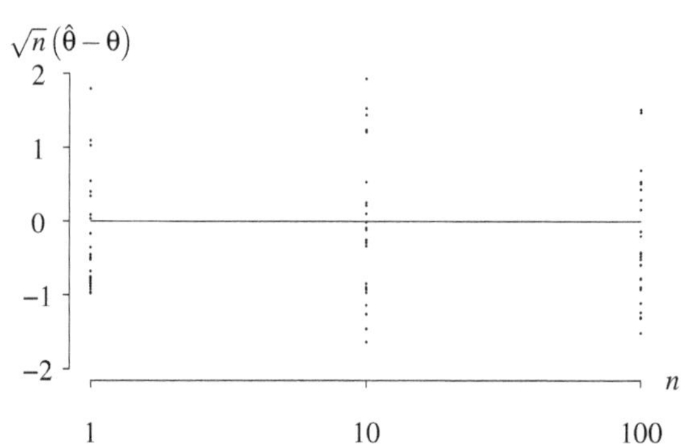

Figure 3.29: Monte Carlo experiment plotting $\sqrt{n}\left(\hat{\theta}-\theta\right)$ for $n=1$, $n=10$, and $n=100$.

Now that we know the expected value and population variance of $\sqrt{n}\,(\hat{\theta}-\theta)$, we might want to know its probability distribution. This question will be answered in two ways—first for finite n and then in the limit as $n \to \infty$. In the case of finite n, we know that since $X_1, X_2, \ldots, X_n$ are a random sample from an exponential population with population mean θ,

$$\sum_{i=1}^{n} X_i \sim \text{Erlang}(1/\theta, n)$$

and

$$\hat{\theta} = \bar{X} = \frac{1}{n}\sum_{i=1}^{n} X_i \sim \text{Erlang}(n/\theta, n),$$

which corresponds to the probability density function

$$f_{\hat{\theta}}(x) = \frac{(n/\theta)^n x^{n-1} e^{-nx/\theta}}{(n-1)!} \qquad x > 0.$$

Since $\hat{\theta}-\theta$ is just a translation of this distribution, it has probability density function

$$f_{\hat{\theta}-\theta}(x) = \frac{(n/\theta)^n (x+\theta)^{n-1} e^{-n(x+\theta)/\theta}}{(n-1)!} \qquad x > -\theta.$$

Finally, consider the product of $\sqrt{n}$ and $\hat{\theta}-\theta$, which is a linear, one-to-one transformation. Leaving the details to the reader, the probability density function of $\sqrt{n}\,(\hat{\theta}-\theta)$ by using the transformation method is

$$f_{\sqrt{n}(\hat{\theta}-\theta)}(x) = \frac{n^n (x/\sqrt{n}+\theta)^{n-1} e^{-n(x/\sqrt{n}+\theta)/\theta}}{\sqrt{n}\,\theta^n (n-1)!} \qquad x > -\sqrt{n}\,\theta.$$

This is the probability density function associated with each point that was plotted in Figure 3.29. These probability density functions are plotted in Figure 3.30 for $\theta = 1$ and $n = 1$, $n = 10$, and $n = 100$. When $n = 1$, the probability distribution of $\sqrt{1}\,(\hat{\theta}-\theta)$ is simply a unit exponential distribution shifted one unit to the left. When $n = 10$, the probability distribution of $\sqrt{10}\,(\hat{\theta}-\theta)$ looks more symmetric and bell-shaped. Finally, for $n = 100$, the probability distribution of $\sqrt{100}\,(\hat{\theta}-\theta)$

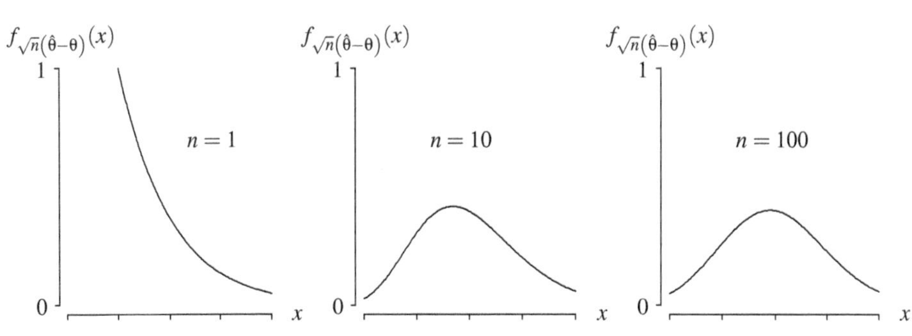

Figure 3.30: Probability density functions of $\sqrt{n}\,(\hat{\theta}-\theta)$ for $n = 1$, $n = 10$, and $n = 100$.

looks nearly symmetric and normally distributed. Each of these three population distributions has expected value 0 and population variance $\theta^2 = 1$.

Now we turn to the probability distribution of $\sqrt{n}\left(\hat{\theta} - \theta\right)$ as $n \to \infty$, which we suspect is normally distributed from the third graph in Figure 3.30. Since the random sampling is from a population distribution with finite population mean and variance, the central limit theorem indicates that $\sum_{i=1}^{n} X_i$ is asymptotically normally distributed. Furthermore, since the transformation $\sqrt{n}\left(\hat{\theta} - \theta\right)$ involves just scalings and translations, it is also asymptotically normally distributed. Hence,

$$\sqrt{n}\left(\hat{\theta} - \theta\right) \xrightarrow{D} N\left(0, \theta^2\right).$$

Replacing θ^2 on the right-hand side of this relationship with its maximum likelihood estimator $\hat{\theta}^2$ is not appropriate for small values of n, but is certainly reasonable as $n \to \infty$ because $\hat{\theta}$ is a consistent estimate of θ. After standardizing, this relationship can be written as

$$\frac{\sqrt{n}\left(\hat{\theta} - \theta\right)}{\hat{\theta}} \xrightarrow{D} N(0, 1).$$

This is the break we have been looking for in the development of our first asymptotically exact confidence interval. Since the standard normal distribution does not involve θ, we can use $\sqrt{n}\left(\hat{\theta} - \theta\right)/\hat{\theta}$ as a pivotal quantity. So using the pivotal method, but this time in the limit as $n \to \infty$,

$$\lim_{n \to \infty} P\left(-z_{\alpha/2} < \frac{\sqrt{n}\left(\hat{\theta} - \theta\right)}{\hat{\theta}} < z_{\alpha/2}\right) = 1 - \alpha,$$

where $z_{\alpha/2}$ is the $100(1-\alpha/2)$th percentile of the standard normal distribution. Rearranging so that θ is isolated in the middle of the inequality,

$$\lim_{n \to \infty} P\left(\hat{\theta} - z_{\alpha/2}\frac{\hat{\theta}}{\sqrt{n}} < \theta < \hat{\theta} + z_{\alpha/2}\frac{\hat{\theta}}{\sqrt{n}}\right) = 1 - \alpha.$$

So an asymptotically exact two-sided $100(1-\alpha)\%$ confidence interval for θ is

$$\hat{\theta} - z_{\alpha/2}\frac{\hat{\theta}}{\sqrt{n}} < \theta < \hat{\theta} + z_{\alpha/2}\frac{\hat{\theta}}{\sqrt{n}}.$$

This is a symmetric confidence interval, so it can also be stated in the more compact form

$$\hat{\theta} \pm z_{\alpha/2}\frac{\hat{\theta}}{\sqrt{n}}.$$

The fact that this confidence interval is symmetric but the population distribution (the exponential distribution) is not symmetric leads us to expect poor performance for small values of n. It is not possible to collect an infinite number of observations in practice, so we are interested in how well this asymptotically exact confidence interval performs for finite values of n. The formula for an asymptotically exact confidence interval is typically used for finite values of n as an approximate confidence interval. Let's investigate the actual coverage of this confidence interval for an arbitrary stated coverage of, say, $\alpha = 0.10$, that is, a 90% confidence interval. The 95th percentile of the standard normal distribution, to four digits, is $z_{\alpha/2} = z_{0.05} = 1.645$. In order to calculate the actual coverage, we need to know the probability density function of $\sqrt{n}\left(\hat{\theta}-\theta\right)/\hat{\theta}$. Leaving the details as an exercise at the end of the chapter, the probability density function of $\sqrt{n}\left(\hat{\theta}-\theta\right)/\hat{\theta}$ is

$$f_{\sqrt{n}(\hat{\theta}-\theta)/\hat{\theta}}(x)=\frac{n^{3n/2}e^{n^{3/2}/(x-\sqrt{n})}}{(n-1)!\left(-x+\sqrt{n}\right)^{n+1}} \qquad x<\sqrt{n},$$

which does not involve θ. The geometry associated with calculating the actual coverage is the area under this probability density function from -1.645 to 1.645. For example, when $n=1$, the probability density function of $\sqrt{1}\left(\hat{\theta}-\theta\right)/\hat{\theta}$ is

$$f_{\sqrt{1}(\hat{\theta}-\theta)/\hat{\theta}}(x)=\frac{e^{1/(x-1)}}{(-x+1)^2} \qquad x<1.$$

So the actual coverage of the asymptotically exact 90% confidence interval when $n=1$ is

$$P\left(-1.645<\frac{\sqrt{1}\left(\hat{\theta}-\theta\right)}{\hat{\theta}}<1.645\right)=\int_{-1.645}^{1}\frac{e^{1/(x-1)}}{(-x+1)^2}\,dx\cong 0.6852,$$

which is not close to the stated coverage $1-\alpha=0.90$. One would expect the actual coverage to improve as n increases. When $n=2$, the probability density function of $\sqrt{2}\left(\hat{\theta}-\theta\right)/\hat{\theta}$ is

$$f_{\sqrt{2}(\hat{\theta}-\theta)/\theta}(x)=\frac{8e^{-2^{3/2}/\left(x-\sqrt{2}\right)}}{\left(-x+\sqrt{2}\right)^3} \qquad x<\sqrt{2}.$$

So the actual coverage of the asymptotically exact 90% confidence interval when $n=2$ is

$$P\left(-1.645<\sqrt{2}\left(\hat{\theta}-\theta\right)/\hat{\theta}<1.645\right)=\int_{-1.645}^{\sqrt{2}}\frac{8e^{-2^{3/2}/\left(x-\sqrt{2}\right)}}{\left(-x+\sqrt{2}\right)^3}\,dx\cong 0.7635.$$

The results of these calculations for $n=1,2,\ldots,100$, are shown in Figure 3.31. The actual coverages approach the stated coverage 0.90 as n increases. The stated coverage is indicated by a horizontal line. Since all of the actual coverages are below the stated coverage, this asymptotically exact confidence interval claims more precision than it actually achieves for finite values of n.

In practice, for a random sample from an exponential population, you would use the exact confidence interval for θ developed in Example 3.7. Using the exponential distribution here illustrates the thinking behind asymptotically exact confidence intervals. The choice of the exponential population was solely to have the ability to calculate the actual coverage exactly because $\sqrt{n}\left(\hat{\theta}-\theta\right)/\hat{\theta}$ has a mathematically-tractable probability distribution that does not depend on θ.

So the mathematics worked out very nicely for random sampling from an exponential population with a maximum likelihood estimator that is a sample mean (relying on the central limit theorem). Do things work out this well in a more general setting? The answer is yes, and the details are given in the theorem that follows.

Theorem 3.1 (Asymptotic normality of maximum likelihood estimators) Let $X_1, X_2, \ldots, X_n$ be a random sample drawn from a population described by $f_X(x)$ with unknown parameter θ. Let $\hat{\theta}$ denote the maximum likelihood estimator of θ. Under certain regularity conditions associated with the population probability distribution,

$$\sqrt{n}\left(\hat{\theta}-\theta\right)\xrightarrow{D}N\left(0,\frac{1}{I(\theta)}\right),$$

where $I(\theta)$ is the information associated with X in estimating θ.

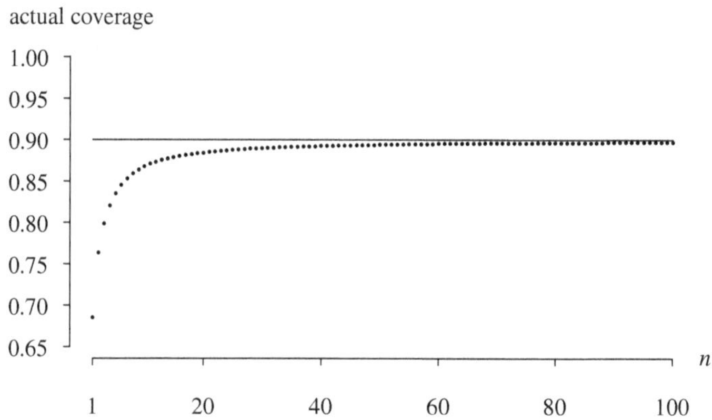

Figure 3.31: Actual coverage for the asymptotically exact 90% confidence interval.

The proof of this result requires enumerating the regularity conditions, which is beyond the scope of this text. The main ideas associated with the proof follow.

- The score can be expanded in a Taylor series about the true parameter value θ.
- The score is zero at the maximum likelihood estimator, that is, $\frac{\partial \ln L(\hat{\theta})}{\partial \theta} = 0$.
- The score can be written as the sum of n random terms as

$$\frac{\partial \ln L(\theta)}{\partial \theta} = \sum_{i=1}^{n} \frac{\partial \ln f(X_i)}{\partial \theta},$$

which is asymptotically normally distributed by the central limit theorem.

Theorem 3.1 states that the maximum likelihood estimators are asymptotically consistent and when $\hat{\theta} - \theta$ is multiplied by $\sqrt{n}$, the resulting random variable is asymptotically normal with a population variance that is the reciprocal of the information. This result is useful in constructing asymptotically exact confidence intervals. One additional concept will be necessary before using this result. Recall from Definition 2.6 and Theorem 2.5 that the information associated with X in estimating θ can be written in two fashions,

$$I(\theta) = E\left[\left(\frac{\partial \ln f(X)}{\partial \theta}\right)^2\right] = E\left[\frac{-\partial^2 \ln f(X)}{\partial \theta^2}\right],$$

under the regularity conditions. Since the information is an expected value, it is a constant. On the other hand, $I\left(\hat{\theta}\right)$ is a random quantity because $\hat{\theta}$ is a random variable as it is a function of the data. The quantity $I\left(\hat{\theta}\right)$ is a consistent estimate of $I(\theta)$ because of the invariance property of maximum likelihood estimators and a result on convergence in probability. This will be used in our development of an asymptotically exact confidence interval for θ next.

All of the exact confidence intervals developed so far in this chapter have been developed from first principles, based on the assumed sampling distribution. They have relied on cleverness and good fortune in identifying an appropriate pivotal quantity. The asymptotically exact confidence interval that we are about to derive is general in nature and applies to sampling from any probability distribution that satisfies the regularity conditions. Using Theorem 3.1, we know that

$$\sqrt{n}\left(\hat{\theta}-\theta\right) \xrightarrow{D} N\left(0, \frac{1}{I(\theta)}\right).$$

Standardizing this random variable by subtracting the population mean and dividing by the population standard deviation, this can be written as

$$\sqrt{nI(\theta)}\left(\hat{\theta}-\theta\right) \xrightarrow{D} N(0, 1).$$

This random variable is an asymptotic pivotal quantity because the standard normal distribution does not involve θ. This means that for $0 < \alpha < 1$,

$$\lim_{n \to \infty} P\left(-z_{\alpha/2} < \sqrt{nI(\theta)}\left(\hat{\theta}-\theta\right) < z_{\alpha/2}\right) = 1 - \alpha.$$

It is not possible to manipulate the inequality in its current form to isolate θ in the middle. However, if we are willing to replace $I(\theta)$ with $I\left(\hat{\theta}\right)$ in the middle of the inequality (using the fact that $I\left(\hat{\theta}\right)$ is a consistent estimator of $I(\theta)$), then this becomes

$$\lim_{n \to \infty} P\left(-z_{\alpha/2} < \sqrt{nI\left(\hat{\theta}\right)}\left(\hat{\theta}-\theta\right) < z_{\alpha/2}\right) = 1 - \alpha.$$

Performing the algebra necessary to isolate θ in the center of the inequality results in

$$\lim_{n \to \infty} P\left(\hat{\theta} - z_{\alpha/2}\frac{1}{\sqrt{nI\left(\hat{\theta}\right)}} < \theta < \hat{\theta} + z_{\alpha/2}\frac{1}{\sqrt{nI\left(\hat{\theta}\right)}}\right) = 1 - \alpha.$$

This effectively constitutes a proof of the following result.

Theorem 3.2 (Asymptotically exact confidence interval based on the normality of maximum likelihood estimators) Let $X_1, X_2, \ldots, X_n$ be a random sample drawn from a population described by $f_X(x)$ with unknown parameter θ. Let $\hat{\theta}$ denote the maximum likelihood estimator of θ. Under certain regularity conditions associated with the population probability distribution, an asymptotically exact two-sided $100(1-\alpha)\%$ confidence interval for θ is

$$\hat{\theta} - z_{\alpha/2}\frac{1}{\sqrt{nI\left(\hat{\theta}\right)}} < \theta < \hat{\theta} + z_{\alpha/2}\frac{1}{\sqrt{nI\left(\hat{\theta}\right)}}.$$

The usefulness of statistical results like Theorem 3.1 in constructing confidence intervals like the one in Theorem 3.2 is one reason that maximum likelihood estimation (popularized by Sir Ronald Fisher between 1912 and 1922) has generally won out over the method of moments (suggested by Karl Pearson in 1894) for parameter estimation. This confidence interval can be applied to most common probability distributions as long as the information and the maximum likelihood estimator can be calculated. The following examples will illustrate the application of the asymptotically exact confidence interval to specific sampling distributions. We begin with the exponential distribution, which was just investigated using first principles.

Example 3.23 Let $X_1, X_2, \ldots, X_n$ be a random sample from an exponential population with positive unknown population mean θ. Use Theorem 3.2 to find an asymptotically exact two-sided $100(1-\alpha)\%$ confidence interval for θ. Apply that confidence interval formula to the air conditioning system failure data from Example 3.7 to give an asymptotically exact two-sided 90% confidence interval for θ.

From Example 2.8 the maximum likelihood estimator of θ is

$$\hat{\theta} = \frac{1}{n}\sum_{i=1}^{n} X_i.$$

For the air conditioning failure data, the maximum likelihood estimate of the population mean time between failures is $\hat{\theta} = 77$ hours. From Example 2.20 the information is

$$I(\theta) = \frac{1}{\theta^2}.$$

Using Theorem 3.2, an asymptotically exact two-sided $100(1-\alpha)\%$ confidence interval for θ is

$$\hat{\theta} - z_{\alpha/2}\frac{\hat{\theta}}{\sqrt{n}} < \theta < \hat{\theta} + z_{\alpha/2}\frac{\hat{\theta}}{\sqrt{n}},$$

which is identical to the confidence interval derived earlier in this subsection. The $n = 27$ times between failures (in hours) for an air conditioning system failure on a Boeing 720 jet airplane from Example 3.7 are

$$97, 51, 11, 4, 141, 18, 142, 68, 77, 80, 1, 16, 106, 206,$$

$$82, 54, 31, 216, 46, 111, 39, 63, 18, 191, 18, 163, 24.$$

The following R code calculates the confidence interval bounds.

```
x    = c(97, 51, 11, 4, 141, 18, 142, 68, 77, 80, 1, 16, 106, 206,
         82, 54, 31, 216, 46, 111, 39, 63, 18, 191, 18, 163, 24)
n    = length(x)
crit = qnorm(0.95)
xbar = mean(x)
xbar - crit * xbar / sqrt(n)
xbar + crit * xbar / sqrt(n)
```

This gives the asymptotically exact 90% confidence interval

$$52 < \theta < 101,$$

which is shifted to the left and is slightly narrower than the exact confidence interval from Example 3.7. Of course one would use the exact confidence interval over this asymptotically exact confidence interval; this example is given just to illustrate the mechanics associated with constructing the asymptotically exact confidence interval. The actual coverage of this confidence interval for the $n = 27$ data values is 0.888 from Figure 3.31, which is slightly less than the stated coverage $1 - \alpha = 0.90$.

The next example considers an asymptotically exact confidence interval for a random sample taken from a special case of the beta distribution.

Example 3.24 Let $X_1, X_2, \ldots, X_n$ be a random sample from a population with probability density function

$$f(x) = \theta x^{\theta-1} \qquad\qquad 0 < x < 1,$$

where θ is a positive unknown parameter. Use Theorem 3.2 to construct an asymptotically exact two-sided $100(1-\alpha)\%$ confidence interval for θ. Also, calculate the actual coverage function for the confidence interval when $\alpha = 0.05$.

The first step is to find the maximum likelihood estimate of θ. For a random sample $X_1, X_2, \ldots, X_n$, the likelihood function is

$$L(\theta) = \theta^n \prod_{i=1}^{n} x_i^{\theta-1}.$$

The log likelihood function is

$$\ln L(\theta) = n \ln \theta + (\theta - 1) \sum_{i=1}^{n} \ln x_i$$

and the score is

$$\frac{\partial \ln L(\theta)}{\partial \theta} = \frac{n}{\theta} + \sum_{i=1}^{n} \ln x_i.$$

Equating the score to zero and solving for θ gives the maximum likelihood estimator

$$\hat{\theta} = -\frac{n}{\sum_{i=1}^{n} \ln x_i}.$$

The negative sign in front of the fraction might look troublesome, but $\hat{\theta}$ is indeed positive because the data values must all fall on the interval $0 < x < 1$, which is the support of the population distribution. The next step is to find the information. The logarithm of the probability density function is

$$\ln f(X) = \ln \theta + (\theta - 1) \ln X.$$

The partial derivative of $\ln f(X)$ with respect to θ is

$$\frac{\partial \ln f(X)}{\partial \theta} = \frac{1}{\theta} + \ln X.$$

The second partial derivative of $\ln f(X)$ with respect to θ is

$$\frac{\partial^2 \ln f(X)}{\partial \theta^2} = -\frac{1}{\theta^2}.$$

So the information is

$$I(\theta) = E\left[-\frac{\partial^2 \ln f(X)}{\partial \theta^2}\right] = E\left[\frac{1}{\theta^2}\right] = \frac{1}{\theta^2}.$$

Finally, plugging the maximum likelihood estimator $\hat{\theta}$ and the information $I(\theta)$ into Theorem 3.2 gives the asymptotically exact two-sided $100(1-\alpha)\%$ confidence interval for θ as

$$\hat{\theta} - z_{\alpha/2}\frac{\hat{\theta}}{\sqrt{n}} < \theta < \hat{\theta} + z_{\alpha/2}\frac{\hat{\theta}}{\sqrt{n}}.$$

Although the signs look awkward, this confidence interval can be written in terms of the original data values $X_1, X_2, \ldots, X_n$ as

$$-\frac{n}{\sum_{i=1}^{n} \ln X_i} + z_{\alpha/2} \frac{\sqrt{n}}{\sum_{i=1}^{n} \ln X_i} < \theta < -\frac{n}{\sum_{i=1}^{n} \ln X_i} - z_{\alpha/2} \frac{\sqrt{n}}{\sum_{i=1}^{n} \ln X_i}.$$

The mathematical tractability of the probability density function associated with the population probability distribution allows us to calculate the actual coverage for finite values of n. Leaving the details as an exercise, the probability density function of the random variable

$$\frac{\sqrt{n}\,(\hat{\theta} - \theta)}{\hat{\theta}}$$

is

$$f_{\sqrt{n}(\hat{\theta}-\theta)/\hat{\theta}}(x) = \frac{n^{n/2}\left(-x+\sqrt{n}\right)^{n-1} e^{\sqrt{n}(x-\sqrt{n})}}{(n-1)!} \qquad x < \sqrt{n},$$

which does not involve θ. To calculate the actual coverage for finite values of n, set $\alpha = 0.05$, which corresponds to an approximate 95% confidence interval. The appropriate percentile from the standard normal distribution is $z_{\alpha/2} = 1.96$. The actual coverage for any particular n is found by integrating the probability density function of $\sqrt{n}\,(\hat{\theta} - \theta)/\hat{\theta}$ with respect to x from -1.96 to 1.96. The results are shown in Figure 3.32. Notice the units on the vertical axis. This asymptotically exact two-sided 95% confidence interval has actual coverage that is quite close to the stated coverage 0.95, even for small values of n. The confidence interval is conservative (that is, the actual coverage exceeds the stated coverage) for $n \geq 2$. The actual coverage deviates furthest from the stated coverage at $n = 6$ with actual coverage 0.9563. The performance of this asymptotically exact confidence interval is far better than that for random sampling from an exponential population from the previous example, with actual coverage shown in Figure 3.31.

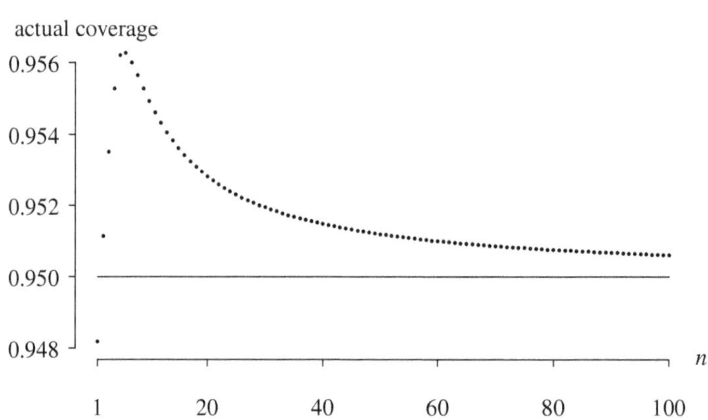

Figure 3.32: Actual coverage for the asymptotically exact 95% confidence interval.

The next example considers the construction of an asymptotically exact confidence interval for a random sample drawn from a probability distribution with positive support.

Example 3.25 Let $X_1, X_2, \ldots, X_n$ be a random sample from a population with probability density function

$$f(x) = \frac{1}{\sqrt{2\pi x^3}} e^{-(x-\mu)^2/(2\mu^2 x)} \qquad x > 0,$$

where μ is a positive unknown parameter, which is the population mean. This population distribution is a special case of the two-parameter inverse Gaussian distribution. Use Theorem 3.2 to find an asymptotically exact two-sided $100(1-\alpha)\%$ confidence interval for μ.

The first step is to find the maximum likelihood estimator of μ. The likelihood function is

$$\begin{aligned} L(\mu) &= \prod_{i=1}^{n} \left(2\pi x_i^3\right)^{-1/2} e^{-(x_i-\mu)^2/\left(2\mu^2 x_i\right)} \\ &= (2\pi)^{-n/2} \left[\prod_{i=1}^{n} x_i\right]^{-3/2} e^{-\sum_{i=1}^{n}(x_i-\mu)^2/\left(2\mu^2 x_i\right)}. \end{aligned}$$

The log likelihood function is

$$\ln L(\mu) = -\frac{n}{2}\ln(2\pi) - \frac{3}{2}\sum_{i=1}^{n} \ln x_i - \frac{1}{2\mu^2}\sum_{i=1}^{n} \frac{(x_i-\mu)^2}{x_i}.$$

The score is the derivative of the log likelihood function with respect to μ, which after simplification is

$$\frac{\partial \ln L(\mu)}{\partial \mu} = \frac{1}{\mu^3}\left[\sum_{i=1}^{n} x_i - n\mu\right].$$

When this equation is equated to zero, the maximum likelihood estimator $\hat{\mu}$ is

$$\hat{\mu} = \frac{1}{n}\sum_{i=1}^{n} x_i,$$

which is the sample mean. The second partial derivative of the log likelihood function is

$$\frac{\partial^2 \ln L(\mu)}{\partial \mu^2} = -\frac{3\sum_{i=1}^{n} x_i}{\mu^4} + \frac{2n}{\mu^3},$$

which is negative at the maximum likelihood estimator, so the maximum likelihood estimator maximizes the log likelihood function. The next step is to find the information. Using the second partial derivative of the log likelihood function when $n = 1$, the information is

$$I(\mu) = E\left[-\frac{\partial^2 \ln f(X)}{\partial \mu^2}\right] = E\left[\frac{3X}{\mu^4} - \frac{2}{\mu^3}\right] = \frac{3\mu}{\mu^4} - \frac{2}{\mu^3} = \frac{1}{\mu^3}$$

because $E[X] = \mu$ for this population distribution. Applying Theorem 3.2, an asymptotically exact two-sided $100(1-\alpha)\%$ confidence interval for μ is

$$\hat{\mu} - z_{\alpha/2}\frac{\hat{\mu}^{3/2}}{\sqrt{n}} < \mu < \hat{\mu} + z_{\alpha/2}\frac{\hat{\mu}^{3/2}}{\sqrt{n}},$$

which can be written in terms of the original data as

$$\bar{X} - z_{\alpha/2}\frac{\bar{X}^{3/2}}{\sqrt{n}} < \mu < \bar{X} + z_{\alpha/2}\frac{\bar{X}^{3/2}}{\sqrt{n}}.$$

The previous three examples have all involved sampling from continuous populations. As will be seen next, the procedure is largely the same for sampling from discrete populations.

Example 3.26 Let $X_1, X_2, \ldots, X_n$ be a random sample from a Poisson(λ) population, where λ is a positive unknown parameter. Find an asymptotically exact two-sided $100(1-\alpha)\%$ confidence interval for λ and apply this confidence interval procedure to the horse kick data set from Example 2.9 to give a 95% confidence interval for λ.

From Example 2.9, the maximum likelihood estimator of λ is

$$\hat{\lambda} = \frac{1}{n}\sum_{i=1}^{n} X_i,$$

which is the sample mean. From Example 2.21, the information associated with the random sample is

$$I(\lambda) = \frac{1}{\lambda}.$$

Applying Theorem 3.2, an asymptotically exact two-sided $100(1-\alpha)\%$ confidence interval for λ is

$$\hat{\lambda} - z_{\alpha/2}\sqrt{\frac{\hat{\lambda}}{n}} < \lambda < \hat{\lambda} + z_{\alpha/2}\sqrt{\frac{\hat{\lambda}}{n}}.$$

Recall that the horse kick data consisted of $n = 200$ corps-years of data, which is listed below.

number of deaths per corps per year	0	1	2	3	4
number of observed values	109	65	22	3	1

The maximum likelihood estimator of λ is

$$\hat{\lambda} = \frac{122}{200} = 0.61.$$

The 95% confidence interval for λ is

$$0.61 - 1.96\sqrt{\frac{0.61}{200}} < \lambda < 0.61 + 1.96\sqrt{\frac{0.61}{200}}$$

or

$$0.50 < \lambda < 0.72,$$

which is shifted slightly to the left of the approximate 95% confidence interval that was calculated in Example 3.12.

The previous four examples have concerned the construction of an asymptotically exact two-sided $100(1-\alpha)\%$ confidence interval for an unknown parameter based on the asymptotic normality of maximum likelihood estimators. But what if an asymptotically exact confidence interval is desired on a function of an unknown parameter such as $g(\theta)$? A method that is used in advanced statistics texts known as the *delta method* can be used to prove the following generalization of Theorem 3.1:

$$\sqrt{n}\left(g(\hat{\theta}) - g(\theta)\right) \xrightarrow{D} N\left(0, \frac{[g'(\theta)]^2}{I(\theta)}\right),$$

where $g'(\theta) = \frac{\partial g(\theta)}{\partial \theta}$. Standardizing this random variable by subtracting the population mean and dividing by the population standard deviation gives

$$\sqrt{\frac{nI(\theta)}{[g'(\theta)]^2}}\left(g(\hat{\theta}) - g(\theta)\right) \xrightarrow{D} N(0, 1).$$

Placing this quantity between the appropriate percentiles of a standard normal distribution, we know that

$$\lim_{n\to\infty} P\left(-z_{\alpha/2} < \sqrt{\frac{nI(\theta)}{[g'(\theta)]^2}}\left(g(\hat{\theta}) - g(\theta)\right) < z_{\alpha/2}\right) = 1-\alpha.$$

Performing the algebra necessary to isolate $g(\theta)$ in the middle of the inequality and replacing values of θ by $\hat{\theta}$ gives an asymptotically exact two-sided $100(1-\alpha)\%$ confidence interval for $g(\theta)$:

$$g(\hat{\theta}) - z_{\alpha/2}\sqrt{\frac{[g'(\hat{\theta})]^2}{nI(\hat{\theta})}} < g(\theta) < g(\hat{\theta}) + z_{\alpha/2}\sqrt{\frac{[g'(\hat{\theta})]^2}{nI(\hat{\theta})}}.$$

This confidence interval will be applied to a random sample from a Bernoulli population in the next example.

Example 3.27 Let $X_1, X_2, \ldots, X_n$ be a random sample from a Bernoulli(p) population with unknown parameter p satisfying $0 < p < 1$.

(a) Find a point estimator and use Theorem 3.2 to find an asymptotically exact two-sided $100(1-\alpha)\%$ confidence interval for p.

(b) Analyze the actual coverage of the asymptotically exact two-sided $100(1-\alpha)\%$ confidence interval for p.

(c) Find a point estimator and an asymptotically exact two-sided $100(1-\alpha)\%$ confidence interval for the population variance $\sigma^2 = p(1-p)$.

(a) From Example 2.16, the maximum likelihood estimator of p is

$$\hat{p} = \frac{1}{n}\sum_{i=1}^{n} X_i.$$

From Example 2.18, the information is

$$I(p) = \frac{1}{p(1-p)}.$$

Using Theorem 3.2, an asymptotically exact two-sided $100(1-\alpha)\%$ confidence interval for p is

$$\hat{p} - z_{\alpha/2}\sqrt{\frac{\hat{p}\,(1-\hat{p}\,)}{n}} < p < \hat{p} + z_{\alpha/2}\sqrt{\frac{\hat{p}\,(1-\hat{p}\,)}{n}}.$$

This can be recognized as the notorious Wald confidence interval that was analyzed in Example 3.14.

As a numerical example, a basketball player takes 100 free throws over the course of a season and makes 78 of those free throws. Assuming each shot is a mutually independent Bernoulli trial (a contested assumption in the world of sports statistics because of the "hot-hand" phenomenon), a point estimate of p is

$$\hat{p} = \frac{78}{100} = 0.78,$$

and an asymptotically exact 90% confidence interval for p is

$$0.78 - 1.645\sqrt{\frac{0.78(1-0.78)}{100}} < p < 0.78 + 1.645\sqrt{\frac{0.78(1-0.78)}{100}}$$

or

$$0.71 < p < 0.85$$

because $z_{0.05} = 1.645$. We are approximately 90% confident that the true probability of making a free throw is between 0.71 and 0.85.

(b) We compute the actual coverage of the asymptotically exact 90% confidence interval from part (a) for sample sizes $n = 5$ and $n = 30$. Other values of n and α yield similar results. Using the formula from the previous section, the actual coverage $c(p)$ of any confidence interval for the binomial proportion is

$$c(p) = P(L < p < U) = \sum_{y=0}^{n} I(y,\, p)\binom{n}{y} p^y(1-p)^{n-y} \qquad 0 < p < 1,$$

where $I(y,\, p)$ is an indicator function that is 0 when a confidence interval does not contain the binomial proportion p and is 1 when a confidence interval contains the binomial proportion p, when the number of successes is $Y = y$. This can be applied to the Wald confidence interval for $n = 5$ and $n = 30$ using the `binomTestCoveragePlot` function from the `conf` package, which results in the graph in Figure 3.33. A horizontal line has been drawn at the stated coverage 0.9. Two conclusions can be drawn from Figure 3.33.

- One should avoid using this confidence interval for p values near $p = 0$ and $p = 1$ because the actual coverage is significantly below the stated coverage.
- One should avoid using this confidence interval for small values of n, again, because of the degraded coverage.

(c) We know that from the invariance property of maximum likelihood estimators that the maximum likelihood estimator of $g(p) = p(1-p)$ is

$$g(\hat{p}\,) = \hat{p}\,(1-\hat{p}\,).$$

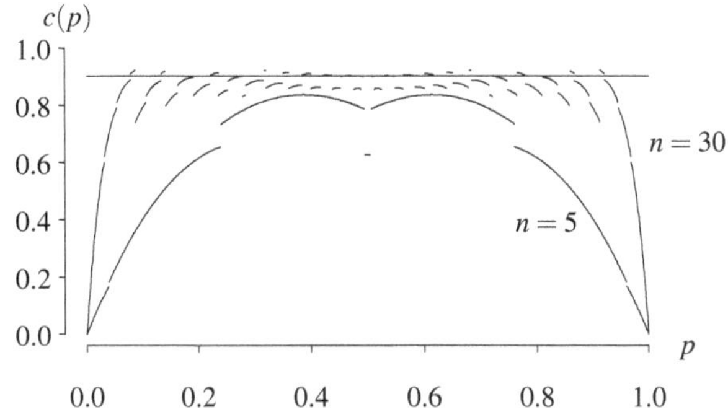

Figure 3.33: Actual coverage $c(p)$ for a 90% Wald confidence interval for $n = 5$ and $n = 30$.

Since $g'(p) = 1 - 2p$, an asymptotically exact two-sided $100(1-\alpha)\%$ confidence interval for $g(p) = p(1-p)$ using the generalization of Theorem 3.2

$$g(\hat{p}) - z_{\alpha/2}\sqrt{\frac{[g'(\hat{p})]^2}{nI(\hat{p})}} < g(p) < g(\hat{p}) + z_{\alpha/2}\sqrt{\frac{[g'(\hat{p})]^2}{nI(\hat{p})}}$$

or

$$\hat{p}(1-\hat{p}) \pm z_{\alpha/2}\sqrt{\frac{(1-2\hat{p})^2\hat{p}(1-\hat{p})}{n}}.$$

(The problems associated with $\hat{p} = 1/2$ are deferred to an advanced statistics text.) Returning to the basketball free throws with $n = 100$ shots and 78 successes, a point estimator of the population variance is

$$\frac{78}{100}\left(1 - \frac{78}{100}\right) = 0.17$$

and an approximate two-sided 90% confidence interval for the population variance $g(p) = p(1-p)$ is

$$0.17 \pm 1.645\sqrt{\frac{(1-2\cdot 0.17)^2 0.17(1-0.17)}{100}}$$

or

$$0.13 < p(1-p) < 0.21.$$

We are approximately 90% confident that the population variance lies between 0.13 and 0.21.

These five examples have illustrated some of the issues which can arise when constructing an asymptotically exact confidence interval based on the asymptotic normality of the maximum likelihood estimator. But there is another way to construct an asymptotically exact confidence interval, which is introduced next.

The likelihood ratio statistic

The second technique for constructing asymptotically exact confidence intervals is based on a quantity known as the *likelihood ratio statistic*. In order to motivate the development of this statistic, we begin with a simple coin flipping experiment that will highlight the importance of the likelihood ratio.

Let's say I have two biased coins. The probability of heads for coin 1 is $1/4$. The probability of heads for coin 2 is $3/4$. I select a coin at random and flip it four times, resulting in X heads. Regardless of the coin chosen, the possible values of X are $x = 0, 1, 2, 3, 4$. Table 3.9 gives the conditional probabilities associated with X for coin 1 and coin 2. These probabilities are the probability mass function values associated with the binomial(4, $1/4$) distribution for coin 1 and the binomial(4, $3/4$) distribution for coin 2. The bottom row of Table 3.9 contains the ratio of these two probabilities for each value of X. This ratio is known as the likelihood ratio. For example, if I select a coin at random, flip it four times and observe $x = 0$ heads, then the likelihood ratio tells me that it is 81 times more likely that the coin that was selected was coin 1. So this likelihood ratio is helpful in the effort to identify which coin was selected based on the number of heads that were observed.

	$x=0$	$x=1$	$x=2$	$x=3$	$x=4$
$P(X=x \mid \text{coin } 1)$	$81/256$	$27/64$	$27/128$	$3/64$	$1/256$
$P(X=x \mid \text{coin } 2)$	$1/256$	$3/64$	$27/128$	$27/64$	$81/256$
$\dfrac{P(X=x \mid \text{coin } 1)}{P(X=x \mid \text{coin } 2)}$	81	9	1	$1/9$	$1/81$

Table 3.9: Conditional distribution of X and likelihood ratio.

Table 3.9 contains the ratio of two probability mass functions. This ratio can be generalized to the ratio of likelihood functions, as defined next.

Definition 3.8 Let $X_1, X_2, \ldots, X_n$ be a random sample from a population whose probability distribution has a single unknown parameter θ and is described by $f(x)$. Let $\hat{\theta}$ be the maximum likelihood estimator of θ. The random variable

$$\Lambda = \frac{L(\theta)}{L\left(\hat{\theta}\right)}$$

is the *likelihood ratio statistic*.

The likelihood ratio statistic is typically introduced in conjunction with hypothesis testing (which is introduced in the next chapter), but it is defined here because it can also be used to construct asymptotically exact confidence intervals. Once again, the notation has been grossly oversimplified because the likelihood function is a function of the values from the random sample $X_1, X_2, \ldots, X_n$ and also θ. So it would be completely reasonable to write the definition of the likelihood ratio statistic as

$$\Lambda(X_1, X_2, \ldots, X_n, \theta) = \frac{L(X_1, X_2, \ldots, X_n, \theta)}{L\left(X_1, X_2, \ldots, X_n, \hat{\theta}\right)},$$

but compactness was chosen in the expression given in Definition 3.8. The likelihood ratio statistic

- is a random variable because the likelihood function and maximum likelihood estimator are functions of the data values, which are themselves random variables;

- is a random variable with positive support because the likelihood function is always positive;
- is less than or equal to one because $L(\theta)$ is maximized at the maximum likelihood estimator $\hat{\theta}$, so the numerator is always less than or equal to the denominator;
- has support on the interval $(0,\ 1]$ because of the previous two facts, so $0 < \Lambda \le 1$.

The probability distribution of the likelihood ratio statistic differs for each population probability distribution, so it alone is not of much use in constructing confidence intervals. However, as the following theorem shows, the opposite of twice the logarithm of the likelihood ratio statistic has a recognizable limiting probability density function—the chi-square distribution with one degree of freedom—which can be used to construct asymptotically exact confidence intervals.

Theorem 3.3 Let $X_1, X_2, \ldots, X_n$ be a random sample from a population whose probability distribution has a single unknown parameter θ and is described by $f(x)$. Let $\hat{\theta}$ be the maximum likelihood estimator of θ. Under certain regularity conditions concerning the population distribution,

$$-2\ln\Lambda = -2\ln\left(\frac{L(\theta)}{L(\hat{\theta})}\right) \xrightarrow{D} \chi^2(1).$$

Proof An outline of the proof follows. For notational simplicity, let $l(\theta) = \ln L(\theta)$ for this proof only, so that

$$l'(\theta) = \frac{\partial \ln L(\theta)}{\partial\theta} \qquad \text{and} \qquad l''(\theta) = \frac{\partial^2 \ln L(\theta)}{\partial\theta^2}.$$

Expanding the log likelihood function $l(\theta)$ about $\hat{\theta}$ in a Taylor series gives

$$l(\theta) = l\left(\hat{\theta}\right) + l'\left(\hat{\theta}\right)\left(\theta - \hat{\theta}\right) + l''\left(\hat{\theta}\right)\frac{\left(\theta - \hat{\theta}\right)^2}{2!} + \cdots.$$

Recognizing that the derivative of the log likelihood function at $\hat{\theta}$ is zero (because $\hat{\theta}$ maximizes the log likelihood function) and ignoring the terms of order three and higher, the log likelihood function is approximately

$$l(\theta) \cong l\left(\hat{\theta}\right) + l''\left(\hat{\theta}\right)\frac{\left(\theta - \hat{\theta}\right)^2}{2}.$$

So $-2\ln\Lambda$ can be written as

$$\begin{aligned}
-2\ln\Lambda &= -2\ln\left(\frac{L(\theta)}{L(\hat{\theta})}\right) \\
&= -2\left(\ln L(\theta) - \ln L\left(\hat{\theta}\right)\right) \\
&= -2\left(l(\theta) - l\left(\hat{\theta}\right)\right) \\
&\cong -2\left(l\left(\hat{\theta}\right) + l''\left(\hat{\theta}\right)\frac{\left(\theta - \hat{\theta}\right)^2}{2} - l\left(\hat{\theta}\right)\right) \\
&= -l''\left(\hat{\theta}\right)\left(\theta - \hat{\theta}\right)^2 \\
&= nI\left(\hat{\theta}\right)\left(\theta - \hat{\theta}\right)^2.
\end{aligned}$$

Since

$$\sqrt{nI(\theta)}\left(\hat{\theta}-\theta\right) \xrightarrow{D} N(0, 1)$$

from Theorem 3.1, $I\left(\hat{\theta}\right) \xrightarrow{D} I(\theta)$, and the square of a standard normal random variable is a chi-square random variable with one degree of freedom,

$$-2\ln\Lambda \xrightarrow{D} \chi^2(1)$$

as desired. □

The regularity conditions in Theorem 3.3 are similar to those listed just after Theorem 2.4. Theorem 3.3 can be used to construct asymptotically exact confidence intervals for θ. Since

$$P\left(-2\ln\left(\frac{L(\theta)}{L\left(\hat{\theta}\right)}\right) < \chi^2_{1,\alpha}\right) = 1-\alpha,$$

an asymptotically exact two-sided confidence interval for θ is the set of all θ satisfying

$$-2\ln\left(\frac{L(\theta)}{L\left(\hat{\theta}\right)}\right) < \chi^2_{1,\alpha}.$$

Calculating the upper and lower bounds for a confidence interval for θ of this type typically requires the use of numerical methods, which is not ideal.

There is some nice geometry associated with this asymptotically exact confidence interval. The confidence interval can be rewritten as

$$\ln L(\theta) - \ln L\left(\hat{\theta}\right) > -\frac{\chi^2_{1,\alpha}}{2}$$

or

$$\ln L(\theta) > \ln L\left(\hat{\theta}\right) - \frac{\chi^2_{1,\alpha}}{2}.$$

So the asymptotically exact two-sided $100(1-\alpha)\%$ confidence interval for θ consists of all θ values associated with the log likelihood function being at a distance of less than $\chi^2_{1,\alpha}/2$ of its peak value at the maximum likelihood estimator. Figures will illustrate this geometry in each of the three examples that follow.

Example 3.28 Construct an asymptotically exact $100(1-\alpha)\%$ confidence interval for θ based on the likelihood ratio statistic for a random sample consisting of a single $(n=1)$ observation X drawn from a population distribution with probability density function

$$f(x) = \theta x^{\theta-1} \qquad 0 < x < 1,$$

where θ is a positive unknown parameter. Apply this confidence interval procedure to the single observation $x = 0.63$ to give a 90% confidence interval for θ.

It might seem pointless to construct an asymptotically exact $100(1-\alpha)\%$ confidence interval for θ from just a single observation here for two reasons: (*a*) we already derived an *exact* confidence interval for θ in Example 3.1, and (*b*) an asymptotically exact confidence interval will probably not work well for $n = 1$. The confidence interval is constructed here so as to see the mechanics and the geometry associated with the process. For a single observation, the likelihood function is the same as the probability density function

$$L(\theta) = f(x) = \theta x^{\theta - 1}.$$

The log likelihood function is

$$\ln L(\theta) = \ln \theta + (\theta - 1) \ln x.$$

The score is

$$\frac{\partial \ln L(\theta)}{\partial \theta} = \frac{1}{\theta} + \ln x.$$

Equating the score to zero gives the maximum likelihood estimate of θ as

$$\hat{\theta} = -\frac{1}{\ln x}.$$

The likelihood ratio statistic is

$$\Lambda = \frac{L(\theta)}{L(\hat{\theta})} = \frac{\theta X^{\theta-1}}{-\frac{1}{\ln X} X^{-1/\ln X - 1}} = -\theta (\ln X) \left(X^{\theta + 1/\ln X}\right).$$

Figure 3.34 shows that the likelihood ratio statistic Λ maps the support of X, which is $\mathcal{A} = \{x \,|\, 0 < x < 1\}$ to the support of Λ, which is $\mathcal{B} = \{\lambda \,|\, 0 < \lambda < 1\}$, although the transformation is not one-to-one. The transformation is plotted for $\theta = 0.5$, $\theta = 1$, and $\theta = 3$.

The asymptotically exact two-sided $100(1-\alpha)\%$ confidence interval for θ consists of all θ satisfying

$$-2 \ln \left(\frac{L(\theta)}{L(\hat{\theta})} \right) < \chi^2_{1,\alpha}.$$

In the case of the single data value $x = 0.63$ and $\alpha = 0.1$, the maximum likelihood estimate is

$$\hat{\theta} = -1/\ln x = -1/\ln 0.63 = 2.2.$$

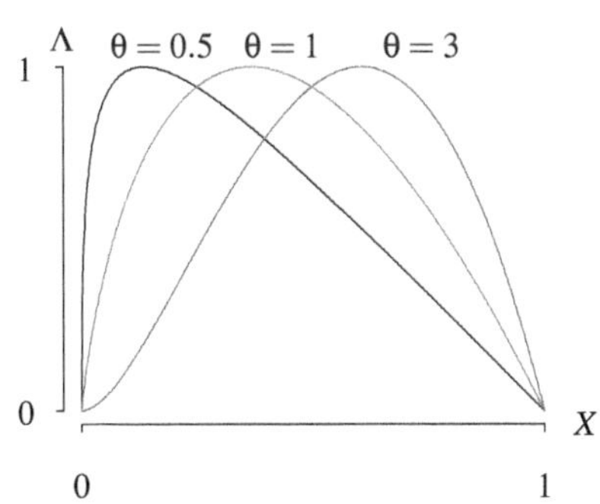

Figure 3.34: Likelihood ratio statistic Λ as a function of X.

Furthermore, the value of the log likelihood function at the maximum likelihood estimator is

$$\ln L(\hat{\theta}) = \ln L(2.2) = 0.23.$$

Next, find the 90th percentile of the chi-square distribution with one degrees of freedom via the R command `qchisq(0.9, 1)`, which gives

$$\chi^2_{1,0.1} = 2.7055.$$

After some algebra, the asymptotically exact 90% confidence interval consists of all values θ such that

$$\ln L(\theta) > \ln L(\hat{\theta}) - \frac{\chi^2_{1,0.1}}{2}$$

or

$$\ln L(\theta) > 0.23 - 1.35.$$

Figure 3.35 contains a plot of the log likelihood function, which is distinctly nonsymmetric because of the small sample size. The asymptotically exact 90% confidence interval corresponds to all of the values of the log likelihood function that are within 1.35 units of the peak value of the log likelihood function, which is indicated by a horizontal dashed line. The R function `uniroot` numerically solves for the confidence interval bounds in the code below. The variable `x` holds the data value, `theta` holds the maximum likelihood estimator, `logl` holds the value of the log likelihood function at the maximum likelihood estimator, and `crit` holds $\chi^2_{1,0.1}/2$.

```
x     = 0.63
theta = -1 / log(x)
logl  = log(theta) + (theta - 1) * log(x)
crit  = qchisq(0.9, 1) / 2
l = uniroot(function(th) log(th) + (th - 1) * log(x) - logl + crit,
            interval = c(0.1, theta))$root
u = uniroot(function(th) log(th) + (th - 1) * log(x) - logl + crit,
            interval = c(theta, 10))$root
```

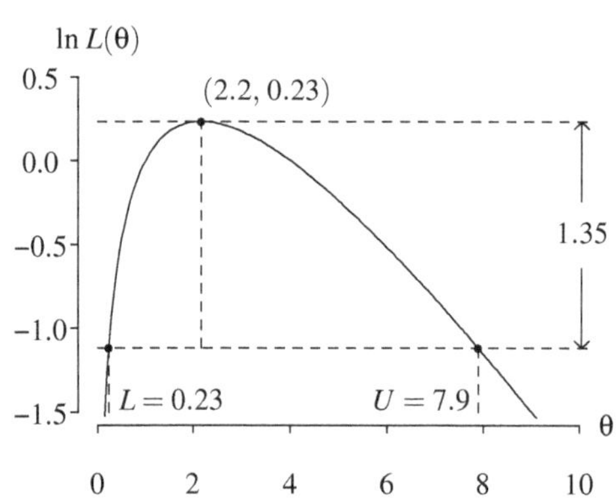

Figure 3.35: The log likelihood function near $\hat{\theta} = 2.2$ for the data set $x = 0.63$.

This code computes the asymptotically exact two-sided 90% confidence interval

$$0.23 < \theta < 7.9,$$

which is particularly nonsymmetric about the maximum likelihood estimator $\hat{\theta} = 2.2$. We are approximately 90% confident that θ lies between 0.23 and 7.9. The wide confidence interval is due to the fact that only a single observation X was collected. As a point of comparison, the exact two-sided 90% confidence interval from Example 3.1 was

$$0.11 < \theta < 6.5.$$

The asymptotically exact confidence interval developed here has approximate actual coverage for any finite value of n, which is always the case in practice. The next example considers a more appropriate application of the asymptotically exact confidence interval procedure, this time with $n = 200$.

Example 3.29 Construct an asymptotically exact $100(1-\alpha)\%$ confidence interval for λ based on the likelihood ratio statistic for a random sample $X_1, X_2, \ldots, X_n$ drawn from a Poisson(λ) population, where λ is a positive unknown parameter. Apply this confidence interval procedure to the horse kick data set from Example 2.9 to give a 95% confidence interval for λ.

From Example 2.9, the log likelihood function is

$$\ln L(\lambda) = -n\lambda + \left(\sum_{i=1}^{n} x_i\right) \ln \lambda - \sum_{i=1}^{n} \ln(x_i!)$$

and the maximum likelihood estimate of λ is the sample mean

$$\hat{\lambda} = \frac{1}{n} \sum_{i=1}^{n} x_i.$$

The asymptotically exact two-sided $100(1-\alpha)\%$ confidence interval for λ consists of all λ values satisfying

$$-2\ln\left(\frac{L(\lambda)}{L(\hat{\lambda})}\right) < \chi^2_{1,\alpha}.$$

Recall that the horse kick data consisted of $n = 200$ corps-years of data, which is listed below. The $n = 200$ data values consist of 109 zeros, 65 ones, 22 twos, 3 threes, and 1

number of deaths per corps per year	0	1	2	3	4
number of observed values	109	65	22	3	1

four. The maximum likelihood estimate of λ is

$$\hat{\lambda} = \frac{109 \cdot 0 + 65 \cdot 1 + 22 \cdot 2 + 3 \cdot 3 + 1 \cdot 4}{200} = \frac{122}{200} = 0.61.$$

The geometry associated with the confidence interval is shown in Figure 3.36. Since the sample size ($n = 200$) is large, the log likelihood function is nearly symmetric. The log likelihood function is maximized at the point

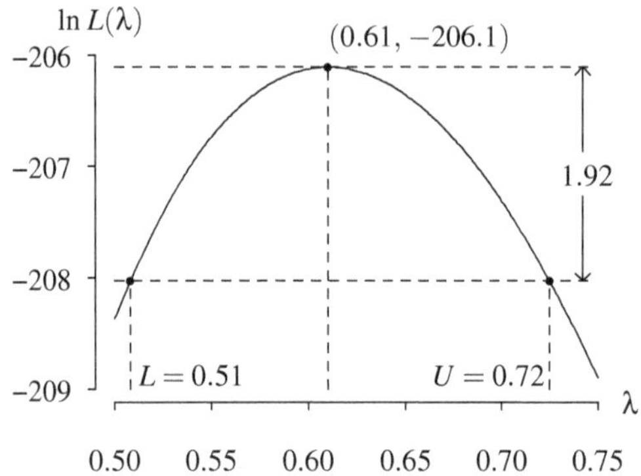

Figure 3.36: The log likelihood function near $\hat{\lambda} = 0.61$ for the horse kick data.

$$\left(\hat{\lambda}, \ln L(\hat{\lambda})\right) = (0.61, -206.1067).$$

The confidence interval bounds are found by drawing a horizontal line

$$\frac{\chi^2_{1,\alpha}}{2} = \frac{\chi^2_{1,0.05}}{2} = \frac{3.84}{2} = 1.92$$

units below the maximum point in Figure 3.36; the points where this horizontal line intersects the log likelihood function corresponds to the 95% confidence interval bounds. The R code below again uses the `uniroot` function to numerically solve for the lower and upper bound of the confidence interval.

```
x   = c(rep(0, 109), rep(1, 65), rep(2, 22), rep(3, 3), 4)
n   = length(x)
lam = mean(x)
t1  = sum(x)
t2  = sum(log(factorial(x)))
t3  = -n * lam + t1 * log(lam) - t2
t4  = qchisq(0.95, 1) / 2
l = uniroot(function(lambda) -n * lambda + t1 * log(lambda) - t2 -
          t3 + t4, lower = 0.1, upper = lam, tol = 1e-9)$root
u = uniroot(function(lambda) -n * lambda + t1 * log(lambda) - t2 -
          t3 + t4, lower = lam, upper = 0.9, tol = 1e-9)$root
```

The asymptotically exact two-sided 95% confidence interval, with bounds rounded to two significant digits, is

$$0.51 < \lambda < 0.72.$$

This approximate confidence interval can be compared to the approximate confidence intervals in Examples 3.12 and 3.26, which are

$$0.51 < \lambda < 0.73 \qquad \text{and} \qquad 0.50 < \lambda < 0.72.$$

All three of these confidence intervals are close to one another because the large sample size, $n = 200$, means that the asymptotic relationship on which two of them are based is quite accurate.

One of the problems associated with constructing asymptotically exact confidence intervals is that you cannot tell how well they perform for a finite sample size. And of course, the rate of convergence depends on the population distribution. In this light, a Monte Carlo simulation is conducted to see how close $-2\ln\Lambda$ comes to the chi-square distribution with one degree of freedom when samples of size $n = 200$ are drawn from a Poisson population with λ arbitrarily set to $\lambda = 3$. The following R code generates 100,000 random values of $-2\ln\Lambda$, which are stored in the vector `stat`.

```
lam  = 3
n    = 200
nrep = 100000
stat = numeric(nrep)
for (i in 1:nrep) {
  x = rpois(n, lam)
  lamhat = mean(x)
  t1   = sum(x)
  t2   = sum(log(factorial(x)))
  num = -n * lam + t1 * log(lam) - t2
  den = -n * lamhat + t1 * log(lamhat) - t2
  stat[i] = -2 * (num - den)
}
```

The empirical cumulative distribution function associated with these values, along with the cumulative distribution function of a chi-square random variable with one degree of freedom are plotted in Figure 3.37. The two cumulative distribution functions are nearly indistinguishable, so at least for a Poisson population with $\lambda = 3$ and a sample size of $n = 200$, using the likelihood ratio statistic will provide a confidence interval whose actual coverage is near the stated coverage.

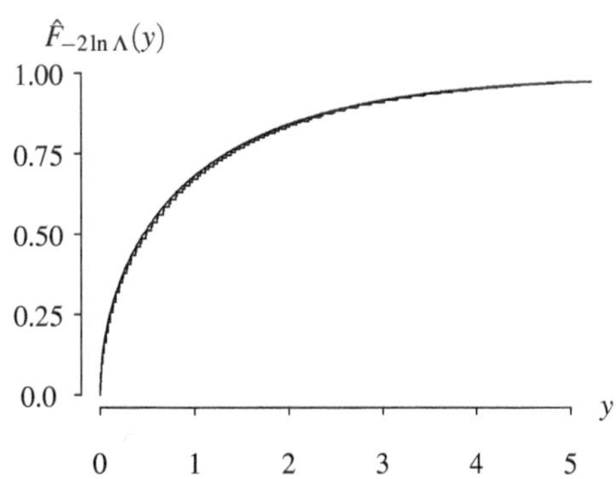

Figure 3.37: Cumulative distribution functions for $-2\ln\Lambda$ for a Poisson population and $\chi^2(1)$.

As a third and final example of asymptotically exact confidence intervals based on the likelihood ratio statistic, we return to sampling from a Bernoulli population.

Example 3.30 Construct an asymptotically exact $100(1-\alpha)\%$ confidence interval for p based on the likelihood ratio statistic for a random sample $X_1, X_2, \ldots, X_n$ drawn from a Bernoulli(p) population, where p is an unknown parameter satisfying $0 < p < 1$. Apply this confidence interval procedure to arrive at a 95% asymptotically exact confidence interval to a data set with $n = 100$ Bernoulli trials (free throw shots) and 78 successes.

From Example 2.16, the likelihood function is

$$L(p) = p^{\sum_{i=1}^n x_i}(1-p)^{n-\sum_{i=1}^n x_i},$$

the log likelihood function is

$$\ln L(p) = \left(\sum_{i=1}^n x_i\right)\ln p + \left(n - \sum_{i=1}^n x_i\right)\ln(1-p),$$

and the maximum likelihood estimator is

$$\hat{p} = \frac{1}{n}\sum_{i=1}^n x_i.$$

From Theorem 3.3, the asymptotically exact two-sided $100(1-\alpha)\%$ confidence interval for p based on the likelihood ratio statistic consists of all p values such that

$$-2\ln\left(\frac{L(p)}{L(\hat{p})}\right) < \chi^2_{1,\alpha}.$$

For the free throw data, the point estimate of p is $\hat{p} = 0.78$. Again using the R function `uniroot` to calculate the confidence interval bounds associated with the geometry in Figure 3.38, an asymptotically exact two-sided 95% confidence interval for p is

$$0.69 < p < 0.85.$$

We are approximately 95% confident that the true free throw percentage lies between 69% and 85%. This is slightly wider than the approximate 95% confidence interval based on the asymptotic normality of the maximum likelihood estimator from Example 3.27, which was $0.71 < p < 0.85$. The confidence interval based on the asymptotic normality of the maximum likelihood estimator must necessarily be symmetric about the maximum likelihood estimator, while the confidence interval based on the fact that $-2\ln\Lambda$ is asymptotically chi-square with one degree of freedom can be nonsymmetric about the maximum likelihood estimator.

This section has presented two ways to compute an asymptotically exact confidence interval for an unknown parameter θ, one based on the asymptotic normality of the maximum likelihood estimator and the other based on the likelihood ratio statistic. One difficulty with having the two options is that the superior choice depends on the population distribution and the sample size. The superior choice must be determined on a case-by-case basis. The next section extends the topic of interval estimation beyond just confidence intervals.

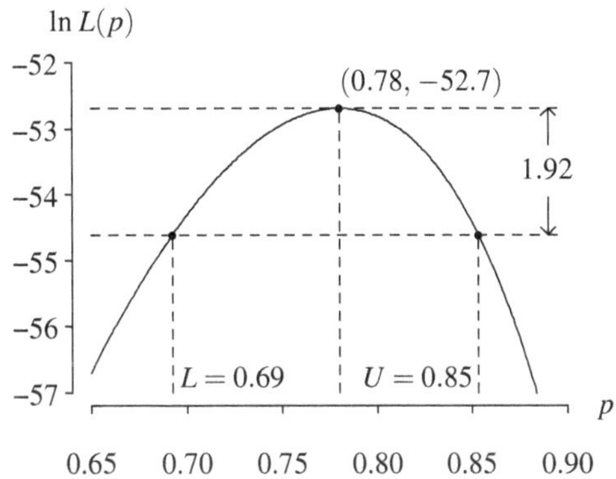

Figure 3.38: The log likelihood function near $\hat{p} = 0.78$ for the free throw data.

3.4 Other Interval Estimators

Confidence intervals are certainly the most prevalent of the interval estimators. But there are other interval estimators that are useful in applications. Four classes of interval estimators are introduced and illustrated in this section: prediction intervals, tolerance intervals, Bayesian credible intervals, and confidence regions.

Prediction intervals

All of the confidence intervals introduced thus far have been used to quantify the precision of the estimate of some unknown parameter θ or some function of θ. The thinking associated with a prediction interval is quite different. As before, there is a data set of n observations that is denoted by $X_1, X_2, \ldots, X_n$ that is randomly sampled from a population distribution. A prediction interval gives bounds associated with a single future observation drawn from the same population denoted by X_{n+1}.

Definition 3.9 Let $X_1, X_2, \ldots, X_n$ be a random sample from some population distribution. Let X_{n+1} denote a single future additional observation from the same population distribution. An exact two-sided $100(1-\alpha)\%$ *prediction interval* for X_{n+1} satisfies

$$P(l < X_{n+1} < u) = 1 - \alpha,$$

for $0 < \alpha < 1$.

Notice that l and u in Definition 3.9 are set in lower case because they are constants. As was the case with confidence intervals, the typical choices for α, that is, 0.1, 0.05, and 0.01, correspond to 90%, 95%, and 99% prediction intervals. The simplest case of a prediction interval for a *single* future observation X_{n+1} will be considered in this subsection. Prediction intervals can also be constructed for the case of m future observations.

The sequence of three examples that follow all consider the construction of an exact two-sided $100(1 - \alpha)\%$ prediction interval for X_{n+1} from a random sample of n data values drawn from a normally distributed population.

Example 3.31 Let $X_1, X_2, \ldots, X_n$ be a random sample from a $N(\mu_0, \sigma_0^2)$ population, where μ_0 and σ_0^2 are known parameters. Find an exact two-sided $100(1-\alpha)\%$ prediction interval for a single future observation X_{n+1} drawn from the same population.

Since the parameters μ_0 and σ_0^2 are known, there is no reason to estimate them from the data values. The data values can be ignored in this case. The prediction interval satisfies

$$\begin{aligned} P(l < X_{n+1} < u) &= P\left(\frac{l-\mu_0}{\sigma_0} < \frac{X_{n+1}-\mu_0}{\sigma_0} < \frac{u-\mu_0}{\sigma_0}\right) \\ &= P\left(\frac{l-\mu_0}{\sigma_0} < Z < \frac{u-\mu_0}{\sigma_0}\right) \\ &= 1-\alpha, \end{aligned}$$

where $Z \sim N(0, 1)$. As was the case with confidence intervals, the probability α is divided between the two tails of the distribution, so the two endpoints of the last inequality are the $100(\alpha/2)$th and $100(1-\alpha/2)$th percentiles of the standard normal distribution. So solving

$$\frac{l-\mu_0}{\sigma_0} = -z_{\alpha/2} \qquad \text{and} \qquad \frac{u-\mu_0}{\sigma_0} = z_{\alpha/2}$$

for the prediction interval bounds l and u results in

$$P\left(\mu_0 - z_{\alpha/2}\sigma_0 < X_{n+1} < \mu_0 + z_{\alpha/2}\sigma_0\right) = 1-\alpha.$$

So an exact two-sided $100(1-\alpha)\%$ prediction interval for X_{n+1} is

$$\mu_0 - z_{\alpha/2}\sigma_0 < X_{n+1} < \mu_0 + z_{\alpha/2}\sigma_0.$$

The previous example considered the ideal case in which the population mean and population variance are known constants. Unfortunately, this is almost never the case in practice. The next example considers the case in which one of these parameters is known.

Example 3.32 Let $X_1, X_2, \ldots, X_n$ be a random sample from a $N(\mu, \sigma_0^2)$ population, where μ is an unknown parameter and σ_0^2 is a known parameter. Find an exact two-sided $100(1-\alpha)\%$ prediction interval for a single future observation X_{n+1} drawn from the same population.

The population mean is unknown, so it must be estimated by the sample mean $\bar{X}$, which is the maximum likelihood estimator as shown in Example 2.13. Since

$$\bar{X} \sim N\left(\mu, \frac{\sigma_0^2}{n}\right) \qquad \text{and} \qquad X_{n+1} \sim N(\mu, \sigma_0^2),$$

the difference between X_{n+1} and $\bar{X}$ is normally distributed:

$$X_{n+1} - \bar{X} \sim N\left(0, \sigma_0^2 + \frac{\sigma_0^2}{n}\right).$$

Standardizing this random variable,

$$\frac{X_{n+1} - \bar{X}}{\sigma_0\sqrt{1+1/n}} \sim N(0, 1).$$

This will serve as the pivotal quantity because it does not involve the unknown parameter μ. So with probability $1-\alpha$,

$$P\left(-z_{\alpha/2} < \frac{X_{n+1}-\bar{X}}{\sigma_0\sqrt{1+1/n}} < z_{\alpha/2}\right) = 1-\alpha.$$

Isolating X_{n+1} in the middle of the inequality results in

$$P\left(\bar{X} - z_{\alpha/2}\sigma_0\sqrt{1+\frac{1}{n}} < X_{n+1} < \bar{X} + z_{\alpha/2}\sigma_0\sqrt{1+\frac{1}{n}}\right) = 1-\alpha.$$

So an exact two-sided $100(1-\alpha)\%$ prediction interval for X_{n+1} is

$$\bar{X} - z_{\alpha/2}\sigma_0\sqrt{1+\frac{1}{n}} < X_{n+1} < \bar{X} + z_{\alpha/2}\sigma_0\sqrt{1+\frac{1}{n}}.$$

The population variance was known in the previous example. In the next example, both the population mean and the population variance are unknown, which is the most common case in practice. All the analyst knows is that the data appears to come from a bell-shaped distribution, presumably determined by the shape of a histogram.

Example 3.33 Let $X_1, X_2, \ldots, X_n$ be a random sample from a $N(\mu, \sigma^2)$ population, where μ and σ^2 are unknown parameters. Find an exact two-sided $100(1-\alpha)\%$ prediction interval for a single future observation X_{n+1} drawn from the same population. Apply this formula to the $n = 78$ IQ scores from Example 3.9 to give an exact two-sided 99% prediction interval for a 79th observation.

Since both parameters are unknown, they must be estimated from data. Using the maximum likelihood estimators from Example 2.13 with an unbiasing constant for the estimator of σ^2, we use the sample mean $\bar{X}$ to estimate the population mean μ and the sample variance S^2 to estimate the population variance σ^2. As in the previous example, we have

$$\bar{X} \sim N\left(\mu, \frac{\sigma^2}{n}\right) \qquad \text{and} \qquad X_{n+1} \sim N(\mu, \sigma^2).$$

The difference between X_{n+1} and $\bar{X}$ is again normally distributed:

$$X_{n+1} - \bar{X} \sim N\left(0, \sigma^2 + \frac{\sigma^2}{n}\right).$$

Standardizing this random variable gives

$$\frac{X_{n+1}-\bar{X}}{\sigma\sqrt{1+1/n}} \sim N(0, 1).$$

But we do not have the value of σ because it is unknown. Replacing σ with S gives

$$\frac{X_{n+1}-\bar{X}}{S\sqrt{1+1/n}} \sim t(n-1).$$

The proof of this result is analogous to the proof of Theorem 1.7. This will serve as the pivotal quantity because it does not involve the unknown parameters μ and σ^2. Since the t distribution with $n-1$ degrees of freedom is symmetric about 0,

$$P\left(-t_{n-1,\alpha/2} < \frac{X_{n+1}-\bar{X}}{S\sqrt{1+1/n}} < t_{n-1,\alpha/2}\right) = 1-\alpha.$$

Isolating X_{n+1} in the middle of the inequality results in

$$P\left(\bar{X} - t_{n-1,\alpha/2}S\sqrt{1+\frac{1}{n}} < X_{n+1} < \bar{X} + t_{n-1,\alpha/2}S\sqrt{1+\frac{1}{n}}\right) = 1-\alpha.$$

So an exact two-sided $100(1-\alpha)\%$ prediction interval for X_{n+1} is

$$\bar{X} - t_{n-1,\alpha/2}S\sqrt{1+\frac{1}{n}} < X_{n+1} < \bar{X} + t_{n-1,\alpha/2}S\sqrt{1+\frac{1}{n}}.$$

The $n = 78$ IQ scores from Example 3.9 are given below.

111	107	100	107	114	115	111	97	100	112	104	89	104
102	91	114	114	103	106	105	113	109	108	113	130	128
128	118	113	120	132	111	124	127	128	136	106	118	119
123	124	126	116	127	119	97	86	102	110	120	103	115
93	72	111	103	123	79	119	110	110	107	74	105	112
105	110	107	103	77	98	90	96	112	112	114	93	106

The sample mean and sample standard deviation are

$$\bar{x} = 109 \qquad \text{and} \qquad s = 13.$$

The following R statements read the data values from the file `iq.d` and compute the bounds of the exact two-sided 99% prediction interval for X_{79}.

```
x    = scan("iq.d")
n    = length(x)
a    = 0.01
xbar = mean(x)
sdev = sd(x)
crit = qt(1 - a / 2, n - 1)
l    = xbar - crit * sdev * sqrt(1 + 1 / n)
u    = xbar + crit * sdev * sqrt(1 + 1 / n)
```

The exact two-sided 99% prediction interval is

$$74 < X_{79} < 144.$$

If an IQ test is given to a 79th seventh grader at this rural midwestern school, the probability that their score will lie between 74 and 144 is 0.99. The wide prediction interval is driven by the high probability (0.99) and the significant variability ($s = 13$) in the data. The distinction between confidence intervals and prediction intervals is crucial. In Example 3.9, we computed the 99% confidence interval for the population mean μ

$$105 < \mu < 113,$$

which is much narrower than the 99% prediction interval. The width of the 99% confidence interval will shrink to zero as n increases; the width of the 99% prediction interval will converge to a non-zero constant as n increases. It would not be reasonable to have a very narrow prediction interval for a new student's IQ score, regardless of the sample size.

As a fourth and final example of constructing a prediction interval, consider a random sample drawn from an exponential population with positive unknown population mean θ.

Example 3.34 Let $X_1, X_2, \ldots, X_n$ be a random sample from an exponential population with positive unknown population mean θ. Find an exact two-sided $100(1-\alpha)\%$ prediction interval for a single future observation X_{n+1} drawn from the same population. Apply this formula to the $n = 27$ air conditioner failure times from Example 3.7 to give a 90% prediction interval for a 28th observed air conditioner failure time.

Recall from Example 3.7 that for $X_1, X_2, \ldots, X_n$ independent and identically distributed exponential random variables each with population mean θ,

$$\frac{2}{\theta}\sum_{i=1}^{n} X_i \sim \chi^2(2n).$$

When this result is applied to the new observation X_{n+1},

$$\frac{2}{\theta} X_{n+1} \sim \chi^2(2).$$

Since the ratio of two independent chi-square random variables divided by their degrees of freedom has the F distribution,

$$\frac{\frac{2}{\theta}\sum_{i=1}^{n} X_i/(2n)}{\frac{2}{\theta}X_{n+1}/2} = \frac{\sum_{i=1}^{n} X_i}{nX_{n+1}} \sim F(2n, 2).$$

This will serve as the pivotal quantity because it does not involve the unknown parameter θ. With probability $1-\alpha$,

$$P\left(F_{2n,2,1-\alpha/2} < \frac{\sum_{i=1}^{n} X_i}{nX_{n+1}} < F_{2n,2,\alpha/2}\right) = 1-\alpha.$$

Isolating X_{n+1} in the middle of the inequality results in

$$P\left(\frac{\sum_{i=1}^{n} X_i}{nF_{2n,2,\alpha/2}} < X_{n+1} < \frac{\sum_{i=1}^{n} X_i}{nF_{2n,2,1-\alpha/2}}\right) = 1-\alpha.$$

So an exact two-sided $100(1-\alpha)\%$ prediction interval for X_{n+1} is

$$\frac{\bar{X}}{F_{2n,2,\alpha/2}} < X_{n+1} < \frac{\bar{X}}{F_{2n,2,1-\alpha/2}}.$$

The $n = 27$ times between failures (in hours) for an air conditioning system on a Boeing 720 jet airplane from Example 3.7 are

$$97, 51, 11, 4, 141, 18, 142, 68, 77, 80, 1, 16, 106, 206,$$

$$82, 54, 31, 216, 46, 111, 39, 63, 18, 191, 18, 163, 24.$$

The sample mean time between failures is

$$\bar{x} = \frac{2074}{27} = 77$$

hours. The following R code computes an exact two-sided 90% prediction interval for the new observation X_{28}.

```
x      = c(97, 51, 11, 4, 141, 18, 142, 68, 77, 80, 1, 16, 106, 206,
           82, 54, 31, 216, 46, 111, 39, 63, 18, 191, 18, 163, 24)
n      = length(x)
a      = 0.1
crithi = qf(1 - a / 2, 2 * n,  2)
critlo = qf(a / 2, 2 * n,  2)
l      = sum(x) / (n * crithi)
u      = sum(x) / (n * critlo)
```

The resulting exact two-sided 90% prediction interval is

$$3.9 < X_{28} < 243.$$

If we were to wait for a 28th failure to occur on this air conditioning system, the probability that the next failure will occur between 3.9 and 243 hours is 0.9. This is much wider than the 90% confidence interval for θ from Example 3.7, which was $57 < \theta < 109$. As in the previous example, the width of the confidence interval shrinks to zero as n increases (because the precision concerning the parameter increases), but the width of the prediction interval stabilizes to a non-zero constant as n increases.

These four examples comprise a brief introduction to prediction intervals. A related topic is a tolerance interval, which is introduced next.

Tolerance intervals

Consider the following experiment. You collect a random sample of $n = 30$ observations denoted by $X_1, X_2, \ldots, X_{30}$ from a continuous population. Consider the random interval that consists of the two extreme observations, that is, the interval $(X_{(1)}, X_{(30)})$. If you then sample a 31st observation from the same continuous population, what is the probability that the new observation will fall in the random interval? If your answer was "pretty likely," then your intuition is correct concerning this experiment. The random interval $(X_{(1)}, X_{(30)})$ is very likely to encompass almost all of the probability distribution associated with the population, so the 31st observation will probably fall in the interval.

Monte Carlo simulation can be used to estimate the probability associated with the question posed in the previous paragraph. Since the continuous population distribution was not specified, the R code below arbitrarily uses the unit exponential population. The code conducts the experiment described in the previous paragraph four million times, then prints the fraction of the intervals that contain the 31st observation.

```
n = 30
nrep = 4000000
count = 0
for (i in 1:nrep) {
  x = rexp(n, 1)
  xnew = rexp(1, 1)
  if (min(x) < xnew && xnew < max(x)) count = count + 1
}
print(count / nrep)
```

After a call to `set.seed(3)` to initialize the random number stream, five runs of the program results in

$$0.9354 \qquad 0.9354 \qquad 0.9355 \qquad 0.9354 \qquad 0.9355.$$

From these results, it is safe to say that to three digits of accuracy the probability that the 31st observation will fall in the random interval $(X_{(1)}, X_{(30)})$ is 0.935. The population distribution, which for this Monte Carlo experiment was a unit exponential distribution, was an arbitrary choice. But surprisingly, when the population distribution is changed to the normal, Weibull, or gamma distributions, the probability remains 0.935. So it appears that the probability depends on n, but the probability does not depend on the continuous population probability distribution.

For a data set $x_1, x_2, \ldots, x_n$, the interval $(x_{(1)}, x_{(n)})$, or more generally $(x_{(i)}, x_{(j)})$, is known as a *tolerance interval*. This subsection investigates properties of tolerance intervals. We begin with an important result concerning the distribution of order statistics drawn from a continuous population.

Theorem 3.4 Let $X_1, X_2, \ldots, X_n$ be a random sample from a continuous population with cumulative distribution function $F_X(x)$. Let $X_{(1)}, X_{(2)}, \ldots, X_{(n)}$ denote the associated order statistics. Let $U_i = F_X\left(X_{(i)}\right)$, for $i = 1, 2, \ldots, n$. Then the joint probability density function of $U_1, U_2, \ldots, U_n$ is

$$f_{U_1,U_2,\ldots,U_n}(u_1, u_2, \ldots, u_n) = n! \qquad 0 < u_1 < u_2 < \cdots < u_n < 1,$$

the marginal probability density function of U_i is

$$f_{U_i}(u) = \frac{n!}{(i-1)!\,(n-i)!}u^{i-1}(1-u)^{n-i} \qquad 0 < u < 1,$$

for $i = 1, 2, \ldots, n$, and the joint probability density function of U_i and U_j is

$$f_{U_i,U_j}(u_i, u_j) = \frac{n!}{(i-1)!\,(j-i-1)!\,(n-j)!}u_i^{i-1}(u_j-u_i)^{j-i-1}(1-u_j)^{n-j} \qquad 0 < u_i < u_j < 1,$$

for indexes $1 \le i < j \le n$.

Proof By the probability integral transformation, $F_X(X_1), F_X(X_2), \ldots, F_X(X_n)$ are mutually independent and identically distributed $U(0, 1)$ random variables. So the random variables $U_1 = F_X\left(X_{(1)}\right), U_2 = F_X\left(X_{(2)}\right), \ldots, U_n = F_X\left(X_{(n)}\right)$ are equivalent to the n order statistics from a $U(0, 1)$ population. Using an order statistic result, and the fact that the probability density function function for $U \sim U(0, 1)$ is $f_U(u) = 1$ for $0 < u < 1$, the joint probability density function of $U_1, U_2, \ldots, U_n$ is

$$f_{U_1,U_2,\ldots,U_n}(u_1, u_2, \ldots, u_n) = n! \qquad 0 < u_1 < u_2 < \cdots < u_n < 1.$$

Using another order statistic result, and the fact that the cumulative distribution function for $U \sim U(0, 1)$ on its support is $F_U(u) = u$ for $0 < u < 1$, the marginal probability density function of U_i is

$$f_{U_i}(u) = \frac{n!}{(i-1)!\,(n-i)!} u^{i-1}(1-u)^{n-i} \qquad 0 < u < 1,$$

for $i = 1, 2, \ldots, n$, which also happens to be the probability density function of a beta$(i, n-i+1)$ random variable. Finally, using yet another order statistic result, the joint probability density function of U_i and U_j is

$$f_{U_i,U_j}(u_i, u_j) = \frac{n!}{(i-1)!\,(j-i-1)!\,(n-j)!} u_i^{i-1}(u_j-u_i)^{j-i-1}(1-u_j)^{n-j}$$

for $0 < u_i < u_j < 1$ and indexes $1 \le i < j \le n$. □

The notation in Theorem 3.4 is illustrated in Figure 3.39 for $n = 3$. The raw data values X_1, X_2, X_3 are randomly sampled from a continuous population with cumulative distribution function $F_X(x)$. Their values are denoted by dots on the upper axis. The associated order statistics are denoted by $X_{(1)}, X_{(2)}, X_{(3)}$. The probability distribution of these order statistics depends on the population probability distribution. So instead of working with these order statistics, we work with the transformed random variables $U_1 = F_X(X_{(1)})$, $U_2 = F_X(X_{(2)})$, and $U_3 = F_X(X_{(3)})$, which have the same probability distribution regardless of the population distribution. Arrows identify transformations from the upper axis to the lower axis. The random variables U_1, U_2, U_3 lie between 0 and 1 and are plotted on the lower axis. As indicated in Theorem 3.4, the probability distribution of these random variables does not depend on the population probability distribution. When we work with the random variables on the lower axis, any results we derive correspond to *all* continuous population distributions.

The question that was posed at the beginning of this subsection concerning the probability that a 31st observation falls in the random interval $\left(X_{(1)}, X_{(30)}\right)$ can be restated in terms of the probability that U_{31} falls in the interval (U_1, U_{30}). But since $U_{31} \sim U(0, 1)$, this is the same as asking the expected distance between U_1 and U_{30}, that is, $E[U_{30} - U_1]$. We have encountered $R = U_{30} - U_1$ in Section 1.4 as the sample range. Leaving the derivation as an exercise, the probability density function of the sample range of n mutually independent $U(0, 1)$ random variables is

$$f_R(r) = n(n-1)r^{n-2}(1-r) \qquad 0 < r < 1.$$

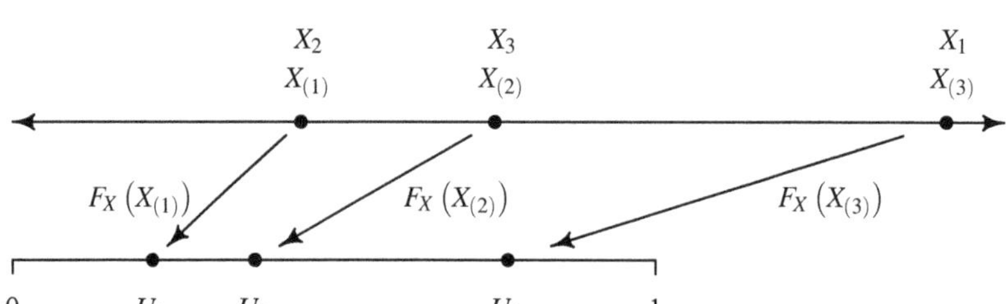

Figure 3.39: Tolerance interval notation.

So the expected value of R is

$$E[R] = \int_0^1 r \cdot n(n-1)r^{n-2}(1-r)\,dr = n(n-1)\left[\frac{r^n}{n} - \frac{r^{n+1}}{n+1}\right]_0^1 = \frac{n-1}{n+1}.$$

This means that the analytic result associated with the Monte Carlo simulation with $n = 30$ is

$$E[R] = E[U_{30} - U_1] = \frac{29}{31} = 0.9355.$$

The graph of the expected value of R for $n = 2$ through $n = 100$ in Figure 3.40 gives the probability that the $(n+1)$st observation falls in the random interval $(X_{(1)}, X_{(n)})$. Not surprisingly, the probabilities asymptotically approach 1. The thinking associated with tolerance intervals concerns how much of the probability associated with a distribution is covered by the interval $(X_{(i)}, X_{(j)})$.

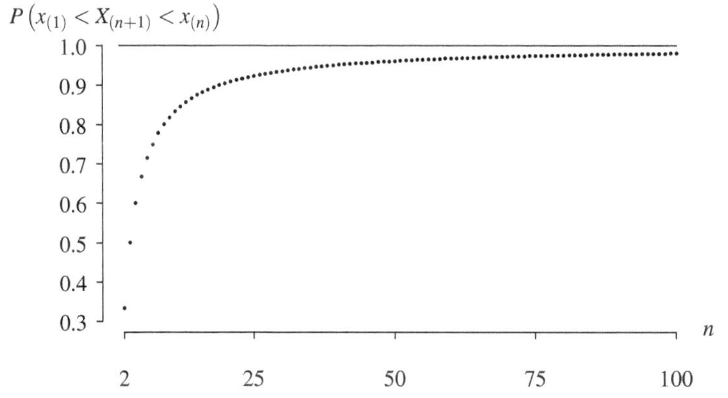

Figure 3.40: Estimated probability that the interval $(x_{(1)}, x_{(n)})$ includes $X_{(n+1)}$.

We begin the process of constructing a tolerance interval by selecting two indexes $i < j$. Although the most common choice would be $i = 1$ and $j = n$, as was the case in the Monte Carlo simulation, the derivation here will consider any indexes i and j. Even though the interest is in the random tolerance interval $(X_{(i)}, X_{(j)})$, Theorem 3.4 indicates that it is equivalent to consider the interval (U_i, U_j). Next consider the random distance between U_i and U_j, that is, $U_j - U_i$, which is a continuous random variable. It is of interest to see how much of the population probability distribution is contained in the interval (U_i, U_j). The probability that the interval (U_i, U_j) captures p or more of the $U(0, 1)$ probability distribution, where $0 < p < 1$, can be written as

$$P(U_j - U_i \geq p) = 1 - \alpha.$$

Using Theorem 3.4, this probability is calculated by

$$\begin{aligned}
1 - \alpha &= P(U_j - U_i \geq p) \\
&= \int_0^{1-p} \int_{p+u_i}^1 f_{U_i,U_j}(u_i, u_j)\,du_j\,du_i \\
&= \int_0^{1-p} \int_{p+u_i}^1 \frac{n!}{(i-1)!\,(j-i-1)!\,(n-j)!} u_i^{i-1}(u_j - u_i)^{j-i-1}(1-u_j)^{n-j}\,du_j\,du_i.
\end{aligned}$$

Since the integrand is a polynomial, this double integral can easily be worked out for specific values of n, i, j, and p using a computer algebra system.

Example 3.35 Let $X_1, X_2, \ldots, X_{30}$ be a random sample of size $n = 30$ from a continuous population distribution. What is the probability that the tolerance interval $(x_{(1)}, x_{(30)})$ covers at least 90% of the probability distribution?

Fortunately, this procedure works regardless of the sampling distribution of the observations, so it is not necessary to specify the population distribution. Since $n = 30$, $i = 1$, $j = 30$, and $p = 0.9$, the probability associated with this particular tolerance interval is

$$\begin{aligned} 1-\alpha &= P(U_{30} - U_1 \geq 0.9) \\ &= \int_0^{1-0.9} \int_{0.9+u_1}^{1} \frac{30!}{28!}(u_{30} - u_1)^{28}\, du_{30}\, du_1 \\ &= 0.8163. \end{aligned}$$

The probability is approximately 0.82 that the tolerance interval $(x_{(1)}, x_{(30)})$ contains at least 90% of the probability for the population distribution.

The interpretation of a tolerance interval is somewhat more complex than the interpretation associated with a confidence interval or a prediction interval because there are two probabilities specified. The p parameter concerns how much of the probability distribution is captured in the tolerance interval, and the $1-\alpha$ parameter gives the probability that the interval captures at least this proportion of the probability distribution.

So when does one choose a tolerance interval over a confidence interval? Here are two such situations.

- Manufacturing companies often face what is known as a "fill problem" when an automatic filling device is used to fill a container with a product. A bottling company, for example, might have an interest in filling a bottle that is labeled as providing 16 ounces of product. The bottle can actually contain 16.3 ounces of product. The bottling company wants to have very few occurrences of the automatic filling device delivering less than 16 ounces (which annoys the customer and the government might allow only a small percentage of such occurrences). On the other hand, the bottling company wants to have very few occurrences of the automatic filling device delivering more than 16.3 ounces (which corresponds to an overfill). If the tolerance interval $(16.1, 16.2)$, for example, contains at least 99% of the probability distribution of the fill amount with probability 0.95, then the bottler can run the operation with certainty that either of the two extremes will be rare. In this case, $p = 0.99$ and $\alpha = 0.05$.

- Let $x_1, x_2, \ldots, x_n$ denote the observed gas mileages, measured in miles per gallon, for n automobiles of the same make, model, and year. A one-sided 95% confidence interval, for example $\mu > 38$, would be of use for projecting the worst-case scenario for the gas consumption of a fleet of such automobiles. But this would not do you much good if you intend to rent such a vehicle and you want to know if you can get to your destination which is 500 miles away on a single tank of gas. In this case, a 95% prediction interval, for example $X_{n+1} > 36$, would be the interval of choice. But for the design engineer who needs to determine the gas tank size, neither the confidence interval nor the prediction interval suffices for determining the capacity of the gas tank required so that 95% of the cars will achieve a cruising distance of 500 miles. This would be the role of a one-sided tolerance interval, for example, the probability that the 95% tolerance interval $X_{(1)} > 37$ captures at least 90% of the probability distribution. In this case, $p = 0.9$ and $\alpha = 0.05$.

Bayesian credible intervals

The confidence intervals discussed so far have been based on the following paradigm. There is an unknown parameter θ which has a fixed, but unknown, value. Furthermore, the analyst has absolutely no information or insight about θ prior to data collection. A random sample is collected in order to compute a point estimate $\hat{\theta}$, which is a best guess for the parameter based on the data, and to compute a confidence interval $L < \theta < U$ that quantifies the precision of the point estimator. This paradigm is generally known as "classical statistics" by statisticians.

There is a group of very passionate statisticians known as "Bayesians" who reject this paradigm. The origins of their philosophy comes from the rule of Bayes formulated by Rev. Thomas Bayes. They believe that in nearly all applications, *something* must be known about θ. Consider the following situations.

- Does a botanist really have *no idea* how many petals are on a certain type of flower?
- Does an engineer really have *no idea* how long the newly designed tire will last?
- Does a surgeon really have *no idea* what the success rate is on a particular type of surgery?
- Does a sociologist really have *no idea* what the social distance is between two friends having a conversation?

Bayesians would argue that the botanist, engineer, surgeon, and sociologist really do have some sense of the probability distribution of their parameter of interest even before the data is collected. They believe that it is prudent to combine their knowledge of the process with the data to arrive at a point estimate and interval estimate for an unknown parameter that is superior to just using the data alone. This perspective implies that we no longer think of θ as a constant that needs to be estimated, but rather a parameter that is itself a random variable. This perspective has resulted in a branch of statistics known as *Bayesian statistics*. The probability distribution of θ that is formulated before the data is collected is known as the *prior distribution*. A crude relationship describing how Bayesians piece together the prior distribution with the data set is

$$\begin{bmatrix} \text{prior} \\ \text{distribution} \end{bmatrix} + \begin{bmatrix} \text{data} \end{bmatrix} \longrightarrow \begin{bmatrix} \text{posterior} \\ \text{distribution} \end{bmatrix}.$$

The individual formulating the prior distribution need not necessarily be a statistician. Even if the subject matter expert is interviewed and is able to make a reasonable guess at the 25th, 50th, and 75th percentiles (and perhaps also the range of the parameter θ), then an analyst can form a probability distribution that conforms to these percentiles. Alternatively, if the subject matter expert can give a subjective estimate of the population mean and population variance of the unknown parameter, this is often enough information to formulate a two-parameter prior distribution. After the data has been collected, there is the step of modifying the prior distribution by the observed data values $x_1, x_2, \ldots, x_n$, which is where the rule of Bayes comes into play. An application of the rule of Bayes applied to the prior distribution and the data results in the *posterior distribution* of θ. This posterior distribution is used to determine a point estimator and an interval estimator for θ. Since the interpretation of the interval estimator for θ differs from that of a standard confidence interval, these intervals are known as *Bayesian credible intervals*.

Let's begin with a simple example involving coin tossing. This will be followed by a development of some general notation for Bayesian statistics.

Example 3.36 Let's say I have two biased coins. The probability of heads for coin 1 is $1/4$. The probability of heads for coin 2 is $3/4$. I select a coin and flip it four times, resulting in X heads. Given your extensive experience observing the way that I select the coin for flipping, you believe that the probability that I will select coin 1 is $2/3$. If $x = 3$ heads are observed on the four flips, what are the posterior probabilities associated with the selection of the two coins?

In this problem, the unknown parameter is θ, the probability of heads for the coin chosen. It is now considered a random variable Θ. The prior distribution for Θ was determined by you observing me select coins in the past, which results in the probability mass function

$$f_{\Theta}(\theta) = \begin{cases} 2/3 & \theta = 1/4 \\ 1/3 & \theta = 3/4. \end{cases}$$

So prior to executing the coin-flipping experiment you believe that it is more likely that I will choose coin 1 for flipping. Regardless of the coin chosen, the possible values of X are $x = 0, 1, 2, 3, 4$. The next step is to find the conditional probability mass function associated with the number of heads X associated with each of the coins. If coin 1 is chosen, X has conditional probability mass function

$$f_{X\,|\,\Theta=1/4}(x\,|\,\Theta = 1/4) = \binom{4}{x}\left(\frac{1}{4}\right)^{x}\left(1-\frac{1}{4}\right)^{4-x} \qquad x = 0, 1, 2, 3, 4.$$

If coin 2 is chosen, X has conditional probability mass function

$$f_{X\,|\,\Theta=3/4}(x\,|\,\Theta = 3/4) = \binom{4}{x}\left(\frac{3}{4}\right)^{x}\left(1-\frac{3}{4}\right)^{4-x} \qquad x = 0, 1, 2, 3, 4.$$

These probability mass functions are associated with the binomial(4, 1/4) distribution for coin 1 and the binomial(4, 3/4) distribution for coin 2. If coin 1 is chosen, the expected number of heads is $E[X\,|\,\Theta = 1/4] = (4)(1/4) = 1$; if coin 2 is chosen, the expected number of heads is $E[X\,|\,\Theta = 3/4] = (4)(3/4) = 3$. Now the experiment is conducted. A coin is selected, flipped four times, and the outcome of the experiment is that $x = 3$ heads appear. This outcome certainly makes it more likely that coin 2 was selected. So now we combine the prior distribution with the data to arrive at posterior probabilities associated with $\Theta = 1/4$ and $\Theta = 3/4$. Using the rule of Bayes,

$$\begin{aligned} f_{\Theta\,|\,X=3}(\theta\,|\,X=3) &= P(\Theta = \theta\,|\,X = 3) \\ &= \frac{P(\Theta = \theta, X = 3)}{P(X = 3)} \\ &= \frac{P(X = 3\,|\,\Theta = \theta)P(\Theta = \theta)}{P(X = 3\,|\,\Theta = 1/4)P(\Theta = 1/4) + P(X = 3\,|\,\Theta = 3/4)P(\Theta = 3/4)}. \end{aligned}$$

When $\theta = 1/4$, this conditional probability is

$$\begin{aligned} f_{\Theta\,|\,X=3}(1/4\,|\,X=3) &= \frac{P(X = 3\,|\,\Theta = 1/4)P(\Theta = 1/4)}{P(X = 3\,|\,\Theta = 1/4)P(\Theta = 1/4) + P(X = 3\,|\,\Theta = 3/4)P(\Theta = 3/4)} \\ &= \frac{(3/64)(2/3)}{(3/64)(2/3) + (27/64)(1/3)} \\ &= \frac{2}{11}. \end{aligned}$$

When $\theta = 3/4$, this conditional probability is

$$\begin{aligned} f_{\Theta\,|\,X=3}(3/4\,|\,X=3) &= \frac{P(X=3\,|\,\Theta=3/4)P(\Theta=3/4)}{P(X=3\,|\,\Theta=1/4)P(\Theta=1/4)+P(X=3\,|\,\Theta=3/4)P(\Theta=3/4)} \\ &= \frac{(27/64)(1/3)}{(3/64)(2/3)+(27/64)(1/3)} \\ &= \frac{9}{11}. \end{aligned}$$

So the posterior distribution of Θ is

$$f_{\Theta\,|\,X=3}(\theta\,|\,X=3) = \begin{cases} 2/11 & \theta = 1/4 \\ 9/11 & \theta = 3/4. \end{cases}$$

Our initial hunch that it was more likely that coin 1 was selected has been revised by the data so that we now believe that it was more likely that coin 2 was selected. This outcome is slightly different than the classical approach in which we have no idea about the coin selection prior to the flipping. In this case we assume that selecting each coin is equally likely; the rule of Bayes indicates that the probability that coin 1 was selected is

$$\frac{(3/64)(1/2)}{(3/64)(1/2)+(27/64)(1/2)} = \frac{1}{10}$$

and the probability that coin 2 was selected is

$$\frac{(27/64)(1/2)}{(3/64)(1/2)+(27/64)(1/2)} = \frac{9}{10}.$$

The prior distribution associated with no prior knowledge concerning the coin selection is known as a *noninformative prior distribution*.

In the previous example, the random variable Θ was the probability of tossing heads on the coin selected. This will now be generalized to a generic Θ, which is the random parameter of interest. Rather than assuming that θ has a constant but unknown value, as was the case in all previous discussions, we now assume that θ is a random variable denoted by Θ. To make the discussion a bit more general, we assume that Θ is a continuous random variable. The probability distribution of Θ is known as the *prior distribution* of Θ. The probability density function of Θ is denoted by $f_\Theta(\theta)$, and is known as the *prior probability density function* of Θ. This probability distribution is defined on the support set Ω. The discrete case is analogous. Now let X be a random variable whose probability distribution depends on the random parameter Θ. For the discussion here, assume that X is also a continuous random variable. Since both X and Θ are random variables, we can consider x to be an experimental value of X and θ to be an experimental value of Θ. The conditional distribution of X given $\Theta = \theta$ is denoted by $f_{X\,|\,\Theta=\theta}(x\,|\,\Theta=\theta)$. These two probability distributions, namely the conditional distribution of X given $\Theta = \theta$ and the prior distribution of Θ, describe the inputs to the Bayesian model.

Let $X_1, X_2, \ldots, X_n$ be a random sample from the conditional distribution of X given $\Theta = \theta$. Since the observations are mutually independent and identically distributed, the joint conditional probability density function of $X_1, X_2, \ldots, X_n$ given $\Theta = \theta$ is

$$f_{\boldsymbol{X}\,|\,\Theta=\theta}(\boldsymbol{x}\,|\,\Theta=\theta) = f_{X\,|\,\Theta=\theta}(x_1\,|\,\Theta=\theta)f_{X\,|\,\Theta=\theta}(x_2\,|\,\Theta=\theta)\ldots f_{X\,|\,\Theta=\theta}(x_n\,|\,\Theta=\theta).$$

The boldface font is used on $\boldsymbol{X}$ and $\boldsymbol{x}$ for notational compactness. The joint probability density function of $\boldsymbol{X}$ and Θ is

$$f_{\boldsymbol{X},\Theta}(\boldsymbol{x}, \theta) = f_{\boldsymbol{X}\,|\,\Theta=\theta}(\boldsymbol{x}\,|\,\Theta=\theta) f_\Theta(\theta).$$

Finally, the posterior distribution of Θ conditioned on the experimental values $x_1, x_2, \ldots, x_n$ of the random variables $X_1, X_2, \ldots, X_n$ is

$$f_{\Theta\,|\,\boldsymbol{X}=\boldsymbol{x}}(\theta\,|\,\boldsymbol{X}=\boldsymbol{x}) = \frac{f_{\boldsymbol{X},\Theta}(\boldsymbol{x}, \theta)}{f_{\boldsymbol{X}}(\boldsymbol{x})},$$

where

$$f_{\boldsymbol{X}}(\boldsymbol{x}) = \int_\Omega f_{\boldsymbol{X},\Theta}(\boldsymbol{x}, \theta)\,d\theta = \int_\Omega f_{\boldsymbol{X}\,|\,\Theta=\theta}(\boldsymbol{x}\,|\,\Theta=\theta) f_\Theta(\theta)\,d\theta$$

is the marginal distribution of $\boldsymbol{X}$. The population mean of the posterior distribution, $E\left[\Theta\,|\,\boldsymbol{X}=\boldsymbol{x}\right]$, is traditionally used as the *Bayesian point estimator* of θ. The $100(\alpha/2)$th and $100(1-\alpha/2)$th percentiles of the posterior distribution can be used as a $100(1-\alpha)\%$ *Bayesian credible interval* for θ. The reason that the term *confidence interval* is not used here is that the Bayesian credible interval has an interpretation that involves probability in a different fashion than that for a confidence interval. The reason that an interpretation involving probability is possible is that Θ is assumed to be a random variable. The interpretation will be illustrated in the next example.

In the three examples that follow, we use θ to denote the parameter of interest rather than the more traditional parameter associated with the sampling distribution in order to be more consistent with the above notation.

Example 3.37 Let $X_1, X_2, \ldots, X_n$ be a random sample drawn from a Poisson distribution with random parameter Θ. Assume that Θ has a prior gamma(λ, κ) distribution. In other words, the Bayesian model is

$$X\,|\,\Theta=\theta \sim \text{Poisson}(\theta)$$

$$\Theta \sim \text{gamma}(\lambda, \kappa).$$

Find a Bayesian point estimate and an exact two-sided $100(1-\alpha)\%$ Bayesian credible interval for θ. Apply these formulas to the horse kick data from Example 2.9 to compute a Bayesian point estimator and an exact two-sided 95% Bayesian credible interval when the prior parameters are $\lambda = 12$ and $\kappa = 6$.

The gamma prior probability density function of Θ is given by

$$f_\Theta(\theta) = \frac{\lambda^\kappa \theta^{\kappa-1} e^{-\lambda\theta}}{\Gamma(\kappa)} \qquad \theta > 0.$$

The joint conditional probability mass function of $X_1, X_2, \ldots, X_n$ given $\Theta = \theta$ is

$$f_{\boldsymbol{X}\,|\,\Theta=\theta}(\boldsymbol{x}\,|\,\Theta=\theta) = \frac{\theta^{x_1} e^{-\theta}}{x_1!} \cdot \frac{\theta^{x_2} e^{-\theta}}{x_2!} \cdot \ldots \cdot \frac{\theta^{x_n} e^{-\theta}}{x_n!} = \frac{\theta^{x_1+x_2+\cdots+x_n} e^{-n\theta}}{x_1!x_2!\ldots x_n!}$$

for nonnegative integers x_i, for $i = 1, 2, \ldots, n$ and $\theta > 0$. The joint distribution of $X_1, X_2, \ldots, X_n$ and Θ is described by

$$f_{\boldsymbol{X},\Theta}(\boldsymbol{x}, \theta) = f_{\boldsymbol{X}\,|\,\Theta=\theta}(\boldsymbol{x}\,|\,\Theta=\theta) f_\Theta(\theta) = \frac{\theta^{x_1+x_2+\cdots+x_n} e^{-n\theta}}{x_1!x_2!\ldots x_n!} \cdot \frac{\lambda^\kappa \theta^{\kappa-1} e^{-\lambda\theta}}{\Gamma(\kappa)}$$

for $\theta > 0$ and nonnegative integers x_i, for $i = 1, 2, \ldots, n$. This is an unusual joint probability distribution because Θ is a continuous random variable, but $X_1, X_2, \ldots, X_n$ are discrete random variables. The marginal probability mass function of $X_1, X_2, \ldots, X_n$ is

$$\begin{aligned}
f_{\boldsymbol{X}}(\boldsymbol{x}) &= \int_{\Omega} f_{\boldsymbol{X}\,|\,\Theta=\theta}(\boldsymbol{x}\,|\,\Theta=\theta) f_{\Theta}(\theta)\, d\theta \\
&= \int_0^{\infty} \frac{\theta^{x_1+x_2+\cdots+x_n} e^{-n\theta}}{x_1!x_2!\ldots x_n!} \cdot \frac{\lambda^{\kappa}\theta^{\kappa-1}e^{-\lambda\theta}}{\Gamma(\kappa)}\, d\theta \\
&= \int_0^{\infty} \frac{\lambda^{\kappa}\theta^{x_1+x_2+\cdots+x_n+\kappa-1} e^{-(n+\lambda)\theta}}{\Gamma(\kappa)x_1!x_2!\ldots x_n!}\, d\theta \\
&= \frac{\lambda^{\kappa}}{\Gamma(\kappa)x_1!x_2!\ldots x_n!} \int_0^{\infty} \left(\frac{u}{n+\lambda}\right)^{x_1+x_2+\cdots+x_n+\kappa-1} e^{-u} \frac{1}{n+\lambda}\, du \\
&= \frac{\lambda^{\kappa}\Gamma(x_1+x_2+\cdots+x_n+\kappa)}{\Gamma(\kappa)x_1!x_2!\ldots x_n!(n+\lambda)^{x_1+x_2+\cdots+x_n+\kappa}}
\end{aligned}$$

by performing the integration using the substitution $u = (n+\lambda)\theta$ and the definition of the gamma function. The support of this marginal distribution is the nonnegative integers x_i, for $i = 1, 2, \ldots, n$. The posterior probability density function of Θ conditioned on the experimental values $x_1, x_2, \ldots, x_n$ of the random variables $X_1, X_2, \ldots, X_n$ is

$$\begin{aligned}
f_{\Theta\,|\,\boldsymbol{X}=\boldsymbol{x}}(\theta\,|\,\boldsymbol{X}=\boldsymbol{x}) &= \frac{f_{\boldsymbol{X},\Theta}(\boldsymbol{x},\theta)}{f_{\boldsymbol{X}}(\boldsymbol{x})} \\
&= \frac{\theta^{x_1+x_2+\cdots+x_n+\kappa-1} e^{-(n+\lambda)\theta}(n+\lambda)^{x_1+x_2+\cdots+x_n+\kappa}}{\Gamma(x_1+x_2+\cdots+x_n+\kappa)}
\end{aligned}$$

for $\theta > 0$. This probability density function is recognized as that of a gamma random variable. More specifically,

$$\Theta\,|\,\boldsymbol{X} \sim \text{gamma}(n+\lambda, x_1+x_2+\cdots+x_n+\kappa).$$

When both the prior distribution and the posterior distribution belong to the same distribution family as they do in this particular example, then this choice of prior distribution is known as a *conjugate prior distribution.* The Bayesian point estimate of θ is the population mean of the posterior distribution, which is

$$\hat{\theta} = \frac{x_1+x_2+\cdots+x_n+\kappa}{n+\lambda}.$$

This point estimator uses the information provided in the prior distribution, given in λ and κ and the information provided in the data $x_1, x_2, \ldots, x_n$. To further emphasize this point, the Bayesian point estimate can be written as

$$\hat{\theta} = \left(\frac{n}{n+\lambda}\right)\left(\frac{x_1+x_2+\cdots+x_n}{n}\right) + \left(\frac{\lambda}{n+\lambda}\right)\left(\frac{\kappa}{\lambda}\right).$$

As seen in Example 2.9, the maximum likelihood estimate of θ using the data alone is the sample mean $\bar{x} = (x_1+x_2+\cdots+x_n)/n$. The prior distribution has population mean κ/λ. So the Bayesian point estimate of θ is a weighted average of the sample mean of the data values and the population mean of the prior distribution. The weight attached to the sample mean is $n/(n+\lambda)$ and the weight attached to the population mean of the prior distribution is $\lambda/(n+\lambda)$. As the sample size increases, the weight attached to the sample mean increases, as it should. In the limit as $n \to \infty$, the prior distribution gets no weight and the point estimator converges to the sample mean.

To construct a Bayesian credible interval for θ requires computing the $100(\alpha/2)$th and $100(1-\alpha/2)$th percentiles of the posterior distribution. Although these Bayesian credible interval bounds cannot be written in closed form because the posterior distribution is the gamma distribution, these percentiles can easily be found with the `qgamma` function in R.

We now turn to analyzing the horse kick data in this Bayesian setting. The prior distribution was chosen to be a gamma(12, 6) distribution. A prior distribution of this type might be formulated by gathering a group of military (and equine) experts and asking them their best guess for the population mean and variance of the number of horse-kick deaths per corps-year. If they estimate a population mean of 1/2 death per year and a population variance of 1/24 death per year squared, then these two values can be equated to the population mean and population variance of the gamma distribution, that is,

$$\frac{\kappa}{\lambda} = \frac{1}{2} \qquad \text{and} \qquad \frac{\kappa}{\lambda^2} = \frac{1}{24}.$$

The small population variance can be attributed to a high degree of certainty that the experts have placed on their estimate of the prior population mean. Solving for λ and κ gives $\lambda = 12$ and $\kappa = 6$, which are the values given in the problem statement. The $n = 200$ observations for the horse kick data are given below.

number of deaths per year	0	1	2	3	4
number of observed values	109	65	22	3	1

Recall that the maximum likelihood estimator of θ from Example 2.9 was the sample mean

$$\hat{\theta} = \frac{x_1 + x_2 + \cdots + x_n}{n} = \frac{122}{200} = 0.61$$

horse kick deaths per corps-year. The Bayesian point estimate of θ is

$$\hat{\theta} = \frac{x_1 + x_2 + \cdots + x_n + \kappa}{n + \lambda} = \frac{122 + 6}{200 + 12} = \frac{128}{212} = 0.60.$$

So the large data set has pulled the expert's estimate of 0.50 nearly up to the classical estimator of 0.61. The 95% Bayesian credible interval can be calculated by finding the appropriate percentiles of the gamma$(n + \lambda, x_1 + x_2 + \cdots + x_n + \kappa)$ distribution. These values are calculated with the R code shown below.

```
k = 6
l = 12
x = 0:4
h = c(109, 65, 22, 3, 1)
y = sum(x * h)
n = sum(h)
a = 0.05
```

```
qgamma(a / 2, y + k, n + 1)
qgamma(1 - a / 2, y + k, n + 1)
```

This returns the 95% Bayesian credible interval $0.50 < \theta < 0.71$. This interval is quite close to the three confidence intervals listed in Example 3.29 because the sample size is so large that the effect of the data overwhelms the prior distribution. Since Θ is assumed to be a random variable in the Bayesian paradigm, the Bayesian credible interval can be interpreted probabilistically. Using the posterior distribution

$$P(0.50 < \Theta < 0.71) = 0.95.$$

Figure 3.41 contains graphs of the prior and posterior probability density functions for the horse kick data using the same vertical and horizontal scales. The population means of the prior and posterior distributions are indicated by vertical dashed lines. The 95% Bayesian credible interval limits are indicated by vertical solid lines. The effect of the large sample size is a large reduction in the population variance of the posterior distribution over the population variance of the prior distribution.

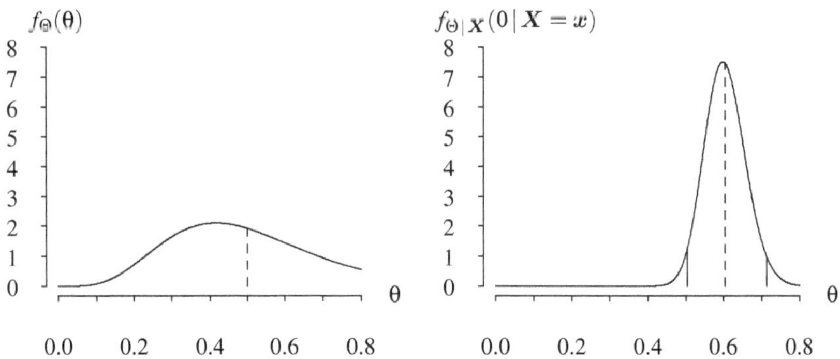

Figure 3.41: Prior and posterior probability density functions for the horse kick data.

Several previous examples have explored sampling data values from a Bernoulli population. We now address that same question in a Bayesian framework. Since the unknown parameter θ in a Bernoulli(θ) population assumes the values on the interval $0 < \theta < 1$, a natural distribution to use as a prior distribution is the beta distribution.

Example 3.38 Let $X_1, X_2, \ldots, X_n$ be a random sample drawn from a Bernoulli population with random parameter Θ. Assume that Θ has a prior beta(α, β) distribution. In other words, the Bayesian model is

$$X \,|\, \Theta = \theta \sim \text{Bernoulli}(\theta)$$

$$\Theta \sim \text{beta}(\alpha,\ \beta).$$

Find a Bayesian point estimate of θ and an exact two-sided $100(1-\alpha)\%$ Bayesian credible interval for θ.

The prior probability density function is given by

$$f_\Theta(\theta) = \frac{\Gamma(\alpha+\beta)}{\Gamma(\alpha)\Gamma(\beta)}\theta^{\alpha-1}(1-\theta)^{\beta-1} \qquad 0<\theta<1.$$

The joint conditional probability mass function of $X_1, X_2, \ldots, X_n$ given $\Theta = \theta$ is

$$\begin{aligned} f_{\boldsymbol{X}\,|\,\Theta=\theta}(\boldsymbol{x}\,|\,\Theta=\theta) &= \theta^{x_1}(1-\theta)^{1-x_1}\theta^{x_2}(1-\theta)^{1-x_2}\ldots\theta^{x_n}(1-\theta)^{1-x_n} \\ &= \theta^{x_1+x_2+\cdots+x_n}(1-\theta)^{n-(x_1+x_2+\cdots+x_n)} \end{aligned}$$

for $0<\theta<1$ and binary x_i, for $i = 1, 2, \ldots, n$. The joint distribution of $X_1, X_2, \ldots, X_n$ and Θ is described by

$$\begin{aligned} f_{\boldsymbol{X},\Theta}(\boldsymbol{x},\theta) &= f_{\boldsymbol{X}\,|\,\Theta=\theta}(\boldsymbol{x}\,|\,\Theta=\theta)f_\Theta(\theta) \\ &= \theta^{x_1+x_2+\cdots+x_n}(1-\theta)^{n-(x_1+x_2+\cdots+x_n)}\cdot\frac{\Gamma(\alpha+\beta)}{\Gamma(\alpha)\Gamma(\beta)}\theta^{\alpha-1}(1-\theta)^{\beta-1} \\ &= \frac{\Gamma(\alpha+\beta)}{\Gamma(\alpha)\Gamma(\beta)}\theta^{x_1+x_2+\cdots+x_n+\alpha-1}(1-\theta)^{n+\beta-(x_1+x_2+\cdots+x_n)-1} \end{aligned}$$

for $0<\theta<1$ and binary x_i, for $i = 1, 2, \ldots, n$. This is again an unusual joint probability distribution because Θ is a continuous random variable, but $X_1, X_2, \ldots, X_n$ are discrete random variables. The marginal probability mass function of $X_1, X_2, \ldots, X_n$ is

$$\begin{aligned} f_{\boldsymbol{X}}(\boldsymbol{x}) &= \int_\Omega f_{\boldsymbol{X}\,|\,\Theta=\theta}(\boldsymbol{x}\,|\,\Theta=\theta)f_\Theta(\theta)\,d\theta \\ &= \int_0^1 \frac{\Gamma(\alpha+\beta)}{\Gamma(\alpha)\Gamma(\beta)}\theta^{x_1+x_2+\cdots+x_n+\alpha-1}(1-\theta)^{n+\beta-(x_1+x_2+\cdots+x_n)-1}\,d\theta \\ &= \frac{\Gamma(\alpha+\beta)}{\Gamma(\alpha)\Gamma(\beta)}\cdot\frac{\Gamma(x_1+x_2+\cdots+x_n+\alpha)\Gamma\big(n+\beta-(x_1+x_2+\cdots+x_n)\big)}{\Gamma(n+\alpha+\beta)} \end{aligned}$$

by using the fact that the probability density function of the beta distribution integrates to one. The support of this marginal distribution is binary x_i, for $i = 1, 2, \ldots, n$. The posterior probability density function of Θ conditioned on the experimental values $x_1, x_2, \ldots, x_n$ of the random variables $X_1, X_2, \ldots, X_n$ is

$$\begin{aligned} f_{\Theta\,|\,\boldsymbol{X}=\boldsymbol{x}}(\theta\,|\,\boldsymbol{X}=\boldsymbol{x}) &= \frac{f_{\boldsymbol{X},\Theta}(\boldsymbol{x},\theta)}{f_{\boldsymbol{X}}(\boldsymbol{x})} \\ &= \frac{\Gamma(n+\alpha+\beta)\theta^{x_1+x_2+\cdots+x_n+\alpha-1}(1-\theta)^{n+\beta-(x_1+x_2+\cdots+x_n)-1}}{\Gamma(x_1+x_2+\cdots+x_n+\alpha)\Gamma\big(n+\beta-(x_1+x_2+\cdots+x_n)\big)} \end{aligned}$$

for $0<\theta<1$. This probability density function is recognized as that of a beta random variable. More specifically,

$$\Theta\,|\,\boldsymbol{X} \sim \text{beta}\big(x_1+x_2+\cdots+x_n+\alpha,\; n+\beta-(x_1+x_2+\cdots+x_n)\big).$$

So the choice of a beta prior distribution when the sampling is from a Bernoulli population distribution is another conjugate prior distribution because the prior and posterior distributions belong to the same probability distribution family. In this case that distribution family is the beta distribution. The Bayesian point estimate of θ is the population mean of the posterior distribution, which is

$$\hat{\theta} = \frac{x_1+x_2+\cdots+x_n+\alpha}{n+\alpha+\beta}.$$

The Bayesian point estimate of θ can be written as

$$\hat{\theta} = \left(\frac{n}{n+\alpha+\beta}\right)\left(\frac{x_1+x_2+\cdots+x_n}{n}\right)+\left(\frac{\alpha+\beta}{n+\alpha+\beta}\right)\left(\frac{\alpha}{\alpha+\beta}\right).$$

As seen in Example 2.16, the maximum likelihood estimator of θ is the sample mean $(x_1+x_2+\cdots+x_n)/n$. Since the population mean of a beta(α, β) random variable is $\alpha/(\alpha+\beta)$, the Bayesian point estimator of θ is a weighted average of the sample mean (the classical estimator of θ) and the population mean of the prior distribution. The weight attached to the sample mean is $n/(n+\alpha+\beta)$ and the weight attached to the population mean of the prior distribution is $(\alpha+\beta)/(n+\alpha+\beta)$. As the sample size increases, the weight attached to the sample mean increases, as it should. In the limit as $n \to \infty$, the prior distribution gets no weight and the point estimator converges to the sample mean.

To construct a Bayesian credible interval for θ requires calculating the $100(\alpha/2)$th and $100(1-\alpha/2)$th percentiles of the posterior distribution. Although these Bayesian credible interval bounds cannot be written in closed form because the posterior distribution is the beta distribution, these percentiles can easily be found with the `qbeta` function in R.

We now illustrate the effect of two different choices for the prior distribution on the posterior distribution. We begin with a *noninformative* prior distribution. That is to say we know that θ is between 0 and 1 but have no idea where it might fall. This situation corresponds to the beta distribution with $\alpha = 1$ and $\beta = 1$, which is a $U(0, 1)$ prior distribution. The population mean of the prior distribution is $1/2$. Assume that the data consists of five free throw attempts, with four successes, that is, $n = 5$ and $x_1+x_2+\cdots+x_5 = 4$. So the posterior distribution is the beta(5, 2) distribution. Figure 3.42 shows a graph of the prior probability density function on the left and the associated posterior probability density function on the right. The population mean of the posterior distribution is $5/(5+2) = 5/7$. The data tugs the prior distribution to the right, toward the classical estimator of $\hat{\theta} = 4/5$. As before, the population means for the prior and posterior distributions are marked with vertical dashed lines and the

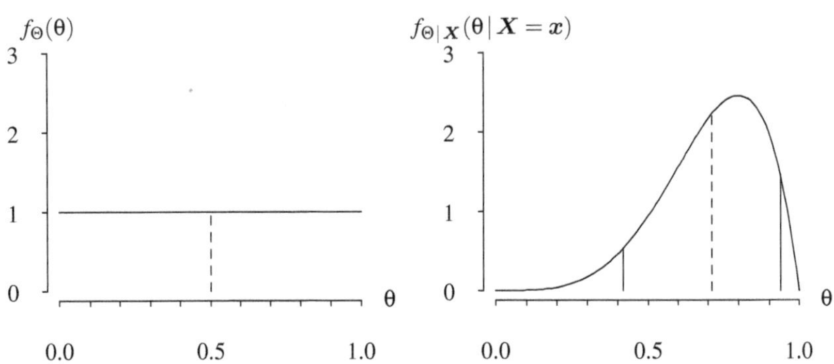

Figure 3.42: Noninformative prior and posterior probability density functions.

90% Bayesian credible interval limits are marked with vertical solid lines. The 90% Bayesian credible interval is

$$0.42 < \theta < 0.94.$$

This should be compared with the wider approximate 90% confidence interval from Example 3.13 based on the data alone, which was $0.34 < \theta < 0.99$.

To contrast the noninformative prior case, now consider a beta prior distribution with $\alpha = 5$ and $\beta = 5$. The population prior mean is again $5/(5+5) = 1/2$, but this time we have some evidence that the probability of success on each trial is indeed near $1/2$ and is not likely to be at the extremes near 0 and 1. With a data set of $n = 5$ observed Bernoulli trials with four successes, the prior and posterior probability density functions are plotted in Figure 3.43. The posterior distribution is a beta(9, 6) distribution, which has population posterior mean $9/(9+6) = 3/5$. This time the tug of the data to the right is not a strong as in the noninformative prior case. Since we now have more information in the prior distribution, the data does not tug the posterior distribution as far to the right as it did in the noninformative prior case. The 90% Bayesian credible interval is

$$0.39 < \theta < 0.79.$$

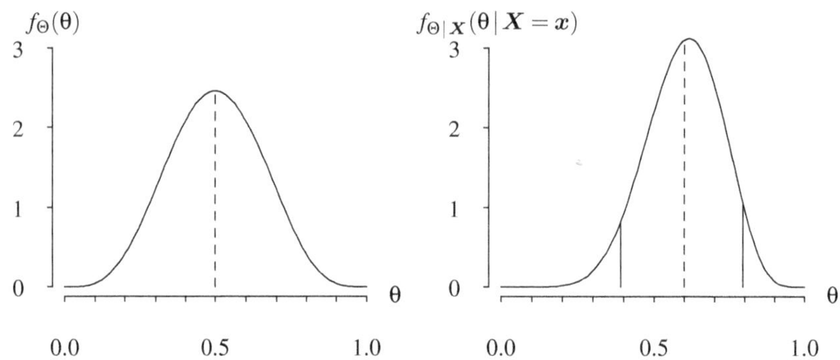

Figure 3.43: Informative prior and posterior probability density functions.

The first two examples of using the Bayesian approach have illustrated random sampling from a discrete population with a random parameter with a continuous distribution. Example 3.37 showed how a large data set ($n = 200$) had a large influence over the prior distribution. Example 3.38 showed how a small data set ($n = 5$) had a lesser influence over the prior distribution. The next example of Bayesian credible intervals illustrates the case of random sampling from a continuous population with a random parameter having a continuous distribution with a moderate-sized data set.

Example 3.39 Let $X_1, X_2, \ldots, X_n$ be a random sample drawn from an exponential distribution with random failure rate Θ. Assume that Θ has a prior exponential(λ) distribution, where λ is a fixed rate parameter that is estimated by subject matter experts. So the Bayesian model is

$$X \mid \Theta = \theta \sim \text{exponential}(\theta)$$

$$\Theta \sim \text{exponential}(\lambda).$$

Find a Bayesian point estimate and an exact two-sided $100(1-\alpha)\%$ Bayesian credible interval for θ. Apply these formula to the $n = 27$ aircraft air conditioner failure times from Example 3.7 to compute a Bayesian point estimator and an exact two-sided 90% Bayesian credible interval for θ when the prior parameter is $\lambda = 100$.

The prior probability density function of Θ is given by

$$f_\Theta(\theta) = \lambda e^{-\lambda\theta} \qquad \theta > 0.$$

The joint conditional probability density function of $X_1, X_2, \ldots, X_n$ given $\Theta = \theta$ is

$$\begin{aligned} f_{\boldsymbol{X}\,|\,\Theta=\theta}(\boldsymbol{x}\,|\,\Theta=\theta) &= \theta e^{-\theta x_1}\theta e^{-\theta x_2}\ldots\theta e^{-\theta x_n} \\ &= \theta^n e^{-(x_1+x_2+\cdots+x_n)\theta} \end{aligned}$$

for $\theta > 0$ and $x_i > 0$, for $i = 1, 2, \ldots, n$. The joint probability density function of $X_1, X_2, \ldots, X_n$ and Θ is

$$\begin{aligned} f_{\boldsymbol{X},\Theta}(\boldsymbol{x},\theta) &= f_{\boldsymbol{X}\,|\,\Theta=\theta}(\boldsymbol{x}\,|\,\Theta=\theta)f_\Theta(\theta) \\ &= \theta^n e^{-(x_1+x_2+\cdots+x_n)\theta}\cdot\lambda e^{-\lambda\theta} \\ &= \lambda\theta^n e^{-(\lambda+x_1+x_2+\cdots+x_n)\theta} \end{aligned}$$

for $\theta > 0$ and $x_i > 0$, for $i = 1, 2, \ldots, n$. The marginal probability density function of $X_1, X_2, \ldots, X_n$ is

$$\begin{aligned} f_{\boldsymbol{X}}(\boldsymbol{x}) &= \int_\Omega f_{\boldsymbol{X}\,|\,\Theta=\theta}(\boldsymbol{x}\,|\,\Theta=\theta)f_\Theta(\theta)\,d\theta \\ &= \int_0^\infty \lambda\theta^n e^{-(\lambda+x_1+x_2+\cdots+x_n)\theta}\,d\theta \\ &= \frac{\lambda\Gamma(n+1)}{(\lambda+x_1+x_2+\cdots+x_n)^{n+1}} \end{aligned}$$

for $x_i > 0$, for $i = 1, 2, \ldots, n$ by using the substitution $u = (\lambda + x_1 + x_2 + \cdots + x_n)\theta$ and the definition of the gamma function. The posterior probability density function of Θ conditioned on the experimental values $x_1, x_2, \ldots, x_n$ of the random variables $X_1, X_2, \ldots, X_n$ is

$$\begin{aligned} f_{\Theta\,|\,\boldsymbol{X}=\boldsymbol{x}}(\theta\,|\,\boldsymbol{X}=\boldsymbol{x}) &= \frac{f_{\boldsymbol{X},\Theta}(\boldsymbol{x},\theta)}{f_{\boldsymbol{X}}(\boldsymbol{x})} \\ &= \frac{\theta^n e^{-(\lambda+x_1+x_2+\cdots+x_n)\theta}(\lambda+x_1+x_2+\cdots+x_n)^{n+1}}{\Gamma(n+1)} \end{aligned}$$

for $\theta > 0$. This probability density function is recognized as that of a gamma random variable. More specifically,

$$\Theta\,|\,\boldsymbol{X} \sim \text{gamma}(\lambda + x_1 + x_2 + \cdots + x_n, n+1).$$

The Bayesian point estimate of θ is the population mean of the posterior distribution, which is

$$\hat{\theta} = \frac{n+1}{\lambda + x_1 + x_2 + \cdots + x_n}.$$

The bounds associated with the Bayesian credible interval are found by calculating the appropriate percentiles of the posterior distribution using the `qgamma` function in R.

The $n = 27$ times between failures (in hours) for an air conditioning system on a Boeing 720 jet airplane from Example 3.7 are

$$97, 51, 11, 4, 141, 18, 142, 68, 77, 80, 1, 16, 106, 206,$$

$$82, 54, 31, 216, 46, 111, 39, 63, 18, 191, 18, 163, 24.$$

The population mean of the prior distribution is 100 hours. Since the sum of the data values is $x_1 + x_2 + \cdots + x_{27} = 2074$, the posterior distribution is

$$\Theta \,|\, \boldsymbol{X} \sim \text{gamma}(2174, 28).$$

The Bayesian point estimate of θ is the population mean of the posterior distribution, which is

$$\hat{\theta} = \frac{28}{2174} = 0.013.$$

The bounds of the 90% Bayesian credible interval are calculated with the R statements.

```
qgamma(0.05, 28, 2174)
qgamma(0.95, 28, 2174)
```

This gives the exact two-sided 90% Bayesian credible interval

$$0.0092 < \theta < 0.017.$$

To summarize the point estimates of this Bayesian analysis in term of population means rather than population failure rates, the air conditioners are estimated to function, on average, for 100 hours prior to data collection by subject-matter experts. The classical (non-Bayesian) analysis gives the point estimate $2074/27 = 77$ hours for the population mean time to failure, which is the sample mean of the data. The population mean of the posterior distribution is $2174/28 = 78$ hours. The data pulled the prior point estimator toward the sample mean. Figure 3.44 shows the prior and posterior distributions, along with the point estimator and associated 90% Bayesian credible interval. Taking reciprocals, the exact two-sided 90% Bayesian credible interval interval for the population mean time to failure is

$$58 < \frac{1}{\theta} < 109,$$

which is shifted slightly to the right of the asymptotically exact 90% confidence interval from Example 3.23, which was $52 < 1/\theta < 101$, and is quite close to the exact two-sided 90% confidence interval from Example 3.7, which was $57 < 1/\theta < 109$.

This concludes the discussion of Bayesian credible intervals. The Bayesian methodology combines a prior distribution, which can be formulated from previous statistical studies or from expert opinion, with a current data set to arrive at a posterior distribution. A Bayesian credible interval can be calculated from the posterior distribution. Three examples have illustrated the construction of the Bayesian credible intervals for relatively tractable choices of prior and conditional distributions. For intractable choices, numerical methods can be used to determine Bayesian point estimators and Bayesian credible intervals. The final topic in this section extends the discussion of interval estimation for a single parameter to interval estimation for multiple parameters.

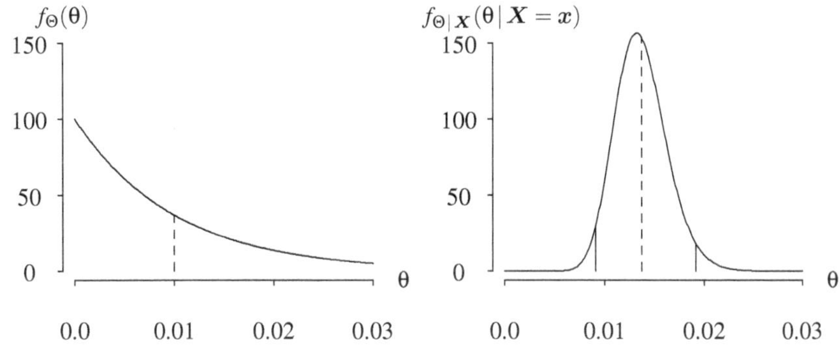

Figure 3.44: Prior and posterior probability density functions.

Confidence regions

Confidence intervals and Bayesian credible intervals have concerned just a single parameter, which has been referred to generically as θ. But what if there is more than one parameter, such as μ and σ in the normal distribution? In this case an *interval* is not the correct term. A set that reflects the precision of a point estimator when there are two or more unknown parameters is typically called a *region*. We consider *confidence regions* in this subsection, which are the multiparameter analog of confidence intervals.

Perhaps the simplest confidence region is a rectangular-shaped confidence region for just $p = 2$ unknown parameters. Not surprisingly, we start with random sampling from a normal population.

Example 3.40 Find a confidence region for the unknown parameters μ and σ for a random sample $X_1, X_2, \ldots, X_n$ drawn from a $N(\mu, \sigma^2)$ population and plot the confidence region for μ and σ for Michelson's $n = 100$ speed of light data values from Example 1.6.

Using the confidence interval formulas from Table 3.1, an exact two-sided $100(1-\alpha)\%$ confidence interval for μ is

$$\bar{X} - t_{n-1,\alpha/2}\frac{S}{\sqrt{n}} < \mu < \bar{X} + t_{n-1,\alpha/2}\frac{S}{\sqrt{n}}$$

and an exact two-sided $100(1-\alpha)\%$ confidence interval for σ is

$$\sqrt{\frac{(n-1)S^2}{\chi^2_{n-1,\alpha/2}}} < \sigma < \sqrt{\frac{(n-1)S^2}{\chi^2_{n-1,1-\alpha/2}}}.$$

Arbitrarily choosing $\alpha = 0.1$ for Michelson's speed of light data, the sample mean is $\bar{x} = 852$, the sample standard deviation is $s = 79$, an exact two-sided 90% confidence interval for μ is

$$839 < \mu < 866,$$

and an exact two-sided 90% confidence interval for σ is

$$71 < \sigma < 90.$$

The interior of the rectangular region in Figure 3.45 that is bounded by these confidence intervals is the confidence region. The coordinates of the point plotted in the interior of the rectangle are the point estimates: $(\bar{x}, s) = (852, 79)$. This point is exactly centered between the vertical sides of the rectangle (because the confidence interval for μ is symmetric) and slightly below the middle of the two horizontal sides of the rectangle (because the confidence interval for σ is not symmetric).

The rectangular-shaped confidence region either contains the point (μ, σ) or it does not. The values μ and σ are the population parameters. In the case of a confidence interval for a single parameter θ, there were three outcomes: the interval either contained θ, missed θ low, or missed θ high. There are more options for missing with two parameters. As indicated by the dashed lines in Figure 3.45, there are $3^2 - 1 = 8$ different ways for the rectangle to miss the point (μ, σ), for example, the confidence interval for μ might miss high and the confidence interval for σ might miss low.

Is there a way to determine whether this particular rectangular-shaped confidence region covers the true population parameters? Certainly the true speed of light in air is now known very precisely, so we can tell if it falls in the interval $839 < \mu < 866$. But determining whether the confidence region covers the population standard deviation is much more difficult. This would require an experiment with a large sample size using the technology available to Michelson in the late 1800's to accurately determine the population standard deviation.

So far, there has not been an actual coverage assigned to this confidence region. Certainly 90% coverage would not be appropriate because $\alpha = 0.1$ for each of the two exact confidence intervals, and we are requiring both parameters to fall in the rectangle. The actual coverage must be less than 90%. If the two exact confidence intervals were independent, then the actual coverage would be exactly 81% because $0.9 \cdot 0.9 = 0.81$. Even though $\bar{X}$ and S^2 are independent in this setting via Theorem 1.5, a careful inspection of the confidence interval bounds for the two exact confidence intervals shows that the value of S impacts the coverage of both confidence intervals, so the actual coverage will not be the product of the two stated coverages for the individual confidence intervals. Determining the actual coverage of this confidence region is nontrivial.

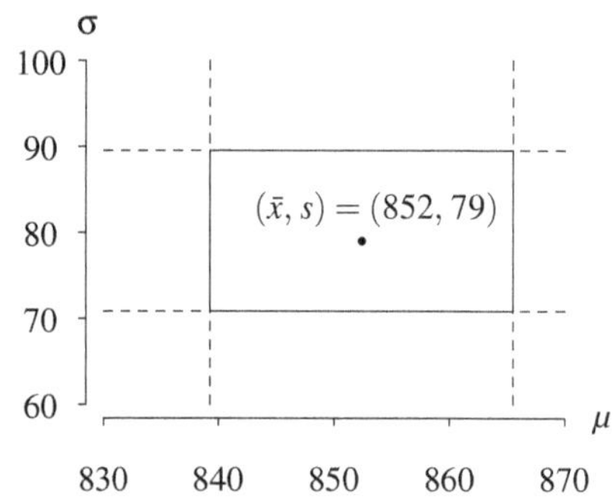

Figure 3.45: Confidence region for μ and σ for Michelson's speed of light data.

The next example describes a Monte Carlo simulation experiment that generates two exact two-sided $100(1-\alpha)\%$ confidence intervals, one for μ and one for σ^2 for a random sample drawn from a normal population, and estimates the actual coverage of the rectangular confidence region created by the two confidence intervals. The sample size and population parameters are chosen arbitrarily.

Example 3.41 Consider a random sample of $n = 8$ observations from a normal population with population mean $\mu = 2$ and population standard deviation $\sigma = 2$. Conduct a Monte Carlo simulation experiment that estimates the actual coverage associated with the two exact 90% confidence intervals given in the previous example. Generalize the results to other values of n, μ, and σ if possible.

The R code below generates 1,000,000 confidence regions for μ and σ and prints the fraction of those confidence regions that cover the true parameter values.

```
mu    = 2
sigma = 2
n     = 8
a     = 0.1
nrep  = 1000000
count = 0
for (i in 1:nrep) {
  x = rnorm(n, mu, sigma)
  m = mean(x)
  s = sd(x)
  l = m - qt(1 - a / 2, n - 1) * s / sqrt(n)
  u = m + qt(1 - a / 2, n - 1) * s / sqrt(n)
  b = sqrt((n - 1) * s ^ 2 / qchisq(1 - a / 2, n - 1))
  h = sqrt((n - 1) * s ^ 2 / qchisq(a / 2, n - 1))
  if (l < mu && mu < u && b < sigma && sigma < h) count = count + 1
}
print(count / nrep)
```

After a call to `set.seed(9)` to initialize the random number stream, five runs of this simulation return

0.8185 0.8189 0.8185 0.8190 0.8188.

The actual coverage for this confidence region, to three digits, is 81.9%. The two sets of confidence interval bounds are dependent, so their coverages cannot be multiplied to arrive at the actual coverage. To provide some additional insight concerning this confidence region, the values of $\bar{x}$ and s for the first 100 confidence intervals in the simulation experiment are plotted in Figure 3.46. A $\bullet$ is associated with values in which the associated confidence region covers the true values $(\mu, \sigma) = (2, 2)$ and a $\circ$ is associated with values that miss. The cross hairs meet at true values of the parameters $(\mu, \sigma) = (2, 2)$. The 82 confidence regions that cover the true values tend to cluster about $(\mu, \sigma) = (2, 2)$. The 18 confidence regions that miss the true value tend to be at the periphery. An unusually small or large value of $\bar{x}$ will tend to make the exact confidence interval for μ miss (and therefore the confidence region will miss). An unusually small or large value of s will tend to make the exact confidence interval for σ miss (and therefore the confidence region will miss).

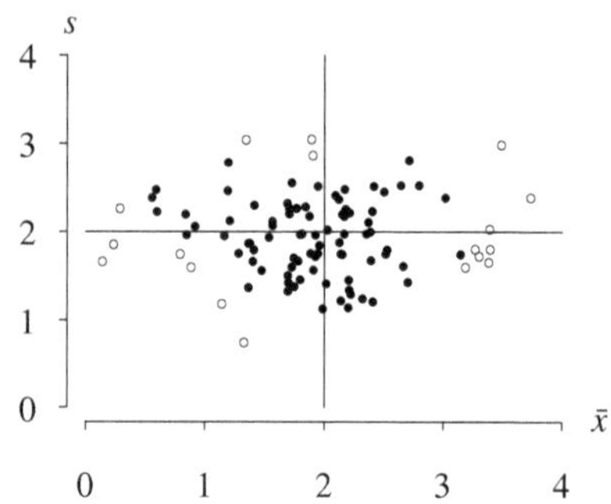

Figure 3.46: Scatterplot of 100 $(\bar{x}, s)$ pairs for a confidence region for μ and σ.

Figure 3.46 is reminiscent of Figure 3.15, which was a plot of the sample mean vs. the confidence interval halfwidth. Since the confidence interval halfwidth is just a constant multiplied by the sample standard deviation, the boundaries associated with the exact confidence interval for μ are linear in Figure 3.46. Furthermore, the exact confidence interval for σ has bounds that do not involve $\bar{x}$, so horizontal lines in the $\bar{x}$ vs. s plot are associated with the exact confidence interval for σ in Figure 3.46. The resulting trapezoidal-shaped area that is associated with the confidence region containing the true values of μ and σ is shown in Figure 3.47. The eight ways that the confidence region can miss the true values of μ and σ are apparent in the figure.

If we had the joint probability density function of $\bar{X}$ and S, we could integrate over the trapezoid and arrive at the actual coverage of the confidence region. By Theorem 1.5, we know that $\bar{X}$ and S are independent for random sampling from a normal population, so all that is necessary is to find the marginal distribution of $\bar{X}$ and the marginal distribution of S. We now switch the notation from $\mu = 2$, $\sigma = 2$, and $n = 8$ to general values for μ, σ, and n. The marginal distribution of $\bar{X}$ is already known. By Theorem 1.3, we

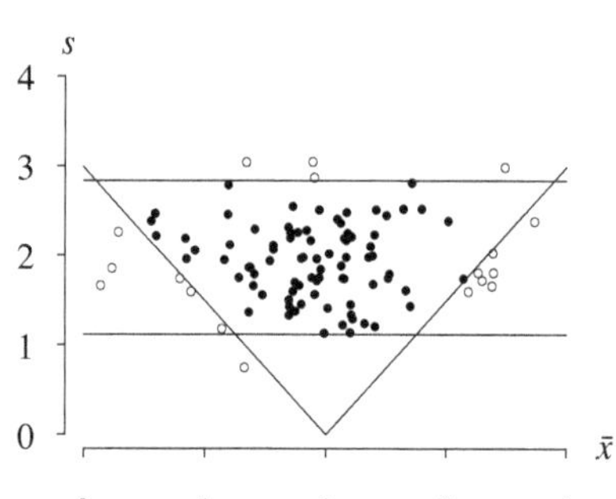

Figure 3.47: Scatterplot of 100 $(\bar{x}, s)$ pairs for a confidence region for μ and σ.

know that

$$\bar{X} \sim N\left(\mu, \frac{\sigma^2}{n}\right),$$

so the probability density function of $\bar{X}$ is

$$f_{\bar{X}}(x) = \frac{1}{\sqrt{2\pi\sigma^2/n}} e^{-(x-\mu)^2/(2\sigma^2/n)} \qquad -\infty < x < \infty.$$

Finding the probability distribution of S will not be as easy. From Theorem 1.6,

$$X = \frac{(n-1)S^2}{\sigma^2} \sim \chi^2(n-1).$$

So the probability density function of X is

$$f_X(x) = \frac{x^{(n-1)/2-1} e^{-x/2}}{\Gamma((n-1)/2) 2^{(n-1)/2}} \qquad x > 0.$$

We would like the probability density function of

$$S = Y = \sqrt{\frac{\sigma^2 X}{n-1}}.$$

The transformation $y = g(x) = \sqrt{\sigma^2 x/(n-1)}$ is a one-to-one transformation from the support of X, which is $\mathcal{A} = \{x \mid x > 0\}$, to the support of Y, which is $\mathcal{B} = \{y \mid y > 0\}$, with inverse

$$x = g^{-1}(y) = (n-1)y^2/\sigma^2$$

and

$$\frac{dx}{dy} = \frac{2y(n-1)}{\sigma^2}.$$

By the transformation technique, the probability density function of Y, which is the sample standard deviation S, after simplification, is

$$f_Y(y) = \frac{(n-1)^{(n-1)/2} y^{n-2} e^{-(n-1)y^2/(2\sigma^2)}}{\sigma^{n-1} 2^{(n-3)/2} \Gamma((n-1)/2)} \qquad y > 0.$$

So finally, the actual coverage of the confidence region is computed as the volume under the joint probability density function above the trapezoid in Figure 3.47:

$$\int_{\sqrt{\chi^2_{n-1,0.95}\sigma^2/(n-1)}}^{\sqrt{\chi^2_{n-1,0.05}\sigma^2/(n-1)}} \int_{\mu - t_{n-1,0.05}y/\sqrt{n}}^{\mu + t_{n-1,0.05}y/\sqrt{n}} f_{\bar{X}}(x) f_Y(y)\, dx\, dy.$$

Fortunately, the value of this double integral does not depend on the values of the population parameters μ and σ. So evaluating it at $n = 8$ gives actual coverage for the confidence region of 0.8189, which is consistent with the Monte Carlo simulation in this example. For the $n = 100$ speed of light data values from Example 3.40, the actual coverage for the confidence region is 0.8107. As n increases, the trapezoid in Figure 3.47 associated with $n = 8$ becomes more rectangular-shaped, as illustrated in Figure 3.48. This provides strong geometric evidence that the actual coverage of the confidence interval approaches $(0.9)(0.9) = 0.81$ as $n \to \infty$. A plot of the actual coverage for the confidence region is shown in Figure 3.49. The horizontal axis begins at $n = 2$ because S does not exist for $n = 1$.

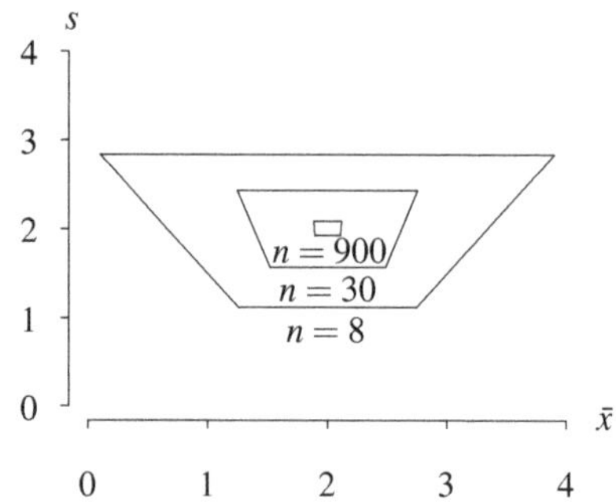

Figure 3.48: Integration region for calculating the actual coverage of the confidence region.

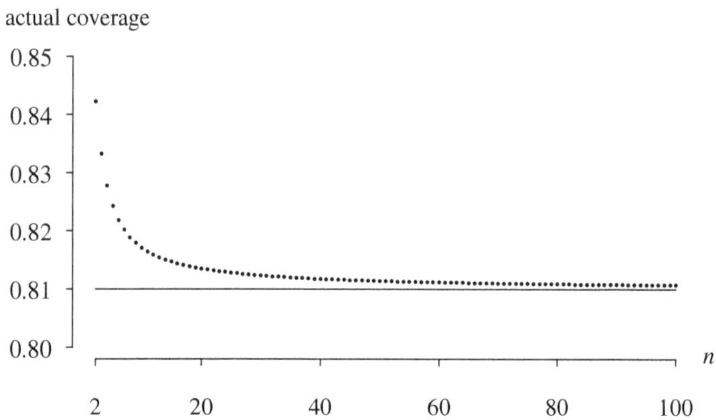

Figure 3.49: Actual coverage for the confidence region.

Although in principle one could calculate the actual coverage for all rectangular-shaped confidence regions as in the previous example, a lower bound on the actual coverage probability can be found using a result known as Bonferroni's inequality.

Theorem 3.5 (Bonferroni's inequality) If $A_1, A_2, \ldots, A_p$ are events, then

$$P(A_1 \cap A_2 \cap \ldots \cap A_p) \geq P(A_1) + P(A_2) + \cdots + P(A_p) - (p-1).$$

Proof The proof uses induction. The first non-trivial case is $p = 2$. The addition rule states that for any two events A_1 and A_2,

$$P(A_1 \cup A_2) = P(A_1) + P(A_2) - P(A_1 \cap A_2).$$

Since the left-hand side of this equation is a probability,

$$P(A_1) + P(A_2) - P(A_1 \cap A_2) \le 1$$

or

$$P(A_1 \cap A_2) \ge P(A_1) + P(A_2) - 1,$$

which proves Bonferroni's inequality for $p = 2$ events. Now assume that Bonferroni's inequality holds for p events, that is,

$$P(A_1 \cap A_2 \cap \ldots \cap A_p) \ge P(A_1) + P(A_2) + \cdots + P(A_p) - (p-1),$$

and show that this implies that the inequality holds for $p+1$ events. Using the $p = 2$ case which has already been proven,

$$P(A_1 \cap A_2 \cap \ldots \cap A_n \cap A_{p+1}) \ge P(A_1 \cap A_2 \cap \ldots \cap A_p) + P(A_{p+1}) - 1.$$

Replacing $P(A_1 \cap A_2 \cap \ldots \cap A_p)$ with the expression in the inductive hypothesis results in

$$P(A_1 \cap A_2 \cap \ldots \cap A_{p+1}) \ge P(A_1) + P(A_2) + \cdots + P(A_p) - (p-1) + P(A_{p+1}) - 1.$$

Rearranging terms results in Bonferroni's inequality for $p+1$ events, which completes the proof. □

Bonferroni's inequality is very useful for assessing the actual coverage of rectangular-shaped confidence regions. In the previous example, the two "events" from Bonferroni's inequality are the confidence interval for μ covering the true population mean and the confidence interval for σ covering the true population standard deviation. If the two 90% confidence intervals were independent (which they are not for finite n), then the actual coverage of the confidence region would be 81% because $(0.9)(0.9) = 0.81$. Bonferroni's inequality gives a lower bound on the actual coverage of the confidence region of 80% because

$$0.9 + 0.9 - (2-1) = 0.8.$$

Bonferroni's inequality can be applied in more than two dimensions. If five 99% confidence intervals for five unknown parameters are assumed to be independent, then the actual coverage of the associated confidence region in five dimensions is 95.1% because $0.99^5 = 0.951$. In most cases, however, there is no justification for the independence assumption. Bonferroni's inequality gives a lower bound on the actual coverage of 95% because

$$0.99 + 0.99 + 0.99 + 0.99 + 0.99 - (5-1) = 0.95.$$

Particularly when α is small for the individual confidence intervals, Bonferroni's inequality gives a lower bound that is surprisingly close to the actual coverage associated with the independence assumption.

Since constructing confidence regions is hampered by mathematical intractability, many confidence regions are based on asymptotic relationships. In order to develop these asymptotic relationships, it is necessary to first define the Fisher information matrix. In Definition 2.6, the information associated with a single unknown parameter was defined as $I(\theta)$. We now define its multiparameter analog.

Definition 3.10 For a random variable X whose probability distribution is a parametric distribution described by $f(x)$ with p unknown parameters $\theta_1, \theta_2, \ldots, \theta_p$, the $p \times p$ *information matrix* associated with X in estimating $\boldsymbol{\theta} = (\theta_1, \theta_2, \ldots, \theta_p)'$ is denoted by $I(\boldsymbol{\theta})$, which has (j, k)th entry

$$E\left[-\frac{\partial^2 \ln f(X)}{\partial \theta_j \partial \theta_k}\right]$$

under certain regularity conditions when the expectation exists.

The information matrix is often referred to as the "Fisher information matrix," to highlight its discovery by Sir Ronald Fisher. The information matrix defined in Definition 3.10 applies to a single random variable X. The information matrix for the random sample $X_1, X_2, \ldots, X_n$ is simply $nI(\boldsymbol{\theta})$. A single random variable X from the normal distribution is used to illustrate the information matrix in the next example.

Example 3.42 Find the information matrix for the unknown parameters μ and σ^2 associated with $X \sim N\left(\mu, \sigma^2\right)$.

Using the derivatives from Example 2.13, the appropriate expected values are

$$E\left[-\frac{\partial^2 \ln f(X)}{\partial \mu^2}\right] = E\left[\frac{1}{\sigma^2}\right] = \frac{1}{\sigma^2},$$

$$E\left[-\frac{\partial^2 \ln f(X)}{\partial \left(\sigma^2\right)^2}\right] = E\left[-\frac{1}{2\sigma^4} + \frac{1}{\sigma^6}(X-\mu)^2\right] = \frac{1}{2\sigma^4},$$

and

$$E\left[-\frac{\partial^2 \ln f(X)}{\partial \mu \partial \sigma^2}\right] = E\left[\frac{1}{\sigma^4}(X-\mu)\right] = 0.$$

Notice that these expected values would differ if σ, rather than σ^2, were used as the second parameter. So the information matrix is

$$I\left(\mu, \sigma^2\right) = \begin{bmatrix} \frac{1}{\sigma^2} & 0 \\ 0 & \frac{1}{2\sigma^4} \end{bmatrix}.$$

In most practical statistical applications, the values of the parameters μ and σ^2 are not known. Since the maximum likelihood estimators

- $\hat{\mu} = \bar{X}$ is a consistent estimator of μ, and
- $\hat{\sigma}^2 = \frac{1}{n}\sum_{i=1}^{n}\left(X_i - \bar{X}\right)^2$ is a consistent estimator of σ^2,

$I\left(\hat{\mu}, \hat{\sigma}^2\right)$ is a consistent estimator of the information matrix.

The information matrix plays an important role in the next result. Recall from Section 2.3 that the score vector consists of the partial derivatives of the log likelihood function with respect to the unknown parameters $\boldsymbol{\theta} = (\theta_1, \theta_2, \ldots, \theta_p)'$. The symbol $\boldsymbol{U}(\boldsymbol{\theta})$ will be used to denote the score vector, that is,

$$\boldsymbol{U}(\boldsymbol{\theta}) = \left(U_1(\boldsymbol{\theta}), U_2(\boldsymbol{\theta}), \ldots, U_p(\boldsymbol{\theta})\right)' = \left(\frac{\partial \ln L(\boldsymbol{\theta})}{\partial \theta_1}, \frac{\partial \ln L(\boldsymbol{\theta})}{\partial \theta_2}, \ldots, \frac{\partial \ln L(\boldsymbol{\theta})}{\partial \theta_p}\right)'.$$

Theorem 3.6 Let $X_1, X_2, \ldots, X_n$ be a random sample from a population probability distribution described by $f(x)$, with unknown parameters $\boldsymbol{\theta} = (\theta_1, \theta_2, \ldots, \theta_p)'$. For a population distribution satisfying certain regularity conditions, the score vector $\boldsymbol{U}(\boldsymbol{\theta})$ has expected value $\mathbf{0}$, that is,

$$E\left[\boldsymbol{U}(\boldsymbol{\theta})\right] = \mathbf{0},$$

and variance–covariance matrix $nI(\boldsymbol{\theta})$, where $I(\boldsymbol{\theta})$ is the information matrix.

Proof Assume that the population is continuous for this proof. The discrete case is analogous. This is an outline of the proof because the regularity conditions have not been listed. Since the likelihood function $L(\boldsymbol{\theta})$ is the joint probability density function for $X_1, X_2, \ldots, X_n$, it must integrate to 1, that is,

$$\int_{-\infty}^{\infty}\int_{-\infty}^{\infty}\cdots\int_{-\infty}^{\infty} L(\boldsymbol{\theta})\,d\boldsymbol{x} = 1,$$

which will be referred to as the normalizing equation. Assuming that the likelihood function is continuous (and hence differentiation and integration can be interchanged), the partial derivative of the left-hand side of the normalizing equation with respect to one of the parameters, θ_j, yields

$$\begin{aligned}
\frac{\partial}{\partial\theta_j}\int_{-\infty}^{\infty}\int_{-\infty}^{\infty}\cdots\int_{-\infty}^{\infty} L(\boldsymbol{\theta})\,d\boldsymbol{x} &= \int_{-\infty}^{\infty}\int_{-\infty}^{\infty}\cdots\int_{-\infty}^{\infty}\frac{\partial}{\partial\theta_j}L(\boldsymbol{\theta})\,d\boldsymbol{x} \\
&= \int_{-\infty}^{\infty}\int_{-\infty}^{\infty}\cdots\int_{-\infty}^{\infty}\frac{\partial\ln L(\boldsymbol{\theta})}{\partial\theta_j}L(\boldsymbol{\theta})\,d\boldsymbol{x} \\
&= E\left[\frac{\partial\ln L(\boldsymbol{\theta})}{\partial\theta_j}\right] \\
&= E\left[U_j(\boldsymbol{\theta})\right] \qquad j = 1, 2, \ldots, p.
\end{aligned}$$

Since the partial derivative of the right-hand side of the normalizing equation with respect to θ_j is zero,

$$E\left[U_j(\boldsymbol{\theta})\right] = 0 \qquad j = 1, 2, \ldots, p,$$

or, in vector form,

$$E\left[\boldsymbol{U}(\boldsymbol{\theta})\right] = \mathbf{0}.$$

This proves the first result. Differentiating the left-hand side of the normalizing equation first with respect to θ_j, then with respect to θ_k, and using the chain rule yields

$$\begin{aligned}
&\frac{\partial}{\partial\theta_k}\int_{-\infty}^{\infty}\int_{-\infty}^{\infty}\cdots\int_{-\infty}^{\infty}\frac{\partial\ln L(\boldsymbol{\theta})}{\partial\theta_j}L(\boldsymbol{\theta})\,d\boldsymbol{x} \\
&= \int_{-\infty}^{\infty}\int_{-\infty}^{\infty}\cdots\int_{-\infty}^{\infty}\left(\frac{\partial\ln L(\boldsymbol{\theta})}{\partial\theta_j}\frac{\partial L(\boldsymbol{\theta})}{\partial\theta_k} + \frac{\partial^2\ln L(\boldsymbol{\theta})}{\partial\theta_j\partial\theta_k}L(\boldsymbol{\theta})\right)d\boldsymbol{x} \\
&= \int_{-\infty}^{\infty}\int_{-\infty}^{\infty}\cdots\int_{-\infty}^{\infty}\left(\frac{\partial\ln L(\boldsymbol{\theta})}{\partial\theta_j}\frac{\partial\ln L(\boldsymbol{\theta})}{\partial\theta_k}L(\boldsymbol{\theta}) + \frac{\partial^2\ln L(\boldsymbol{\theta})}{\partial\theta_j\partial\theta_k}L(\boldsymbol{\theta})\right)d\boldsymbol{x} \\
&= E\left[U_j(\boldsymbol{\theta})U_k(\boldsymbol{\theta})\right] + E\left[\frac{\partial^2\ln L(\boldsymbol{\theta})}{\partial\theta_j\partial\theta_k}\right] \qquad \begin{array}{l} j = 1, 2, \ldots, p, \\ k = 1, 2, \ldots, p. \end{array}
\end{aligned}$$

Since the second partial derivative of the right-hand side of the normalizing equation with respect to θ_j and θ_k is zero,

$$E\left[-\frac{\partial^2 \ln L(\boldsymbol{\theta})}{\partial\theta_j\partial\theta_k}\right] = E\left[U_j(\boldsymbol{\theta})U_k(\boldsymbol{\theta})\right] \qquad \begin{array}{l} j=1,2,\ldots,p, \\ k=1,2,\ldots,p. \end{array}$$

But since $E\left[U_j(\boldsymbol{\theta})\right]=0$ for $j=1,2,\ldots,p$, and $E\left[U_k(\boldsymbol{\theta})\right]=0$ for $k=1,2,\ldots,p$,

$$E\left[-\frac{\partial^2 \ln L(\boldsymbol{\theta})}{\partial\theta_j\partial\theta_k}\right] = \text{Cov}\left(U_j(\boldsymbol{\theta}),\,U_k(\boldsymbol{\theta})\right) \qquad \begin{array}{l} j=1,2,\ldots,p, \\ k=1,2,\ldots,p. \end{array}$$

So the $p\times p$ variance–covariance matrix of the score vector $\boldsymbol{U}(\boldsymbol{\theta})$ is $nI(\boldsymbol{\theta})$, which completes the proof. □

To review the results derived so far, the $p\times 1$ score vector $\boldsymbol{U}(\boldsymbol{\theta})$ has components

$$U_j(\boldsymbol{\theta}) = \frac{\partial \ln L(\boldsymbol{\theta})}{\partial\theta_j} \qquad j=1,2,\ldots,p,$$

which, when equated to zero and solved for the unknown parameters, yields the $p\times 1$ maximum likelihood estimate vector $\hat{\boldsymbol{\theta}}$. The expected value of the jth element of the score vector is

$$E\left[U_j(\boldsymbol{\theta})\right] = 0 \qquad j=1,2,\ldots,p,$$

and the variance–covariance matrix of $\boldsymbol{U}(\boldsymbol{\theta})$ is

$$nI(\boldsymbol{\theta}) = E\left[\boldsymbol{U}(\boldsymbol{\theta})\boldsymbol{U}'(\boldsymbol{\theta})\right],$$

where the $p\times p$ information matrix $I(\boldsymbol{\theta})$ has components

$$E\left[-\frac{\partial^2 \ln f(X)}{\partial\theta_j\partial\theta_k}\right] \qquad \begin{array}{l} j=1,2,\ldots,p, \\ k=1,2,\ldots,p. \end{array}$$

These results are exact and apply to all sample sizes n. In order to construct confidence regions, we need asymptotic results. When there was a single parameter θ, Theorem 3.1 was used to show that $\sqrt{n}\left(\hat{\theta}-\theta\right)$ converged in distribution to a normal distribution, that is,

$$\sqrt{n}\left(\hat{\theta}-\theta\right) \xrightarrow{D} N\left(0,\,\frac{1}{I(\theta)}\right).$$

The next result, which is proved in most advanced mathematical statistics textbooks, states the multiparameter analog.

Theorem 3.7 Let $X_1, X_2, \ldots, X_n$ be a random sample from a population probability distribution described by $f(x)$, with unknown parameters $\boldsymbol{\theta} = (\theta_1, \theta_2, \ldots, \theta_p)'$. Let $\hat{\boldsymbol{\theta}}$ denote the maximum likelihood estimator of $\boldsymbol{\theta}$. Under certain regularity conditions associated with the population probability distribution,

$$\sqrt{n}\left(\hat{\boldsymbol{\theta}}-\boldsymbol{\theta}\right) \xrightarrow{D} N\left(\boldsymbol{0},\,I^{-1}(\boldsymbol{\theta})\right),$$

where $I(\boldsymbol{\theta})$ is the information matrix associated with X in estimating $\boldsymbol{\theta}$.

So if the information matrix (and its inverse) were diagonal, or if the off-diagonal elements in the inverse of the information matrix are small enough to be ignored, one can use this result to construct asymptotically exact confidence intervals for the unknown parameters. More specifically, denoting the (j, j) element of the information matrix by $I_{jj}(\boldsymbol{\theta})$,

$$\sqrt{n}(\hat{\theta}_j - \theta_j) \xrightarrow{D} N\left(0, I_{jj}^{-1}(\boldsymbol{\theta})\right)$$

for $j = 1, 2, \ldots, p$. Subtracting the populations mean and dividing by the population standard deviation, this can be written as

$$\frac{\hat{\theta}_j - \theta_j}{\sqrt{I_{jj}^{-1}(\boldsymbol{\theta})/n}} \xrightarrow{D} N(0, 1)$$

for $j = 1, 2, \ldots, p$. This random variable is an asymptotic pivotal quantity because the standard normal distribution does not involve θ. This implies that for $0 < \alpha < 1$,

$$\lim_{n\to\infty} P\left(-z_{\alpha/2} < \frac{\hat{\theta}_j - \theta_j}{\sqrt{I_{jj}^{-1}(\boldsymbol{\theta})/n}} < z_{\alpha/2}\right) = 1 - \alpha$$

for $j = 1, 2, \ldots, p$. It is not possible to manipulate the inequality in its current form to isolate θ_j in the middle. However, if we are willing to replace $I_{jj}(\boldsymbol{\theta})$ with $I_{jj}\left(\hat{\boldsymbol{\theta}}\right)$ in the middle of the inequality $\left(\text{using the fact that } I_{jj}\left(\hat{\boldsymbol{\theta}}\right) \text{ is a consistent estimator of } I_{jj}(\boldsymbol{\theta})\right)$, then this becomes

$$\lim_{n\to\infty} P\left(-z_{\alpha/2} < \frac{\hat{\theta}_j - \theta_j}{\sqrt{I_{jj}^{-1}\left(\hat{\boldsymbol{\theta}}\right)/n}} < z_{\alpha/2}\right) = 1 - \alpha$$

for $j = 1, 2, \ldots, p$. Performing the algebra necessary to isolate θ_j in the center of the inequality results in

$$\lim_{n\to\infty} P\left(\hat{\theta} - z_{\alpha/2}\sqrt{\frac{I_{jj}^{-1}\left(\hat{\boldsymbol{\theta}}\right)}{n}} < \theta_j < \hat{\theta} + z_{\alpha/2}\sqrt{\frac{I_{jj}^{-1}\left(\hat{\boldsymbol{\theta}}\right)}{n}}\right) = 1 - \alpha$$

for $j = 1, 2, \ldots, p$. So an asymptotically exact two-sided $100(1-\alpha)\%$ confidence interval for θ_j is

$$\hat{\theta}_j - z_{\alpha/2}\sqrt{\frac{I_{jj}^{-1}\left(\hat{\boldsymbol{\theta}}\right)}{n}} < \theta_j < \hat{\theta}_j + z_{\alpha/2}\sqrt{\frac{I_{jj}^{-1}\left(\hat{\boldsymbol{\theta}}\right)}{n}}$$

for $j = 1, 2, \ldots, p$.

The next example illustrates the construction of a confidence region for the parameters for random sampling from the two-parameter inverse Gaussian distribution.

Example 3.43 Let $X_1, X_2, \ldots, X_n$ be a random sample from an *inverse Gaussian* population with $p = 2$ positive unknown parameters λ and μ and probability density function

$$f(x) = \sqrt{\frac{\lambda}{2\pi}} x^{-3/2} e^{-\lambda(x-\mu)^2/(2\mu^2 x)} \qquad x > 0.$$

The population mean is μ. Find the maximum likelihood estimators and a confidence region for λ and μ. Calculate $\hat{\lambda}$, $\hat{\mu}$, and a confidence region for the ball bearing failure times from Example 2.3.

The likelihood function is

$$\begin{aligned} L(\lambda, \mu) &= \prod_{i=1}^{n} \sqrt{\frac{\lambda}{2\pi}} x_i^{-3/2} e^{-\lambda(x_i-\mu)^2/(2\mu^2 x_i)} \\ &= \lambda^{n/2}(2\pi)^{-n/2} \left[\prod_{i=1}^{n} x_i\right]^{-3/2} e^{-\lambda/(2\mu^2)\sum_{i=1}^{n}(x_i-\mu)^2/x_i}. \end{aligned}$$

The log likelihood function is

$$\ln L(\lambda, \mu) = \frac{n}{2}\ln\lambda - \frac{n}{2}\ln(2\pi) - \frac{3}{2}\sum_{i=1}^{n}\ln x_i - \frac{\lambda}{2\mu^2}\sum_{i=1}^{n}\frac{(x_i-\mu)^2}{x_i}.$$

The score vector, $\boldsymbol{U}(\lambda, \mu)$, has two components:

$$U_1(\lambda, \mu) = \frac{\partial \ln L(\lambda, \mu)}{\partial \lambda} = \frac{n}{2\lambda} - \frac{1}{2\mu^2}\sum_{i=1}^{n}\frac{(x_i-\mu)^2}{x_i}$$

and

$$U_2(\lambda, \mu) = \frac{\partial \ln L(\lambda, \mu)}{\partial \mu} = \frac{\lambda}{\mu^3}\left[\sum_{i=1}^{n} x_i - n\mu\right].$$

When the second equation is equated to zero, the maximum likelihood estimator $\hat{\mu}$ is determined. Then using $\hat{\mu}$ as an argument in the first equation and solving for $\hat{\lambda}$ results in the closed-form maximum likelihood estimators

$$\hat{\lambda} = \left[\frac{1}{n}\sum_{i=1}^{n}\frac{1}{x_i} - \frac{n}{\sum_{i=1}^{n} x_i}\right]^{-1} \qquad \text{and} \qquad \hat{\mu} = \frac{1}{n}\sum_{i=1}^{n} x_i.$$

The next step is to compute the information matrix. The second partial derivatives of $\ln f(X)$ are

$$\begin{aligned} \frac{\partial^2 \ln f(X)}{\partial \lambda^2} &= -\frac{1}{2\lambda^2} \\ \frac{\partial^2 \ln f(X)}{\partial \lambda \partial \mu} &= \frac{X}{\mu^3} - \frac{1}{\mu^2} \\ \frac{\partial^2 \ln f(X)}{\partial \mu^2} &= -\frac{3\lambda X}{\mu^4} + \frac{2\lambda}{\mu^3}. \end{aligned}$$

Since $E[X] = \mu$, the information matrix consists of the expected values of the negatives of these derivatives:

$$I(\lambda, \mu) = \begin{pmatrix} E\left[-\frac{\partial^2 \ln f(X)}{\partial \lambda^2}\right] & E\left[-\frac{\partial^2 \ln f(X)}{\partial \lambda \partial \mu}\right] \\ E\left[-\frac{\partial^2 \ln f(X)}{\partial \mu \partial \lambda}\right] & E\left[-\frac{\partial^2 \ln f(X)}{\partial \mu^2}\right] \end{pmatrix} = \begin{pmatrix} \frac{1}{2\lambda^2} & 0 \\ 0 & \frac{\lambda}{\mu^3} \end{pmatrix},$$

where the off-diagonal elements being zero implies that the elements of the score vector are uncorrelated. The inverse of the information matrix is

$$I^{-1}(\lambda, \mu) = \begin{pmatrix} 2\lambda^2 & 0 \\ 0 & \frac{\mu^3}{\lambda} \end{pmatrix}.$$

Since the off-diagonal elements of the information matrix are zero, an asymptotically exact two-sided $100(1-\alpha)\%$ confidence interval for λ is

$$\hat{\lambda} - z_{\alpha/2}\sqrt{\frac{2\hat{\lambda}^2}{n}} < \lambda < \hat{\lambda} + z_{\alpha/2}\sqrt{\frac{2\hat{\lambda}^2}{n}}$$

and an asymptotically exact two-sided $100(1-\alpha)\%$ confidence interval for μ is

$$\hat{\mu} - z_{\alpha/2}\sqrt{\frac{\hat{\mu}^3}{n\hat{\lambda}}} < \mu < \hat{\mu} + z_{\alpha/2}\sqrt{\frac{\hat{\mu}^3}{n\hat{\lambda}}}.$$

A confidence region for λ and μ satisfies both of these inequalities. The $n = 23$ ball bearing failure times (measured in millions of revolutions) from Example 2.3 are given below.

17.88, 28.92, 33.00, 41.52, 42.12, 45.60, 48.48, 51.84, 51.96, 54.12, 55.56, 67.80,

68.64, 68.64, 68.88, 84.12, 93.12, 98.64, 105.12, 105.84, 127.92, 128.04, 173.40.

The maximum likelihood estimators are

$$\hat{\lambda} = 231.7 \qquad \text{and} \qquad \hat{\mu} = 72.22.$$

Figure 3.50 contains a plot of the empirical cumulative distribution function for the data set and the cumulative distribution function for the fitted inverse Gaussian distribution. The cumulative distribution function for the inverse Gaussian distribution cannot be expressed in closed form, so the probability density function was numerically integrated to arrive at the fitted cumulative distribution function in Figure 3.50. It appears that the inverse Gaussian distribution provides a reasonable fit to the ball bearing failure times.

Selecting $\alpha = 0.05$ arbitrarily, an asymptotically exact two-sided 95% confidence interval for λ is

$$231.7 - 1.96\sqrt{\frac{2 \cdot 231.7^2}{23}} < \lambda < 231.7 + 1.96\sqrt{\frac{2 \cdot 231.7^2}{23}}$$

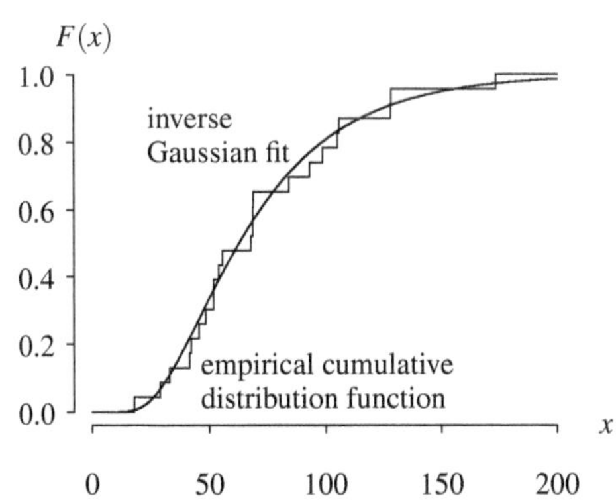

Figure 3.50: Empirical and fitted cumulative distribution functions for the ball bearing data.

or

$$97.78 < \lambda < 365.6.$$

Likewise, an asymptotically exact two-sided 95% confidence interval for μ is

$$72.22 - 1.96\sqrt{\frac{72.22^3}{23 \cdot 231.7}} < \mu < 72.22 + 1.96\sqrt{\frac{72.22^3}{23 \cdot 231.7}}$$

or

$$55.74 < \mu < 88.70.$$

The confidence region associated with these two confidence intervals is plotted in Figure 3.51. The Bonferroni inequality can be used to conclude that the asymptotic actual coverage of this confidence region is at least 90%.

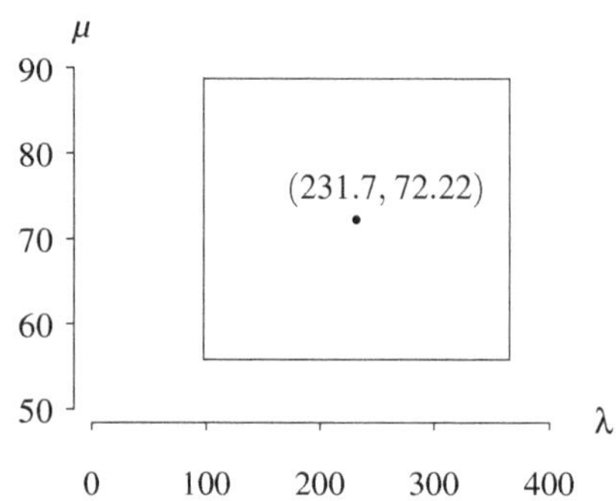

Figure 3.51: Confidence region for λ and μ for the ball bearing failure times.

Using the inverse Gaussian population distribution in the previous example was, in some sense, ideal because

- the maximum likelihood estimators were closed form,
- the expected values required to compute the information matrix could be computed in closed form,
- the information matrix was diagonal.

In general, we will not be so lucky. In the cases in which the expected values in the information matrix cannot be computed in closed form, the observed information matrix can be used to approximate the information matrix. Rather than computing the expected values from Definition 3.10, the maximum likelihood estimators are used directly in the second derivatives of $\ln f(x)$.

Definition 3.11 For a random variable X whose probability distribution is a parametric distribution described by $f(x)$ with p unknown parameters $\theta_1, \theta_2, \ldots, \theta_p$, the *observed information matrix* associated with X in estimating $\boldsymbol{\theta} = (\theta_1, \theta_2, \ldots, \theta_p)'$ is denoted by $O(\hat{\boldsymbol{\theta}})$, which has (j, k)th entry

$$\left[-\frac{\partial^2 \ln f(x)}{\partial \theta_j \partial \theta_k} \right]_{\boldsymbol{\theta} = \hat{\boldsymbol{\theta}}}.$$

Although it was not necessary to use the observed information matrix in the previous example, it works out to

$$O\left(\hat{\lambda}, \hat{\mu}\right) = \begin{pmatrix} -\dfrac{\partial^2 \ln f(x)}{\partial \lambda^2} & -\dfrac{\partial^2 \ln f(x)}{\partial \lambda \partial \mu} \\ -\dfrac{\partial^2 \ln f(x)}{\partial \mu \partial \lambda} & -\dfrac{\partial^2 \ln f(x)}{\partial \mu^2} \end{pmatrix}_{\lambda = \hat{\lambda}, \mu = \hat{\mu}} = \begin{pmatrix} \dfrac{1}{2\hat{\lambda}^2} & 0 \\ 0 & \dfrac{\hat{\lambda}}{\hat{\mu}^3} \end{pmatrix},$$

which is identical to $I\left(\hat{\lambda}, \hat{\mu}\right)$.

Every confidence region that we have encountered thus far has been rectangular shaped. This need not necessarily be the case. Parameter estimators are oftentimes correlated. Asymptotically exact confidence regions can be constructed for random sampling for certain probability distributions based on the asymptotic distribution of $-2\ln\Lambda$, where Λ is the likelihood ratio statistic. For a single unknown parameter θ, Theorem 3.3 showed that the asymptotic distribution of $-2\ln\Lambda$ is the chi-square distribution with one degree of freedom, that is,

$$-2\ln\Lambda = -2\ln\left(\frac{L(\theta)}{L\left(\hat{\theta}\right)}\right) \xrightarrow{D} \chi^2(1),$$

under certain regularity conditions. We now state the multivariate analog, which is useful for constructing asymptotically exact confidence regions. This result is known as *Wilks's lemma* after Samuel S. Wilks, and is proved in most advanced mathematical statistics textbooks.

Theorem 3.8 (Wilks's lemma) Let $X_1, X_2, \ldots, X_n$ be a random sample from a population probability distribution described by $f(x)$, which has p unknown parameters $\theta_1, \theta_2, \ldots, \theta_p$. Let $\boldsymbol{\theta} = (\theta_1, \theta_2, \ldots, \theta_p)'$ be the vector of unknown parameters and $\hat{\boldsymbol{\theta}}$ be the maximum likelihood estimator of $\boldsymbol{\theta}$. Under certain regularity conditions concerning the population distribution,

$$-2\ln\Lambda = -2\ln\left(\frac{L(\boldsymbol{\theta})}{L\left(\hat{\boldsymbol{\theta}}\right)}\right) = -2\left[\ln L(\boldsymbol{\theta}) - \ln L\left(\hat{\boldsymbol{\theta}}\right)\right] \xrightarrow{D} \chi^2(p).$$

This result can be used to construct an asymptotically exact confidence region for $\theta_1, \theta_2, \ldots, \theta_p$. In an analogous fashion to the single-parameter case, an asymptotically exact confidence region for $\boldsymbol{\theta}$ is the set of all $\boldsymbol{\theta}$ satisfying

$$-2\ln\left(\frac{L(\boldsymbol{\theta})}{L\left(\hat{\boldsymbol{\theta}}\right)}\right) < \chi^2_{p,\alpha}.$$

This will be illustrated for random sampling from a Weibull(λ, κ) population in the next example.

Example 3.44 Let $X_1, X_2, \ldots, X_n$ be a random sample from a Weibull(λ, κ) population, where λ and κ are positive unknown scale and shape parameters. Find an asymptotically exact 95% confidence region for the $n = 23$ ball bearing failure times from Example 2.3 using Theorem 3.8.

In Example 2.14, the Weibull distribution was fitted to the $n = 23$ ball bearing failure times (measured in millions of revolutions) given below, yielding the maximum likelihood estimates $\hat{\lambda} = 0.0122$ and $\hat{\kappa} = 2.10$.

17.88	28.92	33.00	41.52	42.12	45.60	48.48	51.84
51.96	54.12	55.56	67.80	68.64	68.64	68.88	84.12
93.12	98.64	105.12	105.84	127.92	128.04	173.40	

The log likelihood function evaluated at the maximum likelihood estimators is

$$\ln L\left(\hat{\lambda}, \hat{\kappa}\right) = -113.691,$$

which is the peak value of the log likelihood function graphed in Figure 3.52.

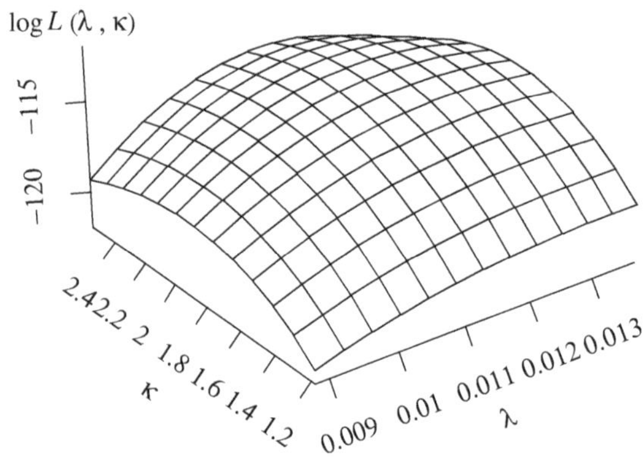

Figure 3.52: Log likelihood function for the ball bearing failure times.

Using the fact that the likelihood ratio statistic, $-2\left[\ln L(\lambda, \kappa) - \ln L\left(\hat{\lambda}, \hat{\kappa}\right)\right]$, is asymptotically $\chi^2(2)$ by Theorem 3.8, a 95% confidence region for the parameters is all λ and κ that satisfy

$$2[-113.691 - \ln L(\lambda, \kappa)] < 5.99,$$

since $\chi^2_{2,0.05} = 5.99$, which can be found with the R statement `qchisq(0.95, 2)`. The geometry associated with the confidence region is analogous to the geometry in one dimension as shown in Figures 3.35, 3.36, and 3.38. Consider a plane that is parallel to the λ, κ plane in Figure 3.52 that is $\chi^2_{2,0.05}/2 = 5.99/2$ units below the peak of the log likelihood function at $\ln L\left(\hat{\lambda}, \hat{\kappa}\right) = -113.691$. The intersection of this plane and the log likelihood function is the boundary of the 95% confidence region. The 95% confidence region is shown in Figure 3.53, and, not surprisingly, the line $\kappa = 1$ is not interior to the region. Since the Weibull distribution collapses to the exponential distribution when $\kappa = 1$, this indicates that the exponential distribution is *not* an appropriate model for this particular data set. This was confirmed visually in Example 2.3 by comparing the empirical and fitted exponential cumulative distribution functions. So the fact that $\kappa = 1$ is not in the 95% confidence region is statistical evidence that the ball bearings are wearing out. The point that is plotted in Figure 3.53 is $\left(\hat{\kappa}, \hat{\lambda}\right)$.

Graphing a confidence region like that in Figure 3.53 requires numerical methods. The confidence region can be plotted in R using the `crplot` function after installing and loading the `conf` package. The `crplot` function can be used for several popular two-parameter continuous distributions and plots confidence regions for complete and right-censored data sets. The following R command, with the ball bearing failure times stored in the vector `x`, plots the 95% confidence region in Figure 3.53.

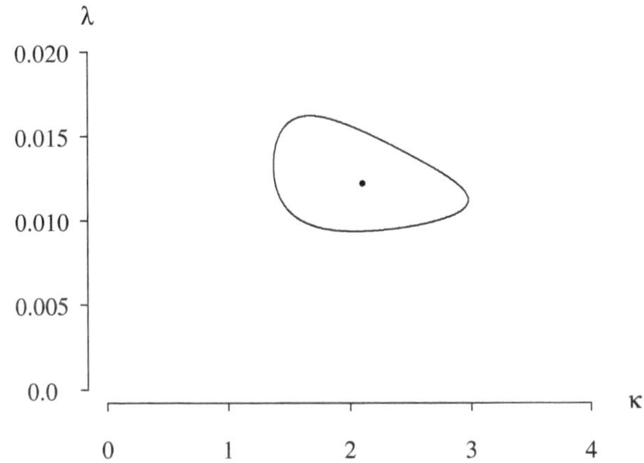

Figure 3.53: Confidence region for λ and κ ($\alpha = 0.05$) for the ball bearing failure times.

```
crplot(x, alpha = 0.05, distn = "weibull", origin = TRUE, pts = FALSE)
```

This ends the chapter on interval estimation. The most common type of statistical interval is a confidence interval. The gold standard for a confidence interval is an *exact* $100(1-\alpha)\%$ *confidence interval*. An exact two-sided $100(1-\alpha)\%$ confidence interval for an unknown parameter θ of the form $L < \theta < U$ satisfies $P(L < \theta < U) = 1 - \alpha$ for all values of θ. Three classifications of other types of confidence intervals are given below.

- A confidence interval that is not exact is an *approximate confidence interval.*
- A confidence interval that satisfies $P(L < \theta < U) \geq 1 - \alpha$ for all values of θ is a *conservative confidence interval.* These confidence intervals never claim more precision than they should.
- A confidence interval that satisfies $\lim_{n \to \infty} P(L < \theta < U) = 1 - \alpha$ for all values of θ is an *asymptotically exact confidence interval.* These confidence intervals are usually based on the asymptotic normality of the maximum likelihood estimator of θ or the asymptotic distribution of the likelihood ratio statistic.

The four other interval estimators presented in this chapter are prediction intervals, tolerance intervals, Bayesian credible intervals, and confidence regions.

- Prediction intervals assess the precision of a new observation X_{n+1}.
- Tolerance intervals capture at least a fraction p of the population probability distribution with a prescribed probability.
- Bayesian analysis treats the unknown parameter θ as a random variable, and combines a prior distribution and the data to produce a posterior distribution. This posterior distribution is used to derive a Bayesian point estimator and a Bayesian credible interval estimator for θ.
- Confidence regions are the multiparameter analog of confidence intervals.

3.5 Exercises

3.1 Which of the following statements are true of an exact 90% confidence interval for the population mean μ?

(a) In the long run, 90% of the intervals constructed in this manner will include the sample mean.

(b) In the long run, 90% of the intervals constructed in this manner will include the population mean.

(c) In the long run, 90% of the intervals constructed in this manner will include one particular sample value.

(d) In the long run, 90% of the intervals constructed in this manner will include the population variance.

(e) In the long run, 90% of the intervals constructed in this manner will include zero.

3.2 Kristina constructs three independent exact 90% confidence intervals for an unknown parameter θ. What is the probability that all three of the confidence intervals contain θ?

3.3 In a Monte Carlo simulation experiment, 100 exact 99% confidence intervals for θ are generated. Find the probability that all 100 of the confidence intervals cover the true parameter θ.

3.4 A reliability engineer has a single observation X from an exponential distribution with probability density function

$$f(x) = \lambda e^{-\lambda x} \qquad x > 0,$$

where λ is a positive unknown rate parameter. Find an exact two-sided 95% confidence interval for λ.

3.5 Let X be a single observation from a population with probability density function

$$f(x) = \theta x^{\theta - 1} \qquad 0 < x < 1,$$

where θ is a positive unknown parameter. Find an exact two-sided $100(1-\alpha)\%$ confidence interval for θ.

3.6 Let X be a single observation from a special case of the log logistic distribution with cumulative distribution function

$$F(x) = 1 - \frac{1}{1 + \lambda x} \qquad x > 0,$$

where λ is a positive unknown scale parameter. Find an exact two-sided 90% confidence interval for λ.

3.7 Let X be a single observation from a Rayleigh population with probability density function

$$f(x) = \frac{2x}{\theta^2} e^{-(x/\theta)^2} \qquad x > 0,$$

where θ is a positive unknown scale parameter. Show that the method of moments estimator of θ derived in Example 3.3 is not an efficient estimator.

3.8 Let X be a single observation from a population with probability density function

$$f(x) = \frac{\theta}{(1+x)^{\theta+1}} \qquad x > 0,$$

where θ is a positive unknown parameter. Construct an exact two-sided 90% confidence interval for θ.

3.9 Let $X_1, X_2, \ldots, X_n$ be a random sample from a Rayleigh population with probability density function

$$f(x) = \frac{2x}{\theta^2} e^{-(x/\theta)^2} \qquad x > 0,$$

where θ is a positive unknown scale parameter.

(a) Find the maximum likelihood estimator of θ.

(b) Find an exact two-sided $100(1-\alpha)\%$ confidence interval for θ.

(c) Conduct a Monte Carlo simulation experiment to estimate the actual coverage of the confidence interval from part (b) for a population with $\theta = 4$, $n = 7$, and $\alpha = 0.12$.

(d) Calculate the maximum likelihood estimator and an exact two-sided 95% confidence interval for θ for the aircraft air conditioning data from Example 3.7:

$$97, 51, 11, 4, 141, 18, 142, 68, 77, 80, 1, 16, 106, 206,$$

$$82, 54, 31, 216, 46, 111, 39, 63, 18, 191, 18, 163, 24.$$

3.10 Let X_1, X_2, X_3 be mutually independent random variables such that X_i is exponentially distributed with population mean $i\theta$, for $i = 1, 2, 3$, where θ is a positive unknown parameter.

(a) Find the maximum likelihood estimator $\hat{\theta}$.

(b) Find the probability density function of the maximum likelihood estimator $\hat{\theta}$.

(c) Is $\hat{\theta}$ an unbiased estimator of θ?

(d) Find an exact two-sided $100(1-\alpha)\%$ confidence interval for θ.

(e) Perform a Monte Carlo simulation experiment to evaluate the coverage of the confidence interval for $\theta = 10$ and $\alpha = 0.1$.

3.11 Let $X_1, X_2, \ldots, X_7$ be a random sample of size $n = 7$ from an exponential population with positive unknown population mean θ.

(a) Find an exact two-sided 90% confidence interval for the population median by finding a pivotal quantity based on the sample median $X_{(4)}$.

(b) Give an exact two-sided 90% confidence interval for the population median for the $n = 7$ rat survival times in the treatment group from Table 3.5:

$$16, 23, 38, 94, 99, 141, 197.$$

(c) Conduct a Monte Carlo simulation experiment to provide convincing numerical evidence that the exact two-sided 90% confidence interval for the population median is indeed an exact two-sided 90% confidence interval for an exponential population when θ is arbitrarily set to 1.

3.12 Let $X_1, X_2, \ldots, X_n$ be a random sample from an exponential(λ_X) population, where λ_X is a positive unknown parameter. Let $Y_1, Y_2, \ldots, Y_m$ be a random sample from an exponential(λ_Y) population, where λ_Y is a positive unknown parameter. The two random samples are independent. Use the results

$$2\lambda_X \sum_{i=1}^{n} X_i \sim \chi^2(2n) \qquad \text{and} \qquad 2\lambda_Y \sum_{i=1}^{m} Y_i \sim \chi^2(2m)$$

to form an exact two-sided $100(1-\alpha)\%$ confidence interval for λ_X/λ_Y.

3.13 Let $X_1, X_2, \ldots, X_n$ be a random sample from an exponential(λ) population, where λ is a positive unknown parameter. An exact two-sided 90% confidence interval for λ is

$$2 < \lambda < 5.$$

A client is not concerned about small values of λ. Only large values of λ are of concern. What is an exact one-sided 95% confidence interval of the form $\lambda < k$, for some constant k?

3.14 Let $X_1, X_2, \ldots, X_n$ be a random sample from a population with probability density function

$$f(x) = \frac{1}{\theta_1} e^{-(x-\theta_2)/\theta_1} \qquad x > \theta_2,$$

where θ_1 is a positive parameter and θ_2 is known as a shift, guarantee, or threshold parameter. This population distribution is known as the shifted exponential distribution, and the shift parameter θ_2 can be used, for example, to model warranty times for a manufactured product.

(a) Find an exact two-sided $100(1-\alpha)\%$ confidence interval for θ_1 when θ_1 is an unknown positive parameter and θ_2 is a known parameter.

(b) Find an exact two-sided $100(1-\alpha)\%$ confidence interval for θ_2 when θ_2 is an unknown parameter and θ_1 is a known parameter.

(c) Conduct a Monte Carlo simulation experiment that provides convincing numerical evidence that the two confidence intervals given in parts (a) and (b) are indeed exact confidence intervals for the arbitrary setting of the parameters $\theta_1 = 3$, $\theta_2 = 10$, and $\alpha = 0.05$, and sample size $n = 10$.

3.15 Let $X_1, X_2, \ldots, X_n$ be a random sample from a population with probability density function

$$f(x) = \frac{2x}{\theta^2} \qquad 0 < x < \theta,$$

where θ is a positive unknown parameter.

(a) Find the maximum likelihood estimator of θ.

(b) Find an unbiasing constant c_n, which is a function of n, such that $E\left[c_n\hat{\theta}\right] = \theta$.

(c) Find the maximum likelihood estimator of the 95th percentile of the distribution.

(d) Find an exact two-sided 95% confidence interval for θ when $n = 8$ and $\hat{\theta} = 7$.

3.16 Let $X_1, X_2, \ldots, X_n$ be a random sample from a continuous, symmetric probability distribution with unknown, finite population mean θ. An unbiased and efficient point estimate of θ is $\hat{\theta} = 10$. If an exact one-sided 99% confidence interval for θ is

$$\theta < 14$$

give an exact two-sided 98% confidence interval for θ.

3.17 Let $X_1, X_2, \ldots, X_n$ be a random sample from a $U(0, \theta)$ population, where θ is a positive unknown parameter. In Example 3.8, an exact two-sided $100(1-\alpha)\%$ confidence interval for θ was given by

$$\frac{X_{(n)}}{(1-\alpha/2)^{1/n}} < \theta < \frac{X_{(n)}}{(\alpha/2)^{1/n}}.$$

Conduct a Monte Carlo simulation experiment that provides convincing numerical evidence that this is indeed an exact two-sided confidence interval for θ for $n = 15$, $\theta = 10$, and $\alpha = 0.05$.

3.18 Let X be a single observation from a $N(\mu, 100)$ population, where μ is an unknown parameter. Find an exact two-sided 95% confidence interval for μ.

3.19 Let $X_1, X_2, \ldots, X_n$ be a random sample from a $N(\mu, \sigma^2)$ population, where μ and $\sigma^2 > 0$ are known parameters. Give an expression for the smallest sample size n that is required to assure that

$$P\left(\left|\bar{X} - \mu\right| < d\right) > 1 - \alpha,$$

for some real positive constant d and some real constant α satisfying $0 < \alpha < 1$.

3.20 Let $X_1, X_2, \ldots, X_9$ be a random sample from a $N(\mu, \sigma^2)$ population, where μ and $\sigma^2 > 0$ are unknown parameters. The associated experimental values are $x_1, x_2, \ldots, x_9$. If

$$\sum_{i=1}^{9} (x_i - \bar{x})^2 = 15.8,$$

give an exact two-sided 90% confidence interval for σ^2.

3.21 Let X_1 and X_2 be a random sample from a $N(\mu, \sigma^2)$ population, where μ and $\sigma^2 > 0$ are unknown parameters.

(a) Express the lower bound of an exact two-sided $100(1-\alpha)\%$ confidence interval for μ in terms of $\bar{X}$ and S.

(b) Write this formula in terms of X_1 and X_2 in the simplest possible algebraic terms.

(c) Write a paragraph describing how you would find the probability distribution of this lower bound.

3.22 Let X_1 and X_2 denote a random sample from a $N(\mu, \sigma^2)$ population, where μ and $\sigma^2 > 0$ are unknown parameters. Consider the random variables

$$Y_1 = \frac{X_1 + X_2}{2} - k\sqrt{\frac{(X_1 - X_2)^2}{2}} \qquad \text{and} \qquad Y_2 = \frac{X_1 + X_2}{2} + k\sqrt{\frac{(X_1 - X_2)^2}{2}},$$

where k is a positive real constant. Notice that when k is an appropriately selected percentile from the t distribution with 1 degree of freedom, these are the lower and upper bounds of an exact two-sided confidence interval for μ. Graphically and/or analytically examine the properties of this transformation.

3.23 Let $X_1, X_2, \ldots, X_n$ be a random sample from a $N(\mu, 81)$ population, where μ is an unknown parameter. Find the smallest sample size n required to assure that the expected width of an exact two-sided 95% confidence interval for μ is less than 10.

3.24 Let $X_1, X_2, \ldots, X_n$ be a random sample from a $N(\mu, 17)$ population, where μ is an unknown parameter. Find the smallest sample size n required to assure that the expected width of an exact two-sided 99% confidence interval for μ is less than 1.

3.25 Let $X_1, X_2, \ldots, X_n$ be a random sample from a normal population with population variance $\sigma^2 = 5$. Find the sample size n if

$$P\left(\sum_{i=1}^{n} (X_i - \bar{X})^2 < 160\right) = 0.99.$$

3.26 Let $X_1, X_2, \ldots, X_5$ be a random sample from a $N\left(\mu, \sigma^2\right)$ population, where μ and $\sigma^2 > 0$ are unknown parameters. The associated experimental values are $x_1, x_2, \ldots, x_5$. Find an exact two-sided 95% confidence interval for μ when

$$\bar{x} = 17 \qquad \text{and} \qquad \sum_{i=1}^{5} (x_i - \bar{x})^2 = 23.$$

3.27 Let $X_1, X_2, \ldots, X_5$ be a random sample from a $N\left(\mu, \sigma^2\right)$ population, where μ and $\sigma^2 > 0$ are unknown parameters. The associated experimental values are $x_1, x_2, \ldots, x_5$. Given that

$$\sum_{i=1}^{5} x_i = 10 \qquad \text{and} \qquad \sum_{i=1}^{5} (x_i - \bar{x})^2 = 20,$$

for what value of c is the interval

$$2 - c < \mu < 2 + c$$

an exact two-sided 90% confidence interval for μ?

3.28 Let X_1, X_2 and Y_1, Y_2, Y_3 and Z_1, Z_2, Z_3, Z_4 be three mutually independent random samples from three normal populations that share a common positive population variance σ^2. The associated sample variances from these three populations are S_1^2, S_2^2, and S_3^2.

(a) Show that the sample mean of the sample variances

$$\frac{S_1^2 + S_2^2 + S_3^2}{3}$$

is an unbiased estimator of σ^2.

(b) Suggest another unbiased estimator of σ^2 which has a smaller variance than the point estimator in part (a).

(c) Construct an exact two-sided 95% confidence interval for σ^2.

(d) Conduct a Monte Carlo simulation experiment that provides convincing numerical evidence that the actual coverage of the exact two-sided 95% confidence interval matches the stated coverage for parameter values of your choice.

3.29 Mr. Trump and Mrs. Clinton are in a very close race for U.S. President. You are going to conduct a poll that estimates p, the true proportion of likely voters supporting Mr. Trump. Find the appropriate sample size necessary to be 99% certain that your estimate of p is within 0.01 of the true value. You may assume that these are the only two candidates running for President.

3.30 Sophie spins a penny 100 times and obtains 32 heads. Find approximate 95% confidence intervals for p, the true probability of obtaining a head for a single spin of a penny, using the Wald, Clopper–Pearson, Wilson–score, Jeffreys, Agresti–Coull, and arcsine transformation confidence interval procedures.

3.31 Show that the Clopper–Pearson confidence interval for the Bernoulli parameter p can be written in terms of percentiles of the F distribution.

3.32 Consider the acceptance curves for a fixed positive integer n concerning a confidence interval for a binomial proportion p:

$$b(p, y_0, y_1) = \sum_{y=y_0}^{y_1} \binom{n}{y} p^y (1-p)^{n-y},$$

where p satisfies $0 < p < 1$ and the integers y_0 and y_1 satisfy $0 \le y_0 \le y_1 \le n$. Show that $b(p, y_0, y_1)$ on $0 < p < 1$ is

(a) a decreasing function of p when $y_0 = 0$ and $0 \le y_1 < n$,

(b) an increasing function of p when $y_1 = n$ and $0 < y_0 \le n$,

(c) equal to 1 when $y_0 = 0$ and $y_n = n$, and

(d) a unimodal function that achieves a maximum on $0 < p < 1$ when $0 < y_0 \le y_1 < n$ (find the value of p where the maximum occurs),

as given in Wang, H. (2007), "Exact Confidence Coefficients of Confidence Intervals for a Binomial Proportion," *Statistica Sinica*, Volume 17, Number 1, 361–368.

3.33 Let $X_1, X_2, \ldots, X_5$ be a random sample from a Bernoulli(p) population, where p is an unknown parameter satisfying $0 < p < 1$. Conduct a Monte Carlo simulation experiment to estimate the actual coverage of an approximate two-sided 90% Wald confidence interval for p when the true value of p is $p = 0.35$.

3.34 What are the degrees of freedom for the percentile of the t distribution that appears in the confidence interval for $\mu_X - \mu_Y$ using Welch's approximation when $s_X = s_Y$?

3.35 What are the degrees of freedom for the percentile of the t distribution that appears in the confidence interval for $\mu_X - \mu_Y$ using Welch's approximation when $n = m$ and $s_X = s_Y$?

3.36 Let X be a single observation from a discrete uniform population with probability mass function

$$f(x) = \frac{1}{\theta} \qquad x = 1, 2, \ldots, \theta,$$

where θ is a positive unknown integer parameter.

(a) Find the method of moments estimator of θ and show that it is an unbiased estimator of θ.

(b) Find an approximate two-sided $100(1-\alpha)\%$ confidence interval for θ.

(c) Calculate the method of moments estimator and the approximate two-sided 90% confidence interval for θ when the single observation is $x = 4$.

(d) Find the actual coverage of the approximate two-sided 90% confidence interval when $\theta = 97$.

3.37 Let $X_1, X_2, \ldots, X_n$ be a random sample from a discrete uniform population with probability mass function

$$f(x) = \frac{1}{\theta} \qquad x = 1, 2, \ldots, \theta,$$

where θ is an unknown parameter which is a positive integer.

(a) Find an approximate two-sided $100(1-\alpha)\%$ confidence interval for θ.

(b) Calculate an approximate two-sided 95% confidence interval for the small data set of $n = 4$ observations: 155, 803, 243, 606.

(c) Perform a Monte Carlo simulation to estimate the actual coverage for the arbitrary parameter settings $\theta = 1000$, $n = 4$ and $\alpha = 0.05$. Use enough replications to achieve three digits of accuracy.

3.38 Let $X_1, X_2, \ldots, X_n$ be a random sample from a Poisson(λ) population, where λ is a positive unknown parameter. An approximate two-sided $100(1-\alpha)\%$ confidence interval for λ is

$$\frac{1}{2n}\chi^2_{2Y,1-\alpha/2} < \lambda < \frac{1}{2n}\chi^2_{2(Y+1),\alpha/2},$$

where $Y = \sum_{i=1}^{n} X_i$. Run two Monte Carlo simulations (one for $n = 20$ and another for $n = 200$) that estimate the actual coverage for the confidence interval for the parameters $\lambda = 0.61$ and $\alpha = 0.05$. Use enough replications to achieve three digits of accuracy.

3.39 In the United States in the 1960s, 16 states owned the retail liquor stores while in 26 of the states the stores were privately owned. (Some states were omitted for technical reasons.) The prices for a fifth of Seagram 7 Crown Whisky in the two sets of states are given below in dollars. The prices for the 16 monopoly states are (Hand, et al., *Small Data Sets*, 1994, Chapman & Hall, page 318):

$$4.65, 4.55, 4.11, 4.15, 4.20, 4.55, 3.80, 4.00, 4.19, 4.75, 4.74, 4.50, 4.10, 4.00, 5.05, 4.20.$$

The prices for the 26 private-ownership states are:

$$4.82, 5.29, 4.89, 4.95, 4.55, 4.90, 5.25, 5.30, 4.29, 4.85, 4.54, 4.75, 4.85,$$

$$4.85, 4.50, 4.75, 4.79, 4.85, 4.79, 4.95, 4.95, 4.75, 5.20, 5.10, 4.80, 4.29.$$

(a) Give an exact two-sided 99% confidence interval for the difference between the population means of the two populations assuming that the two data sets are independent random samples from normal populations with equal population variances.

(b) Give an approximate two-sided 99% confidence interval for the difference between the population means of the two populations assuming that the two data sets are independent random samples from normal populations.

(c) What conclusion do you draw from your confidence intervals?

3.40 Let $X_1, X_2, \ldots, X_9$ be a random sample from a continuous population with population 75th percentile $x_{0.75}$. Find the coverage of the confidence interval $X_{(2)} < x_{0.75} < X_{(9)}$.

3.41 Let $X_1, X_2, \ldots, X_9$ be a random sample from a continuous population with population median $x_{0.5}$. Find the coverage of the confidence interval $X_{(1)} < x_{0.5} < X_{(9)}$.

3.42 Let $X_1, X_2, \ldots, X_6$ be a random sample from a continuous population with population median $x_{0.5}$. Find $P\left(X_{(3)} < x_{0.5} < X_{(5)}\right)$.

3.43 Consider the $n = 23$ ball bearing failure times (in millions of revolutions)

17.88	28.92	33.00	41.52	42.12	45.60	48.48	51.84
51.96	54.12	55.56	67.80	68.64	68.64	68.88	84.12
93.12	98.64	105.12	105.84	127.92	128.04	173.40	

from Lieblein, J. and Zelen, M. (1956), "Statistical Investigation of the Fatigue Life of Deep-Groove Ball Bearings," *Journal of Research of the National Bureau of Standards*, Volume 57, Number 5, pages 273–316.

(a) Find a point estimator of the population median.

(b) Find an approximate nonparametric two-sided 90% confidence interval for the population median.

(c) Find a point estimator of the 75th percentile of the population distribution.

(d) Find an approximate nonparametric two-sided 90% confidence interval for the 75th percentile of the population distribution.

(e) Use bootstrapping to compute an approximate two-sided 90% confidence interval for the population median and compare it to the confidence interval from part (b).

3.44 The interest in this question is a 90% parametric bootstrap confidence interval for the population median survival time using maximum likelihood estimation to estimate the exponential distribution for the $n = 9$ rat survival times (in days) from the control group:

$$10, 27, 30, 40, 46, 51, 52, 104, 146.$$

Example 3.19 illustrates the use of the parametric bootstrap.

(a) Use R to generate a 90% parametric bootstrap confidence interval for the population median survival time.

(b) Use APPL to confirm the upper and lower limits of the confidence interval.

3.45 Let $X_1, X_2, \ldots, X_n$ be a random sample from an exponential population with probability density function

$$f(x) = \frac{1}{\theta} e^{-x/\theta} \qquad x > 0,$$

where θ is a positive unknown parameter. Example 2.8 showed that the maximum likelihood estimator of θ is

$$\hat{\theta} = \bar{X}.$$

Example 3.7 showed that an exact two-sided $100(1-\alpha)\%$ confidence interval for θ is

$$\frac{2\sum_{i=1}^{n} X_i}{\chi^2_{2n,\alpha/2}} < \theta < \frac{2\sum_{i=1}^{n} X_i}{\chi^2_{2n,1-\alpha/2}}.$$

Calculate the exact values of the actual coverage of this confidence interval if the random sample values $X_1, X_2, \ldots, X_n$ are not drawn from an exponential population with population mean θ, but rather are drawn from a $N\left(\theta, \theta^2\right)$ population, where θ is a positive unknown parameter, when $\alpha = 0.05$ for sample sizes $n = 1, 2, \ldots, 10$.

3.46 Let $X_1, X_2, \ldots, X_n$ be a random sample from an exponential(λ) population, where λ is a positive unknown rate parameter. When all of the data values are observed, the sample is known as a *complete* data set. Situations arise in practice, however, in which only a lower bound is known for a particular data value. Such a data value is known as a *right censored* data value. Any data set containing one or more right censored data values is known as a *right censored data set*. Placing n light bulbs on test, for example, and ending the test when r (for a prescribed integer $r < n$) of the bulbs have failed constitutes a *Type II right censored data set*. In this case an exact confidence interval for the failure rate λ can be constructed based on the fact that $2\lambda T \sim \chi^2(2r)$, where T is the *total time on test* (that is, the total burning time for all bulbs over the duration of the test). On the other hand, placing n light bulbs on test and ending the test at time c (for a positive real number c) constitutes a *Type I right censored data set*. In this case, an approximate confidence interval for the failure rate λ can be constructed based on the fact that $2\lambda T$ is approximately $\chi^2(2r+1)$, where T is the total time on test and r is the number of observed failures.

(a) A life test is conducted that consists of placing n light bulbs with mutually independent exponential(λ) lifetimes on test, and terminating the test when the rth light bulb burns out. What is the expected time to complete this test?

(b) A life test is conducted that consists of placing n light bulbs on test and terminating the test at time $c > 0$. What is the expected number of light bulbs that will fail during this test?

(c) Calculate an exact two-sided 95% confidence interval for λ for the Type II right censored automotive switch failure data set (from Kapur, K.C., and Lamberson, L.R. (1977), *Reliability in Engineering Design*, John Wiley & Sons, Inc., pages 253–254) with $n = 15$ switches on test and $r = 5$ observed failures (measured in cycles) at

$$1410, 1872, 3138, 4218, 6971.$$

(d) Give the rationale associated with the result that $2\lambda T$ is approximately $\chi^2(2r+1)$ for a Type I right censored data set with n items on test, r observed failures, and total time on test T.

(e) Conduct a Monte Carlo simulation experiment that estimates the actual coverage of an approximate two-sided 95% confidence interval for a Type I right-censored data set with the arbitrary parameter settings $\alpha = 0.05$, $n = 15$, $\lambda = 1/4$, and $c = 3$. Use enough replications to achieve three digits of accuracy.

3.47 Consider the bootstrap estimate of the difference between the population median survival time (in days), m_X, for the $n = 7$ rats in the treatment group,

$$16, 23, 38, 94, 99, 141, 197,$$

and the population median survival time, m_Y for the $n = 9$ rats in the control group,

$$10, 27, 30, 40, 46, 51, 52, 104, 146.$$

In Example 3.20 an approximate 90% bootstrap confidence interval for the difference via bootstrapping was determined to be

$$-28 < m_X - m_Y < 95.$$

But if the R code in the example is run several times, different approximate confidence intervals result. Devise a procedure for determining the best of these confidence intervals.

3.48 Margarita collects the following data values:

$$3, 1, 2.$$

She then creates a *bootstrap sample* by randomly sampling $n = 3$ of the data values *with replacement*. What is the probability mass function of the sample range of the bootstrap sample?

3.49 Let $X_1, X_2, \ldots, X_n$ be a random sample from a population with probability density function

$$f(x) = \frac{\theta}{(1+x)^{\theta+1}} \qquad x > 0,$$

where θ is a positive unknown parameter. Calculate an asymptotically exact two-sided $100(1-\alpha)\%$ confidence interval for θ based on the asymptotic normality of the maximum likelihood estimator.

3.50 Let $X_1, X_2, \ldots, X_n$ be a random sample from a Poisson(λ) population, where λ is a positive unknown parameter. Example 2.9 showed that the maximum likelihood estimator of λ is

$$\hat{\lambda} = \bar{X}.$$

Three approximate confidence intervals for λ have been derived in this chapter. Example 3.12 derived an approximate two-sided $100(1-\alpha)\%$ confidence interval for λ, which is

$$\frac{1}{2n}\chi^2_{2Y,1-\alpha/2} < \lambda < \frac{1}{2n}\chi^2_{2(Y+1),\alpha/2},$$

where $Y = \sum_{i=1}^{n} X_i$. Example 3.26 derived an approximate two-sided $100(1-\alpha)\%$ confidence interval for λ based on the asymptotic normality of the maximum likelihood estimator, which is

$$\hat{\lambda} - z_{\alpha/2}\sqrt{\frac{\hat{\lambda}}{n}} < \lambda < \hat{\lambda} + z_{\alpha/2}\sqrt{\frac{\hat{\lambda}}{n}}.$$

Example 3.29 derived an approximate two-sided $100(1-\alpha)\%$ confidence interval for λ based on the fact that the asymptotic distribution of the likelihood ratio statistic is chi-square with one degree of freedom, which consists of all λ values satisfying

$$-2\left[\ln L(\lambda) - \ln L(\hat{\lambda})\right] < \chi^2_{1,\alpha},$$

where the log likelihood function is

$$\ln L(\lambda) = -n\lambda + \left(\sum_{i=1}^{n} X_i\right)\ln\lambda - \sum_{i=1}^{n}\ln(X_i!).$$

In a practical problem, which of these approximate confidence intervals should be used? Conduct a Monte Carlo simulation experiment to estimate the actual coverages of these three approximate confidence intervals for $\lambda = 3$, $\alpha = 0.05$ and sample sizes $n = 10, 20, \ldots, 50$. Comment on which of the three confidence intervals should be used based on the estimated actual coverages.

3.51 Let $X_1, X_2, \ldots, X_n$ be a random sample from a population with probability density function

$$f(x) = \frac{\theta}{2\sqrt{x}} e^{-\theta\sqrt{x}} \qquad x > 0,$$

where θ is a positive unknown parameter.

(a) Find the method of moments estimator of θ.

(b) Find the maximum likelihood estimator of θ.

(c) Find the method of moments and the maximum likelihood estimates of θ for the $n = 7$ rat survival times (in days) of the treatment group:

$$16, 23, 38, 94, 99, 141, 197.$$

(d) Find an asymptotically exact two-sided 95% confidence interval for θ based on the asymptotic normality of the maximum likelihood estimator of θ for the rat survival times from part (c).

(e) Find an asymptotically exact two-sided 95% confidence interval for θ based on the likelihood ratio statistic for the rat survival times from part (c).

(f) Which of the two asymptotically exact two-sided 95% confidence intervals is superior?

3.52 Let $X_1, X_2, \ldots, X_n$ be a random sample from a population with probability density function

$$f(x) = \theta x^{\theta-1} \qquad 0 < x < 1,$$

where θ is a positive unknown parameter.

(a) Show that the probability density function of $\sqrt{n}\left(\hat{\theta} - \theta\right)/\hat{\theta}$ is

$$f_{\sqrt{n}(\hat{\theta}-\theta)/\hat{\theta}}(x) = \frac{n^{n/2}\left(-x + \sqrt{n}\right)^{n-1} e^{\sqrt{n}(x-\sqrt{n})}}{(n-1)!} \qquad x < \sqrt{n},$$

where $\hat{\theta}$ is the maximum likelihood estimator.

(b) For the asymptotically exact 95% confidence interval for θ

$$\hat{\theta} - 1.96 \cdot \frac{\hat{\theta}}{\sqrt{n}} < \theta < \hat{\theta} + 1.96 \cdot \frac{\hat{\theta}}{\sqrt{n}},$$

find the actual coverage for $n = 6$ by integrating the probability density function from part (a).

(c) Conduct a Monte Carlo simulation experiment to provide convincing numerical evidence that the actual coverage you computed in part (b) for $n = 6$ is correct.

3.53 Let $X_1, X_2, \ldots, X_n$ be a random sample from a population with probability density function

$$f(x) = \sqrt{\frac{1}{2\pi x^3}}\, e^{-(x-\theta)^2/(2x\theta^2)} \qquad x > 0,$$

where θ is a positive unknown parameter. This population distribution is a special case of the inverse Gaussian distribution. Calculate an asymptotically exact two-sided $100(1-\alpha)\%$ confidence interval for θ based on the asymptotic normality of the maximum likelihood estimator. *Hint*: The population mean is $E[X] = \theta$.

3.54 Let $X_1, X_2, \ldots, X_n$ be a random sample from an exponential population with positive unknown population mean θ.

(a) Show that the probability density function of $\sqrt{n}\left(\hat{\theta}-\theta\right)/\hat{\theta}$ is

$$f_{\sqrt{n}(\hat{\theta}-\theta)/\hat{\theta}}(x) = \frac{n^{3n/2}e^{n^{3/2}/(x-\sqrt{n})}}{(n-1)!\left(-x+\sqrt{n}\right)^{n+1}} \qquad x < \sqrt{n},$$

where $\hat{\theta}$ is the maximum likelihood estimator.

(b) For the asymptotically exact two-sided 95% confidence interval for θ

$$\hat{\theta} - 1.96 \cdot \frac{\hat{\theta}}{\sqrt{n}} < \theta < \hat{\theta} + 1.96 \cdot \frac{\hat{\theta}}{\sqrt{n}},$$

find the actual coverage for $n = 8$ by integrating the probability density function from part (a).

(c) Conduct a Monte Carlo simulation experiment to provide convincing numerical evidence that the actual coverage you computed in part (b) for $n = 8$ is correct.

3.55 Let $X_1, X_2, \ldots, X_n$ be a random sample from a population with probability density function

$$f(x) = \frac{\theta}{x^{\theta+1}} \qquad x > 1,$$

where θ is a positive unknown parameter.

(a) Find the maximum likelihood estimator of θ.

(b) Find the maximum likelihood estimate of θ for the $n = 7$ rat survival times (in days) of the treatment group:

$$16, 23, 38, 94, 99, 141, 197.$$

(c) Find an asymptotically exact two-sided 95% confidence interval for θ based on the likelihood ratio statistic for the rat survival times from part (b).

3.56 Let $X_1, X_2, \ldots, X_n$ be a random sample from a population with probability density function

$$f(x) = \frac{\theta}{(1+\theta x)^2} \qquad x > 0,$$

where θ is a positive unknown parameter. This is a special case of the log logistic distribution.

(a) Find the maximum likelihood estimator of θ. *Hint*: The maximum likelihood estimator cannot be expressed in closed form.

(b) Find the maximum likelihood estimate of θ for the $n = 7$ rat survival times (in days) of the treatment group:

$$16, 23, 38, 94, 99, 141, 197.$$

(c) Find an asymptotically exact two-sided 95% confidence interval for θ based on the likelihood ratio statistic for the rat survival times from part (b).

3.57 Let $X_1, X_2, \ldots, X_n$ be a random sample from a continuous population. Another independent observation, X_{n+1}, is collected from the same population. Calculate

$$P\left(X_{n+1} > \max\left\{X_1, X_2, \ldots, X_n\right\}\right).$$

3.58 Let $X_1, X_2, \ldots, X_n$ be a random sample from an exponential population with a positive unknown population mean θ. Conduct a Monte Carlo simulation experiment that provides convincing numerical evidence that the $100(1-\alpha)\%$ prediction interval for X_{n+1}

$$\frac{\bar{X}}{F_{2n,2,\alpha/2}} < X_{n+1} < \frac{\bar{X}}{F_{2n,2,1-\alpha/2}}$$

is an *exact* prediction interval for the arbitrary parameter settings $n = 11$, $\alpha = 0.05$, and $\theta = 19$.

3.59 Let $X_1, X_2, \ldots, X_n$ be a random sample from an exponential population with a positive unknown population mean θ. Find the asymptotic expected width of the exact two-sided $100(1-\alpha)\%$ prediction interval for X_{n+1},

$$\frac{\bar{X}}{F_{2n,2,\alpha/2}} < X_{n+1} < \frac{\bar{X}}{F_{2n,2,1-\alpha/2}},$$

for the arbitrary parameter setting $\alpha = 0.1$.

3.60 Let $X_1, X_2, \ldots, X_n$ be a random sample from a $N\left(\mu, \sigma^2\right)$ population, where μ and $\sigma^2 > 0$ are unknown parameters. Show that for a single future observation X_{n+1} from the same population,

$$\frac{X_{n+1} - \bar{X}}{S\sqrt{1+1/n}} \sim t\,(n-1).$$

3.61 Let $X_1, X_2, \ldots, X_n$ be a random sample from a $N\left(\mu, \sigma^2\right)$ population, where μ and $\sigma^2 > 0$ are unknown parameters. The experimental values of the random variables $X_1, X_2, \ldots, X_n$ are denoted by $x_1, x_2, \ldots, x_n$. The sample mean and sample standard deviation of the experimental values are denoted by $\bar{x}$ and s. Now consider a random sample of m future observations $X_{n+1}, X_{n+2}, \ldots, X_{n+m}$ from the same population. Let $\bar{Y}$ denote the sample mean of the future observations. Derive an exact two-sided $100(1-\alpha)\%$ prediction interval for $\bar{Y}$.

3.62 In the development of a tolerance interval, we needed the distribution of the sample range of n mutually independent $U(0, 1)$ random variables. This exercise provides that derivation. Let $X_1, X_2, \ldots, X_n$ be mutually independent $U(0, 1)$ random variables. Find the distribution of the sample range $X_{(n)} - X_{(1)}$.

3.63 Let $X_1, X_2, \ldots, X_n$ be a random sample from a continuous population with cumulative distribution function $F(x)$. Let $X_{(1)}, X_{(2)}, \ldots, X_{(n)}$ be the associated order statistics. Find the smallest value of n for which

$$P\left(F\left(X_{(n)}\right) - F\left(X_{(1)}\right) \geq 0.8\right)$$

is at least 0.9.

3.64 Let $X_1, X_2, \ldots, X_{100}$ be a random sample from a continuous population with cumulative distribution function $F(x)$. Let $X_{(1)}, X_{(2)}, \ldots, X_{(100)}$ be the associated order statistics, and let $x_{(1)}, x_{(2)}, \ldots, x_{(100)}$ be their experimental values. The interval $\left(x_{(9)}, x_{(93)}\right)$ is used as a tolerance interval to capture at least 80% of the distribution.

(a) Find the coverage of this tolerance interval.

(b) Conduct a Monte Carlo simulation experiment using a continuous probability distribution of your choice that provides convincing numerical evidence that your coverage from part (a) is correct.

3.65 Let $X_1, X_2, \ldots, X_n$ be a random sample from a Bernoulli population distribution with random parameter Θ. Assuming that Θ has a prior beta(2, 2) distribution, $n = 100$, and $\sum_{i=1}^{100} x_i = 54$, find the Bayesian point estimate of θ and an exact 95% Bayesian credible interval for θ.

3.66 Let $X_1, X_2, \ldots, X_n$ be a random sample from a Geometric(P) distribution, where P is a random variable with a noninformative uniform prior distribution. Find a Bayesian point estimate for p and a $100(1-\alpha)\%$ Bayesian credible interval for p.

3.67 Consider a parallel system of two independent components, each with an exponential(λ) time to failure, where λ is a positive unknown failure rate. You collect n *system* failure times $X_1, X_2, \ldots, X_n$, and would like to estimate the *component* failure rate λ via maximum likelihood.

(a) Find the log likelihood function and score statistic.

(b) Find the maximum likelihood estimate $\hat{\lambda}$ for the $n = 3$ system failure times

$$2.5 \qquad 1.2 \qquad 6.3.$$

(c) Find an approximate two-sided 95% confidence interval for λ based on the observed information matrix for the three data values from part (b).

3.68 Let $X_1, X_2, \ldots, X_n$ be a random sample from an exponential population with random rate parameter Λ. Assume that Λ has a prior gamma(α, β) distribution.

(a) What is the posterior probability density function of Λ?

(b) What is the population mean of the posterior distribution of Λ, that is, what is the Bayesian point estimator of λ?

(c) Calculate the Bayesian point estimator of λ and construct an exact two-sided 95% Bayesian credible interval for λ for the $n = 27$ times between failures (in hours) for an air conditioning system on a Boeing 720 collected by Frank Proschan (Proschan, F. (1963), "Theoretical Explanation of Observed Decreasing Failure Rate," *Technometrics*, Volume 5, Number 3, pages 375–383):

$$97, 51, 11, 4, 141, 18, 142, 68, 77, 80, 1, 16, 106, 206,$$

$$82, 54, 31, 216, 46, 111, 39, 63, 18, 191, 18, 163, 24,$$

assuming that the prior distribution of Λ is gamma(200, 2).

3.69 Janice collects 12 data values $x_1, x_2, \ldots, x_{12}$. She performs an exploratory data analysis (for example, plotting a histogram of the data values) and concludes that the normal distribution is a reasonable model for her data. She then calculates a point estimate and an exact two-sided confidence interval for the population mean μ. She is unhappy about the large width of the confidence interval due to the small sample size of $n = 12$. To tighten the confidence interval, she devises the following scheme.

- She fits the data values to a normal model using maximum likelihood estimation.
- She generates 100,000 observations from the fitted normal distribution, and then calculates an improved (much narrower) confidence interval from these simulated data values.

Choose one of the following options and justify your choice.

(a) I think that this is a most excellent and clever scheme.

(b) I think that this scheme has some pros and cons.

(c) I think that this is a very bad scheme.

3.70 Consider the $n = 23$ ball bearing failure times given by (in millions of revolutions)

17.88	28.92	33.00	41.52	42.12	45.60	48.48	51.84
51.96	54.12	55.56	67.80	68.64	68.64	68.88	84.12
93.12	98.64	105.12	105.84	127.92	128.04	173.40	

from Lieblein, J. and Zelen, M. (1956), "Statistical Investigation of the Fatigue Life of Deep-Groove Ball Bearings," *Journal of Research of the National Bureau of Standards*, Volume 57, Number 5, pages 273–316. Use the `crplot` function in the `conf` package in R to find point estimates and a 95% confidence region for the point estimates based on the likelihood ratio statistic for the following two-parameter probability distributions.

(a) Weibull(λ, κ).

(b) gamma(λ, κ).

(c) log normal(μ, σ).

(d) log logistic(λ, κ).

(e) inverse Gaussian(λ, μ).

Give any insights concerning the ball bearings that can be gleaned from the confidence regions.

Chapter 4

Hypothesis Testing

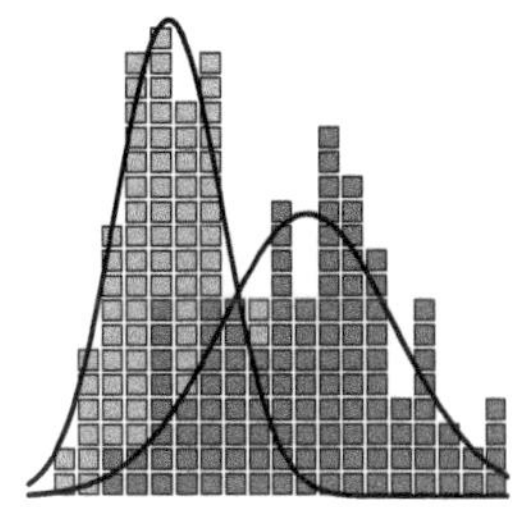

Obtaining a point estimate of an unknown parameter using the techniques in Chapter 2, and an associated interval estimate using the techniques in Chapter 3, is one type of statistical inference. Using these methods, the experimenter obtains a best guess for the unknown parameter (the point estimate) and a measure of the precision of the point estimate (the interval estimate). This approach is adequate in some settings, but inadequate in other settings.

There is a second broad type of statistical inference known as *hypothesis testing* which is introduced in this chapter. Hypothesis testing is a formal approach that has been devised by statisticians for testing whether data supports or refutes a conjecture (that is, a hypothesis) concerning an unknown population parameter. Over the years it has become the de facto standard method for drawing conclusions from a data set. Hypothesis testing contains more elements than point and interval estimation; these elements will be defined and illustrated in Section 4.1.

4.1 Elements of Hypothesis Testing

Hypothesis testing is used to answer a question that is posed concerning an unknown parameter. We begin with the definition of a hypothesis.

Definition 4.1 A *hypothesis* is a statement concerning an unknown parameter.

As a simple illustration, consider a coin with probability p of turning up heads when flipped, where p is an unknown but fixed parameter satisfying $0 < p < 1$. The hypothesis associated with the coin being fair is

$$p = \frac{1}{2}.$$

The hypothesis associated with the coin being biased is

$$p \neq \frac{1}{2}.$$

The hypothesis associated with the coin being biased in favor of turning up heads is

$$p > \frac{1}{2}.$$

These hypotheses could be tested with a coin flipping experiment.

Now consider a somewhat more complex example. A researcher has developed a new drug that allegedly extends the average survival time of a patient with liver cancer more than the drug currently in use. Let μ_X be the population mean survival time of a liver cancer patient who uses the current drug. Let μ_Y be the population mean survival time of a liver cancer patient who uses the new drug. The hypothesis that the two drugs perform equally well in terms of the population mean survival time of a liver cancer patient is

$$\mu_X = \mu_Y.$$

The hypothesis that the new drug performs better than the current drug in terms of the population mean survival time is

$$\mu_X < \mu_Y.$$

This second hypothesis is often known as the *research worker's hypothesis* because this is the hypothesis that the experimenter is trying to show. Hypothesis testing parallels the scientific method. The two pillars of the scientific method are theory and experimentation. Formulating a hypothesis, as we have in the liver cancer drug example, corresponds to the development of a theory. In order to test the hypothesis, or theory, an experiment is conducted. In the case of the two drugs, the survival times $X_1, X_2, \ldots, X_n$ of n liver cancer patients taking the current drug are collected; the survival times $Y_1, Y_2, \ldots, Y_m$ of m liver cancer patients taking the new drug are also collected. Care should be taken in the way that the patients are placed into the two groups so that the survival times in the two groups are not influenced by other factors such as age, gender, or the progress of the disease. This process is known to statisticians as *stratified sampling*. In addition, hypothesis testing in medicine often uses a *double-blind experiment* in which both the patient and the physician are unaware of whether the current drug or the new drug is being used. This procedure is used to ward off the *placebo effect*, which can bias the results of the experiment. A hypothesis test can then be conducted to assess the effectiveness of the new drug relative to the effectiveness of the current drug.

The time-tested approach in hypothesis testing is to test one hypothesis, known as the null hypothesis, versus a second hypothesis, known as the alternative hypothesis. These two competing hypotheses are defined next.

Definition 4.2 A *hypothesis test* is a statistical inference technique for deciding between two competing propositions concerning an unknown parameter. The first proposition, known as the *null hypothesis*, is denoted by H_0. The second proposition, known as the *alternative hypothesis*, is denoted by H_1. The decision between H_0 and H_1 is based on a set of data values and is stated as either "reject H_0" or "fail to reject H_0."

Stating the null and alternative hypothesis is one of the first steps in conducting a hypothesis test. Some further details associated with H_0 and H_1 are given below.

- Some authors use the notation H_a or H_A for the alternative hypothesis. The more traditional H_1 is used here because of some helpful notational advantages.
- The null hypothesis has an inherently different nature than the alternative hypothesis because it is assumed to be true until enough evidence in the data causes it to be rejected. The null hypothesis is hardly ever true in any practical problem; it is almost always an approximation.
- The null hypothesis H_0 tends to be "no effect," "no difference," or "no change." The alternative hypothesis, on the other hand, is often the hoped-for conclusion of the test.
- The language used in the decision associated with the conclusion of the hypothesis test, that is, "reject H_0" or "fail to reject H_0," is important. There is a subtle but important difference

between the conclusions "fail to reject H_0" and "accept H_0." The words "fail to reject H_0" imply that there is not enough statistical evidence in the sample to reject H_0 in favor of H_1. The null hypothesis H_0 has not been confirmed—there just simply isn't enough evidence in the data to reject it.

- Some authors refer to H_0 and H_1 as "statistical hypotheses."
- In the examples considered in this chapter, the data values will always be a random sample.

The first example of a hypothesis test is used to decide whether an unknown parameter θ assumes one value or another. To keep the mathematics simple, only a single observation, denoted by X, is collected to decide between the null and alternative hypotheses.

Example 4.1 A single observation X is collected to test the hypothesis

$$H_0 : \theta = 1$$

versus

$$H_1 : \theta = 3,$$

where θ is a positive unknown parameter from the Rayleigh population distribution described by the probability density function

$$f(x) = 2\theta^{-2} x e^{-(x/\theta)^2} \qquad x > 0.$$

Conduct a hypothesis test associated with the data value $x = 2.2$.

The null and alternative hypotheses completely define the population probability distribution of X. The population probability density functions associated with H_0 and H_1 are graphed in Figure 4.1. Let x be a realization of the random variable X. Because of the support of the population probability distribution, x will always be positive. It is clear from Figure 4.1 that a smaller value of x is evidence in favor of H_0 and a larger value of x is evidence in favor of H_1. Now consider the particular value of x collected in this case, that is, $x = 2.2$. The data value is out in the far right-hand tail of the

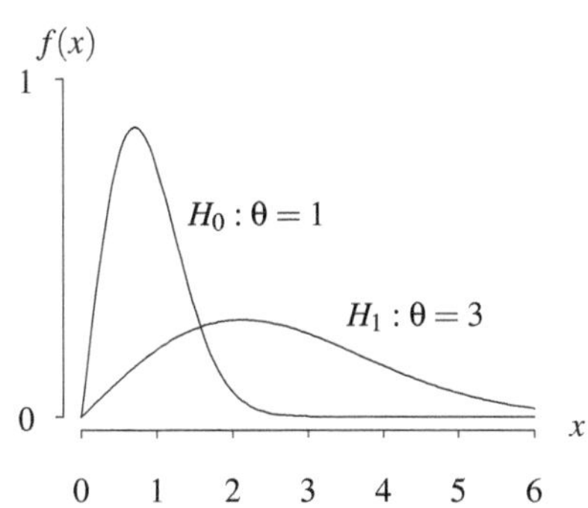

Figure 4.1: Population probability density functions for $\theta = 1$ and $\theta = 3$.

probability density function associated with $\theta = 1$, but in the center of the probability density function associated with $\theta = 3$, so it is reasonable to conclude that H_0 should be rejected in this case. The probability distribution associated with $\theta = 1$ is unlikely to have produced $x = 2.2$. This conclusion was drawn in a rather intuitive and unscientific manner by appealing to Figure 4.1, so the next example establishes a cutoff value for x for deciding between H_0 and H_1.

The null and alternative hypotheses in the previous example completely defined the probability density function of X. A hypothesis that completely defines a probability distribution is known as a simple hypothesis, which is defined next.

Definition 4.3 A hypothesis that completely specifies a probability distribution is called a *simple hypothesis*. A hypothesis that does not completely specify a probability distribution is called a *composite hypothesis*.

The next example considers a simple null hypothesis and a simple alternative hypothesis to test a hypothesis concerning the parameter θ from a special case of the beta distribution.

Example 4.2 A single observation X is collected to test the simple null hypothesis

$$H_0 : \theta = 1/2$$

versus the simple alternative hypothesis

$$H_1 : \theta = 3,$$

where θ is a positive unknown parameter from the continuous population distribution described by the probability density function

$$f(x) = \theta x^{\theta - 1} \qquad 0 < x < 1.$$

Analyze an appropriate decision rule based on the single data value.

The null and alternative hypotheses are both simple because they completely define the population probability distribution of X. The population probability density functions associated with H_0 and H_1 are graphed in Figure 4.2. Let x be a realization of the random variable X. Because of the support of the population probability distribution, x will always fall in the interval $0 < x < 1$. It is clear from Figure 4.2 that a smaller value of x is evidence in favor of H_0 and a larger value of x is evidence in favor of H_1. Using terminology that will be defined formally subsequently, x is known as the *test statistic*. Since we do not have any criteria yet for selecting which values of x correspond to the two decisions, let's arbitrarily choose the cut-off value of $c = 0.64$. If $x > 0.64$, then we reject H_0; if $x \le 0.64$, then we fail to reject H_0. Using terminology that will be defined formally subsequently, $c = 0.64$ is known as the *critical value* for the hypothesis test and $\{x \mid x > 0.64\}$ is known as the *critical region* for the hypothesis test. It is clear from Figure 4.2 that this is not a perfect hypothesis test in the sense that it will not always provide the correct conclusion with respect to which of the two distributions produced the data value. In that light, it is useful to calculate the probability that the test will draw the wrong conclusion. There are two types of errors that can be made. The first, known as a *Type I error*, corresponds to rejecting H_0 when H_0 is true. The second, known as

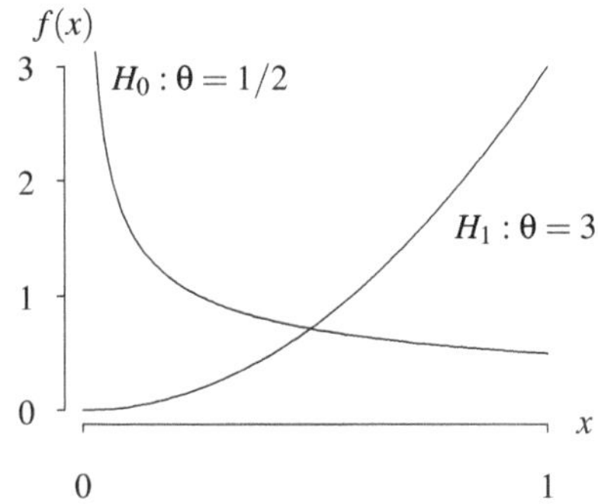

Figure 4.2: Population probability density functions for $\theta = 1/2$ and $\theta = 3$.

a *Type II error*, corresponds to failing to reject H_0 when H_1 is true. For this particular hypothesis test, the probability of a Type I error is

$$\begin{aligned}
\alpha &= P(\text{Type I error}) \\
&= P(\text{reject } H_0 \,|\, H_0 \text{ true}) \\
&= P(X > 0.64 \,|\, \theta = 1/2) \\
&= \int_{0.64}^{1} \frac{1}{2} x^{-1/2}\, dx \\
&= \left[\sqrt{x}\,\right]_{0.64}^{1} \\
&= 0.2,
\end{aligned}$$

as illustrated in Figure 4.3. The curve is the probability density function of X when H_0 is true, that is, when $\theta = 1/2$. The shaded area is the probability that X exceeds 0.64, which is the probability of a Type I error. The probability of a Type II error, on the other hand, is

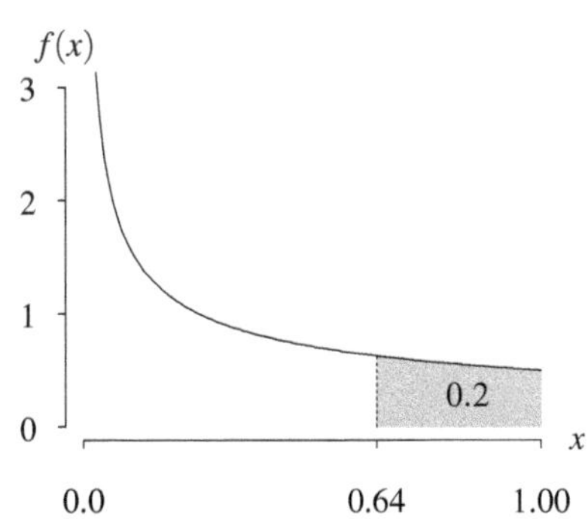

Figure 4.3: The probability of a Type I error $\alpha = P(X > 0.64 \,|\, \theta = 1/2) = 0.2$.

$$\begin{aligned}
\beta &= P(\text{Type II error}) \\
&= P(\text{fail to reject } H_0 \,|\, H_1 \text{ true}) \\
&= P(X \le 0.64 \,|\, \theta = 3) \\
&= \int_0^{0.64} 3x^2\,dx \\
&= \left[x^3\right]_0^{0.64} \\
&\cong 0.2621,
\end{aligned}$$

as illustrated in Figure 4.4. The curve is the probability density function of X when H_1 is true, that is, when $\theta = 3$. The shaded area is the probability that X is less than or equal to 0.64, which is the probability of a Type II error.

It is worth considering the impact of the critical value $c = 0.64$ on α and β. What is the impact of increasing c? Appealing to Figures 4.3 and 4.4, this would result in a smaller α, but a larger β. On the flip side, what is the impact of decreasing c? Again appealing to Figures 4.2 and 4.3, this would result in a larger α, but a smaller β. This see-saw relationship between α and β is typically a characteristic of hypothesis testing.

What if the α and β values calculated in this example are unacceptably high? As was the case with confidence intervals that were unacceptably wide, more data values could be collected in order to bring both of these probabilities down.

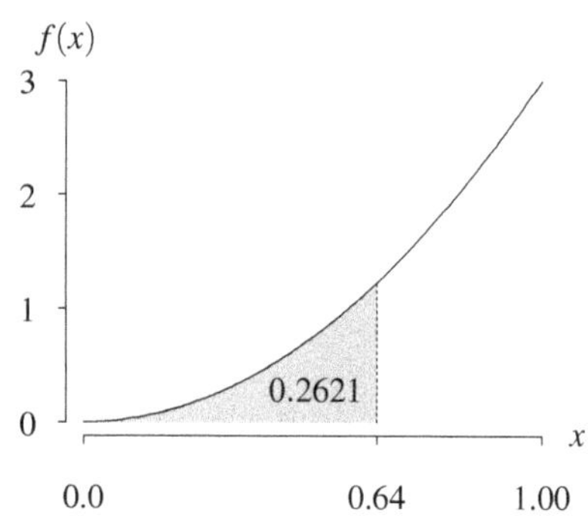

Figure 4.4: The probability of a Type II error is $\beta = P(X \le 0.64 \,|\, \theta = 3) \cong 0.2621$.

Several new terms used in the previous example are now formally defined. A hypothesis test concerns an unknown parameter θ defined on the parameter space Ω. In the previous example, the parameter space was

$$\Omega = \{\theta \,|\, \theta > 0\}.$$

The null and alternative hypotheses can be written generically as

$$H_0 : \theta \in \Omega_0$$

and

$$H_1 : \theta \in \Omega_1,$$

where Ω_0 and Ω_1 are subsets of Ω. In the previous example, $\Omega_0 = \{\theta \mid \theta = 1/2\}$ and $\Omega_1 = \{\theta \mid \theta = 3\}$ are sets containing just a single point because the null hypothesis and the alternative hypothesis are both simple hypotheses. The set of all data values that lead to rejecting H_0 is known as the critical region, which is defined next.

Definition 4.4 For a hypothesis test of a null hypothesis H_0 versus an alternative hypothesis H_1 based on the data values $X_1, X_2, \ldots, X_n$, the set of data values that lead to rejecting H_0 is known as the *critical region* C.

The critical region is sometimes referred to as the *rejection region*. In the previous example, the critical region was $C = \{X \mid X > 0.64\}$. So in general, the decision rule for the test is:

- if $(X_1, X_2, \ldots, X_n) \in C$, then reject H_0,
- if $(X_1, X_2, \ldots, X_n) \notin C$, then fail to reject H_0.

The data values $X_1, X_2, \ldots, X_n$ assume values in an n-dimensional support, so specifying C can be tedious. The usual approach in hypothesis testing is to define a test statistic that is a function of the data values and an associated critical value to implicitly define the critical region C.

Definition 4.5 A *test statistic* is a real-valued function of the data values $X_1, X_2, \ldots, X_n$ that is used in a hypothesis test. The *critical value(s)* associated with the test statistic are cutoff value(s) that determine the boundary of the critical region.

The test statistic in the previous example was x, an experimental value of the random variable X, and the critical value was $c = 0.64$. Determining an appropriate test statistic for a hypothesis test is a nontrivial matter that will be postponed for later in the chapter. For the next few examples that follow, the test statistic will be given in the problem statement. The two types of errors that were introduced in the previous example, along with their associated probabilities, are defined next.

Definition 4.6 Consider a hypothesis test of a simple null hypothesis H_0 versus a simple alternative hypothesis H_1. Rejecting H_0 when H_0 is true is a *Type I error*. Failing to reject H_0 when H_1 is true is a *Type II error*. The probability of a Type I error, also known as the *significance level*, is

$$\alpha = P(\text{Type I error}) = P(\text{reject } H_0 \mid H_0 \text{ true}).$$

The probability of a Type II error is

$$\beta = P(\text{Type II error}) = P(\text{fail to reject } H_0 \mid H_1 \text{ true}).$$

Since α and β are probabilities that are associated with errors, every effort should be made to keep them as small as possible subject to the cost and time constraints associated with data collection. The pre-specified, constant significance level α goes by several aliases:

- level of significance,
- size of the critical region,
- power of the test when H_0 is true,
- the probability of committing a Type I error.

As was the case with confidence intervals, popular values of α are 0.1, 0.05, 0.01. The probability $1-\beta$, which represents

$$1-\beta = P(\text{reject } H_0 \mid H_1 \text{ true}),$$

is known as the *power* of the test. Table 4.1 shows the four outcomes to any hypothesis test. The columns of the 2×2 table denote the true state of nature with respect to H_0 and H_1; the rows of the table denote the decision to the hypothesis test that is made by the experimenter. The associated probabilities are given in parentheses. Two of the four cases correspond to a correct decision, and the other two correspond to an error.

		Reality	
		H_0 true	H_1 true
Decision	Fail to reject H_0	Correct $(1-\alpha)$	Type II error (β)
	Reject H_0	Type I error (α)	Correct $(1-\beta)$

Table 4.1: Four outcomes of a hypothesis test.

Most of the elements of a hypothesis test have now been defined. This leads us to the following algorithm, or method, for conducting a hypothesis test via the following steps.

1. Develop a hunch, or theory, concerning a problem of interest.
2. Translate the theory into a question concerning an unknown parameter θ.
3. State the null hypothesis H_0 in terms of θ.
4. State the alternative hypothesis H_1 in terms of θ.
5. Determine the significance level α of the test.
6. Determine the sample size n.
7. Determine a test statistic, and derive its probability distribution when H_0 is true.
8. Determine the associated critical value(s).
9. Determine the associated critical region.
10. Collect the data values $x_1, x_2, \ldots, x_n$.
11. Compute the observed value of the test statistic.
12. Decide between H_0 and H_1 based on whether the test statistic falls in the critical region.

Many of the elements of a statistical test, for example the null hypothesis, the alternative hypothesis, the significance level, the sample size, and the test statistic, are all defined *before* any data is collected. To define these elements after the data is collected could result in manipulation of the final conclusion. Looking at the data prior to formulating the steps in a hypothesis test is known as "data snooping" and is considered a violation of the process.

The choice of the level of significance α is important in a hypothesis test because it determines the critical value c, which is the key value determining the final outcome of the hypothesis test. The value of the level of significance α should reflect the consequences of rejecting H_0 when H_0 is true. The more dire the consequences (for example, loss of life), the smaller α should be.

The final step in the process will result in "reject H_0" or "fail to reject H_0," but should also be translated back into the language associated with the problem of interest from the first step. For example, a hypothesis test might result in the conclusion that "the coin is not fair" or "the new drug prolongs the population mean survival time for liver cancer patients."

In the previous example, we defined a test statistic X, a critical value $c = 0.64$, and an associated critical region $C = \{X \mid X > 0.64\}$, which led to the calculation of α, namely $\alpha = 0.2$. In the next example, we collect $n = 2$ data values and set α initially, as in the hypothesis testing algorithm, to determine a critical value c.

Example 4.3 Let X_1 and X_2 be a random sample of size $n = 2$ from a $N(\theta, 1)$ population, where the population mean θ is an unknown parameter. Use the test statistic $X_1 + X_2$ to test the simple null hypothesis

$$H_0 : \theta = 5$$

against the simple alternative hypothesis

$$H_1 : \theta = 3$$

at $\alpha = 0.1$. Find c, the critical value, and β, the probability of a Type II error.

The population variance has been set to the constant $\sigma^2 = 1$ in the problem statement so that there is just one unknown parameter θ. Since X_1 and X_2 are independent $N(\theta, 1)$ random variables, small values of the test statistic $X_1 + X_2$ are evidence in favor of rejecting H_0. The decision rule for the hypothesis test is to reject H_0 if $X_1 + X_2 < c$. In other words, the critical region is $C = \{(X_1, X_2) \mid X_1 + X_2 < c\}$. The first step is to find the critical value c that provides a threshold for the value of $X_1 + X_2$ to reject H_0. Since the sum of two independent normal random variables is also normally distributed,

- $X_1 + X_2 \sim N(10, 2)$ under H_0, and
- $X_1 + X_2 \sim N(6, 2)$ under H_1.

The phrase "under H_0" is equivalent to "when H_0 is true." Thus, the critical value c is the 10th percentile of a $N(10, 2)$ random variable because $\alpha = 0.1$. This can be calculated with the R command

```
qnorm(0.1, 10, sqrt(2))
```

which returns $c = 8.1876$. So the decision rule for the hypothesis test is to reject H_0 whenever $X_1 + X_2 < 8.1876$. The geometry behind this choice of critical value is shown in Figure 4.5. The curve is the probability density function of a $N(10, 2)$ random variable, which is the probability distribution of $X_1 + X_2$ under H_0. The shaded area under the curve to the left of the critical value $c = 8.1876$ is the level of significance $\alpha = 0.1$. We now turn to the calculation of β. Using the definition of a Type II error,

$$\begin{aligned}
\beta &= P(\text{Type II error}) \\
&= P(\text{fail to reject } H_0 \mid H_1 \text{ true}) \\
&= P(X_1 + X_2 \geq 8.1876 \mid \theta = 3) \\
&= 0.0609.
\end{aligned}$$

This probability was calculated with the following R statements.

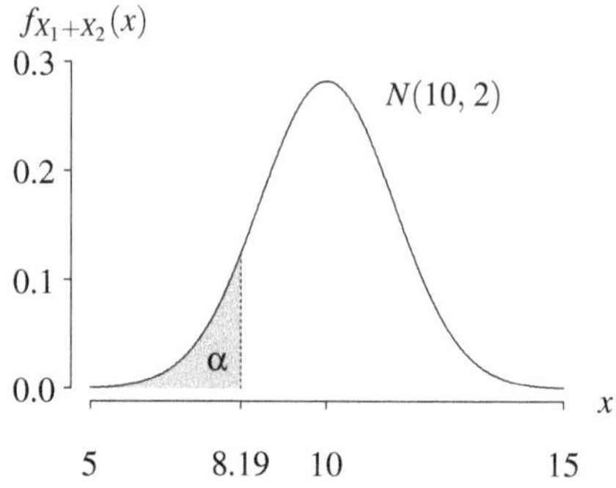

Figure 4.5: The sampling distribution of $X_1 + X_2$ when $\theta = 5$ with $\alpha = 0.1$.

```
critical.value = qnorm(0.1, 10, sqrt(2))
1 - pnorm(critical.value, 6, sqrt(2))
```

The probability of a Type I error α and the probability of a Type II error β are shown in Figure 4.6. The area under the $N(10, 2)$ probability density function to the left of $c = 8.1876$ is $\alpha = P(\text{Type I error}) = 0.1$. The area under the $N(6, 2)$ probability density function to the right of $c = 8.1876$ is $\beta = P(\text{Type II error}) = 0.0609$. The see-saw relationship between α and β is again apparent from this figure as c moves along the horizontal axis. As the critical value $c = 8.1876$ moves to the left, α decreases and β increases. As the critical value $c = 8.1876$ moves to the right, α increases and β decreases. The only way to decrease both α and β simultaneously is to increase the sample size n.

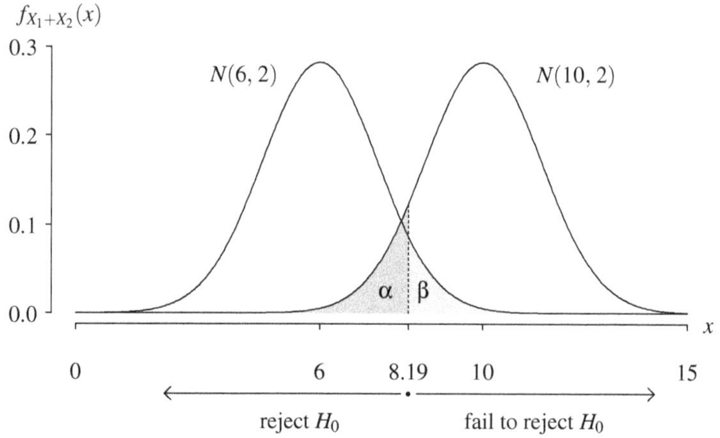

Figure 4.6: The sampling distribution of $X_1 + X_2$ when $\theta = 3$ and $\theta = 5$.

The next example also considers a sample of size $n = 2$, but this time drawn from a $U(0, \theta)$ population rather than a $N(\theta, 1)$ population.

Example 4.4 Let X_1 and X_2 be a random sample of size $n = 2$ from a $U(0, \theta)$ population, where θ is a positive unknown parameter. Chico uses the test statistic $X_1 + X_2$ to test the simple null hypothesis

$$H_0 : \theta = 5$$

against the simple alternative hypothesis

$$H_1 : \theta = 2$$

at $\alpha = 0.08$. Find the critical value c and the probability of a Type II error β.

Since X_1 and X_2 are independent $U(0, \theta)$ random variables, small values of the test statistic $X_1 + X_2$ are evidence in favor of rejecting H_0. The decision rule for the hypothesis test is to reject H_0 when $X_1 + X_2 < c$. In other words, the critical region is $C = \{(X_1, X_2) \,|\, X_1 + X_2 < c\}$. The first step is to find the critical value c that provides a threshold for the value of $X_1 + X_2$ to reject H_0. The marginal probability density functions for $X_1 \sim U(0, \theta)$ and $X_2 \sim U(0, \theta)$ are

$$f_{X_1}(x_1) = \frac{1}{\theta} \qquad 0 < x_1 < \theta \qquad \text{and} \qquad f_{X_2}(x_2) = \frac{1}{\theta} \qquad 0 < x_2 < \theta.$$

Since X_1 and X_2 are independent, their joint probability density function is

$$f_{X_1,X_2}(x_1, x_2) = \frac{1}{\theta^2} \qquad 0 < x_1 < \theta, 0 < x_2 < \theta.$$

To determine the value of c associated with $\alpha = 0.08$,

$$\begin{aligned}
\alpha &= P(\text{Type I error}) \\
&= P(\text{reject } H_0 \,|\, H_0 \text{ true}) \\
&= P(X_1 + X_2 < c \,|\, \theta = 5) \\
&= \int_0^c \int_0^{c-x_1} \frac{1}{25}\, dx_2\, dx_1 \\
&= \frac{c^2}{50}.
\end{aligned}$$

Since α was set to 0.08 in the problem statement, solving $c^2/50 = 0.08$ for c results in the critical value $c = 2$. So Chico should reject H_0 when the sum of the data values is less than $c = 2$. Under H_0, the support for X_1 and X_2 is a 5×5 square, and the rejection region is the shaded triangle in the lower-left portion of that square, as illustrated in Figure 4.7.

The next step is to determine the probability of a Type II error, which is

$$\begin{aligned}
\beta &= P(\text{Type II error}) \\
&= P(\text{fail to reject } H_0 \,|\, H_1 \text{ true}) \\
&= P(X_1 + X_2 \geq 2 \,|\, \theta = 2) \\
&= \int_0^2 \int_{2-x_1}^2 \frac{1}{4}\, dx_2\, dx_1 \\
&= \frac{1}{2}.
\end{aligned}$$

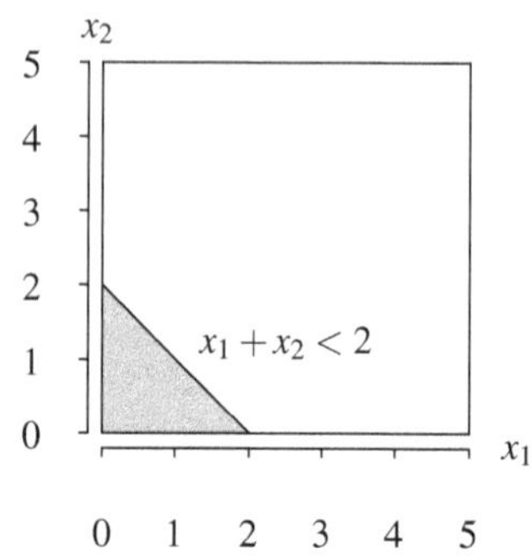

Figure 4.7: Critical region under H_0 ($\theta = 5$) for the test statistic $X_1 + X_2$.

The region associated with failing to reject H_0 under H_1, that is, when $\theta = 2$, is the shaded triangle in Figure 4.8. Instead of calculating the values of the critical value c and the probability of a Type II error β using calculus as has been done here, the following APPL code can also be used to perform these calculations.

```
n := 2;
a := 0.08;
X := UniformRV(0, 5);
Y := ConvolutionIID(X, n);
c := IDF(Y, a);
X := UniformRV(0, 2);
Y := ConvolutionIID(X, n);
b := SF(Y, c);
```

The probability of a Type II error, $\beta = 1/2$, is quite high. If this probability is unac-

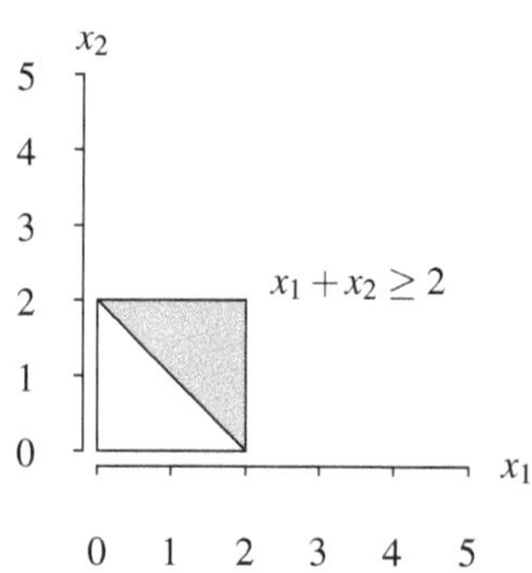

Figure 4.8: Fail to reject region under H_1 ($\theta = 2$) for the test statistic $X_1 + X_2$.

ceptably high, the analyst has two options: increase α or increase n. The first option—increasing the significance level α—will result in an increase in the critical value c. Geometrically, this corresponds to the slanted line in Figure 4.7 moving away from the origin, with an associated larger shaded area. Taking another perspective, the test statistic $Y = X_1 + X_2$ has a triangular(0, 5, 10) distribution when $\theta = 5$ and a triangular(0, 2, 4) distribution when $\theta = 2$. The probability density functions for these two distributions are graphed in Figure 4.9, along with the two shaded areas associated with α and β. This provides another illustration of the see-saw relationship between α and β. As the critical value c is shifted to the left from $c = 2$, α decreases and β increases; as the critical value c is shifted to the right from $c = 2$, α increases and β decreases.

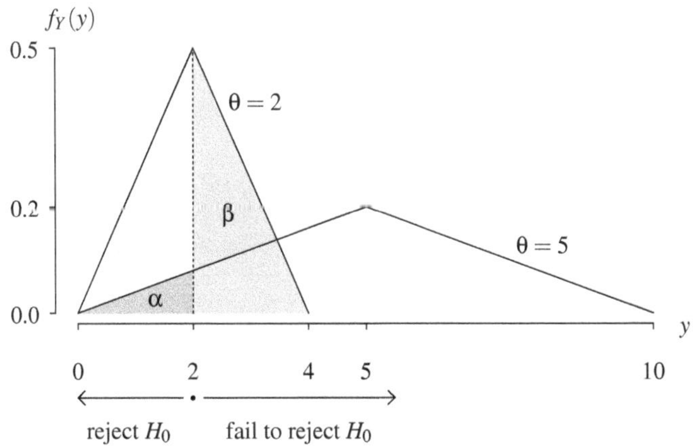

Figure 4.9: Probability density functions of the test statistic $Y = X_1 + X_2$ under H_0 and H_1.

A plot of the probability of a Type II error β for values of the significance level α on the interval $0 \le \alpha \le 0.32$ is given in Figure 4.10. Three important values of the critical value c, namely $c = 0$, $c = 2$, and $c = 4$, are highlighted as points on this graph.

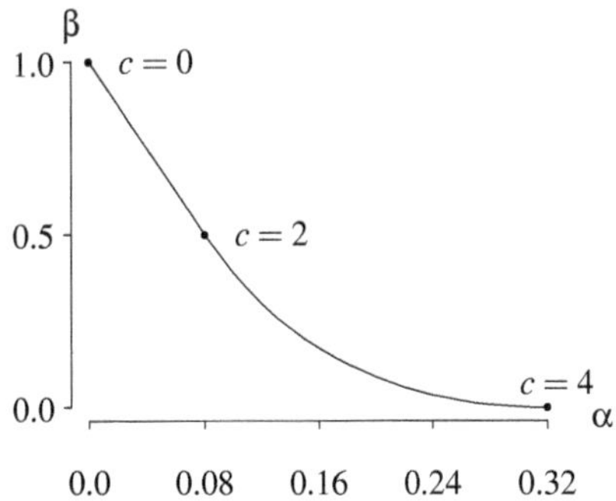

Figure 4.10: Probability of a Type I error versus the probability of a Type II error.

The second option for decreasing β—increasing the sample size n—also results in an increase in the critical value c. If Chico has the time and money to collect more data values, he can reduce β to an acceptable level. The APPL code given earlier can be used to calculate the critical value c and the probability of a Type II error β associated with various sample sizes n. These values are listed in Table 4.2, where the value of α is fixed at $\alpha = 0.08$. The probability of a Type II error decreases rapidly for increasing values of n. For example, when $n = 4$ and $\alpha = 0.08$, Chico should reject H_0 when $X_1 + X_2 + X_3 + X_4 < 5.89$, resulting in the probability of a Type II error $\beta = 0.05$.

n	1	2	3	4	5	6
c	0.4	2	3.91	5.89	7.91	9.99
β	0.8	0.5	0.19	0.05	0.01	0.001

Table 4.2: Critical values c and β values for various sample sizes for $\alpha = 0.08$.

Power functions

One inescapable aspect of hypothesis testing is that errors, that is, incorrect conclusions, can be drawn. One would like to minimize the probability that these errors occur. The choice of the test statistic can influence the likelihood of these errors. The test statistic in the previous example, $X_1 + X_2$, was chosen arbitrarily. Is this the best possible choice? Chico certainly could have instead used the product X_1X_2, the quotient X_1/X_2, or the minimum $\min\{X_1, X_2\}$ as a test statistic. Do any of these statistics give him a lower probability of a Type II error β for the same α? Questions of this nature can be answered with a power function, which is defined next.

Definition 4.7 For a hypothesis test of a null hypothesis H_0 against an alternative hypothesis H_1 involving an unknown parameter θ, associated parameter space Ω, and a prescribed level of significance α, the *power function* is $q(\theta) = P(\text{reject } H_0 \,|\, \theta)$ for all $\theta \in \Omega$.

Notice that for the simple hypotheses $H_0 : \theta = \theta_0$ and $H_1 : \theta = \theta_1$, the power function at θ_0 and θ_1 is $q(\theta_0) = \alpha$ and $q(\theta_1) = 1 - \beta$. The power function is useful for comparing competing test statistics. A test statistic with higher power is always preferred, as illustrated in the next example.

Example 4.5 Chico has two brothers, Harpo and Groucho. They do not like Chico's hypothesis test from the previous example. They suggest replacing the simple alternative hypothesis $H_1 : \theta = 2$ with a composite alternative hypothesis $H_1 : \theta < 5$. All agree to this. But the brothers disagree as to which test statistic to use, so

- Chico continues to use the test statistic $X_1 + X_2$,
- Harpo uses the test statistic $\min\{X_1, X_2\}$,
- Groucho uses the test statistic $\max\{X_1, X_2\}$.

The random sample X_1 and X_2 is drawn from a $U(0, \theta)$ population, where θ is a positive unknown parameter. Each brother tests the hypothesis

$$H_0 : \theta = 5$$

versus

$$H_1 : \theta < 5$$

at $\alpha = 0.08$ with their test statistic. Draw the power functions associated with these three test statistics and decide which of the three is the superior test statistic.

As in the previous example, the marginal probability density functions for X_1 and X_2 are

$$f_{X_1}(x_1) = \frac{1}{\theta} \qquad 0 < x_1 < \theta \qquad \text{and} \qquad f_{X_2}(x_2) = \frac{1}{\theta} \qquad 0 < x_2 < \theta.$$

Since X_1 and X_2 are independent random variables, their joint probability density function is

$$f_{X_1,X_2}(x_1, x_2) = \frac{1}{\theta^2} \qquad 0 < x_1 < \theta, 0 < x_2 < \theta.$$

We begin by first calculating the critical values associated with the three test statistics. Let c_1, c_2, and c_3 denote the critical values for Chico, Harpo, and Groucho, respectively. For Chico's test statistic $X_1 + X_2$, the critical value associated with significance level $\alpha = 0.08$ from the previous example is $c_1 = 2$.

Now consider Harpo's test statistic $\min\{X_1, X_2\}$. As with Chico's test statistic, small values of $\min\{X_1, X_2\}$ lead to rejecting H_0, so the decision rule is to reject H_0 whenever $\min\{X_1, X_2\} < c_2$. To determine c_2 at $\alpha = 0.08$,

$$\begin{aligned}
\alpha &= P(\text{Type I error}) \\
&= P(\text{reject } H_0 \,|\, H_0 \text{ true}) \\
&= P(\min\{X_1, X_2\} < c_2 \,|\, \theta = 5) \\
&= 1 - P(\min\{X_1, X_2\} \ge c_2 \,|\, \theta = 5) \\
&= 1 - P(X_1 \ge c_2, X_2 \ge c_2 \,|\, \theta = 5) \\
&= 1 - \int_{c_2}^{5} \int_{c_2}^{5} \frac{1}{25}\, dx_2\, dx_1 \\
&= 1 - \frac{(5 - c_2)^2}{25}.
\end{aligned}$$

Thus, $1 - (5 - c_2)^2/25 = 0.08$, so Harpo's critical value is $c_2 = 5 - \sqrt{23} \cong 0.2042$.

Finally, consider Groucho's test statistic $\max\{X_1, X_2\}$. As with his brothers' test statistics, small values of $\max\{X_1, X_2\}$ lead to rejecting H_0, so the decision rule is to reject H_0 if $\max\{X_1, X_2\} < c_3$. To determine c_3 at $\alpha = 0.08$,

$$\begin{aligned}
\alpha &= P(\text{Type I error}) \\
&= P(\text{reject } H_0 \,|\, H_0 \text{ true}) \\
&= P(\max\{X_1, X_2\} < c_3 \,|\, \theta = 5) \\
&= P(X_1 < c_3, X_2 < c_3 \,|\, \theta = 5) \\
&= \int_{0}^{c_3} \int_{0}^{c_3} \frac{1}{25}\, dx_2\, dx_1 \\
&= \frac{c_3^2}{25}.
\end{aligned}$$

So $c_3^2/25 = 0.08$, which implies that Groucho's critical value is $c_3 = \sqrt{2} \cong 1.4142$.

To summarize the critical values and decision rules for the three brothers,

- Chico rejects H_0 when $X_1 + X_2 < 2$,
- Harpo rejects H_0 when $\min\{X_1, X_2\} < 5 - \sqrt{23}$,
- Groucho rejects H_0 when $\max\{X_1, X_2\} < \sqrt{2}$.

Figure 4.11 shows the critical regions (shaded) for the three test statistics, along with the 5×5 square associated with the support of X_1 and X_2 under the null hypothesis H_0. For each of the three graphs, the significance level $\alpha = 0.08$ is the ratio of the shaded area to the area of the square.

Now that the critical values for the three tests have been established, we can find the power functions for each of the three test statistics. Geometrically, the power function can be viewed as altering the size of the 5×5 squares from Figure 4.11. For Chico's test statistic, the power function for $\theta > 2$, that is, for larger squares on the left graph in Figure 4.11, is

$$\begin{aligned} q(\theta) &= P(\text{reject } H_0 \,|\, \theta) \\ &= P(X_1 + X_2 < 2 \,|\, \theta) \\ &= \int_0^2 \int_0^{2-x_1} \frac{1}{\theta^2}\, dx_2\, dx_1 \\ &= \frac{2}{\theta^2}. \end{aligned}$$

For $1 < \theta \le 2$, the square on the left graph in Figure 4.11 intersects the triangular-shaped critical region, so the power function is

$$\begin{aligned} q(\theta) &= P(\text{reject } H_0 \,|\, \theta) \\ &= P(X_1 + X_2 < 2 \,|\, \theta) \\ &= 1 - P(X_1 + X_2 \ge 2 \,|\, \theta) \\ &= 1 - \int_{2-\theta}^{\theta} \int_{2-x_1}^{\theta} \frac{1}{\theta^2}\, dx_2\, dx_1 \\ &= 1 - \frac{2(\theta - 1)^2}{\theta^2}. \end{aligned}$$

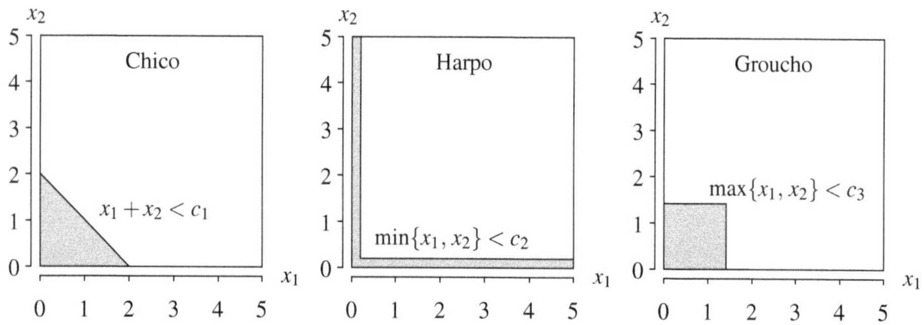

Figure 4.11: Critical regions under H_0 for $X_1 + X_2$, $\min\{X_1, X_2\}$, and $\max\{X_1, X_2\}$.

Finally, for $0 < \theta \le 1$, the power function is always 1 because the square on the left graph in Figure 4.11 falls inside the triangular-shaped critical region. Putting these three pieces together, Chico's power function is

$$q(\theta) = \begin{cases} 1 & 0 < \theta \le 1 \\ 1 - \dfrac{2(\theta-1)^2}{\theta^2} & 1 < \theta \le 2 \\ \dfrac{2}{\theta^2} & \theta > 2. \end{cases}$$

We now turn to Harpo's power function. To minimize clutter, Harpo's critical value $c_2 = 5 - \sqrt{23}$ will be written as just c_2 in the derivation that follows. For $\theta > c_2$, his power function is

$$\begin{aligned} q(\theta) &= P(\text{reject } H_0 \,|\, \theta) \\ &= P(\min\{X_1, X_2\} < c_2 \,|\, \theta) \\ &= 1 - P(\min\{X_1, X_2\} \ge c_2 \,|\, \theta) \\ &= 1 - P(X_1 \ge c_2, X_2 \ge c_2 \,|\, \theta) \\ &= 1 - \int_{c_2}^{\theta} \int_{c_2}^{\theta} \frac{1}{\theta^2}\, dx_2\, dx_1 \\ &= 1 - \frac{(\theta - c_2)^2}{\theta^2}. \end{aligned}$$

For $0 < \theta \le c_2$, Harpo's power function is always 1 because the square associated with the support of X_1 and X_2 has shrunk to fall completely within the L-shaped critical region on the middle graph in Figure 4.11. Putting these two pieces together, Harpo's power function is

$$q(\theta) = \begin{cases} 1 & 0 < \theta \le c_2 \\ 1 - \dfrac{(\theta - c_2)^2}{\theta^2} & \theta > c_2, \end{cases}$$

where $c_2 = 5 - \sqrt{23}$. Finally, we now turn to Groucho's power function. Groucho's critical value was determined earlier to be $c_3 = \sqrt{2}$. For $\theta > \sqrt{2}$, his power function is

$$\begin{aligned} q(\theta) &= P(\text{reject } H_0 \,|\, \theta) \\ &= P\left(\max\{X_1, X_2\} < \sqrt{2} \,\middle|\, \theta\right) \\ &= P\left(X_1 < \sqrt{2}, X_2 < \sqrt{2} \,\middle|\, \theta\right) \\ &= \int_0^{\sqrt{2}} \int_0^{\sqrt{2}} \frac{1}{\theta^2}\, dx_2\, dx_1 \\ &= \frac{2}{\theta^2}. \end{aligned}$$

For $0 < \theta \le \sqrt{2}$, Groucho's power function is always 1 because the square associated with the support of X_1 and X_2 has shrunk to fall completely within the square-shaped critical region on the right graph in Figure 4.11. Putting these two pieces together, Groucho's power function is

$$q(\theta) = \begin{cases} 1 & 0 < \theta \le \sqrt{2} \\ \dfrac{2}{\theta^2} & \theta > \sqrt{2}. \end{cases}$$

The power functions for the three brothers are graphed in Figure 4.12. The first thing to notice is that all three of the power functions pass through the point $(5, 0.08)$, where $\theta = 5$ is associated with the null hypothesis H_0, and $\alpha = 0.08$ is the significance level. This was expected. The hypothesis tests associated with the three statistics will all reject H_0 with probability $\alpha = 0.08$ when $\theta = 5$. In some sense, passing through the point $(5, 0.08)$ allows us to compare the three test statistics on an equal footing. A second thing to notice is that Harpo's power function is significantly lower than the other brothers on the interval of interest $0 < \theta < 5$. Poor Harpo. Why did his test statistic perform so poorly relative to his brothers? The answer can be seen using a toy data set. Let's say the observed data values x_1 and x_2 are

$$x_1 = 5.1 \qquad \text{and} \qquad x_2 = 0.1.$$

Since the population distribution is $U(0, \theta)$, you can immediately conclude from this data set that $\theta > 5.1$ in order to have produced the data value $x_1 = 5.1$. But not Harpo's hypothesis test. His test statistic is $\min\{x_1, x_2\} = \min\{5.1, 0.1\} = 0.1$, which is less than the associated critical value $c_2 = 5 - \sqrt{23} \cong 0.2042$. So Harpo's test rejects H_0 and concludes that $\theta < 5$ even though it is certain that $\theta > 5.1$. The other brothers would fail to reject H_0 for the toy data set because

- Chico's test statistic $x_1 + x_2 = 5.1 + 0.1 = 5.2$ is much greater than his critical value $c_1 = 2$, and
- Groucho's test statistic $\max\{x_1, x_2\} = \max\{5.1, 0.1\} = 5.1$ is also much greater than his critical value $c_3 = \sqrt{2} = 1.4142$.

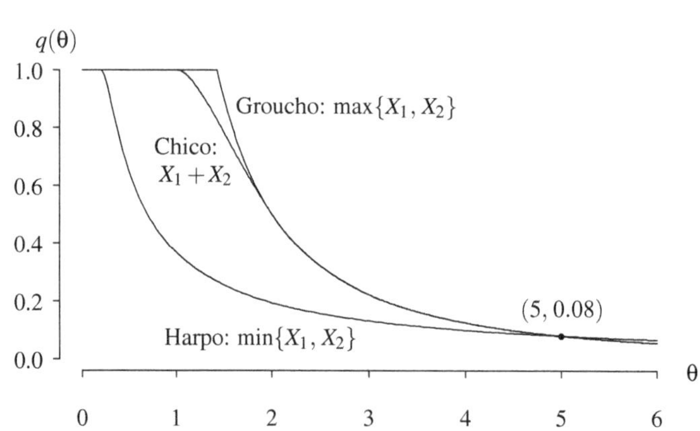

Figure 4.12: Power functions for the test statistics $X_1 + X_2$, $\min\{X_1, X_2\}$, and $\max\{X_1, X_2\}$.

Harpo's test statistic is clearly the worst choice of the three. Groucho's test statistic has the highest power of the three, even though his power function coincides with Chico's for $\theta > 2$. Is there a theoretical reason why Groucho's is the best? This question will be addressed later in the chapter, but it is noted in passing that Groucho's test statistic is a sufficient statistic for θ, while the other two brothers' test statistics were not sufficient statistics for θ.

The null and alternative hypotheses in the previous example were $H_0 : \theta = 5$ and $H_1 : \theta < 5$. This instance of a simple null hypothesis and a composite alternative hypothesis is common in hypothesis testing. The constant 5 for this particular hypothesis test can be treated generically as a real constant θ_0. So the most general description of a null and alternative hypothesis, namely

$$H_0 : \theta \in \Omega_0$$

and

$$H_1 : \theta \in \Omega_1,$$

can be described in terms of a special case consisting of a simple null hypothesis and a composite alternative hypothesis that occurs frequently in practice in the definition that follows.

Definition 4.8 For a hypothesis test involving an unknown parameter θ which can assume a real constant value θ_0 with simple null hypothesis

$$H_0 : \theta = \theta_0,$$

a hypothesis test with an alternative hypothesis of the form

$$H_1 : \theta > \theta_0$$

or

$$H_1 : \theta < \theta_0$$

is known as a *one-sided hypothesis test* and a hypothesis test with an alternative hypothesis of the form

$$H_1 : \theta \neq \theta_0$$

is known as a *two-sided hypothesis test*.

One-sided hypothesis tests are also known as one-tailed hypothesis tests; two-sided hypothesis tests are also known as two-tailed hypothesis tests. Choosing whether to conduct a one-sided or a two-sided test is dictated by the nature of the question being posed. In the previous example, Groucho had the superior test statistic for the one-sided test. We now revisit that example using Groucho's test statistic in a two-sided hypothesis test.

Example 4.6 The random sample X_1 and X_2 is drawn from a $U(0, \theta)$ population, where θ is a positive unknown parameter. Groucho tests the simple null hypothesis

$$H_0 : \theta = 5$$

versus the composite alternative hypothesis

$$H_1 : \theta \neq 5$$

at significance level $\alpha = 0.08$ using the test statistic $\max\{X_1, X_2\}$. Find the critical values and draw the associated power function for the test.

The form of the alternative hypothesis tells us that this is a two-sided hypothesis test. Since small and large values of the test statistic $\max\{X_1, X_2\}$ lead to rejecting H_0, the critical region is of the form

$$C = \{(X_1, X_2) \mid \max\{X_1, X_2\} < c_l \text{ or } \max\{X_1, X_2\} > c_u\},$$

where c_l and c_u are critical values. (The l and u subscripts are for "lower" and "upper.") Their values can be determined by using the definition of the significance level α:

$$\begin{aligned}\alpha &= P(\text{Type I error}) \\ &= P(\text{reject } H_0 \mid H_0 \text{ true}) \\ &= P(\max\{X_1, X_2\} < c_l \text{ or } \max\{X_1, X_2\} > c_u \mid \theta = 5).\end{aligned}$$

When α is set to 0.08, it can be seen that there are an infinite number of pairs of c_l and c_u values that satisfy this equation. In a similar fashion to the construction of confidence intervals, the common practice is to split α evenly between the two events. Under H_0, that is, when $\theta = 5$, the joint probability density function of X_1 and X_2 is

$$f_{X_1, X_2}(x_1, x_2) = \frac{1}{25} \qquad 0 < x_1 < 5, 0 < x_2 < 5,$$

whose support is illustrated by the large 5×5 square in Figure 4.13. Rather than using double integrals to calculate α, its value can be determined via geometry. Since the portion of the critical region associated with $\max\{X_1, X_2\} < c_l$ is a square,

$$c_l^2 \cdot \frac{1}{25} = 0.04,$$

so $c_l = 1$. This portion of the critical region corresponds to the shaded unit square in the lower-left portion of Figure 4.13. The portion of the critical region associated with

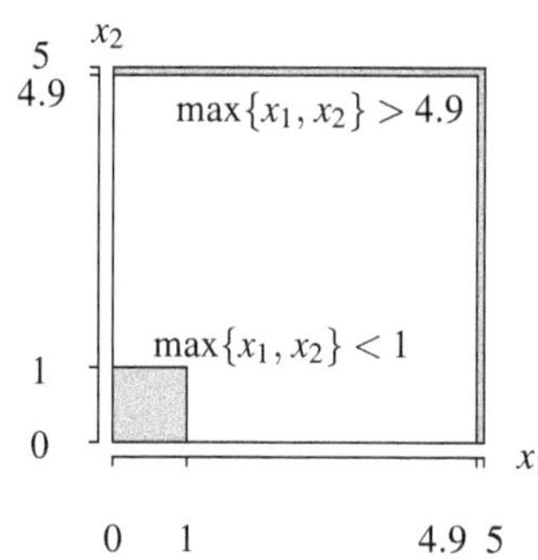

Figure 4.13: Critical region under H_0 ($\theta = 5$) for the test statistic $\max\{X_1, X_2\}$.

$\max\{X_1, X_2\} > c_u$ has a more complex shape. The probability of falling in this portion of the critical region is

$$\left[2 \cdot 5 \cdot (5 - c_u) - (5 - c_u)^2\right] \cdot \frac{1}{25} = 0.04.$$

Solving for c_u gives $c_u = \sqrt{24} = 2\sqrt{6} \cong 4.9$. This portion of the critical region is the thin strip along the top and right-hand side of the support of X_1 and X_2 under H_0 in Figure 4.13. The ratio of the sum of the shaded areas to the area of the square is $\alpha = 0.08$. To summarize, Groucho collects the experimental values of the random variables X_1 and X_2, that is, x_1 and x_2, then he calculates the test statistic $\max\{x_1, x_2\}$ and rejects H_0 if the value of this test statistic is less than the lower critical value $c_l = 1$ or greater than the upper critical value $c_u = \sqrt{24} = 2\sqrt{6} \cong 4.9$. The critical region under H_0 is shaded in Figure 4.13.

The next step is to calculate the power function $q(\theta)$. By definition,

$$\begin{aligned} q(\theta) &= P(\text{reject } H_0 \,|\, \theta) \\ &= P\left(\max\{X_1, X_2\} < 1 \text{ or } \max\{X_1, X_2\} > 4.9 \,\middle|\, \theta\right) \end{aligned}$$

for all $\theta > 0$. Geometrically, this can be viewed as a plot of the ratio of the shaded areas from Figure 4.13 to the area of the large square as the length of the side of the large square θ varies. For $0 < \theta \le c_l$, Groucho's power function is always 1 because the square associated with the support of X_1 and X_2 has shrunk to fall completely within the square-shaped portion of the critical region in Figure 4.13. For $c_l < \theta \le c_u$, Groucho's power function is just $1/\theta^2$ because the area of the shaded square is 1 and the area of the larger square is θ^2. Finally, for $\theta > c_u$, Groucho's power function is the sum of the area of the shaded square, that is, $1/\theta^2$, and the other area, which is

$$\left[2 \cdot \theta \cdot (\theta - c_u) - (\theta - c_u)^2\right] \cdot \frac{1}{\theta^2}.$$

Putting these three pieces together, Groucho's power function is

$$q(\theta) = \begin{cases} 1 & 0 < \theta \le c_l \\ \dfrac{1}{\theta^2} & c_l < \theta \le c_u \\ \dfrac{1}{\theta^2} + \left[2 \cdot \theta \cdot (\theta - c_u) - (\theta - c_u)^2\right] \cdot \dfrac{1}{\theta^2} & \theta > c_u, \end{cases}$$

where $c_l = 1$ and $c_u = \sqrt{24} = 2\sqrt{6} \cong 4.9$. This power function is plotted in Figure 4.14. By design, the power function passes through the point $(5, 0.08)$, or more generally, the point (θ_0, α). The power curve asymptotically approaches 1 as $\theta \to \infty$.

All of the hypothesis testing examples encountered thus far have involved sample sizes $n = 1$ or $n = 2$ observations. Such small sample sizes are rare in statistics. The behavior of Groucho's power function for general values of n will be investigated next.

Example 4.7 The random sample $X_1, X_2, \ldots, X_n$ is drawn from a $U(0, \theta)$ population, where θ is a positive unknown parameter. Groucho tests the simple null hypothesis

$$H_0 : \theta = 5$$

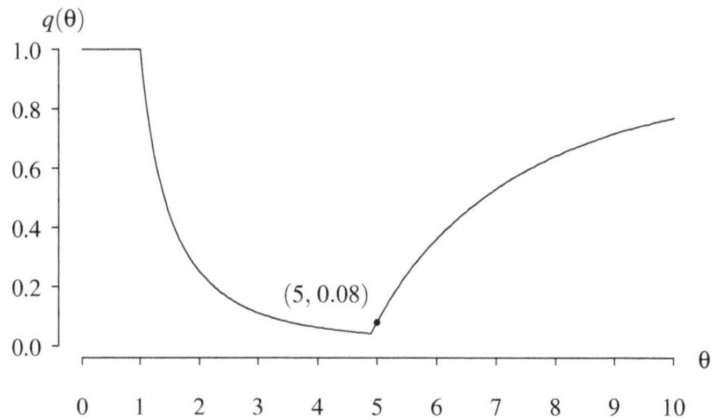

Figure 4.14: Power function for the test statistic $\max\{X_1, X_2\}$.

versus the composite alternative hypothesis

$$H_1 : \theta \neq 5$$

at significance level $\alpha = 0.08$ using the test statistic $X_{(n)} = \max\{X_1, X_2, \ldots, X_n\}$. Find the critical values and draw the associated power functions for the hypothesis test for several values of n.

This is again a two-sided test based on the form of the alternative hypothesis. Since small and large values of the test statistic $X_{(n)} = \max\{X_1, X_2, \ldots, X_n\}$ lead to rejecting H_0, the critical region is of the form

$$C = \{(X_1, X_2, \ldots, X_n) \,|\, X_{(n)} < c_l \text{ or } X_{(n)} > c_u\},$$

where c_l and c_u are critical values. We can no longer appeal to geometric arguments to calculate α and β for general values of n. Calculus must be employed. Since the random sample is drawn from a $U(0, \theta)$ population, the population probability density function is

$$f_X(x) = \frac{1}{\theta} \qquad 0 < x < \theta,$$

and the population cumulative distribution function on the support of X is

$$F_X(x) = \int_0^x \frac{1}{\theta} dw = \frac{x}{\theta} \qquad 0 < x < \theta.$$

Using an order statistic result from probability theory, the probability density function of the test statistic $X_{(n)} = \max\{X_1, X_2, \ldots, X_n\}$ is

$$f_{X_{(n)}}(x) = n \left(\frac{x}{\theta}\right)^{n-1} \frac{1}{\theta} \qquad 0 < x < \theta.$$

The associated cumulative distribution function on the support of $X_{(n)}$ is

$$F_{X_{(n)}}(x) = \int_0^x n \left(\frac{w}{\theta}\right)^{n-1} \frac{1}{\theta} dw = \left(\frac{x}{\theta}\right)^n \qquad 0 < x < \theta.$$

The value of the significance level α can be determined by using its definition:

$$\begin{aligned}\alpha &= P(\text{Type I error})\\ &= P(\text{reject } H_0 \,|\, H_0 \text{ true})\\ &= P(X_{(n)} < c_l \text{ or } X_{(n)} > c_u \,|\, \theta = 5).\end{aligned}$$

Again splitting $\alpha = 0.08$ equally between the two tails of the distribution of $X_{(n)}$, the lower critical value c_l can be found by solving

$$P(X_{(n)} < c_l \,|\, \theta = 5) = F_{X_{(n)}}(c_l \,|\, \theta = 5) = \left(\frac{c_l}{5}\right)^n = 0.04,$$

for c_l, resulting in $c_l = 5(0.04)^{1/n}$. The upper critical value c_u can be found by solving

$$P(X_{(n)} > c_u \,|\, \theta = 5) = 1 - F_{X_{(n)}}(c_u \,|\, \theta = 5) = 1 - \left(\frac{c_u}{5}\right)^n = 0.04,$$

for c_u, resulting in $c_u = 5(0.96)^{1/n}$. So Groucho collects the experimental values of the random variables $X_1, X_2, \ldots, X_n$, that is, $x_1, x_2, \ldots, x_n$, then he calculates the test statistic $x_{(n)} = \max\{x_1, x_2, \ldots, x_n\}$, and rejects H_0 if the value of this test statistic is less than the lower critical value $c_l = 5(0.04)^{1/n}$ or greater than the upper critical value $c_u = 5(0.96)^{1/n}$.

The next step is to calculate the power function $q(\theta)$. By definition,

$$\begin{aligned}q(\theta) &= P(\text{reject } H_0 \,|\, \theta)\\ &= P\left(X_{(n)} < c_l \text{ or } X_{(n)} > c_u \,\middle|\, \theta\right).\end{aligned}$$

For $0 < \theta \le c_l$, the power function is always 1 because the largest value sampled will always be smaller than c_l. For $c_l < \theta \le c_u$, the power function is

$$P(X_{(n)} < c_l \,|\, \theta) = F_{X_{(n)}}(c_l \,|\, \theta) = \left(\frac{c_l}{\theta}\right)^n.$$

Finally, for $\theta > c_u$, the power function is

$$P(X_{(n)} < c_l \text{ or } X_{(n)} > c_u \,|\, \theta) = F_{X_{(n)}}(c_l \,|\, \theta) + 1 - F_{X_{(n)}}(c_u \,|\, \theta) = \left(\frac{c_l}{\theta}\right)^n + 1 - \left(\frac{c_u}{\theta}\right)^n.$$

Putting these three pieces together, Groucho's power function is

$$q(\theta) = \begin{cases} 1 & 0 < \theta \le c_l \\ \left(\frac{c_l}{\theta}\right)^n & c_l < \theta \le c_u \\ \left(\frac{c_l}{\theta}\right)^n + 1 - \left(\frac{c_u}{\theta}\right)^n & \theta > c_u, \end{cases}$$

where $c_l = 5(0.04)^{1/n}$ and $c_u = 5(0.96)^{1/n}$. Power functions are plotted in Figure 4.15 for several values of n. By design, the power functions all pass through the point $(5, 0.08)$. In addition, the power functions tend to be steeper as n increases. This makes sense because the more data values Groucho collects, the more likely he is to detect departures from the null hypothesis $\theta = 5$. What does the power function look like in the limit as $n \to \infty$? If Groucho could collect an unlimited number of observations, then

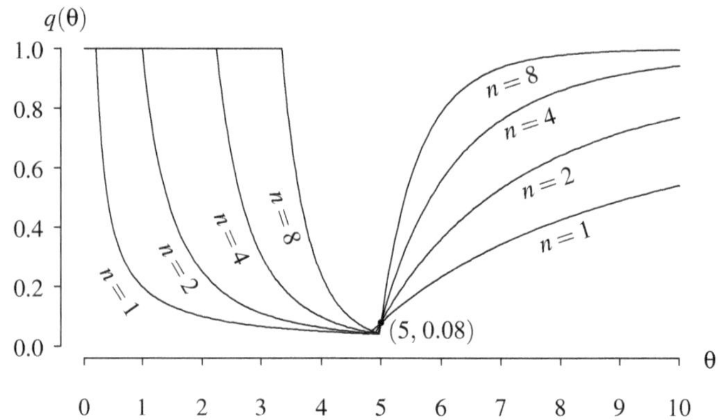

Figure 4.15: Power functions for the test statistic $X_{(n)} = \max\{X_1, X_2, \ldots, X_n\}$.

he would have a test that can detect departures from H_0 with certainty. As was the case with finite n, the power function passes through the point $(5, 0.08)$. Both c_l and c_u are approaching $\theta_0 = 5$ from below, resulting in a hypothesis test that detects θ values other than $\theta_0 = 5$ with probability 1. This "ideal" power function is plotted in Figure 4.16 and can be considered the limiting power function as $n \to \infty$.

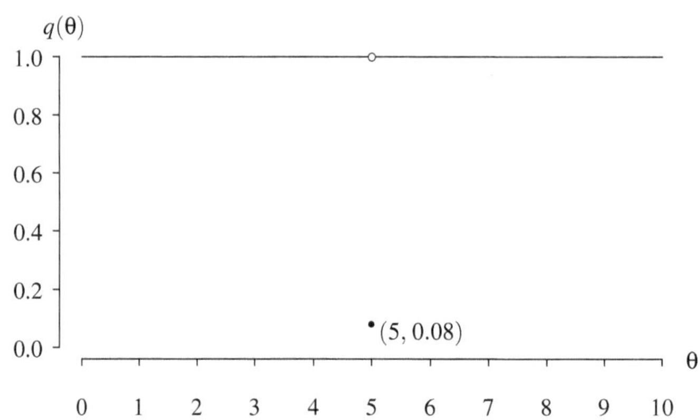

Figure 4.16: Ideal power function for the test statistic $X_{(n)} = \max\{X_1, X_2, \ldots, X_n\}$.

One remaining loose end concerning the level of significance α can now be tied up. In Definition 4.6, it was assumed that the null hypothesis was a simple hypothesis. But this need not necessarily be the case. Testing

$$H_0 : \theta \geq 5$$

versus

$$H_1 : \theta < 5,$$

for example, is a hypothesis test involving a composite null hypothesis. The definition of α can easily be tweaked in order to accommodate such a test, as shown in the next definition. A hypothesis test with a composite null hypothesis is rare in practice, so such a test will not be illustrated here. In the example involving the composite null hypothesis $H_0 : \theta \geq 5$, the test statistic, critical region, and decision rule are all identical to the hypothesis test with the simple null hypothesis $H_0 : \theta = 5$, so there is essentially no new concept to be presented.

Definition 4.9 Consider a hypothesis test of a null hypothesis $H_0 : \theta \in \Omega_0$ versus an alternative hypothesis $H_1 : \theta \in \Omega_1$. The *significance level* is

$$\alpha = \sup_{\theta \in \Omega_0} P(\text{rejecting } H_0 \,|\, \theta).$$

The maximum operator, rather than the supremum operator, would have been adequate for the null hypothesis $H_0 : \theta \geq 5$ in Definition 4.9. The definition has been kept more general, however, because the supremum operator would have been necessary for computing α for the null hypothesis $H_0 : \theta > 5$.

The next example is a non-quantitative example that draws some analogies between hypothesis testing and the Western legal system.

Example 4.8 A person who has been arrested and accused of committing a crime typically undergoes a trial involving evidence that results in a verdict. Much of the process associated with the trial runs parallel to hypothesis testing. This example examines the common threads between these two processes.

In most Western legal systems, the null hypothesis in the courtroom is innocence. The defendant is considered

"innocent until proven guilty."

The "presumed innocence" assumption dates back to the second or third centuries when it was known as "proof lies on him who asserts, not on him who denies." Having innocence as the null hypothesis tips the process in favor of the defendant who is accused of committing the crime. Thus, the null and alternative hypotheses are

$$H_0 : \text{defendant is innocent}$$

versus

$$H_1 : \text{defendant is guilty}.$$

The alternative hypothesis is the reason that the police arrested the defendant in the first place. The task for the judge or jury in the courtroom is to see if there is enough evidence present to reject H_0 and declare the defendant guilty.

As in the case of hypothesis testing, the process is not foolproof and there are two types of mistakes that can be made by the judge or jury, and both are very serious. The first mistake corresponds to a Type I error in hypothesis testing and consists of declaring an innocent person guilty. The probability of a Type I error is

$$\begin{aligned} \alpha &= P(\text{Type I error}) \\ &= P(\text{reject } H_0 \,|\, H_0 \text{ is true}) \\ &= P(\text{declare defendant guilty} \,|\, \text{defendant is innocent}). \end{aligned}$$

A Type I error results in a person being convicted of a crime, with the associated consequences, when in fact the person did not commit the crime. In a perfect world, we would like this probability to be zero. But in order to do so we would have to declare the defendant innocent, which would drive the probability of a Type II error to

$$\begin{aligned}\beta &= P(\text{Type II error}) \\ &= P(\text{fail to reject } H_0 \,|\, H_1 \text{ is true}) \\ &= P(\text{declare defendant not guilty} \,|\, \text{defendant is guilty}) \\ &= 1.\end{aligned}$$

This would mean that a defendant, innocent or guilty, is declared innocent, which is also quite bad for society, particularly in terms of public safety. It is in society's interest to keep the probability of a Type I error, that is, declaring an innocent person guilty, as low as reasonably possible. Although it is a rather vague description of this particular probability, the notion of

"beyond a reasonable doubt"

has been used since medieval times to describe the standard of evidence required to convict a defendant. So for a reasonable person on a jury to vote to declare a person guilty of a crime, there must be a high probability $1-\alpha$ that the defendant committed the crime based on the evidence presented. Technicalities, such as a defendant not being read their Miranda rights in the United States upon arrest, tend to further decrease α, but, of course, at the cost of increasing β.

In hypothesis testing, the final result is either "reject H_0" or "fail to reject H_0." The final result of a courtroom trial is either

"guilty" or "not guilty."

Notice that the defendant is not being declared "innocent," which would be exceedingly difficult to prove, but rather "not guilty," which means that there has not been enough evidence presented to conclude that the defendant is indeed guilty. The legal system allows for a much smaller fraction of innocent people to be declared guilty (that is, convicted) than the fraction of guilty people to be declared not guilty (that is, acquitted). This is a societal choice—the first error is considered more harmful than the second error. To prove guilt or innocence absolutely is nearly impossible. Even if there are dozens of eyewitnesses to a murder, for example, the murder could have been committed by an identical twin, and the eyewitnesses might be testifying against an innocent defendant.

The concept of a sample size also applies to a trial. In some trials there is little evidence; in other trials there is overwhelming evidence. This evidence is analogous to the sample size n, although more difficult to quantify. Is there a way to increase the sample size, and thereby decrease both α and β? As technology has advanced, the sample size has effectively been increasing. Technological advances over the years, such as fingerprinting, polygraph testing, DNA evidence, and advanced forensics, serve to increase the evidence that can be used in a trial. Effectively, these advances make the power function associated with a trial steeper. In summary, some strong parallels exist between hypothesis testing and the legal system in most Western countries, including:

- the adversarial legal system effectively creates H_0 and H_1,
- the presumed innocence assumption means that H_0 corresponds to an innocent defendant,
- choosing H_0 as innocence and using the criterion "beyond a reasonable doubt" results in α being much smaller than β,
- the verdict "guilty" corresponds to rejecting H_0; the verdict "not guilty" corresponds to failing to reject H_0, and
- technology advances that are used in a trial correspond to a steeper power function.

All of the hypothesis testing examples considered thus far have involved sampling from continuous population distributions. As was the case when constructing a confidence interval, random sampling from a discrete population causes complications. The next example illustrates some of these difficulties.

Example 4.9 Roulette is a gambling game that involves a wheel which is spun by a croupier in one direction and a ball which is spun in the opposite direction on a tilted track at the circumference of the wheel. On a French wheel, there are 37 positions where the ball can land once it loses momentum, and they are numbered $0, 1, 2, \ldots, 36$. A wheel is considered "fair" if all 37 of these outcomes are equally likely. One bet that can be made on a roulette wheel is called a bet on the "third column," which wins if the ball falls in one of the following 12 positions: $3, 6, 9, \ldots, 36$. Devise a hypothesis test for determining whether or not the wheel is fair based on the results of 74 bets on the third column at level of significance $\alpha = 0.1$.

This example is another illustration of "missing information" in statistics. If we were testing for the fairness of the wheel, we would prefer the actual numbers that were generated by the 74 spins rather than the outcomes of the 74 bets on the third column. Sometimes the data is given in a form that is less than optimal, as in this case, and an analyst needs to work with the data as presented. We can not conclude that the wheel is fair with just the outcomes of the 74 bets on the third column. (For example, if the probability that the ball falls in the 0 position is $2/37$, the probability that the ball falls in the 1 position is 0, and the probability that the ball falls in all other positions is $1/37$, then this unfair wheel could not be detected by the number of wins associated with bets on the third column, regardless of the sample size.) But it is possible to conclude that the wheel is unfair if there are too many or too few wins on bets on the third column. The null hypothesis corresponds to a fair roulette wheel, which corresponds to the probability of winning a single bet on the third column of $p = 12/37$. Since an unfair roulette wheel would correspond to a lower or higher probability of winning, the hypothesis test is two-sided:

$$H_0 : p = 12/37$$

versus

$$H_1 : p \neq 12/37.$$

Let $X_1, X_2, \ldots, X_{74}$ be the observed outcomes of the $n = 74$ bets, where $X_i = 0$ corresponds to a loss and $X_i = 1$ corresponds to a win, for $i = 1, 2, \ldots, 74$. These constitute a random sample because each spin of the wheel produces a mutually independent random variable. The total number of wins is

$$Y = \sum_{i=1}^{74} X_i,$$

which is a binomial(74, 12/37) random variable under H_0. The random variable Y will serve as the test statistic for this hypothesis test. If the roulette wheel is fair, we expect $E[Y] = (74)(12/37) = 24$ wins. Having this expected value come out evenly was the motivation for the choice of a sample size of $n = 74$. Small and large values of the test statistic Y are evidence in favor of rejecting H_0. Since $\alpha = 0.1$ and the hypothesis test is two-sided, we assign $\alpha/2 = 0.05$ to each tail of the binomial(74, 12/37) distribution. That is, we want to find a lower critical value c_l and an upper critical value c_u such that

$$P(Y \le c_l) = \sum_{y=0}^{c_l} \binom{74}{y} \left(\frac{12}{37}\right)^y \left(\frac{25}{37}\right)^{74-y} = 0.05$$

and

$$P(Y \ge c_u) = \sum_{y=c_u}^{74} \binom{74}{y} \left(\frac{12}{37}\right)^y \left(\frac{25}{37}\right)^{74-y} = 0.05.$$

This is the point at which the random sampling from a discrete distribution causes problems in the hypothesis test. The discrete nature of the population probability distribution makes it extremely unlikely that we can find critical values c_l and c_u that achieve these probabilities exactly. The usual approach that is adopted by statisticians in this case is to choose the critical values c_l and c_u so that the tail probabilities are as large as possible without exceeding $\alpha/2 = 0.05$. The lower critical value is $c_l = 16$ because

$$\sum_{y=0}^{16} \binom{74}{y} \left(\frac{12}{37}\right)^y \left(\frac{25}{37}\right)^{74-y} \cong 0.028 < 0.05 \text{ and } \sum_{y=0}^{17} \binom{74}{y} \left(\frac{12}{37}\right)^y \left(\frac{25}{37}\right)^{74-y} \cong 0.0502 > 0.05.$$

In other words, choosing $c_l = 17$ results in a left-hand tail probability of 0.0502 that exceeds 0.05. The upper critical value is $c_u = 32$ because

$$\sum_{y=32}^{74} \binom{74}{y} \left(\frac{12}{37}\right)^y \left(\frac{25}{37}\right)^{74-y} \cong 0.033 < 0.05 \text{ and } \sum_{y=31}^{74} \binom{74}{y} \left(\frac{12}{37}\right)^y \left(\frac{25}{37}\right)^{74-y} \cong 0.055 > 0.05.$$

So the hypothesis test counts the number of wins Y associated with bets on the third column for $n = 74$ spins of the roulette wheel and rejects H_0 if $Y \le 16$ or $Y \ge 32$. Figure 4.17 shows the probability mass function of $Y = \sum_{i=1}^{74} X_i$ under H_0 $(p = 12/37)$, which is binomial(74, 12/37), along with the critical values $c_l = 16$ and $c_u = 32$. The probability mass function is somewhat bell shaped by the central limit theorem associated with the large $(n = 74)$ sample size. Although the test requested a significance level of $\alpha = 0.1$, this test delivers a significance level of approximately $0.028 + 0.033 = 0.061$, which is the sum of the probability mass function values from 0 to 16 and the probability mass function values from 32 to 74. The exact value of the significance level can be calculated with the R statement

```
pbinom(16, 74, 12 / 37) + 1 - pbinom(31, 74, 12 / 37)
```

The power function for this hypothesis test is

$$\begin{aligned} q(p) &= P(\text{reject } H_0 \,|\, p) \\ &= P(Y \le 16 \text{ or } Y \ge 32 \,|\, p) \\ &= \sum_{y=0}^{16} \binom{74}{y} p^y (1-p)^{74-y} + \sum_{y=32}^{74} \binom{74}{y} p^y (1-p)^{74-y} \end{aligned}$$

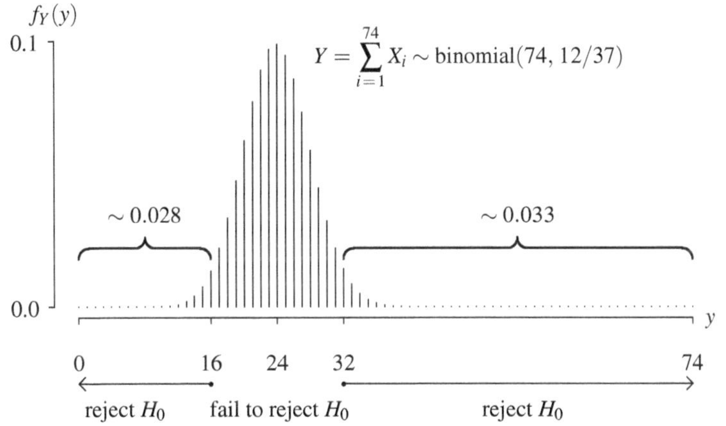

Figure 4.17: The sampling distribution of $Y = \sum_{i=1}^{74} X_i$ under H_0 (when $p = 12/37$).

for $0 < p < 1$. The power function can be plotted with the following R statements.

```
p = seq(0, 1, length = 200)
power = pbinom(16, 74, p) + 1 - pbinom(31, 74, p)
plot(p, power, type = "l")
```

The power function is plotted in Figure 4.18. As expected, the power function passes through the point $(12/37, 0.061)$ which is associated with the null hypothesis. Improving the power function for this hypothesis test involves increasing the number of bets

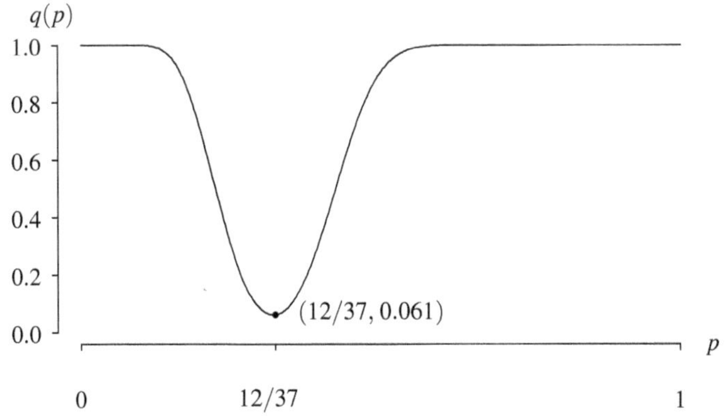

Figure 4.18: Power function for the test statistic $Y = \sum_{i=1}^{74} X_i$.

on the third column, which would make the power function steeper. The discrete nature of the binomial distribution means that not all of the power functions will pass through the same point on the graph, as we found was the case with sampling from continuous populations.

In one of the unlikely quirks of U.S. history, the first two signers of the Declaration of Independence who went on to become President, namely John Adams and Thomas Jefferson, both died on July 4, 1826, the 50th anniversary of the signing. (James Monroe, the fifth president of the U.S. and last president who was also a founding father, died on July 4, 1831, the 55th anniversary of the signing.) Did this highly improbable event occur by chance, or do people have the ability to rally their strength to hang on for some important event? There has been a lot of speculation among sociologists because of conflicting studies concerning a person's ability to influence the length of their life based on the anticipation of a positive (or negative) event. Some other conjectures surrounding this phenomenon involving a positive event include

- a person's ability to stay alive for the arrival of a loved one,
- a person's ability to stay alive to celebrate their birthday,
- a Jewish person's ability to stay alive to celebrate the Passover,
- a Chinese person's ability to stay alive to celebrate the Chinese new year,
- in the United States, a person's ability to stay alive to vote in a Presidential election and see the results in early November of a year divisible by four.

Hypothesis tests have been conducted to assess each of these conjectures, often with contradictory results. The next example contains a hypothesis test concerning the ability of a person to prolong their life in anticipation of their birthday. Since birthdays occur fairly uniformly throughout the year, any possible seasonal effect that might impact the results is hopefully neutralized.

Example 4.10 The Bruton Parish Church on Duke of Gloucester Street in Williamsburg, Virginia has dozens of tombstones in the church yard. There are $n = 72$ of these tombstones that contain a date of birth and a date of death. Use this data set to conduct a hypothesis test at $\alpha = 0.1$ to determine whether people are able to prolong their lives in order to celebrate a birthday.

The earliest tombstone with both the date of birth and the date of death is that of Michael Archer, who was born on September 29, 1681 and died on February 10, 1726. Mr. Archer died 134 days after his birthday. Computing the number of days after their birthday that a person passes away results in the $n = 72$ data values given in Table 4.3. The data values in Table 4.3 can range from 0 (for a person who dies on their birthday) to 365 (for a person born on February 29 who dies on February 28). A strip chart of the data values on a number line is given in Figure 4.19, where tied observations have been "jittered" to more accurately reflect the probability distribution of the data values. Upon cursory inspection, the data values seem to be uniformly distributed between 0 and 365, with some slight clustering on the left-hand end of the axis. If each of the 12 months following the birthday were equally likely to be the month in which an individual dies, then we would expect $72/12 = 6$ deaths during each one-month period. But this is not the case for this data set; the first two months contain 9 deaths each, so perhaps there is some evidence to indicate that people are more likely to die soon after celebrating a

7	14	18	20	22	23	25	25	27	34	34	36
43	51	55	59	60	60	63	66	67	70	75	99
111	113	121	129	129	130	130	134	134	136	136	137
137	138	140	156	169	174	176	189	198	208	218	222
225	231	240	245	262	262	271	278	278	278	279	279
280	290	294	318	321	343	343	348	349	354	354	358

Table 4.3: Days survived after birthday.

birthday. A formal hypothesis test is necessary to determine if the additional deaths in these two months are statistically significant. We now need to formulate an appropriate hypothesis test. Let p be the probability that an individual dies during the two months after their birthday. (The two-month period is chosen arbitrarily, so choosing a different time period could possibly alter the conclusion of the hypothesis test.) If the birthday has no influence on the date of death, then the probability of dying during that two-month period is $p = 2/12 = 1/6$. This will correspond to the null hypothesis. The alternative hypothesis, which is what we are trying to prove, is that a person is *more* likely to die in the two-month period after their birthday, that is, $p > 1/6$. So this corresponds to the one-sided hypothesis test

$$H_0 : p = 1/6$$

versus

$$H_1 : p > 1/6.$$

The wording of the problem, that is, whether the anticipation of a birthday *prolongs* a person's life, is what makes the test one-sided. The next step is to select a test statistic. As in the previous example involving the fairness of a roulette wheel, the sampling distribution is Bernoulli: either the individual dies in the two months after their birthday or in the other ten months. Let $X_1, X_2, \ldots, X_{72}$ be the observed outcomes, where $X_i = 1$ corresponds to dying in the two months following a birthday and $X_i = 0$ corresponds to dying in the other ten months, for $i = 1, 2, \ldots, 72$. Assuming that the Bernoulli random variables are mutually independent, the total number of people who die in the two months following their birthday

$$Y = \sum_{i=1}^{72} X_i$$

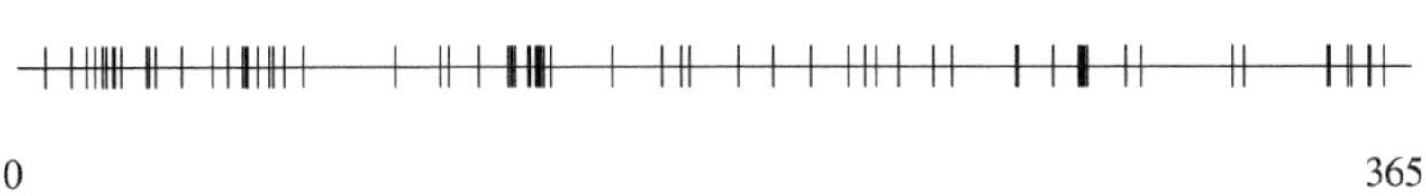

Figure 4.19: Strip chart of days survived after birthday.

will serve as a test statistic. Large values of the test statistic lead to rejecting H_0. Under H_0, $Y \sim \text{binomial}(72, 1/6)$. If a person's birthday has no influence on their time of death, we expect that there will be $E[Y] = (72)(1/6) = 12$ deaths during the first two months after their birthday. Since we observe $y = 18$ deaths during this time period, there is some evidence in favor of H_1. The next step is to find the critical value c. The probability of a Type I error is

$$\begin{aligned}
\alpha &= P(\text{Type I error}) \\
&= P(\text{reject } H_0 \,|\, H_0 \text{ is true}) \\
&= P(Y \geq c \,|\, p = 1/6) \\
&= \sum_{y=c}^{72} \binom{72}{y} \left(\frac{1}{6}\right)^y \left(\frac{5}{6}\right)^{72-y}.
\end{aligned}$$

Since the level of significance is $\alpha = 0.1$, we want to find the critical value c such that

$$\sum_{y=c}^{72} \binom{72}{y} \left(\frac{1}{6}\right)^y \left(\frac{5}{6}\right)^{72-y} = 0.1.$$

Due to the discrete nature of the binomial distribution, there is no value of the critical value c that achieves the probability $\alpha = 0.1$ exactly. Choosing the value that comes closest to $\alpha = 0.1$ without exceeding it results in the critical value $c = 17$, as illustrated in Figure 4.20. The achieved right-hand tail probability associated with $c = 17$ for the probability mass function of a binomial(72, 1/6) random variable is 0.082. The achieved level of significance can be calculated with the R statement

```
1 - pbinom(16, 72, 1 / 6)
```

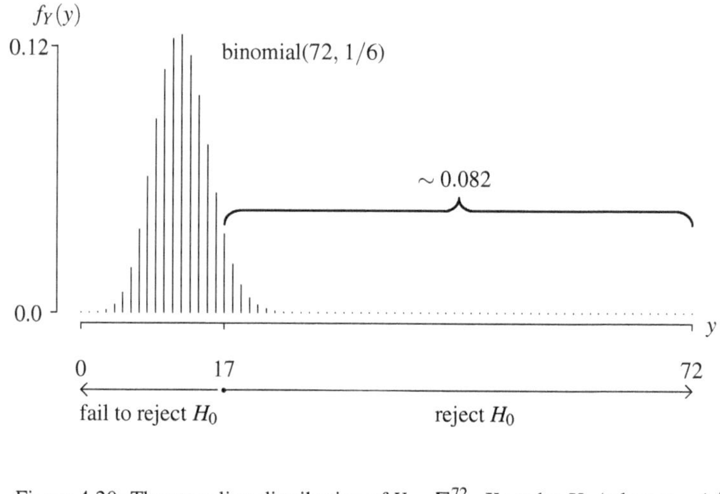

Figure 4.20: The sampling distribution of $Y = \sum_{i=1}^{72} X_i$ under H_0 (when $p = 1/6$).

So the decision rule is to calculate the experimental value y of the random variable Y, and reject H_0 if $y \geq 17$. The power function for this hypothesis test is

$$\begin{aligned} q(p) &= P(\text{reject } H_0 \,|\, p) \\ &= P(Y \geq 17 \,|\, p) \\ &= \sum_{y=17}^{72} \binom{72}{y} p^y (1-p)^{72-y} \end{aligned}$$

for $0 < p < 1$. The power function can be plotted with the following R statements.

```
p = seq(0, 1, length = 200)
power = 1 - pbinom(16, 72, p)
plot(p, power, type = "l")
```

The power function is plotted in Figure 4.21. As expected, the power function passes through the point $(1/6, 0.082)$ which is associated with the null hypothesis.

For our particular data set, $y = 18$ people died within the two-month period following their birthdays, so the null hypothesis is rejected, and it is concluded that people are capable of extending their lives based on the anticipation of a birthday.

The conclusion that was drawn for this data set should not be taken too seriously for the following four reasons. First, the tombstones in the Bruton Parish church yard were quite old, which means that the infant mortality during that time period was high. This would mean that several of the smaller data values might correspond to an early infant death, which could not possibly correspond to a baby extending their life to celebrate their next birthday. The data should be modified to include only people that have achieved a particular age. Second, it is possible that the tombstones might correspond to people that died on one particular day, such as the Battle of Williamsburg on May 5, 1862. If this were the case, then it would be appropriate to exclude such data values. Third, the two-month period after an individual's birthday was chosen arbitrarily.

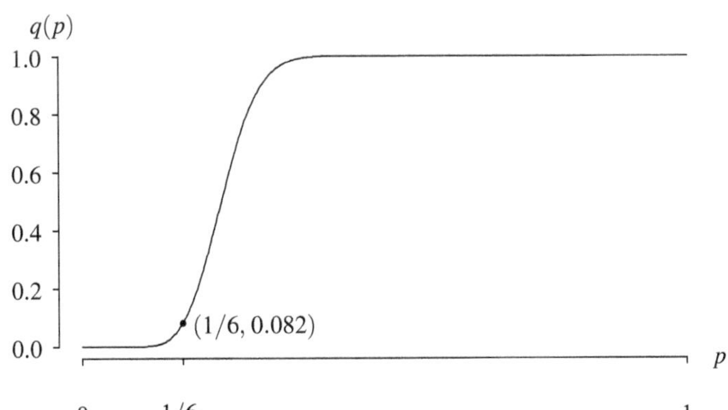

Figure 4.21: Power function for the test statistic $Y = \sum_{i=1}^{72} X_i$.

Changing the length of this period to be longer or shorter could potentially alter the conclusion drawn by the hypothesis test. Finally, the significance level α was set arbitrarily to $\alpha = 0.1$ in the problem statement. Changing the significance level to $\alpha = 0.01$, for example, results in a critical value of $c = 21$, and the test statistic of $y = 18$ now results in failing to reject H_0. One data set can result in two different conclusions based on the choice of the value of α.

The question addressed in this example has been addressed using a larger data set of $n = 2{,}745{,}149$ individuals who died of natural causes. (The reference is given in the preface.) For a huge data set like this one, the power function becomes very steep. In addition, because each probability mass function value associated with a binomial(2745149, 1/6) random variable (the distribution of Y under H_0) is very small, it is possible to come very close to achieving a level of significance equal to the specified $\alpha = 0.1$. The ideal power function graphed in Figure 4.22 is the power function in the limit as $n \to \infty$. This hypothesis test always fails to reject H_0 for $p < 1/6$ and always rejects H_0 for $p > 1/6$, and still passes through the point $(1/6, 0.1)$. The conclusion of the study involving the huge sample size was that women were able to extend their lives in the anticipation of a birthday, but men were not able to do so.

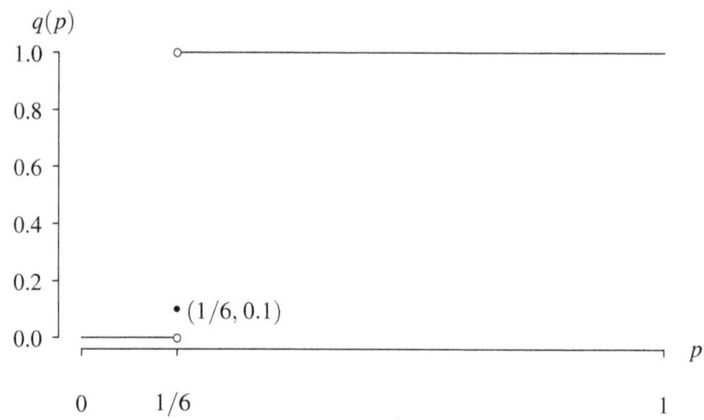

Figure 4.22: Ideal power function for the test statistic $Y = \sum_{i=1}^{n} X_i$.

The discussion at the end of the previous example about the arbitrary nature of α in controlling the outcome of a hypothesis is troubling. One realizes that devious, unscrupulous individuals could skip the step of setting α in advance, collect the data, calculate the test statistic, then choose an α which gives them whichever outcome serves their diabolical purposes. This problem is addressed in the next section.

4.2 Significance Tests

In traditional hypothesis testing, which has been the subject of the chapter thus far, the experimenter is required to select a significance level α which determines the critical value(s), which in turn impacts the final outcome of the hypothesis test. A variant of a hypothesis test, which is often known as a "significance test," releases the experimenter from specifying α and instead reports an achieved significance level, which is also known as a "p-value."

Definition 4.10 For a hypothesis test of a null hypothesis H_0 versus an alternative hypothesis H_1 based on the data values $X_1, X_2, \ldots, X_n$, the *p-value*, also known as the achieved, observed, or attained significance level, is the probability of obtaining the observed test statistic or one more extreme in favor of H_1, assuming that H_0 is true, in a future study with identical sample size and sampling assumptions.

The p-value is the smallest significance level at which the observed data values cause rejection of H_0. A small p-value is evidence against H_0. The process involved with a significance test is largely the same as that of a hypothesis test. The key difference is that the significance level α is not set in advance; rather, an achieved significance level is determined by the test and reported as a p-value. So the algorithm for conducting a significance test involves fewer steps because there is no significance level α, critical value c, or critical region C. The algorithm for conducting a significance test consists of the following steps.

1. Develop a hunch, or theory, concerning a problem of interest.
2. Translate the theory into a question concerning an unknown parameter θ.
3. State the null hypothesis H_0 in terms of θ.
4. State the alternative hypothesis H_1 in terms of θ.
5. Determine the sample size n.
6. Determine a test statistic, and derive its probability distribution when H_0 is true.
7. Collect the data values $x_1, x_2, \ldots, x_n$.
8. Compute the observed value of the test statistic.
9. Compute the p-value, which is the achieved significance level.

The previous example concerning extending a lifetime in the anticipation of a birthday based on a sample of $n = 72$ subjects of whom $y = 18$ died in the two months after their birthday rejected H_0 for a significance level $\alpha = 0.1$, but failed to reject H_0 for $\alpha = 0.01$. An associated significance test will result in a p-value, or achieved significance level, that falls between 0.01 and 0.1.

Example 4.11 Let $X_1, X_2, \ldots, X_{72}$ be the results of whether a person dies in the two months following their birthday, where $X_i = 1$ corresponds to a death within the two months following a person's birthday and $X_i = 0$ corresponds to a death in the other ten months. Conduct the significance test

$$H_0 : p = 1/6$$

versus

$$H_1 : p > 1/6$$

to determine whether the anticipation of a birthday prolongs a person's life using the test statistic $Y = \sum_{i=1}^{72} X_i$. Calculate the p-value associated with the test results from the tombstones in the Bruton Parish Church graveyard with $y = 18$.

As in the previous example, large values of the test statistic lead to rejecting H_0. So the p-value is the probability of getting the observed test statistic $y = 18$ or one more extreme (y greater than 18) under H_0. When H_0 is true, $Y \sim \text{binomial}(72, 1/6)$, so

$$p = P(Y \geq 18 \,|\, H_0 \text{ true}) = \sum_{y=18}^{72} \binom{72}{y} \left(\frac{1}{6}\right)^y \left(\frac{5}{6}\right)^{72-y} \cong 0.04621.$$

This p-value can be calculated with the R statement

```
1 - pbinom(17, 72, 1 / 6)
```

The p-value is shown graphically in Figure 4.23, which shows the probability mass function of a binomial(72, 1/6) distribution, and the sum of the probability mass function values from 18 to 72. Since large values of y lead to rejecting H_0, the p-value is the sum of the values of the probability mass function of Y under H_0 from 18 to 72. The interpretation of the p-value $p = 0.046$ is that if H_0 is true (in this application, the anticipation of a birthday does not prolong life), then the probability is 0.046 of obtaining the observed test statistic $y = 18$ or a test statistic more extreme (that is, 18 or larger). The reader is then required to determine whether this probability is low enough for them to conclude that the anticipation of a birthday prolongs life. The p-value is known as the "achieved significance level" because it indicates the significance level achieved by the test statistic. Setting a significance level α before collecting data has been eliminated from the process.

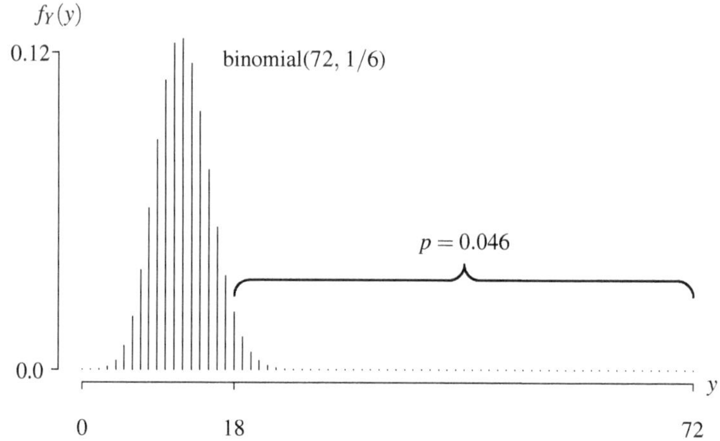

Figure 4.23: Calculating the p-value.

The next question concerns whether it is better to use a hypothesis test or a significance test in practice. In nearly all settings, a significance test is more appropriate for the following reasons.

- Significance testing shifts the burden of drawing a conclusion concerning a test from the experimenter to the reader. If a significance test results in the p-value $p = 0.19$, for example, the reader concludes that there is nearly a one-in-five chance that this test statistic or one more extreme could have been produced by chance when H_0 is true and hence the reader fails to reject H_0. On the other hand, if a significance test results in the p-value $p = 0.003$, the reader concludes that there is only a three-in-a-thousand chance that this test statistic could have been produced by chance when H_0 is true and hence the reader rejects H_0. The p-value quantifies the evidence in a sample for rejecting H_0.

- Significance testing removes the problem associated with stating one level of α in a problem statement, then achieving a lesser value of α for random sampling from discrete distributions. As seen in the previous example, sampling from a discrete population poses no problem over sampling from a continuous population.
- Conducting a significance test is equivalent to conducting an infinite number of traditional hypothesis tests. The reader knows that for all values of α less than the p-value, the associated hypothesis test would fail to reject H_0. Alternatively, for all values of α greater than the p-value, the associated hypothesis test would reject H_0.

Although significance testing is typically the preferred approach for working with actual data sets, it is important to understand hypothesis testing because many of the theoretical results given in the next section require an understanding of hypothesis testing.

The previous example concerned sampling from a discrete distribution—the Bernoulli distribution. The next example concerns sampling from a continuous distribution. In particular, we return to the first example from this chapter, which was solved entirely from intuition. We now conduct a significance test to determine a p-value. A summation was used to calculate the p-value in the previous example; an integral will be used to calculate the p-value in the next example.

Example 4.12 A single observation X is collected to test the hypothesis

$$H_0 : \theta = 1$$

versus

$$H_1 : \theta = 3,$$

where θ is a positive unknown parameter from the continuous population distribution described by the probability density function

$$f(x) = 2\theta^{-2} x e^{-(x/\theta)^2} \qquad x > 0.$$

Conduct a significance test associated with the data value $x = 2.2$.

The population distribution is the Rayleigh distribution. The population probability density functions associated with H_0 and H_1 are graphed in Figure 4.24. It is clear from Figure 4.24 that a large value of x is evidence in favor of rejecting H_0. So the p-value is the area under the probability density function associated with H_0 to the right of $x = 2.2$:

$$p = \int_{2.2}^{\infty} 2xe^{-x^2}\, dx = \left[-e^{-x^2}\right]_{2.2}^{\infty} = e^{-2.2^2} = 0.007907.$$

This small p-value means that there is only an eight-in-a-thousand probability that the value $x = 2.2$ or one more extreme would be produced from the probability distribution associated with H_0. Since $x = 2.2$ corresponds to a very unlikely observation when H_0 is true and also is in the middle of the probability distribution associated with H_1, we decide to reject H_0. The geometry associated with the p-value is shown in Figure 4.25.

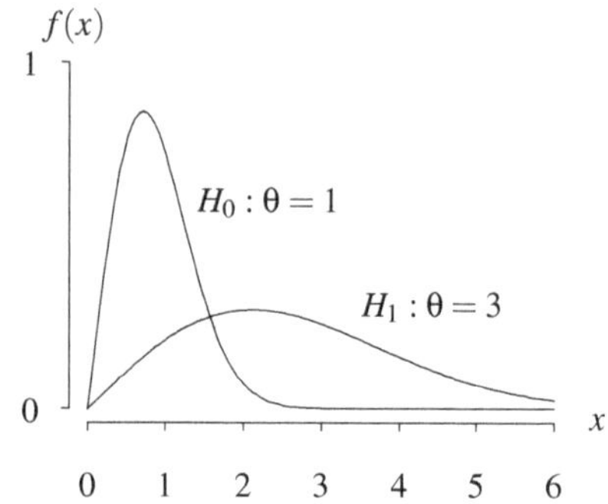

Figure 4.24: Population probability density functions for $\theta = 1$ and $\theta = 3$.

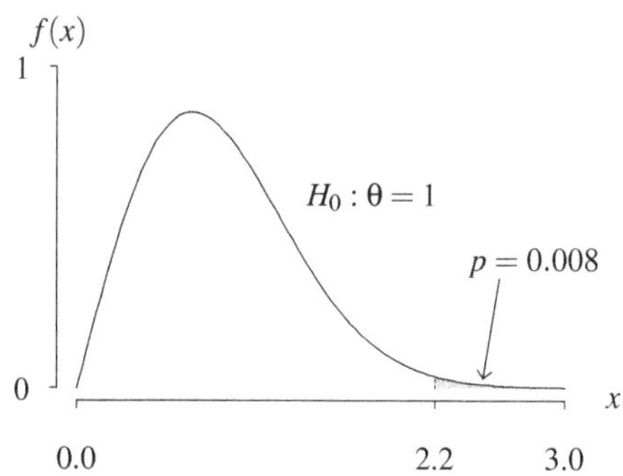

Figure 4.25: Calculating the p-value.

The third and final example of significance testing considers a two-sided test associated with major hurricane strikes on the mainland United States.

Example 4.13 The Saffir–Simpson Hurricane Wind Scale rates hurricanes based on their sustained wind speed and places them into one of five categories. Category 1 hurricanes have 74–95 MPH sustained wind speed and Category 5 hurricanes have greater than 157 MPH sustained wind speed. Hurricanes that reach Categories 3–5 are considered "major hurricanes" because of their potential for significant property damage and loss of life. The National Hurricane Center has determined that there were 79 major hurricanes that struck the mainland United States from 1851–1980. The number of major hurricanes during each of those 13 decades is:

$$6 \quad 1 \quad 7 \quad 5 \quad 8 \quad 4 \quad 7 \quad 5 \quad 8 \quad 10 \quad 8 \quad 6 \quad 4.$$

This corresponds to $79/13 = 6.1$ major hurricanes per decade. (The decade with the lowest number of major hurricanes was the 1860s with just one major hurricane, and the decade with the highest number of major hurricanes was the 1940s with ten major hurricanes.) The purpose of this example is to conduct a significance test to determine whether there has been a *change* in the number of major hurricanes during the 1980s and 1990s relative to the 1851–1980 time period. There were five major hurricanes in the 1980s and five major hurricanes in the 1990s.

The question posed is whether there is a *change* in the number of major hurricanes during this 20-year period. Since this change might correspond to an increase or decrease, the hypothesis test will be two-sided. The next step is to determine an appropriate sampling distribution for this data set. This is a nontrivial step because the data set is so small that plotting a histogram is not meaningful. Each data value, however, is the sum of ten hurricane seasons worth of data because each data point represents a decade. Since the support of the population distribution is $\mathcal{A} = \{x \mid x = 0, 1, 2, \ldots\}$ and the central limit theorem can be invoked, the Poisson distribution is a potential population model. The Poisson distribution is asymptotically normal as $\lambda \to \infty$. One approach is to test the equality of the population means for the first 13 decades and the subsequent two decades. We will take a different approach here and make a further assumption that the sample mean for the first 13 decades is in fact the fixed parameter λ in the Poisson distribution. This assumption means that the significance test will be approximate, rather than exact. So the hypothesis test becomes

$$H_0 : \lambda = 6.1$$

versus

$$H_1 : \lambda \neq 6.1,$$

where λ is a positive unknown parameter in the Poisson distribution which represents the population mean number of major hurricanes per decade. The $n = 2$ experimental values of X_1 and X_2, namely $x_1 = 5$ and $x_2 = 5$, show that the number of major hurricanes has *decreased* from $\lambda = 6.1$ major hurricanes per decade, but it is not clear whether or not the decrease is statistically significant.

The next step is to determine a test statistic. Since the sum of independent Poisson random variables is also Poisson, the test statistic $X_1 + X_2$ has a Poisson(12.2) distribution under H_0. Small and large values of the test statistic lead to rejecting H_0. The value of the test statistic for the two data values is $x_1 + x_2 = 10$. Since the probability mass function for a Poisson(λ) random variable is

$$f(x) = \frac{\lambda^x e^{-\lambda}}{x!} \qquad x = 0, 1, 2, \ldots,$$

the probability that a Poisson(12.2) random variable is ten or less is

$$P(X_1 + X_2 \leq 10) = \sum_{x=0}^{10} \frac{12.2^x e^{-12.2}}{x!} = 0.3313,$$

as illustrated in Figure 4.26. This probability would be the p-value if there were a one-sided alternative hypothesis ($H_1 : \lambda < 6.1$) to see if the number of major hurricanes is decreasing. But this test is two-sided, so this quantity must be doubled; the p-value for the two-sided test is

$$p = 0.6626.$$

This p-value can be calculated with the following R statements.

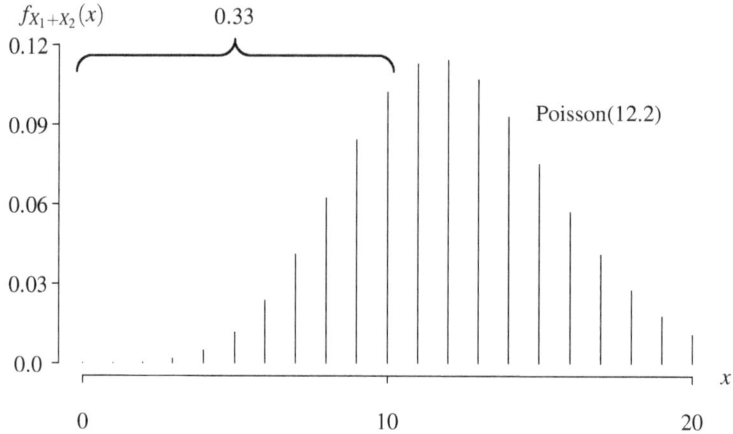

Figure 4.26: Calculating the p-value for the major hurricane data.

```
lambda = 2 * 79 / 13
2 * ppois(10, lambda)
```

Under H_0, the probability is 0.66 that this particular data set or one more extreme could have occurred if there were no change in the number of major hurricanes. This p-value is so high that the conclusion in this case is obvious: we fail to reject H_0 and conclude that the data for the number of major hurricanes in the 1980s and 1990s shows no statistically significant change from the major hurricane data from 1851–1980.

Controversy

Hypothesis tests and significance tests have come under scrutiny recently. In 2016, the American Statistical Association issued a statement concerning the use of p-values in scientific studies. The statement argued against the use of "mechanical bright-line rules (such as $p < 0.05$)" in the application of p-values in significance testing, and, more generally, the approach presented in this chapter of Null Hypothesis Significance Testing (NHST). The statement emphasizes the difficulty interpreting the p-value:

> "While the p-value can be a useful statistical measure, it is commonly misused and misinterpreted."

Since p-values are difficult to interpret cleanly, many non-statisticians misinterpret them, and the results of these misinterpretations are scattered throughout the scientific literature.

To reiterate the proper interpretation of a p-value, it is the probability of getting, in a future study with the same sample size and sampling assumptions, a statistical summary (usually the test statistic) that is equal to or more extreme than the observed value obtained in the current study, conditioned on the null hypothesis and any associated conditions on the null hypothesis (for example, a random sample was drawn from a particular probability distribution) being true. Briefly, a p-value is a measure of the compatibility between a statistical model and a data set.

There is a long list of critiques of significance testing and p-values. Here are a few of them.

- A significance test is used to argue *against* the null hypothesis, but it does not argue *for* the null hypothesis.
- The calculation of a p-value conditions on the postulated null hypothesis rather than conditioning on the known data values.
- Except for random samples from mathematically simple probability distributions, p-values can be difficult to calculate.
- Conditioning on a null hypothesis on a single population parameter, for example, "the coin is fair" is $H_0 : p = 1/2$, is straightforward. This is not the case for a more complicated compound null hypothesis, for example, "the coin and the die are both fair" is

$$H_0 : p_0 = 1/2, p_1 = 1/6, p_2 = 1/6, p_3 = 1/6, p_4 = 1/6, p_5 = 1/6, p_6 = 1/6.$$

- When a significance test has a small p-value, it is not clear which of the assumptions associated with H_0 has been violated. From the previous example concerning hurricanes, if H_0 had been rejected, it could have been that
  - the number of hurricanes per decade in the 1980s and 1990s differed from $\lambda = 6.1$,
  - the Poisson population assumption was violated (perhaps a different population distribution was appropriate),
  - the random sampling assumption was violated (perhaps a major hurricane occurring late in a decade might influence the probability that a hurricane will occur early in the next decade).

Although there are many criticisms of the use of p-values, it is still important to understand them and consider alternatives. Two that have been suggested are (*a*) to replace the usual threshold in the scientific literature $p < 0.05$ with $p < 0.005$, effectively requiring considerably more statistical evidence to reject H_0, and (*b*) replace the frequentist approach to statistics with the Bayesian approach. At this time, it is not clear whether significance testing will persist or will be replaced by an alternative. Stay tuned.

4.3 Sampling from Normal Populations

The most widely used probability distribution in statistical inference is the normal distribution. Even the choice of its name as "normal" implies that some early statisticians found that the normal distribution arose so often in applications that other distributions were considered "abnormal." The reason that the normal distribution is so pervasive is that so many phenomena in nature, science, engineering, economics, business, sociology, etc. follow a bell-shaped probability distribution. Sir Francis Galton (1822–1911), who was one of the early pioneers in statistics, waxed poetic concerning the normal distribution (which at the time was known as the "law of frequency of error"):

> I know of scarcely anything so apt to impress the imagination as the wonderful form of cosmic order expressed by the "Law of Frequency of Error." The law would have been personified by the Greeks and deified, if they had known of it. It reigns with serenity and in complete self-effacement, amidst the wildest confusion. The huger the mob, and the greater the apparent anarchy, the more perfect is its sway. It is the supreme law of Unreason. Whenever a large sample of chaotic elements are taken in hand and marshaled in the order of their magnitude, an unsuspected and most beautiful form of regularity proves to have been latent all along.

Section 1.5 contained derivations of several results concerning random sampling from normal populations which will be used here.

Let's begin with what is known as a *one-sample test*, that is, the question of interest involves a random sample of n data values $X_1, X_2, \ldots, X_n$ drawn from a single normal population with population mean μ and population variance σ^2. Furthermore, assume that the interest is in a hypothesis test concerning the population mean of that normal distribution, that is,

$$H_0 : \mu = \mu_0$$

versus one of the three possible alternative hypotheses, namely, the two-sided alternative hypothesis $H_1 : \mu \neq \mu_0$, or one of the one-sided alternative hypotheses $H_1 : \mu > \mu_0$ or $H_1 : \mu < \mu_0$, where μ_0 is a constant that is determined from the question of interest. A hypothesis test of this kind can be broken into two cases. First, if the population variance σ^2 is known, which is seldom the case in practice, then Theorem 1.3 can be used with the test statistic

$$Z = \frac{\bar{X} - \mu_0}{\sigma/\sqrt{n}},$$

which has the standard normal distribution under H_0. Second, if the population variance σ^2 is not known, which is the much more common case in practice, then Theorem 1.7 can be used with the test statistic

$$T = \frac{\bar{X} - \mu_0}{S/\sqrt{n}},$$

which has the t distribution with $n-1$ degrees of freedom under H_0. The two test statistics, Z and T, have nearly the same probability distribution for large values of n under H_0 because S will typically be quite close to σ. But the two differ significantly for small values of n, particularly in the sense that the probability density function of T has heavier tails than that of Z. The more common case of a test concerning the population mean for an unknown population variance will be illustrated in the next example for a small sample size.

Example 4.14 An automobile tire manufacturing company has a new tire design which it rates as a 50,000 mile tire. Engineers at the company place $n = 5$ randomly selected new tires on various vehicles operating under a variety of operating conditions. The tires fail at the following mileages (in thousands of miles):

38.8 52.4 51.1 33.6 41.6.

These five tires had sample mean lifetime $\bar{x} = 43.5$ thousand miles. Is there enough evidence in this small sample to conclude that the population mean lifetime of the automobile tires is less than 50 thousand miles?

Since there is no significance level α specified, we will conduct a significance test. The problem statement makes it clear that the hypothesis test is

$$H_0 : \mu = 50$$

versus

$$H_1 : \mu < 50.$$

The test on the five tires was conducted under "encountered" or "field" conditions rather than laboratory conditions, which gives the engineers insight on how the tires will perform on the road. But how can we be sure that the normal distribution is an appropriate population model for this hypothesis test? The small number of observations does not make a histogram a viable option for assessing the shape of the population probability density function. Intuition would suggest that the lifetime is bell-shaped because tire treads are wearing away over time, but this is not enough to establish normality. The only hope in this case is that there is previous test data on tire designs that are reasonably close to the current design that have exhibited bell-shaped behavior in the past. Assuming that such data exist and exhibit bell-shaped behavior, we can proceed with the test. Since $n = 5$, $\mu_0 = 50$, $\bar{x} = 43.5$, and $s = 8.1$, the test statistic is

$$t = \frac{\bar{x} - \mu_0}{s/\sqrt{n}} = \frac{43.5 - 50}{8.1/\sqrt{5}} = -1.8.$$

As illustrated in Figure 4.27, the p-value for this significance test is the area to the left of -1.8 under the probability density function of a t random variable with $n - 1 = 4$ degrees of freedom, which is the distribution of the test statistic under H_0.

The R code below calculates the p-value for the hypothesis test.

```
x = c(38.8, 52.4, 51.1, 33.6, 41.6)
n = length(x)
m = mean(x)
s = sd(x)
pt((m - 50) / (s / sqrt(n)), n - 1)
```

The code returns the p-value $p = 0.07$. Since this p-value is greater than a threshold value of, say, 0.05, we fail to reject H_0 and (barely) conclude that there is not enough statistical evidence to conclude that the population mean tire lifetime is less than 50,000 miles. Even though the sample mean lifetime is 6,500 miles below the stated population mean lifetime, a combination of a small sample size ($n = 5$) and significant spread to the data values ($s = 8.1$) results in failing to reject H_0. Since problems of this nature arise often in hypothesis testing, there is a built-in function `t.test` in R that can save a few keystrokes over the code given above. The following R code returns identical results to the code given above.

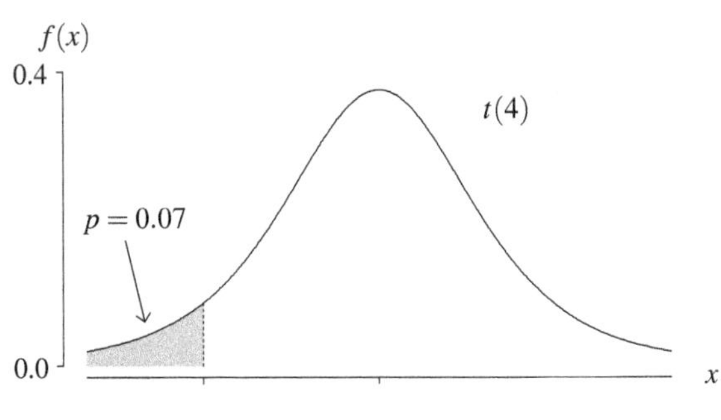

Figure 4.27: Probability density function of a $t(4)$ random variable.

```
x = c(38.8, 52.4, 51.1, 33.6, 41.6)
t.test(x, mu = 50, alternative = "less")
```

There is a second class of tests concerning sampling from normal populations known as *two-sample tests*, where two samples are drawn from normal populations. A long example of a two-sample test is given next.

Example 4.15 An experiment was conducted to determine the influence of caffeine on the performance of a simple physical task known as finger tapping. Twenty male college students were trained in the art of finger tapping. Ten of the students were selected at random to receive a placebo containing no caffeine and the other ten received 200 mg of caffeine. Two hours after the administration of the placebo or caffeine, the students performed a finger tapping experiment in which the number of finger taps per minute was recorded. The experiment was double-blind in the sense that neither the physician nor the student receiving the placebo or caffeine knew which treatment was being administered. The experimenters, however, did know this information. The number of finger taps per minute for the ten students given the placebo were

$$242, 245, 244, 248, 247, 248, 242, 244, 246, 242.$$

The number of finger taps per minute for the ten students given 200 mg of caffeine were

$$246, 248, 250, 252, 248, 250, 246, 248, 245, 250.$$

Conduct a hypothesis test to determine whether students who get caffeine tap faster (that is, an increased number of taps per minute) than students without caffeine at significance level $\alpha = 0.05$.

Let $X_1, X_2, \ldots, X_n$ denote the random sample from the group receiving the placebo and $Y_1, Y_2, \ldots, Y_m$ denote the random sample from the caffeinated group. For this experiment, $n = m = 10$. Some sample statistics associated with these two data sets are

$$\bar{x} = 244.8 \qquad \bar{y} = 248.3 \qquad s_X = 2.39 \qquad s_Y = 2.21.$$

One might wonder how we failed to detect a difference of 6,500 miles in the automobile tire test from Example 4.14, but now we hope to detect a difference of just $\bar{y} - \bar{x} = 248.3 - 244.8 = 3.5$ taps per minute in this example. The explanation lies in the spread of the data values in the data sets. The automobile tire lifetimes ranged from 33,600 miles to 52,400 miles, but the finger tapping data sets vary on a very small range—their values are tightly clustered as evidenced by the small values of the sample standard deviations s_X and s_Y. Let μ_X denote the population mean number of finger taps per minute of the students who were given the placebo. Let μ_Y denote the population mean number of finger taps per minute of the students who were given caffeine. Since the question posed in the problem statement is whether caffeinated students tap faster, the appropriate one-sided hypothesis test, commonly known as a one-sided t test, is

$$H_0 : \mu_X = \mu_Y$$

versus

$$H_1 : \mu_X < \mu_Y.$$

The next step is to determine whether the data came from a normal population. Since the two data sets are only of size ten, drawing a histogram is not a viable option. On the other hand, an empirical cumulative distribution function can be drawn for both data sets, and they are superimposed over one another in Figure 4.28. The classic S-shape associated with a normal population (this was apparent with the speed of light data set in Figure 1.13) is not apparent with either of these data sets. They might have indeed been drawn from normal populations, but that is not immediately apparent from Figure 4.28. Furthermore, the discrete nature of the two data sets consisting of the number of finger taps per minute does not help to establish normality. We will, for now, assume that the normality assumption is justified, then revisit that assumption at the end of this example. The normality assumption would be justified if we had previous data that produced bell-shaped histograms for the placebo and caffeine groups.

So we assume that $X_1, X_2, \ldots, X_n$ is a random sample from a $N(\mu_X, \sigma_X^2)$ population, where μ_X and σ_X^2 are fixed but unknown parameters. Likewise, $Y_1, Y_2, \ldots, Y_m$ is assumed to be a random sample from a $N(\mu_Y, \sigma_Y^2)$ population, where μ_Y and σ_Y^2 are fixed but unknown parameters. One of the first questions is whether the two population variances, that is, σ_X^2 and σ_Y^2, are equal. Answering this question will steer us toward using the pooled sample variance in the hypothesis test or using Welch's approximation (illustrated in Example 3.15). The fact that the two sample standard deviations are nearly the same would lead us to believe that the two population variances are equal, but let's conduct a significance test to assure ourselves that this is indeed the case. So we conduct the two-sided significance test

$$H_0 : \sigma_X^2 = \sigma_Y^2$$

versus

$$H_1 : \sigma_X^2 \neq \sigma_Y^2$$

in order to check for equality of population variances prior to conducting the hypothesis test of interest concerning the population means. The reason that the alternative hypothesis is two-sided is that prior to collecting the data, the experimenter would have no sense of which population would have a larger population variance. In other words,

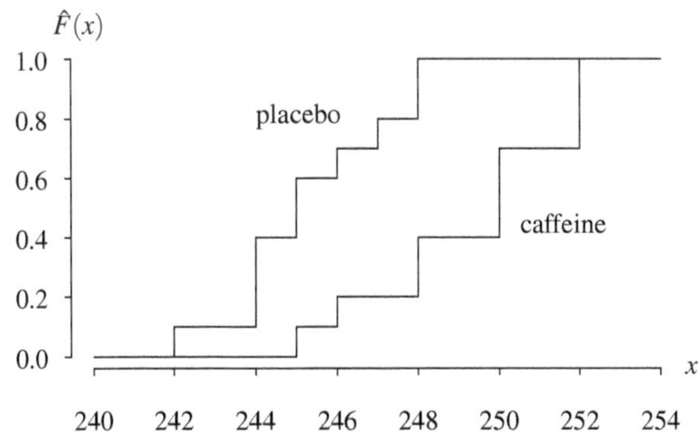

Figure 4.28: Empirical cumulative distribution functions for the finger tapping data.

there should be no data snooping here to determine the alternative hypothesis. Theorem 1.10 indicates that the test statistic

$$F = \frac{S_X^2}{S_Y^2}$$

has the F distribution with $n-1$ and $m-1$ degrees of freedom under H_0. For the finger tapping data for the two groups, the value of the test statistic is

$$f = \frac{s_X^2}{s_Y^2} = \frac{2.39^2}{2.21^2} = 1.17$$

and the associated p-value is

$$p = 2P(F > 1.17) = 0.82,$$

where F is an F random variable with $n-1 = 10-1 = 9$ and $m-1 = 10-1 = 9$ degrees of freedom. The R code below calculates the value of the test statistic and the associated p-value for the significance test.

```
x = c(242, 245, 244, 248, 247, 248, 242, 244, 246, 242)
y = c(246, 248, 250, 252, 248, 250, 246, 248, 245, 250)
f = var(x) / var(y)
p = 2 * (1 - pf(f, 9, 9))
```

Tests for the equality of population variances for sampling from normal populations occur so frequently in practice that they are coded into the R function `var.test`. The R code below produces identical results to those above.

```
x = c(242, 245, 244, 248, 247, 248, 242, 244, 246, 242)
y = c(246, 248, 250, 252, 248, 250, 246, 248, 245, 250)
var.test(x, y, alternative = "two.sided")
```

With the large p-value of $p = 0.82$, we fail to reject H_0. There is not a statistically significant difference between the two population variances based on the data values. This conclusion allows us to now move on to testing the original hypothesis test concerning the population means using the pooled sample variance approach.

So we now return to the original hypothesis test

$$H_0 : \mu_X = \mu_Y$$

versus

$$H_1 : \mu_X < \mu_Y.$$

Theorem 1.9 states that

$$T = \frac{\bar{X} - \bar{Y}}{S_P\sqrt{\frac{1}{n} + \frac{1}{m}}}$$

has the t distribution with $n+m-2$ degrees of freedom under H_0. The pooled sample variance is

$$s_p^2 = \frac{(n-1)s_X^2 + (m-1)s_Y^2}{n+m-2} = \frac{9 \cdot 2.39^2 + 9 \cdot 2.21^2}{10+10-2} = \frac{95.7}{18} = 5.32.$$

As a check, the pooled sample standard deviation $s_p = \sqrt{5.32} = 2.31$ lies between $s_X = 2.39$ and $s_Y = 2.21$, as expected. The test statistic is

$$t = \frac{\bar{x} - \bar{y}}{s_p\sqrt{\frac{1}{n} + \frac{1}{m}}} = \frac{244.8 - 248.3}{2.31\sqrt{\frac{1}{10} + \frac{1}{10}}} = \frac{-3.5\sqrt{5}}{2.31} = -3.39.$$

The associated critical value for $\alpha = 0.05$, which can be found with a table lookup or the `qt` function in R, is $c = -1.73$. Since the test statistic falls way to the left of the critical value, H_0 is rejected and it is concluded that 200 mg of caffeine is effective for superior finger tapping performance in male college students. Figure 4.29 shows that the evidence for rejecting H_0 derived from the test statistic $t = -3.39$ is overwhelming.

The following R code performs the calculations to compute the test statistic and critical value.

```
x = c(242, 245, 244, 248, 247, 248, 242, 244, 246, 242)
y = c(246, 248, 250, 252, 248, 250, 246, 248, 245, 250)
n = length(x)
m = length(y)
s = sqrt(((n - 1) * var(x) + (m - 1) * var(y)) / (n + m - 2))
test.statistic = (mean(x) - mean(y)) / (s * sqrt(1 / n + 1 / m))
critical.value = qt(0.05, n + m - 2)
```

As before, the R function `t.test`, with appropriate parameters, can also be used to conduct the hypothesis test using fewer keystrokes. The appropriate statements are given below.

```
x = c(242, 245, 244, 248, 247, 248, 242, 244, 246, 242)
y = c(246, 248, 250, 252, 248, 250, 246, 248, 245, 250)
t.test(x, y, alternative = "less", var.equal = TRUE, conf.level = 0.95)
```

The concepts of *signal* and *noise* apply naturally to this example. These two terms are familiar to electrical engineers and astronomers who need to separate a signal from

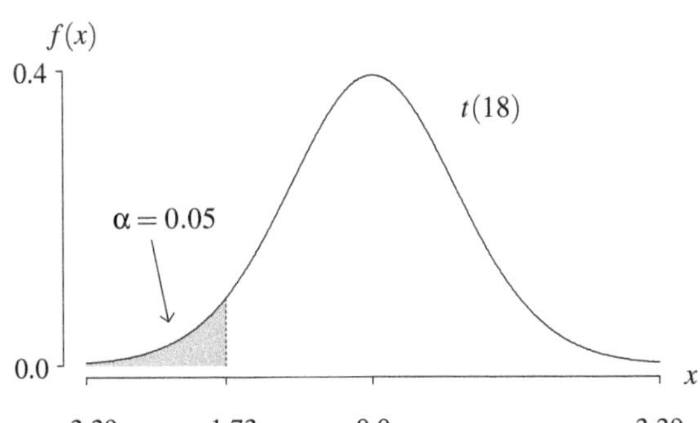

Figure 4.29: Probability density function of a $t(18)$ random variable.

inherent noise. In this example, the signal is the effect of caffeine on finger tapping. The desire is to measure this signal. Unfortunately, the signal can get lost in the noise associated with the process. The noise results in increased variability which can mask the signal. The experimenters have taken the following steps to tamp down the natural random sampling variability associated with the noise.

- The experimenters limited their study to just men. A quick scan of some internet articles seems to indicate that women have a slight advantage over men in terms of dexterity (and one would assume that this dexterity advantage influences finger tapping ability). Thus, including women in the study would have increased the variability in the two groups, resulting in an experiment with more noise that might obscure the signal. A follow-on study involving just female subjects would be appropriate.
- The experimenters limited their study to just college students. If instead they opened the study to anyone between the ages of 5 and 95, this would have resulted in increased variability due to the wide age range. The increased variability due to the children and elderly subjects would make it more difficult to detect the signal.
- The experimenters used a double-blind study. The dose of caffeine was low enough so that the subjects were probably not aware of whether they had taken the placebo or the caffeine. This decision was made to ward off any placebo effects.
- The experimenters waited exactly two hours between the administration of the placebo or caffeine before conducting the finger tapping experiment. Using a different time for each student would have added unnecessary noise to the experiment.
- The experimenters chose a process, namely finger tapping, that was easy to measure. This also results in a less noisy experiment. The experiment should be conducted in the same room with identical lighting conditions, an identical finger tapping measuring device, identical temperature, etc. to decrease the noise infused into the experiment.
- The experimenters split the subjects into the two groups at random. This should be done via a random number generator to avoid any bias.

As this list indicates, it is important to carefully consider every detail associated with the experiment so as to minimize the harmful impact that noise has on detecting the signal of interest. Did the experimenters do all that was possible to reduce the noise? There are two more steps that they could have taken. Although the process of assigning ten subjects to each group was done at random, it is still possible that, by chance, one group contains naturally better finger tappers than the other group. An alternative that would reduce the noise associated with the group assignment is to conduct a test with ten subjects on one day, then conduct a test with *the same ten subjects* the next day. One day a subject would have the placebo or caffeine, based on the flip of a fair coin, then the other day the same subject would receive the other treatment. The analysis would be slightly different from the analysis given above in that the data is now given in pairs, and the difference between the pairs of finger tapping measurements is of interest. Statisticians refer to this as a *paired t test* and the `paired = TRUE` parameter on the R function `t.test` can be set to conduct the test. The test boils down to a one-sample test based on normal sampling because the difference of two independent normal random variables is itself normal. A second additional way to reduce the noise is to drill down deeper into the population of male college students. Selecting a subset of these students, say right-handed, male members of the soccer team, would limit the applicability of the experimental results, but would also decrease the noise. In summary, an important part of conducting a statistical experiment is to eliminate all unnecessary noise so that only random sampling remains and it is easier to detect the signal of interest.

The hypothesis test conducted on the finger tapping data set overwhelmingly indicated that caffeine improves finger tapping. But there is some lingering doubt about our conclusion because it was based on the assumption of normal sampling. The empirical cumulative distribution functions for the two groups in Figure 4.28 did not indicate normality, so going ahead with the analysis with this assumption in place might not have been appropriate. Is there an alternative approach that does not depend on any parametric assumptions? The bootstrapping methodology that was introduced in Chapter 3 can be adapted for hypothesis testing and significance testing. The bootstrapping algorithm given below makes no parametric assumptions on the population distribution.

1. Collect $n = 10$ finger tapping counts from the placebo group and $m = 10$ finger tapping counts from the caffeine group.
2. Calculate the point estimator $\bar{x} - \bar{y} = 244.8 - 248.3 = -3.5$ taps per minute of the parameter of interest $\mu_X - \mu_Y$.
3. Repeat the following steps B times, where B is a large integer.
   a. Generate a *bootstrap sample* of $n = 10$ observations drawn *with replacement* from just the placebo group.
   b. Generate a *bootstrap sample* of $m = 10$ observations drawn *with replacement* from just the caffeine group.
   c. Calculate and store a bootstrap point estimator of $\mu_X - \mu_Y$, which is the difference between the sample means of the two bootstrap samples.
4. Sort the B bootstrap point estimators of $\mu_X - \mu_Y$.
5. Report the fraction of the bootstrap point estimators that exceed zero as an estimate of the p-value of the associated significance test.

The algorithm makes no assumptions concerning the population distribution being the normal distribution, or any other parametric distribution. It simply resamples the data that was collected. Using bootstrapping is appropriate when there is a small sample size, there are no previous similar data sets to provide guidance, or the experimenter is nervous about making a parametric assumption. The following R code implements the bootstrap algorithm for $B = 10,000$ bootstrap replicates.

```
B  = 10000
z  = numeric(B)
x = c(242, 245, 244, 248, 247, 248, 242, 244, 246, 242)
y = c(246, 248, 250, 252, 248, 250, 246, 248, 245, 250)
for (i in 1:B) {
  xx = sample(x, replace = TRUE)
  yy = sample(y, replace = TRUE)
```

```
  z[i] = mean(xx) - mean(yy)
}
print(sum(z > 0) / B)
```

After a call to `set.seed(16)` to initialize the random number stream, the code returns 0.0002, which can be interpreted as an estimated nonparametric p-value for the associated significance test. What is necessary for the difference between the sample means of the two bootstrap samples to be positive? The large values of the placebo group, such as 247 and 248, must be sampled frequently, and the small values of the caffeine group, such as 245 and 246, must also be sampled frequently. This will not occur very often. In the bootstrap experiment, it only happened twice in ten thousand replications. So the estimated p-value for the bootstrapping significance test is $p = 0.0002$. This is fairly consistent with the p-value associated with the significance test for random sampling from two normal populations, which returns $p = 0.002$ via the R statement `pt(-3.39, 18)`. Both approaches overwhelmingly reject H_0. There is strong evidence in the data set that caffeine improves finger tapping performance. The lingering doubt is resolved.

Sometimes there are multiple ways to address a hypothesis test by bootstrapping, and this example provides an opportunity to devise such an alternative approach. Consider the algorithm below.

1. Collect $n = 10$ finger tapping counts from the placebo group and $m = 10$ finger tapping counts from the caffeine group.
2. Calculate the point estimator $\bar{x} - \bar{y} = 244.8 - 248.3 = -3.5$ taps per minute of the parameter of interest $\mu_X - \mu_Y$.
3. Pool the $n + m = 20$ observations into a single combined sample, which corresponds to the 20 observations coming from the same distribution under H_0.
4. Repeat the following steps B times, where B is a large integer.
   a. Generate a *bootstrap sample* of $n = 10$ observations drawn *with replacement* from the combined sample.
   b. Calculate and store the difference between the sample mean of the $n = 10$ observations selected in step (a) minus the sample mean of those not selected.
5. Sort the B bootstrap point estimators.
6. Report the fraction of the bootstrap point estimators that are less than -3.5 as an estimate of the p-value of the associated significance test.

The following R code implements this bootstrap algorithm for $B = 10{,}000$ bootstrap replicates.

```
B = 10000
z = numeric(B)
x = c(242, 245, 244, 248, 247, 248, 242, 244, 246, 242)
y = c(246, 248, 250, 252, 248, 250, 246, 248, 245, 250)
xy = c(x, y)
binary = c(rep(TRUE, 10), rep(FALSE, 10))
for (i in 1:B) {
  index = sample(binary)
```

```
    z[i] = mean(xy[index]) - mean(xy[!index])
}
print(sum(z < -3.5) / B)
```

After a call to `set.seed(16)` to initialize the random number stream, the code returns 0.0016, which can be interpreted as an estimated nonparametric p-value for the associated significance test. This value is fairly consistent with the parametric analysis using the t test and the initial bootstrapping approach.

In the previous example, bootstrapping was given as an alternative to a traditional parametric-based hypothesis test or significance test. Bootstrapping's popularity stems from its lack of parametric assumptions, but also from its versatility. Instead of conducting a test concerning population means, the bootstrapping approach easily adapts to other measures of a probability distribution, such as the population median, the population 95th percentile, the population skewness, etc.

Conducting a test with two different sets of assumptions (in the finger tapping example, an assumption of normal sampling, and no assumption at all concerning the population distribution for bootstrapping) is a very reasonable approach for small data sets. If the results agree, then there is statistical confirmation under the two sets of assumptions. If the results disagree, then the conclusion depends upon the assumptions, and this might indicate that it is wise to collect more data to determine which assumptions are more appropriate for the setting.

The previous two examples have illustrated a one-sample test (on the automobile tire lifetime data) and a two-sample test (on the finger tapping data). But there are numerous other examples of tests that could be conducted on random samples drawn from normal populations. The application of the results from Section 1.5 in the context of hypothesis testing are summarized in Table 4.4. The first column gives the theorem number from Section 1.5 that provides the basis for the test. The second column gives a generic form of the null hypothesis H_0. The third column gives the test statistic and its probability distribution under H_0. The fourth column gives the two-sided and one-sided alternative hypotheses H_1. The fifth column gives the critical region associated with the test. Not counting the headings, the first four rows of the table concern one-sample problems. Then, following the double line, the last three rows concern two-sample problems. The two rows that correspond to the t distribution under H_0 are commonly known as t-tests. The paired t test is not explicitly in the table because it involves taking differences in observations, so the analysis reverts to that of a one-sample problem. The assumptions associated with a random sample drawn from normal populations that allow the results in Table 4.4 to be valid are given in Section 1.5.

The next example uses an entry in Table 4.4 to conduct a hypothesis test for a sample drawn from a normal population. We revisit Example 4.3 in order to show that using Table 4.4 is consistent with solving the problem from first principles.

Example 4.16 Let X_1 and X_2 be a random sample of size $n = 2$ from a $N(\mu, 1)$ population, where μ is an unknown parameter. Use the appropriate entry from Table 4.4 to conduct the hypothesis test

$$H_0 : \mu = 5$$

against the one-sided alternative hypothesis

$$H_1 : \mu < 5$$

at $\alpha = 0.1$. Find the critical value c and plot the power function for this test.

Thm	H_0	Test statistic and distribution under H_0	H_1	Critical region
1.3	$\mu = \mu_0$	$Z = \frac{\bar{X}-\mu_0}{\sigma/\sqrt{n}} \sim N(0, 1)$	$\mu \neq \mu_0$	$z < -z_{\alpha/2}$ or $z > z_{\alpha/2}$
			$\mu > \mu_0$	$z > z_\alpha$
			$\mu < \mu_0$	$z < -z_\alpha$
1.4	$\sigma^2 = \sigma_0^2$	$Y = \sum_{i=1}^{n} \left(\frac{X_i-\mu}{\sigma_0}\right)^2 \sim \chi^2(n)$	$\sigma^2 \neq \sigma_0^2$	$y < \chi^2_{n,1-\alpha/2}$ or $y > \chi^2_{n,\alpha/2}$
			$\sigma^2 > \sigma_0^2$	$y > \chi^2_{n,\alpha}$
			$\sigma^2 < \sigma_0^2$	$y < \chi^2_{n,1-\alpha}$
1.6	$\sigma^2 = \sigma_0^2$	$Y = \frac{(n-1)S^2}{\sigma_0^2} \sim \chi^2(n-1)$	$\sigma^2 \neq \sigma_0^2$	$y < \chi^2_{n-1,1-\alpha/2}$ or $y > \chi^2_{n-1,\alpha/2}$
			$\sigma^2 > \sigma_0^2$	$y > \chi^2_{n-1,\alpha}$
			$\sigma^2 < \sigma_0^2$	$y < \chi^2_{n-1,1-\alpha}$
1.7	$\mu = \mu_0$	$T = \frac{\bar{X}-\mu_0}{S/\sqrt{n}} \sim t(n-1)$	$\mu \neq \mu_0$	$t < -t_{n-1,\alpha/2}$ or $t > t_{n-1,\alpha/2}$
			$\mu > \mu_0$	$t > t_{n-1,\alpha}$
			$\mu < \mu_0$	$t < -t_{n-1,\alpha}$
1.8	$\mu_X = \mu_Y$	$Z = \frac{\bar{X}-\bar{Y}}{\sqrt{\frac{\sigma_X^2}{n}+\frac{\sigma_Y^2}{m}}} \sim N(0, 1)$	$\mu_X \neq \mu_Y$	$z < -z_{\alpha/2}$ or $z > z_{\alpha/2}$
			$\mu_X > \mu_Y$	$z > z_\alpha$
			$\mu_X < \mu_Y$	$z < -z_\alpha$
1.9	$\mu_X = \mu_Y$	$T = \frac{\bar{X}-\bar{Y}}{S_p\sqrt{\frac{1}{n}+\frac{1}{m}}} \sim t(n+m-2)$	$\mu_X \neq \mu_Y$	$t < -t_{n+m-2,\alpha/2}$ or $t > t_{n+m-2,\alpha/2}$
			$\mu_X > \mu_Y$	$t > t_{n+m-2,\alpha}$
			$\mu_X < \mu_Y$	$t < -t_{n+m-2,\alpha}$
1.10	$\sigma_X^2 = \sigma_Y^2$	$F = \frac{S_X^2}{S_Y^2} \sim F(n-1, m-1)$	$\sigma_X^2 \neq \sigma_Y^2$	$f < F_{n-1,m-1,1-\alpha/2}$ or $f > F_{n-1,m-1,\alpha/2}$
			$\sigma_X^2 > \sigma_Y^2$	$f > F_{n-1,m-1,\alpha}$
			$\sigma_X^2 < \sigma_Y^2$	$f < F_{n-1,m-1,1-\alpha}$

Table 4.4: Common hypothesis tests associated with random sampling from normal populations.

The first row of Table 4.4 associated with Theorem 1.3 indicates that the appropriate test statistic is

$$Z = \frac{\bar{X}-\mu_0}{\sigma/\sqrt{n}} = \sqrt{2}\,(\bar{X}-5)$$

which has the standard normal distribution under H_0. The critical value is

$$c = -z_\alpha = -z_{0.1} = -1.28,$$

which is calculated with the R statement `qnorm(0.1)`. So the test statistic $z = \sqrt{2}\,(\bar{x}-5)$ is calculated from the data values x_1 and x_2, and H_0 is rejected if $z < -1.28$. This is illustrated in Figure 4.30, which is, not surprisingly, very reminiscent of Figure 4.5 except for the location and scale. In Example 4.3, we used the test statistic $x_1 + x_2$ and conducted the test by first principles. Here we use the test statistic $z = \sqrt{2}\,(\bar{x}-5)$ and use Table 4.4. Both tests are equivalent and will reject H_0 or fail to reject H_0 for the same data set.

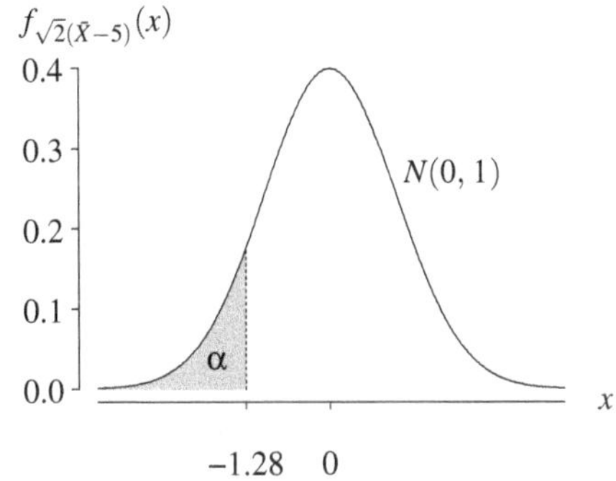

Figure 4.30: The sampling distribution of $\sqrt{2}(\bar{X}-5)$ when $\mu = 5$ with $\alpha = 0.1$.

Now consider the power function for this hypothesis test. Since $\bar{X} \sim N\left(\mu, \sigma^2/n\right)$, the power function is

$$\begin{aligned} q(\mu) &= P(\text{reject } H_0 \mid \mu) \\ &= P\left(\sqrt{2}(\bar{X}-5) < -1.28 \,\middle|\, \mu\right) \\ &= P\left(\bar{X} < 5 - 1.28/\sqrt{2} \,\middle|\, \mu\right) \\ &= P\left(\frac{\bar{X}-\mu}{\sqrt{1/2}} < \frac{5-1.28/\sqrt{2}-\mu}{\sqrt{1/2}} \,\middle|\, \mu\right) \\ &= P\left(Z < 5\sqrt{2} - 1.28 - \sqrt{2}\mu \,\middle|\, \mu\right), \end{aligned}$$

where $Z \sim N(0, 1)$. The power function can be plotted in R on the interval $0 < \mu < 6$ with the following statements.

```
mu    = seq(0, 6, by = 0.1)
crit  = qnorm(0.1)
power = pnorm(5 * sqrt(2) + crit - sqrt(2) * mu)
plot(mu, power, type = "l")
points(5, 0.1, pch = 42, cex = 0.9)
```

The power function is plotted in Figure 4.31. As expected, the power function passes through the point $(\mu_0, \alpha) = (5, 0.1)$ because the probability of rejecting H_0 under H_0 is $\alpha = 0.1$. The power function is not very steep because of the small sample size of $n = 2$.

One of the problems that arises in statistics is that you are not always sure that you have properly identified the population probability distribution. So there is some interest in the impact of misspecifying the population probability distribution. The next example illustrates the impact of using an entry from Table 4.4 when in fact the population distribution is not normally distributed. It is difficult to state the impact of misspecifying the population probability distribution in general, so this example focuses on one specific population distribution.

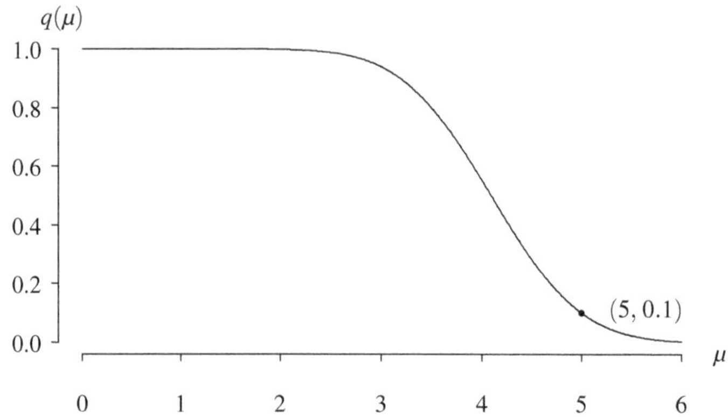

Figure 4.31: Power function for the test statistic $Z = \sqrt{2}(\bar{X} - 5)$ with $n = 2$ and $\alpha = 0.1$.

Example 4.17 Let $X_1, X_2, \ldots, X_6$ denote a sample of size $n = 6$ assumed to be drawn from a $N(\mu, \sigma^2)$ population, where μ and σ^2 are unknown parameters. Draw the power functions associated with testing

$$H_0 : \mu = 4$$

against the one-sided alternative hypothesis

$$H_1 : \mu > 4$$

at $\alpha = 0.05$ for a $N(4, 8)$ population distribution under H_0 and an Erlang(1/2, 2) population distribution under H_0. The power functions should consider the population distribution for varying population mean μ with a constant population variance $\sigma^2 = 8$. Also, draw the power functions associated with testing

$$H_0 : \sigma^2 = 8$$

against the one-sided alternative hypothesis

$$H_1 : \sigma^2 > 8$$

at $\alpha = 0.05$ for a $N(4, 8)$ population distribution under H_0 and an Erlang(1/2, 2) population distribution under H_0. The power functions should consider the population distribution for varying population variances σ^2 with a constant population mean $\mu = 4$. The choice for these particular population distributions is because both of these have population mean 4 and population variance 8 under H_0. Coincidentally, the Erlang(1/2, 2) population is equivalent to a chi-square random variable with 4 degrees of freedom.

Figure 4.32 shows that although these two population distributions have the same population mean $\mu = 4$ and variance $\sigma^2 = 8$, they are quite different in both their support and the shape of their probability density functions. For a small data set, however, one could easily see how an analyst could mistakenly choose one distribution as a population probability distribution when in fact the other probability distribution more accurately

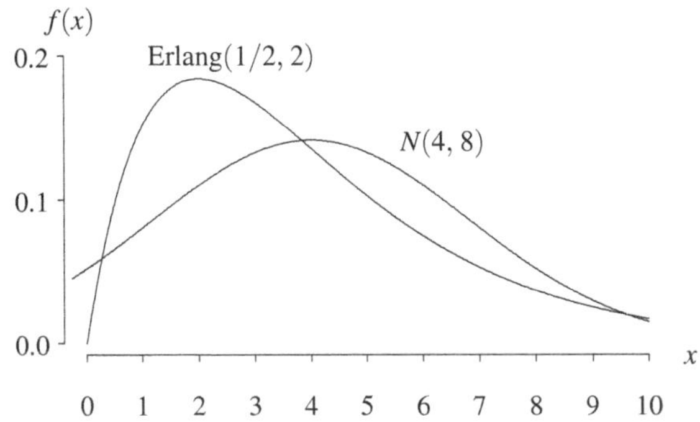

Figure 4.32: Probability density functions of $N(4, 8)$ and Erlang$(1/2, 2)$ random variables.

reflects the population. The presence of random sampling variability often results in misspecified population probability distributions—particularly for small sample sizes. For the first hypothesis test

$$H_0 : \mu = 4$$

versus the alternative hypothesis

$$H_1 : \mu > 4$$

at $\alpha = 0.05$ under the assumption that the random sampling is from a normal population, Table 4.4 indicates that the appropriate test statistic is

$$\frac{\bar{X} - \mu_0}{S/\sqrt{n}} = \frac{\bar{X} - 4}{S/\sqrt{6}},$$

which has the t distribution with $n - 1 = 6 - 1 = 5$ degrees of freedom when the random sample is drawn from a $N(4, 8)$ distribution. The appropriate critical value is

$$c = t_{n-1,\alpha} = t_{5,0.05} = 2.0150,$$

which is calculated with the R statement `qt(0.95, 5)`. Thus, the null hypothesis is rejected in favor of H_1 whenever the experimental value of the test statistic exceeds $c = 2.1050$. The next step is to compute the power function, which is

$$q(\mu) = P(\text{reject } H_0 \,|\, \mu) = P\left(\frac{\bar{X} - 4}{S/\sqrt{6}} > 2.0150 \,\middle|\, \mu\right).$$

But the problem here is that the random variable $\frac{\bar{X}-4}{S/\sqrt{6}}$ has the t distribution for random sampling from a normal population when $\mu = 4$, but not for $\mu \neq 4$. Rather than completing the derivation analytically for the power function (which involves a distribution known as the *noncentral t distribution*) for values of μ other than $\mu = 4$, we instead use Monte Carlo simulation to approximate the power function. Even if we did complete the derivation, it is nearly hopeless to derive the power function for random sampling from an Erlang population. The following R code conducts this hypothesis test 100,000 times using random sampling from the normal distribution with $\mu = 4$ and $\sigma^2 = 8$ via the `rnorm` function, and prints the fraction of times that H_0 is rejected.

```
mu    = 4
sig   = sqrt(8)
alpha = 0.05
n     = 6
crit  = qt(1 - alpha, n - 1)
nrep  = 100000
count = 0
for (i in 1:nrep) {
  x = rnorm(n, mu, sig)
  test.stat = (mean(x) - 4) / sqrt(var(x) / n)
  if (test.stat > crit) count = count + 1
}
print(count / nrep)
```

After a call to `set.seed(3)` to initialize the random number stream, five runs of the program yield

0.0495 0.0500 0.0498 0.0497 0.0508.

Not surprisingly, these fractions are very close to $\alpha = 0.05$. A loop in μ wrapped around the R code above is used to estimate the probability of rejecting H_0 for μ values other than $\mu = 4$. The results are shown in the estimated power function in Figure 4.33. The geometric interpretation is that this function gives the probability of rejecting H_0 as the normal population distribution in Figure 4.32 shifts horizontally. Next, we execute the same R code given above, but instead of sampling from a normal population, we sample from a shifted Erlang population. The geometric interpretation is that the Erlang$(1/2, 2)$ population probability distribution is shifted to the left and right in Figure 4.32, then this curve gives the impact on the probability of rejecting H_0. For both curves, the population variance has been kept constant at $\sigma^2 = 8$.

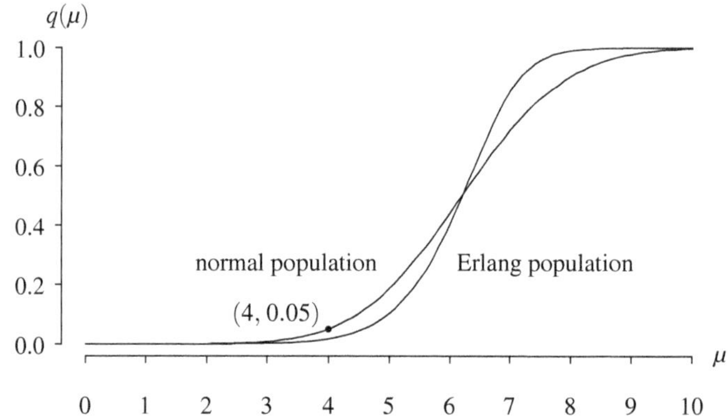

Figure 4.33: Power function estimates for the test statistic $\frac{\bar{X}-4}{S/\sqrt{n}}$ with $n = 6$ and $\alpha = 0.05$.

Considering the differences between the two populations, the two power curves in Figure 4.33 are surprisingly close to one another, which is good news. Their trajectories are similar. This good news generalizes to population distributions other than the Erlang distribution. Statisticians use the term *robust* to indicate an insensitivity to assumptions about the population distribution. This analysis of the two power functions for the hypothesis test concerning the population mean can be generalized to:

> Hypothesis tests concerning population means based on random sampling from a normal population are fairly robust (insensitive) to violations concerning the population distribution.

Of course, the hypothesis test becomes approximate when the true population is Erlang, but as the power functions in Figure 4.33 indicate, the error introduced by misspecifying the population distribution is not extreme.

We now turn to testing the null hypothesis concerning population variances

$$H_0 : \sigma^2 = 8$$

versus the alternative hypothesis

$$H_1 : \sigma^2 > 8$$

at $\alpha = 0.05$. Under the assumption that the random sampling is from a normal population, Table 4.4 indicates that the appropriate test statistic is

$$\frac{(n-1)S^2}{\sigma^2} = \frac{5S^2}{8},$$

which has the chi-square distribution with $n-1 = 6-1 = 5$ degrees of freedom when the random sample is drawn from a $N(4, 8)$ distribution. So the appropriate critical value is

$$c = \chi^2_{n-1,\alpha} = \chi^2_{5,0.05} = 11.0705,$$

which is calculated with the R statement `qchisq(0.95, 5)`. So the null hypothesis is rejected in favor of H_1 whenever the experimental value of the test statistic exceeds $c = 11.0705$. The next step is to compute the power function, which is

$$q(\sigma^2) = P\left(\text{reject } H_0 \,|\, \sigma^2\right) = P\left(\frac{5S^2}{8} > 11.0705 \,\middle|\, \sigma^2\right).$$

But the problem here is that the random variable $\frac{5S^2}{8}$ has the χ^2 distribution with 5 degrees of freedom for random sampling from a normal population when $\sigma^2 = 8$, but not for $\sigma^2 \neq 8$. As before, rather than completing the derivation analytically for the power function, for values of σ^2 other than $\sigma^2 = 8$, we instead use Monte Carlo simulation to approximate the power function. The following R code conducts this hypothesis test 100,000 times using random sampling from the normal distribution via the `rnorm` function, and prints the fraction of times that H_0 is rejected.

```
mu    = 4
sig   = sqrt(8)
alpha = 0.05
n     = 6
```

```
crit  = qchisq(1 - alpha, n - 1)
nrep  = 100000
count = 0
for (i in 1:nrep) {
  x = rnorm(n, mu, sig)
  test.stat = (n - 1) * var(x) / sig ^ 2
  if (test.stat > crit) count = count + 1
}
print(count / nrep)
```

After a call to `set.seed(3)` to initialize the random number stream, five runs of the program yield

0.0490 0.0490 0.0505 0.0511 0.0497.

Not surprisingly, these fractions are very close to $\alpha = 0.05$. This R code allows us to wrap a loop in σ^2 around the existing code to estimate the probability of rejecting H_0 for σ^2 values other than $\sigma^2 = 8$. The results are shown in the estimated power function in Figure 4.34. The geometric interpretation is that this function gives the probability of rejecting H_0 as the normal population distribution in Figure 4.32 as σ^2 varies. Next, we execute the same R code given above, but instead of sampling from a normal population, we sample from an Erlang population with population mean 4 and varying σ^2. For both curves, the population mean has been kept constant at $\mu = 4$.

Unfortunately, the two power curves in Figure 4.34 follow substantially different trajectories, which are diverging. This bad news generalizes to other population distributions than the Erlang distribution. Hypothesis tests concerning population variances based on random sampling assumed to come from a normal population are not robust to violations concerning the population distribution.

This example has provided just one alternative to the normal population distribution,

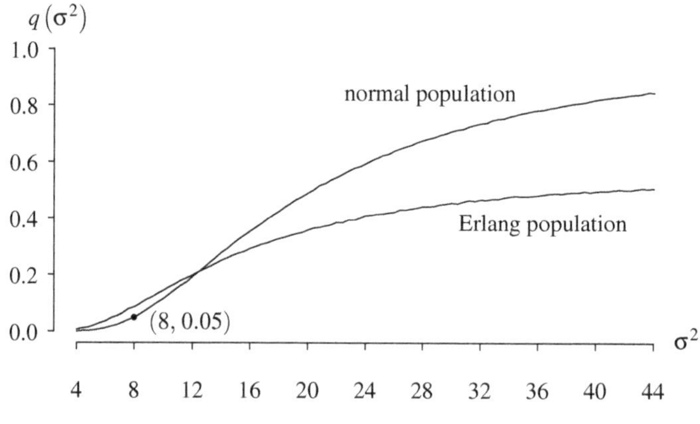

Figure 4.34: Power function estimates for the test statistic $5S^2/8$ with $n = 6$ and $\alpha = 0.05$.

and, of course, the story will be slightly different for every alternative distribution that is considered.

This ends the discussion of the testing methods associated with random sampling from a normal population. Until now, the sample size n for a hypothesis test has been given. The next subsection gives three examples of how n can be calculated in order to give a prescribed power for three types of sampling that occur often in statistical practice.

4.4 Sample Size Determination

In some settings, formulas can be developed that give the sample size necessary to achieve a particular steepness associated with the power function for a hypothesis test or significance test. The first example of such a case concerns the relationship between a producer and a consumer of a product.

Example 4.18 A small start-up company known as Green Motors is marketing their new Sloth, which features excellent gas mileage, which they advertise as 50 miles per gallon, along with dreadful acceleration. You have purchased a Sloth, but are unhappy with the gas mileage results, which have consistently fallen below 50 miles per gallon for you. You have contacted the CEO of Green Motors, who has agreed to conduct the hypothesis test

$$H_0 : \mu = 50$$

versus

$$H_1 : \mu < 50$$

at significance level $\alpha = 0.1$, where μ is the population mean miles per gallon for a Sloth. What is an appropriate sample size n for the test, that is, how many Sloths should be tested?

Let $X_1, X_2, \ldots, X_n$ be the number of miles per gallon for n different Sloths used on the road. The probability of a Type I error is

$$\alpha = P(\text{reject } H_0 \,|\, H_0 \text{ true}) = 0.1.$$

Who is affected by this error? If Green Motors is telling the truth about the gas mileage of their Sloths, then rejecting H_0 is an error that negatively impacts the company. For this reason, α is known in this setting as the "producer's risk."

But you as the consumer also take a risk in conducting the hypothesis test. The parameters associated with this risk must be defined. The probability of a Type II error is

$$\beta = P(\text{fail to reject } H_0 \,|\, H_1 \text{ true}).$$

But there are an infinite number of μ values associated with H_1 being true. You must define a probability β and an associated μ value μ_1 in order to define what is known as the "consumer's risk." For this particular hypothesis test, you decide to let $\beta = 0.05$ when $\mu_1 = 47$ miles per gallon so that

$$\beta = P(\text{fail to reject } H_0 \,|\, \mu = 47) = 0.05.$$

You as the consumer are willing to accept a probability of 0.05 that the hypothesis test will erroneously fail to reject the null hypothesis when in fact Sloths get only 47 miles per gallon.

Still more assumptions are needed before a sample size can be determined. Let's say that a trade journal on gas mileage has indicated that gas mileage for vehicles used on the road has historically been normally distributed, and, furthermore, that the variability about the population mean value, regardless of the population mean value, is $\sigma^2 = 16$ $(\text{miles/gallon})^2$. From this information, we know that the appropriate test statistic from the first row of Table 4.4 is

$$Z = \frac{\bar{X} - \mu}{\sigma/\sqrt{n}} = \frac{\bar{X} - 50}{4/\sqrt{n}},$$

which has the standard normal distribution under H_0. Table 4.4 also indicates that H_0 should be rejected whenever the observed value of Z is less than $-z_\alpha = -z_{0.1} = -1.28$, which can be calculated in R with the statement `qnorm(0.1)`. So H_0 is rejected when

$$z = \frac{\bar{x} - 50}{4/\sqrt{n}} < -1.28.$$

The power function for this hypothesis test is

$$\begin{aligned}
q(\mu) &= P(\text{reject } H_0 \,|\, \mu) \\
&= P\left(\frac{\bar{X} - 50}{4/\sqrt{n}} < -1.28 \,\middle|\, \mu\right) \\
&= P\left(\frac{\bar{X} - 50}{4/\sqrt{n}} + \frac{50}{4/\sqrt{n}} - \frac{\mu}{4/\sqrt{n}} < -1.28 + \frac{50}{4/\sqrt{n}} - \frac{\mu}{4/\sqrt{n}} \,\middle|\, \mu\right) \\
&= P\left(\frac{\bar{X} - \mu}{4/\sqrt{n}} < -1.28 + \frac{50 - \mu}{4/\sqrt{n}} \,\middle|\, \mu\right) \\
&= P\left(Z' < -1.28 + \frac{50 - \mu}{4/\sqrt{n}}\right),
\end{aligned}$$

where $Z' \sim N(0, 1)$. (The symbol Z' is used rather than Z so that it is not mistaken for the test statistic.) The power function for testing $n = 1$, $n = 2$, and $n = 3$ Sloths calculated using the R `pnorm` function is shown in Figure 4.35. All three power curves pass through the point $(\mu_0, \alpha) = (50, 0.1)$. The power function for $n = 2$ is steeper than the power function for $n = 1$. The purpose here is to continue to increase n until the power function intersects the point $(47, 0.95)$. Since n is an integer, a curve is unlikely to pass directly through the point $(47, 0.95)$. So problems of this type are typically stated as "find the smallest n such that $\beta \le 0.05$" rather than "$\beta = 0.05$." As n is increased, the first power function that is steep enough to pass through the point $(47, 0.95)$ or fall above that point determines the sample size to be used.

Returning to the problem of finding the number of Sloths n to test, we want the smallest value of n to satisfy $q(47) \ge 0.95$, or

$$q(47) = P\left(Z' < -1.28 + \frac{50 - 47}{4/\sqrt{n}}\right) = P\left(Z' < -1.28 + \frac{3\sqrt{n}}{4}\right) \ge 0.95.$$

Since $Z' \sim N(0, 1)$, and the 95th percentile of the standard normal distribution is 1.64 (via the R statement `qnorm(0.95)`), the resulting equation is

$$-1.28 + \frac{3\sqrt{n}}{4} = 1.64.$$

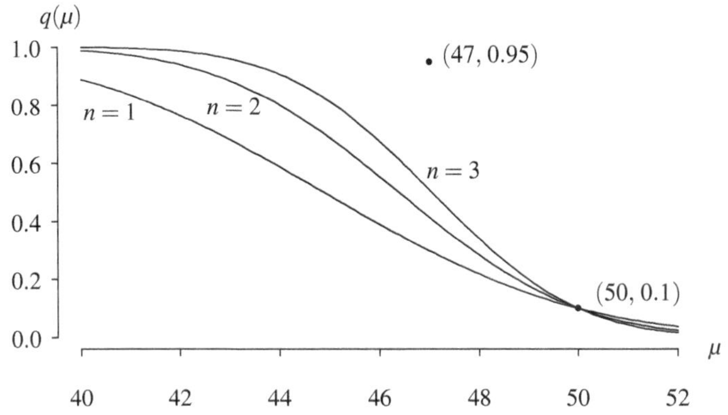

Figure 4.35: Power functions for the test statistic $Z = \frac{\bar{X}-50}{4/\sqrt{n}}$ with $n = 1, 2, 3$.

Solving for n results in

$$n = \frac{(1.64 + 1.28)^2 4^2}{3^2} = 15.2.$$

So using $n = 16$ results in a probability of a Type II error β which is less than 0.05. The power function for $n = 16$ is plotted in Figure 4.36. This power function passes through the point $(\mu_0, \alpha) = (50, 0.1)$ and passes just above the point $(\mu_1, 1-\beta) = (47, 0.95)$. Under the assumption of random sampling from a normal population with known population variance, placing $n = 16$ Sloths on test and gathering the miles per gallon data values $X_1, X_2, \ldots, X_{16}$ will meet both the producer's risk $\alpha = 0.1$ and the consumer's risk $\beta = 0.05$.

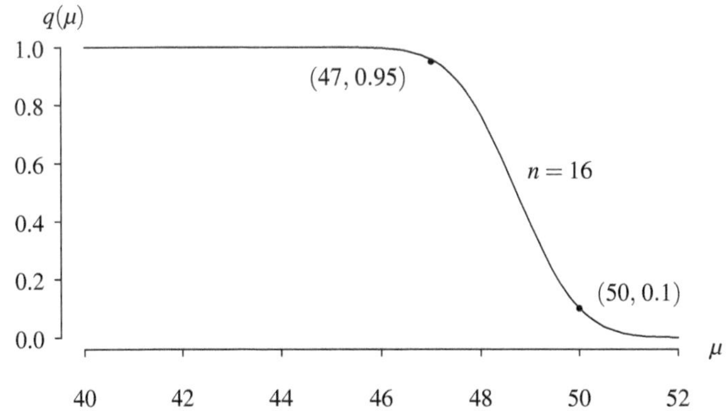

Figure 4.36: Power function for the test statistic $Z = \frac{\bar{X}-50}{4/\sqrt{n}}$ with $n = 16$.

The previous example resulted in a hypothesis test that required a sample size of $n = 16$. But what if conducting such a test is too expensive? Can anything be done? First, the $\sigma = 4$ value in the example might have been based on the miles per gallon from a single tank of gas. If these calculations were made for multiple tanks of gas, the observations would remain normally distributed, but with a smaller population variance. This would allow n to be reduced. Second, the test was conducted in the field. If instead the Sloths were tested on a track, this would effectively reduce σ. Test track results will almost certainly be more consistent although less realistic. Applying these test results to actual driving on the road, however, would be dubious, but the required sample size would be smaller because of the reduced population variance.

The calculations from the previous example can be generalized for determining the sample size n for testing

$$H_0 : \mu = \mu_0$$

versus

$$H_1 : \mu = \mu_1$$

at significance level α, probability of a Type II error β, and $\mu_1 < \mu_0$. Let $X_1, X_2, \ldots, X_n$ be a random sample drawn from a $N(\mu, \sigma^2)$ population, where μ is unknown and σ^2 is known. The appropriate test statistic from the first row of Table 4.4 is

$$Z = \frac{\bar{X} - \mu}{\sigma/\sqrt{n}},$$

which has the standard normal distribution under H_0. Table 4.4 also indicates that H_0 should be rejected whenever the observed value of Z is less than $-z_\alpha$. The power function for this hypothesis test is

$$\begin{aligned}
q(\mu) &= P(\text{reject } H_0 \,|\, \mu) \\
&= P\left(\frac{\bar{X} - \mu_0}{\sigma/\sqrt{n}} < -z_\alpha \,\middle|\, \mu\right) \\
&= P\left(\frac{\bar{X} - \mu_0}{\sigma/\sqrt{n}} + \frac{\mu_0}{\sigma/\sqrt{n}} - \frac{\mu}{\sigma/\sqrt{n}} < -z_\alpha + \frac{\mu_0}{\sigma/\sqrt{n}} - \frac{\mu}{\sigma/\sqrt{n}} \,\middle|\, \mu\right) \\
&= P\left(\frac{\bar{X} - \mu}{\sigma/\sqrt{n}} < -z_\alpha + \frac{\mu_0 - \mu}{\sigma/\sqrt{n}} \,\middle|\, \mu\right) \\
&= P\left(Z' < -z_\alpha + \frac{\mu_0 - \mu}{\sigma/\sqrt{n}}\right),
\end{aligned}$$

where $Z' \sim N(0, 1)$. The power function when $\mu = \mu_1$ is

$$q(\mu_1) = P\left(Z' < -z_\alpha + \frac{\mu_0 - \mu_1}{\sigma/\sqrt{n}}\right) = 1 - \beta.$$

Since the $1 - \beta$ percentile of a standard normal random variable is denoted by z_β, this is equivalent to

$$z_\beta = -z_\alpha + \frac{\mu_0 - \mu_1}{\sigma/\sqrt{n}}.$$

Solving for n gives

$$n = \frac{(z_\alpha + z_\beta)^2 \sigma^2}{(\mu_0 - \mu_1)^2}.$$

If this expression is an integer, then the power function passes through the point $(\mu_1, 1-\beta)$. If this expression is not an integer, then the proper sample size is the next higher integer, and the power function passes just above the point $(\mu_1, 1-\beta)$. These two cases can be combined by using the ceiling function:

$$n = \left\lceil \frac{(z_\alpha + z_\beta)^2 \sigma^2}{(\mu_0 - \mu_1)^2} \right\rceil.$$

One important factor from the previous example that allowed us to find a closed-form expression for n was the fact that σ^2 was known. The reader is encouraged to envision the complications that would result if σ^2 were unknown. If n is large, then σ^2 can be replaced with S^2 with minimal effect, but if n is small, then the mathematics get much stickier and it is not possible to arrive at a closed-form solution for the sample size n.

The next example concerns finding the appropriate sample size n associated with a random sample drawn from a Bernoulli population with unknown parameter p. This type of problem arises frequently in consumer preference tests, go/no-go testing in engineering, and political polls.

Example 4.19 Smith and Jones are running for governor. Several polls show that Smith has about 55% of the vote, but, to be on the safe side, Smith wants to conduct an internal poll. What is an appropriate hypothesis test and sample size associated with $\alpha = 0.1$ and $\beta = 0.05$?

Political polls designed to predict the outcome of an election are very difficult to conduct. A poll of the general population might not accurately reflect who will vote on election day. A poll of registered voters has the same problem. The most accurate political polls identify and poll individuals known as "likely voters." Once these individuals have been identified, how to get their preference is also nontrivial. Errors such as the *Chicago Tribune's* famous "Dewey Defeats Truman" headline are often the result of surveys that oversample a non-representative group. The polls that are conducted today are very sophisticated and are often able to predict elections with considerable precision.

In the case of Smith and Jones, let p be the population fraction of voters that support Smith. Assume that there are no "undecided" voters; everyone has a preference for one of the two candidates. One possible hypothesis test is

$$H_0 : p = 0.55$$

versus

$$H_1 : p = 0.5.$$

If H_0 is rejected for this hypothesis test, then there is reason for Smith to be concerned about the upcoming election. Let $X_1, X_2, \ldots, X_n$ denote the random sample. A voter that supports Smith corresponds to $X_i = 1$; a voter that supports Jones corresponds to $X_i = 0$. As in previous Bernoulli random sampling examples, use the test statistic $Y = X_1 + X_2 + \cdots + X_n$. Since small values of the test statistic lead to rejecting H_0, the test rejects H_0 whenever $Y \le c$, where c is chosen based on the left-hand tail of the binomial distribution with parameters n and 0.55. Since the sum of n mutually independent and identically distributed Bernoulli(p) random variables has the binomial(n, p)

distribution, the power function $q(p)$ for this test is

$$\begin{aligned} q(p) &= P(\text{reject } H_0 \,|\, p) \\ &= P(Y \le c \,|\, p) \\ &= P(X_1 + X_2 + \cdots + X_n \le c \,|\, p) \\ &= P(\text{binomial}(n,\, p) \le c \,|\, p) \\ &= \sum_{x=0}^{c} \binom{n}{x} p^x (1-p)^{n-x} \end{aligned}$$

for $0 < p < 1$. Part of the power function is drawn for the arbitrarily-chosen sample sizes $n = 50$ and $n = 200$ in Figure 4.37. The critical value associated with $n = 50$ is $c = 22$. The null hypothesis is rejected whenever $Y = X_1 + X_2 + \cdots + X_{50} \le 22$. The discrete nature of the binomial distribution means that the power function falls just below the point $(0.55, 0.1)$. The power curve associated with $n = 50$ falls well below the point $(0.5, 0.95)$. So the method of finding the appropriate sample size n for this hypothesis test is to increase n until the power curve is steep enough so that it falls above the point $(0.5, 0.95)$.

Our preference would be to pass through both points on the power function, that is,

$$q(p_0) = q(0.55) = \alpha \qquad \text{and} \qquad q(p_1) = q(0.5) = 1 - \beta.$$

But the discrete nature of the binomial distribution and the discrete nature of n require that the following two steps be placed in a loop in n. First, for a fixed n, find the largest critical value c such that

$$q(0.55) \le \alpha$$

by finding the appropriate percentile of the binomial distribution. Second, for these values of n and c, calculate the power at $p = 0.5$, that is, $q(0.5)$. This process terminates whenever the power function falls above the point $(0.5, 0.95)$, that is,

$$q(0.5) \ge 1 - \beta.$$

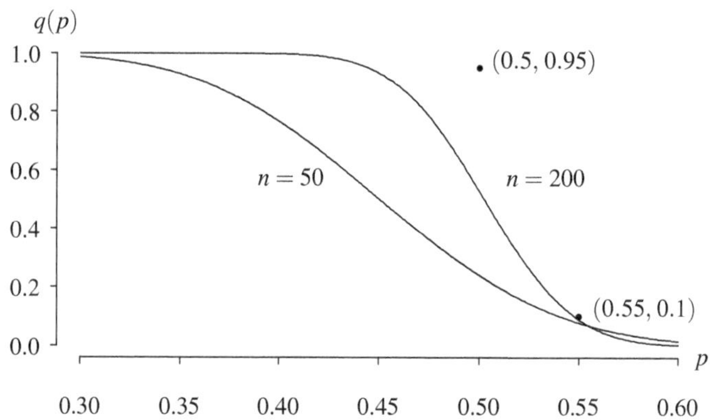

Figure 4.37: Power functions for the test statistic $Y = \sum_{i=1}^{n} X_i$ for $n = 50$ and $n = 200$.

The R code below loops through n beginning at $n = 50$ (which we know from Figure 4.37 is too small), calculating a critical value c, which is stored in `crit`, and then checking to see if

$$q(0.5) = \sum_{x=0}^{c} \binom{n}{x} 0.5^x 0.5^{n-x} \geq 0.95.$$

When this condition is satisfied, the program prints the sample size and critical value required to satisfy the constraints of the problem.

```
p0 = 0.55
p1 = 0.50
a  = 0.10
b  = 0.05
n  = 50
crit = qbinom(a, n, p0) - 1
while (pbinom(crit, n, p1) < 1 - b) {
  n = n + 1
  crit = qbinom(a, n, p0) - 1
}
print(c(n, crit))
```

This code returns sample size $n = 866$ and associated critical value $c = 457$. So Smith's internal poll should randomly sample 866 likely voters and reject H_0 if 457 or fewer say that they will cast their vote for Smith. The power function associated with $n = 866$ and $c = 457$ is shown in Figure 4.38. This is the smallest value of n such that the power function is steep enough so that it simultaneously falls just below the point $(0.55, 0.1)$ and just above the point $(0.5, 0.95)$.

The actual level of significance and the probability of a Type II error for this test with sample size $n = 866$ and critical value $c = 457$ can be calculated with the additional R statements given below.

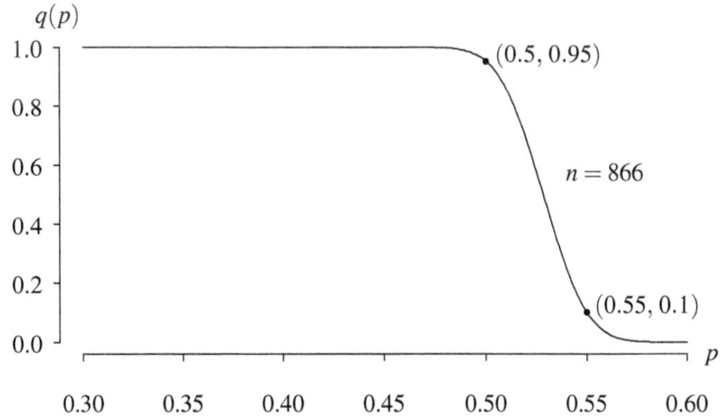

Figure 4.38: Power function for the test statistic $Y = \sum_{i=1}^{n} X_i$ for $n = 866$.

```
pbinom(457, 866, 0.55)
1 - pbinom(457, 866, 0.5)
```

These yield $\alpha = 0.0997$ (just below 0.10) and $\beta = 0.0479$ (just below 0.05).

The third and final setting for determining a sample size comes from an application area known as *life testing*. Large corporations and large government agencies often purchase thousands of items, such as paper clips, fuses, and light bulbs. Since the order size is typically massive, the consumer would like some statistical evidence that the product that they are purchasing in bulk will perform as claimed by the producer. Oftentimes the producer and consumer agree to have a third party test a small sample of the items prior to finalizing the large purchase. The paragraphs below develop a technique for determining an appropriate sample size for a life test concerning items with lifetimes that are exponentially distributed.

A producer ships a large number of items with exponential(λ) lifetimes to a consumer, where λ is a positive unknown failure rate. A random sample of n of these items is placed on test. Let $X_1, X_2, \ldots, X_n$ denote their lifetimes. The appropriate hypothesis test is

$$H_0 : \lambda = \lambda_0$$

versus

$$H_1 : \lambda > \lambda_0,$$

where λ_0 is the failure rate of the items claimed by the producer. The consumer is obviously concerned if H_0 is rejected because this indicates that there is a high probability that the failure rate is higher than λ_0. Neither party is concerned if the population failure rate λ is smaller than λ_0, which is why this is formulated as a one-sided hypothesis test.

As is the case with all hypothesis testing, two different types of errors can be made. The first type of error, which adversely affects the producer, corresponds to rejecting H_0 when H_0 is true (that is, not shipping a good lot of items). The probability that this type of error occurs is called the *producer's risk* and is defined to be

$$\alpha = P(\text{rejecting } H_0 \,|\, \lambda = \lambda_0).$$

The second type of error, which adversely affects the consumer, corresponds to not rejecting H_0 when H_1 is true (that is, shipping a bad lot of items). In this case the consumer will probably be shipped items with a higher failure rate than claimed by the producer. The probability that this type of error will occur is called the *consumer's risk* and is defined to be

$$\beta = P(\text{not rejecting } H_0 \,|\, \lambda = \lambda_1).$$

The consumer must specify λ_1 greater than λ_0, where λ_1 is a failure rate that is unacceptably high.

Recall from Example 3.7 that for the random sample $X_1, X_2, \ldots, X_n$ drawn from an exponential(λ) population,

$$2\lambda \sum_{i=1}^{n} X_i \sim \chi^2(2n).$$

So when H_0 is true,

$$2\lambda_0 \sum_{i=1}^{n} X_i \sim \chi^2(2n).$$

Smaller values of $\sum_{i=1}^{n} X_i$ indicate a higher failure rate and hence will lead to rejecting H_0. Therefore, the null hypothesis is rejected when

$$2\lambda_0 \sum_{i=1}^{n} x_i < \chi^2_{2n,1-\alpha},$$

where α is the level of significance for the test. The power function $q(\lambda)$ gives the probability of rejecting H_0 for various values of the failure rate λ. When $\lambda = \lambda_0$, the probability of rejecting H_0 is α, the producer's risk, that is, $q(\lambda_0) = \alpha$. It is unusual to find an n value such that the power curve will *exactly* pass through the points (λ_0, α) and $(\lambda_1, 1-\beta)$, so the problem is usually formulated as: find the smallest n such that $q(\lambda_0) = \alpha$ and $q(\lambda_1) \geq 1-\beta$. In particular, when $\lambda = \lambda_1$, it is now the case that $2\lambda_1 \sum_{i=1}^{n} X_i \sim \chi^2(2n)$, so

$$\begin{aligned}
q(\lambda_1) &= P\left(2\lambda_0 \sum_{i=1}^{n} X_i < \chi^2_{2n,1-\alpha} \,\middle|\, \lambda = \lambda_1\right) \\
&= P\left(2\lambda_1 \sum_{i=1}^{n} X_i < \frac{\lambda_1}{\lambda_0}\chi^2_{2n,1-\alpha} \,\middle|\, \lambda = \lambda_1\right) \\
&= P\left(\chi^2(2n) < \frac{\lambda_1}{\lambda_0}\chi^2_{2n,1-\alpha} \,\middle|\, \lambda = \lambda_1\right) \\
&\geq 1-\beta.
\end{aligned}$$

Thus, the constraints of the test are satisfied when

$$\chi^2_{2n,\beta} \leq \frac{\lambda_1}{\lambda_0}\chi^2_{2n,1-\alpha}.$$

Once the producer's risk α, the consumer's risk β, and the failure rates λ_0 and λ_1 are defined, the smallest sample size n satisfying the above inequality can be determined. There is no closed-form expression for n. This is solved by trial and error using χ^2 tables or a statistical language like R.

Example 4.20 Glow-Rite lighting company claims that their light bulbs last 1000 hours on average. A federal agency would like to purchase several thousand of these bulbs and their policies require that a life test be conducted so that the lot will be rejected with a probability of 0.9 when the true population mean time to failure is 400 hours. Assuming that the producer's risk is 0.05 and the lifetimes of the bulbs are exponentially distributed, what is the smallest sample size n necessary to satisfy both the producer and the consumer?

Converting to failure rates, the hypothesis being tested here is (since $\lambda_0 = 0.001$)

$$H_0 : \lambda = 0.001$$

versus

$$H_1 : \lambda > 0.001$$

and the other parameters are $\lambda_1 = 0.0025$, $\alpha = 0.05$, and $\beta = 0.1$. To find the smallest value of n that satisfies

$$\frac{\chi^2_{2n,0.1}}{\chi^2_{2n,0.95}} \leq \frac{\lambda_1}{\lambda_0} = \frac{0.0025}{0.001} = 2.5$$

requires trial and error for different quantiles of the chi-square distribution. The following R code loops through n and prints the smallest value of n satisfying the inequality.

```
n = 1
while (qchisq(0.9, 2 * n) / qchisq(0.05, 2 * n) > 2.5) n = n + 1
print(n)
```

The smallest value of n satisfying the inequality is $n = 11$. The life test consists of placing 11 items on test which yield failure times $x_1, x_2, \ldots, x_{11}$. The hypothesis test rejects H_0 if

$$(2)(0.001)\sum_{i=1}^{n} x_i < 12.3,$$

since $\chi^2_{22,0.95} = 12.3380$ via the R statement `qchisq(0.05, 22)`. Figure 4.39 shows the power function

$$q(\lambda) = P\left((2)(0.001)\sum_{i=1}^{n} X_i < 12.3 \,\middle|\, \lambda\right)$$

for this particular test with parameters $\lambda_0 = 0.001$, $\lambda_1 = 0.0025$, $\alpha = 0.05$, $\beta = 0.1$, and $n = 11$. This power function passes through the point $(\lambda_0, \alpha) = (0.001, 0.05)$ and passes just above the point $(\lambda_1, 1 - \beta) = (0.0025, 0.9)$. The actual probability of a Type II error β can be found with the R statements shown below.

```
crit = qchisq(0.05, 22)
beta = 1 - pchisq(0.0025 * crit / 0.001, 22)
```

These return $\beta = 0.09933$ (just below 0.1).

This concludes the discussion of sample size determination. The three examples presented here have illustrated the technique for determining the sample size n that meets a criteria set by the experimenter for random samples drawn from normal, Bernoulli, and exponential populations. Each population distribution requires a custom approach, and two of the populations did not yield a closed-form expression for n. The geometric interpretation in each case is that of choosing the smallest sample size n so as to require the power function to be of a particular prescribed steepness.

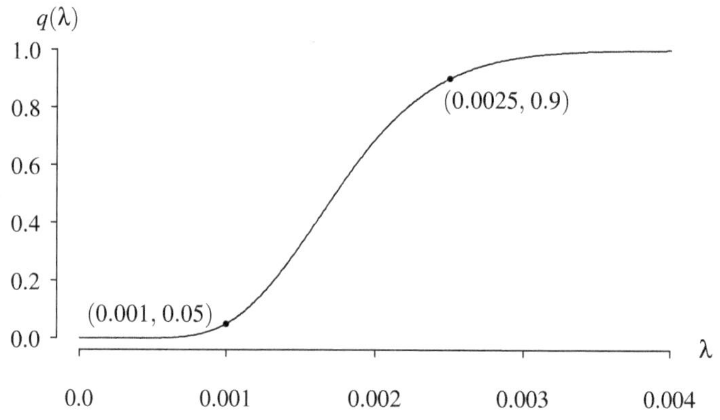

Figure 4.39: Power function for the life test with $n = 11$.

4.5 Confidence Intervals, Hypothesis Tests, & Significance Tests

The use of the symbol α in the development of both confidence intervals and hypothesis testing is not a coincidence.

In confidence intervals, the value of α determines the value of the stated coverage of the confidence interval, which is $1-\alpha$. If the confidence interval $L < \theta < U$ is an exact confidence interval, for instance, then $P(L < \theta < U) = 1-\alpha$.

In hypothesis testing, the value of the significance level α is the probability of committing a Type I error, that is, $\alpha = P(\text{Type I error}) = P(\text{rejecting } H_0 \,|\, H_0 \text{ true})$.

This subsection links the α that is used in constructing a confidence interval with the α used in hypothesis testing. Assume that the goal associated with the analysis of a data set is to conduct a test. As seen in Table 4.5, such a test associated with the null hypothesis $H_0 : \theta = \theta_0$ can be conducted in three different ways: using confidence intervals, using traditional hypothesis testing, and using significance testing. Each column in Table 4.5 is associated with one of the three methods of conducting the test. The rows in Table 4.5 show the sequential steps associated with the three methods. The first three steps in each of the three methods are identical. The methods differ in the last two steps. Regardless of the methods taken, the conclusion that is drawn, either reject H_0 or fail to reject H_0, will be identical for the same data set which is based on the same pivotal value/test statistic.

Confidence interval method	**Hypothesis testing method**	**Significance testing method**
Formulate H_0 and H_1	Formulate H_0 and H_1	Formulate H_0 and H_1
Determine significance level α	Determine significance level α	Determine significance level α
Determine sample size n	Determine sample size n	Determine sample size n
Construct confidence interval	Compute test statistic	Compute test statistic
Reject H_0 if θ_0 does not fall in the confidence interval	Reject H_0 if the test statistic is in the critical region	Reject H_0 if the p-value is less than α

Table 4.5: Steps associated with confidence intervals, hypothesis tests, and significance tests.

The next example illustrates these three methods for conducting a test based on a single data set. The three methods all result in the same conclusion being drawn.

Example 4.21 IQ scores are traditionally centered around a population mean of 100. An experimenter would like to test the simple null hypothesis

$$H_0 : \mu = 100$$

versus the two-sided alternative hypothesis

$$H_1 : \mu \neq 100$$

based on a random sample of IQ scores that is assumed to come from a normally distributed population. The significance level is set to $\alpha = 0.01$ prior to collecting the data. It is also decided that a sample of $n = 78$ IQ scores will be collected in order to conduct the test. More specifically, the $n = 78$ IQ scores of seventh-grade students at a rural school in the midwestern United States from Example 3.9, and also listed below, will be used to conduct the test. This completes the first three common steps for the three methods given in Table 4.5.

111	107	100	107	114	115	111	97	100	112	104	89	104
102	91	114	114	103	106	105	113	109	108	113	130	128
128	118	113	120	132	111	124	127	128	136	106	118	119
123	124	126	116	127	119	97	86	102	110	120	103	115
93	72	111	103	123	79	119	110	110	107	74	105	112
105	110	107	103	77	98	90	96	112	112	114	93	106

The test will now be completed using each of the three methods. To illustrate the use of a confidence interval to complete the test, begin by computing the sample mean and sample standard deviation of the data values, which are approximately

$$\bar{x} = 109 \qquad \text{and} \qquad s = 13.$$

Using Table 3.1, an exact two-sided 99% confidence interval for the population mean IQ score μ is

$$\bar{x} - t_{n-1,\alpha/2}\frac{s}{\sqrt{n}} < \mu < \bar{x} + t_{n-1,\alpha/2}\frac{s}{\sqrt{n}}$$

or

$$109 - 2.6412\frac{13}{\sqrt{78}} < \mu < 109 + 2.6412\frac{13}{\sqrt{78}}$$

or

$$105 < \mu < 113.$$

Since this 99% confidence interval does not contain $\mu_0 = 100$, we reject H_0 and conclude that these IQ scores are not drawn from a population centered around 100.

Shifting to traditional hypothesis testing to complete the test, the appropriate test statistic from Table 4.4 is

$$T = \frac{\bar{X} - \mu}{S/\sqrt{n}} = \frac{\bar{X} - 100}{S/\sqrt{n}},$$

which has the $t(n-1) = t(77)$ distribution under H_0. For the $n = 78$ IQ scores, the value of this test statistic is

$$\frac{109 - 100}{13/\sqrt{78}} = 5.98.$$

The critical values for the test are the 0.5th and 99.5th percentile of the t distribution with 77 degrees of freedom, which are

$$c_l = -t_{77,0.005} = -2.6412 \qquad \text{and} \qquad c_u = t_{77,0.005} = 2.6412,$$

which are values that appear in the confidence interval formula given earlier. So H_0 is rejected if the test statistic is less than $c_l = -2.6412$ or greater than $c_u = 2.6412$. Since the test statistic 5.98 is much greater than the upper critical value $c_u = 2.6412$, H_0 is overwhelmingly rejected. We again conclude that these IQ scores are not drawn from a population centered around population mean 100.

Finally, using a significance test for analysis, the choice of the appropriate test statistic from Table 4.4 remains the same:

$$T = \frac{\bar{X} - \mu}{S/\sqrt{n}} = \frac{\bar{X} - 100}{S/\sqrt{n}}.$$

For the $n = 78$ IQ scores, the value of this test statistic is again

$$\frac{109 - 100}{13/\sqrt{78}} = 5.98.$$

The p-value for the two-sided significance test is twice the area under the probability density function of a $t(77)$ random variable to the right of 5.98, that is,

$$p = 2P(t(77) > 5.98) = 0.00000006.$$

Since this p-value is much smaller than the significance level $\alpha = 0.01$, we again decisively reject H_0 and conclude that the IQ scores are not drawn from a population that is centered around 100.

As indicated in Table 4.5, all three methods for conducting the test, namely confidence intervals, traditional hypothesis testing, and significance testing, draw the same conclusion.

Both confidence intervals and traditional hypothesis testing address the question of consistency between a population parameter and a sample statistic. In the previous example, the confidence interval gives a range of parameter values, in this case $105 < \mu < 113$, that are consistent with the sample statistic $\bar{x} = 109$. On the other hand, the traditional hypothesis test fixes the parameter according to the null hypothesis, in this case $H_0 : \mu = 100$, and asks which values of the sample statistic are consistent with H_0.

The geometry associated with these two points of view for the $n = 78$ IQ data values and $\alpha = 0.01$ is illustrated in Figure 4.40, which contains a plot of the sample mean $\bar{x}$ on the horizontal axis and the population mean μ on the vertical axis. The equation of the top solid diagonal line is

$$\mu = \bar{x} + t_{n-1,\alpha/2}\frac{s}{\sqrt{n}}$$

for $n = 78$, $\alpha = 0.01$, and $s = 13$. The equation of the bottom solid diagonal line is

$$\mu = \bar{x} - t_{n-1,\alpha/2}\frac{s}{\sqrt{n}}.$$

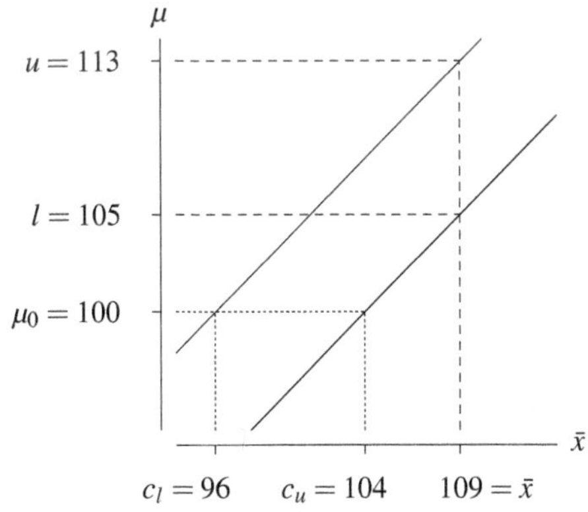

Figure 4.40: Equivalence between confidence intervals and traditional hypothesis testing.

The dashed lines illustrate the construction of a confidence interval. Beginning on the horizontal axis at $\bar{x} = 109$, the vertical dashed line intersects the diagonal solid lines at heights that correspond to the lower confidence limit $l = 105$ and upper confidence limit $u = 113$. The resulting exact two-sided 99% confidence interval for μ is $105 < \mu < 113$. The dotted lines illustrate a traditional hypothesis test. Beginning on the vertical axis at $\mu_0 = 100$ (associated with the null hypothesis $H_0 : \mu = 100$), the horizontal dotted line intersects the diagonal solid lines at the lower critical value $c_l = 96$ and the upper critical value $c_u = 104$. These two critical values are associated with using the test statistic $\bar{x}$. (In the previous example, $(\bar{x} - 100)/(s/\sqrt{n})$ was used as a test statistic, but the geometric interpretation given here uses $\bar{x}$ as the test statistic.) Since the slopes of the two solid lines both equal 1, the confidence interval width $u - l$ equals the width of the region associated with failing to reject H_0, which is $c_u - c_l$, thus establishing the equivalence between the confidence interval and hypothesis testing methods for conducting the test for this particular data set.

The three methods for conducting a test, namely confidence intervals, traditional hypothesis testing, and significance testing, have been shown to be equivalent for one particular test, one particular data set, and one particular choice of α. This is promising, but it does not establish the equivalence of the three methods in general.

The equivalence between traditional hypothesis testing and significance testing should be clear from the development of the two methods given earlier in this section. But the equivalence between using confidence intervals and traditional hypothesis testing to conduct a test is not so obvious.

We illustrate the equivalence for the confidence interval from the previous example for a general value of α and any random sample $X_1, X_2, \ldots, X_n$ assumed to be drawn from a normal population with unknown population mean and variance. The derivation below will show that the confidence interval method for conducting a test is equivalent to traditional hypothesis testing in this slightly more general context. Consider testing

$$H_0 : \mu = \mu_0$$

versus

$$H_1 : \mu \neq \mu_0.$$

Using the confidence interval method to conduct this test, Table 4.5 indicates that we fail to reject H_0 when μ_0 falls in the exact two-sided confidence interval for μ (from Table 3.1), that is, when

$$\bar{X} - t_{n-1,\alpha/2}\frac{S}{\sqrt{n}} < \mu_0 < \bar{X} + t_{n-1,\alpha/2}\frac{S}{\sqrt{n}}.$$

In order to show the equivalence of this method with traditional hypothesis testing, some algebra is performed on the inequality given above resulting in

$$-t_{n-1,\alpha/2} < \frac{\bar{X} - \mu_0}{S/\sqrt{n}} < t_{n-1,\alpha/2}.$$

The random variable in the center of this inequality happens to be the appropriate test statistic for the hypothesis test in this particular setting (from Table 4.4), and the values on the left-hand side and right-hand side of the inequality are the associated critical values c_l and c_u. Thus, failing to reject H_0 using the confidence interval method in this setting is equivalent to failing to reject H_0 using traditional hypothesis testing.

We have now shown the duality between hypothesis testing and interval estimation in one particular statistical setting. The good news is that duality generalizes, as established next.

Theorem 4.1 Let $X_1, X_2, \ldots, X_n$ be a random sample drawn from a population distribution with unknown parameter θ which is defined on the parameter space Ω.

(a) If there exists a critical region C for testing $H_0 : \theta = \theta_0$ versus the one- or two-tailed alternative hypothesis at significance level α for all $\theta_0 \in \Omega$, then the associated set

$$I(X_1, X_2, \ldots, X_n) = \{\theta \,|\, (X_1, X_2, \ldots, X_n) \in C'\}$$

is an exact $100(1-\alpha)\%$ confidence interval for θ.

(b) If $I(X_1, X_2, \ldots, X_n)$ is an exact $100(1-\alpha)\%$ confidence interval for θ, then the region for failing to reject $H_0 : \theta = \theta_0$ at significance level α for all $\theta_0 \in \Omega$ is

$$C' = \{(X_1, X_2, \ldots, X_n) \,|\, \theta \in I(X_1, X_2, \ldots, X_n)\}.$$

Proof

(a) For the hypothesis test $H_0 : \theta = \theta_0$ at significance level α,

$$P\big((X_1, X_2, \ldots, X_n) \in C' \,|\, \theta = \theta_0\big) = 1 - \alpha,$$

which implies that

$$P\big(\theta \in I(X_1, X_2, \ldots, X_n) \,|\, \theta = \theta_0\big) = P\big((X_1, X_2, \ldots, X_n) \in C' \,|\, \theta = \theta_0\big) = 1 - \alpha.$$

(b) Since $I(X_1, X_2, \ldots, X_n)$ is an exact $100(1-\alpha)\%$ confidence interval for θ,

$$P\big(\theta \in I(X_1, X_2, \ldots, X_n) \,|\, \theta = \theta_0\big) = 1 - \alpha,$$

which implies that

$$P\big((X_1, X_2, \ldots, X_n) \in C' \,|\, \theta = \theta_0\big) = P\big(\theta \in I(X_1, X_2, \ldots, X_n) \,|\, \theta = \theta_0\big) = 1 - \alpha. \quad \square$$

Theorem 4.1 establishes the equivalence between using confidence intervals and hypothesis testing for conducting a test: θ_0 falling in a confidence interval is equivalent to the data values falling in C'. Three important aspects of the theorem are listed below.

- Although the term confidence "interval" is used in the statement of Theorem 4.1, the term confidence "set" would be more accurate and more general because occasions arise when these are not intervals. It is for this reason that the notation $I(X_1, X_2, \ldots, X_n)$ is used for the confidence interval instead of the usual $L < \theta < U$.

- The alternative hypothesis has not been explicitly stated in Theorem 4.1. When the alternative hypothesis is one tailed, the associated confidence interval is one sided; when the alternative hypothesis is two tailed, the associated confidence interval is two sided.

- Although the equivalence between confidence intervals and hypothesis testing is interesting mathematically, there is also a practical use for Theorem 4.1. Occasions arise in which a critical region can be determined for a test, but it is quite difficult to determine a confidence interval. Theorem 4.1 indicates that C' can be transformed to arrive at a confidence interval.

Due to the large number of new concepts that have been introduced so far in this chapter, we pause to summarize some of the concepts and notation associated with hypothesis testing.

Summary

Hypothesis testing is one of the bedrocks of inferential statistics. It is used in numerous settings to draw a conclusion concerning an unknown parameter based on statistical evidence. The outline given below is a brief summary of the concepts presented in this section. The bulleted items correspond to the subsections.

- Elements of hypothesis testing
  - A *hypothesis* is a statement concerning an unknown parameter.
  - The unknown parameter is typically referred to generically as θ and the associated parameter space is typically referred to generically as Ω.
  - A *hypothesis test* consists of the following two competing propositions.
    - The *null hypothesis* H_0 is a belief that is assumed to be correct until enough statistical evidence (from the data) causes it to be rejected.
    - The *alternative hypothesis* H_1 is typically what the experimenter is trying to prove.
  - The decision between H_0 and H_1 is based on a set of n data values that are typically denoted by $X_1, X_2, \ldots, X_n$. The realizations of these random variables are typically denoted by $x_1, x_2, \ldots, x_n$.
  - The decision between H_0 and H_1 is stated as "reject H_0" or "fail to reject H_0."
  - There are two subsets of the parameter space Ω, which are Ω_0 corresponding to H_0 and Ω_1 corresponding to H_1.
  - Hypotheses can be classified as simple or composite.
    - A *simple hypothesis* completely specifies a probability distribution.
    - A *composite hypothesis* does not completely specify a probability distribution.
  - The set of all data values leading to rejecting H_0 is known as the *critical region* C. The critical region should be chosen so that there is significant (statistical) evidence to reject H_0.
  - Some hypothesis tests are concluded based on the value of a *test statistic*, which is a real-valued function of the data values.
  - The *critical value(s)* associated with a test statistic are cutoff value(s) that determine the boundary of the critical region.
  - A hypothesis test can result in one of the two following types of errors.
    - Rejecting H_0 when H_0 is true is a *Type I error*.
    - Failing to reject H_0 when H_1 is true is a *Type II error*.
  - The probabilities of committing the two types of errors are
    - $\alpha = P(\text{Type I error}) = P(\text{reject } H_0 \,|\, H_0 \text{ true})$,
    - $\beta = P(\text{Type II error}) = P(\text{fail to reject } H_0 \,|\, H_1 \text{ true})$.
  - The probability of committing a Type I error α is also known as the *significance level*, the *level of significance*, or the *size* of the critical region.
  - For a simple null hypothesis $H_0 : \theta = \theta_0$,

    * an alternative hypothesis of the form $H_1 : \theta > \theta_0$ or $H_1 : \theta < \theta_0$ is known as a *one-sided hypothesis test*,
    * an alternative hypothesis of the form $H_1 : \theta \neq \theta_0$ is known as a *two-sided hypothesis test*.
  - A hypothesis test is considered to be *robust* if it is insensitive to assumptions about the population distribution.

- Power functions
  - The power function for a test is $q(\theta) = P(\text{reject } H_0 \,|\, \theta)$ for all $\theta \in \Omega$.
  - Power functions can be used to compare test statistics for a particular hypothesis test. If one test statistic's power function is greater than or equal to another statistic's power function for all values of θ associated with H_1, then the first test statistic is preferred over the second test statistic for conducting the hypothesis test.
  - Power functions tend to become steeper as the sample size n increases.

- Significance tests
  - Significance tests shift the burden of drawing a conclusion concerning a test from the experimenter to the reader.
  - The *p-value* associated with a significance test is the probability of obtaining the observed test statistic or one more extreme in favor of H_1 in a future experiment with identical sample size n and identical sampling assumptions, assuming that H_0 is true.

- Random sampling from normal populations
  - Hypothesis tests and significance tests associated with random samples drawn from normal populations can be classified into four groups using the following two criteria.
    * Tests concerning random sampling from one normal population or two normal populations.
    * Tests concerning population means and tests concerning population variances.
  - Critical values for tests concerning random sampling from normal populations are typically percentiles of the normal, t, chi-square, and F distributions.

- Sample size determination
  - For some hypothesis tests, formulas or algorithms can be developed that give the sample size n necessary to achieve a particular steepness associated with the power function.

- Equivalence of confidence intervals, hypothesis tests, and significance tests
  - A test can be conducted via confidence intervals, traditional hypothesis testing, and significance testing.
  - Under certain conditions, these three methods are equivalent.

There are two loose ends that need to be tied up concerning hypothesis testing. One of these loose ends is related to the three test statistics used by Chico, Harpo, and Groucho in Example 4.5 for sampling from a $U(0, \theta)$ population. By examining the power functions for the three test statistics in Figure 4.12, we concluded that Groucho's test statistic, $\max\{X_1, X_2\}$, outperformed the others. Although it is the best of the three, is there another test statistic that would perform better still?

The second loose end concerns how to determine an appropriate test statistic. For every example in this section the test statistic has simply been provided in the problem statement. Is there a general way to determine a test statistic given a particular population distribution and hypothesis test? These two loose ends will be addressed in the final two sections.

4.6 Most Powerful Tests

As seen in the examples from the previous section, given a random sample $X_1, X_2, \ldots, X_n$ and an associated hypothesis test, there are several choices of both a test statistic and an associated critical region. A *best critical region of size* α can be defined based on having the highest power of all critical regions associated with level of significance α. This definition only considers a simple null hypothesis and a simple alternative hypothesis.

Definition 4.11 Let the data values $X_1, X_2, \ldots, X_n$ have a probability distribution defined on support $\mathcal{A}$. The hypothesis test

$$H_0 : \theta = \theta_0$$

versus

$$H_1 : \theta = \theta_1,$$

where θ is an unknown parameter, is to be conducted at significance level α. The critical region $C \subset \mathcal{A}$ is known as a *best critical region of size* α if

$$P\big((X_1, X_2, \ldots, X_n) \in C \,|\, \theta = \theta_0\big) = \alpha$$

and for every other potential critical region $C^* \subset \mathcal{A}$ satisfying

$$P\big((X_1, X_2, \ldots, X_n) \in C^* \,|\, \theta = \theta_0\big) = \alpha,$$

it must be the case that

$$P\big((X_1, X_2, \ldots, X_n) \in C \,|\, \theta = \theta_1\big) \ge P\big((X_1, X_2, \ldots, X_n) \in C^* \,|\, \theta = \theta_1\big).$$

The final inequality in this definition is the key. The power is the highest for the best critical region C. The fact that C and C^* have the same size α under H_0 assures that the two critical regions are compared on an equal footing. This definition is illustrated next for a coin flipping example.

Example 4.22 Hayeon is given a coin which is either fair (a flip comes up heads with probability $1/2$) or biased (a flip comes up heads with probability $2/3$). She is not told which type of coin she has been given. Let θ denote the probability that the coin comes up heads on a single flip. Hayeon would like to test

$$H_0 : \theta = \frac{1}{2}$$

versus

$$H_1 : \theta = \frac{2}{3}.$$

She decides to flip the coin four times and observe the random variable X, which denotes the total number of heads that she observes. So the hypothesis test will be executed based on the observation of a single binomial random variable X. Find the best critical region for the test at

(a) level of significance $\alpha = 1/16$,

(b) level of significance $\alpha = 5/16$.

Since both the null and alternative hypotheses are simple, they completely define the distribution of X, the number of heads that appear in the four coin flips:

- under H_0, $X \sim \text{binomial}(4, 1/2)$,
- under H_1, $X \sim \text{binomial}(4, 2/3)$.

Using the R commands `dbinom(0:4, 4, 1 / 2)` and `dbinom(0:4, 4, 2 / 3)`, Table 4.6 can be constructed. The first row gives the probability mass function values of X associated with $H_0 : \theta = 1/2$; the second row gives the probability mass function values of X associated with $H_1 : \theta = 2/3$; the third row gives the ratio of the first two rows.

	$x=0$	$x=1$	$x=2$	$x=3$	$x=4$
$P\left(X=x \,\middle|\, \theta=\frac{1}{2}\right)$	$\frac{1}{16}$	$\frac{1}{4}$	$\frac{3}{8}$	$\frac{1}{4}$	$\frac{1}{16}$
$P\left(X=x \,\middle|\, \theta=\frac{2}{3}\right)$	$\frac{1}{81}$	$\frac{8}{81}$	$\frac{24}{81}$	$\frac{32}{81}$	$\frac{16}{81}$
$\frac{P(X=x \mid \theta=1/2)}{P(X=x \mid \theta=2/3)}$	$\frac{81}{16}$	$\frac{81}{32}$	$\frac{81}{64}$	$\frac{81}{128}$	$\frac{81}{256}$

Table 4.6: Conditional distribution of X and likelihood ratio.

Now Hayeon turns to the question of finding the best critical region for the test.

(a) When the level of significance is $\alpha = 1/16$, Hayeon has just two choices for potential critical regions:

$$C_1 = \{x \mid x=0\} \qquad \text{and} \qquad C_2 = \{x \mid x=4\},$$

because

$$P\left(X \in C_1 \,\middle|\, \theta=\frac{1}{2}\right) = P\left(X \in C_2 \,\middle|\, \theta=\frac{1}{2}\right) = \frac{1}{16}.$$

She is looking for the best of these two critical regions. To apply Definition 4.11, she must calculate the power associated with these two potential critical regions. From Table 4.6, the power associated with these two potential critical regions is

$$P\left(X \in C_1 \,\middle|\, \theta=\frac{2}{3}\right) = \frac{1}{81}$$

and

$$P\left(X \in C_2 \,\middle|\, \theta=\frac{2}{3}\right) = \frac{16}{81}.$$

Since $P(X \in C_2 \mid \theta = 2/3)$ is the larger of the two, she selects C_2 as the best critical region C. So Hayeon's decision rule for a test of size $\alpha = 1/16$ is to flip her coin four times, record the number of heads that appear x, and reject H_0 when $x = 4$.

This decision is consistent with intuition because the biased coin with a higher probability of heads is more likely to produce more heads in the four flips than the fair coin. The power associated with this test is $q(2/3) = 16/81$. The probability of a Type II error is $\beta = 1 - 16/81 = 65/81$. Notice that the choice of the best critical region of size $\alpha = 1/16$, $C = \{x \,|\, x = 4\}$, corresponds to the smallest value in the third row of Table 4.6. We will return to this observation later.

(b) When the level of significance is $\alpha = 5/16$, Hayeon now has four choices for potential critical regions:

$$C_1 = \{x \,|\, x = 0, 1\}, \quad C_2 = \{x \,|\, x = 0, 3\}, \quad C_3 = \{x \,|\, x = 1, 4\}, \quad C_4 = \{x \,|\, x = 3, 4\},$$

because

$$P\left(X \in C_i \,\middle|\, \theta = \frac{1}{2}\right) = \frac{5}{16},$$

for $i = 1, 2, 3, 4$ from Table 4.6. She is looking for the best of these four critical regions. From Table 4.6, the power associated with these four potential critical regions is

$$P\left(X \in C_1 \,\middle|\, \theta = \frac{2}{3}\right) = \frac{9}{81},$$

$$P\left(X \in C_2 \,\middle|\, \theta = \frac{2}{3}\right) = \frac{33}{81},$$

$$P\left(X \in C_3 \,\middle|\, \theta = \frac{2}{3}\right) = \frac{24}{81},$$

and

$$P\left(X \in C_4 \,\middle|\, \theta = \frac{2}{3}\right) = \frac{48}{81}.$$

Since $P(X \in C_4 \,|\, \theta = 2/3)$ is the largest of the four, she selects C_4 as the best critical region C. So Hayeon's decision rule for a test of size $\alpha = 5/16$ is to flip her coin four times, record the number of heads that appear x, and reject H_0 when $x = 3$ or $x = 4$. This decision is also consistent with intuition because the biased coin is more likely to produce more heads in the four flips than the fair coin. The power associated with this test is $q(2/3) = 48/81$. The probability of a Type II error is $\beta = 1 - 48/81 = 33/81$. Notice that the choice of the best critical region of size $\alpha = 5/16$, $C = \{x \,|\, x = 3, 4\}$, corresponds to the two smallest values in the third row of Table 4.6. We will return to this observation later.

Parts (a) and (b) of the previous example have once again illustrated the see-saw relationship between α and β for the best critical regions. The probabilities of the two types of errors are

$$\alpha = P(\text{Type I error}) = \frac{1}{16} \qquad \text{and} \qquad \beta = P(\text{Type II error}) = \frac{65}{81}$$

for part (a) and

$$\alpha = P(\text{Type I error}) = \frac{5}{16} \qquad \text{and} \qquad \beta = P(\text{Type II error}) = \frac{33}{81}$$

for part (b). As α increases from $\alpha = 1/16$ to $\alpha = 5/16$, β decreases from $\beta = 65/81$ to $\beta = 33/81$.

Hayeon was "lucky" in some sense in the previous example because there were only two potential critical regions associated with $\alpha = 1/16$, and only four potential critical regions associated with $\alpha = 5/16$. This allowed her to enumerate all of the potential critical regions and choose the best critical region from among them. The next example returns us to the very first hypothesis test presented in this chapter, where we will not be so lucky.

Example 4.23 Graham collects a single observation X to test the hypothesis

$$H_0 : \theta = 1$$

versus

$$H_1 : \theta = 3,$$

where θ is a positive unknown parameter from the continuous population distribution described by the probability density function

$$f(x) = 2\theta^{-2} x e^{-(x/\theta)^2} \qquad x > 0.$$

Find the best critical region associated with the test of size $\alpha = 0.1$.

The population distribution is the Rayleigh distribution. The null and alternative hypotheses are simple, so they completely define the population probability distribution of the single ($n = 1$) observation X. The population probability density functions associated with H_0 and H_1 are graphed in Figure 4.41. One problem that arises in Graham's quest to find the best critical region of size α is that there are an infinite number of critical regions of size $\alpha = 0.1$, so it is impossible to enumerate them all. This puts him at a significant disadvantage relative to Hayeon in the previous example. In this example, Graham will compare just two such potential critical regions. The cumulative distribution function of the population under H_0 ($\theta = 1$) on its support is

$$F(x) = \int_{-\infty}^{x} f(w)\,dw = \int_{0}^{x} 2we^{-w^2}\,dw = \left[-e^{-w^2}\right]_0^x = 1 - e^{-x^2} \qquad x > 0.$$

The 90th percentile of this probability distribution, $x_{0.9}$, can be found by solving

$$F(x_{0.9}) = 1 - e^{-x_{0.9}^2} = 0.9$$

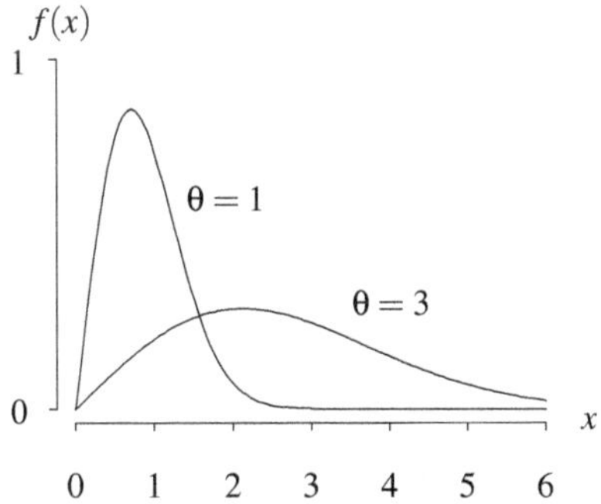

Figure 4.41: Population probability density functions for $\theta = 1$ and $\theta = 3$.

for $x_{0.9}$, yielding $x_{0.9} = \sqrt{-\ln(0.1)} \cong 1.5174$. So one potential critical region of size $\alpha = 0.1$ is

$$C_1 = \{x \,|\, x > 1.5174\}.$$

There are infinitely many ways to devise a second potential critical region C_2 to compare with C_1. Another potential critical region of size $\alpha = 0.1$ consists of X values that lie between some value c and 1. The value of c can be found by solving

$$P(c < X < 1) = F(1) - F(c) = \left(1 - e^{-1}\right) - \left(1 - e^{-c^2}\right) = e^{-c^2} - e^{-1} = 0.1$$

for c yielding $c = \sqrt{-\ln\left(0.1 + e^{-1}\right)} \cong 0.8715$. So a second potential critical region of size $\alpha = 0.1$ is

$$C_2 = \{x \,|\, 0.8715 < x < 1\}.$$

The areas under the probability density function of X associated with $H_0 : \theta = 1$, both equal to 0.1 for the two critical regions, are shown in Figure 4.42. Although C_1 and C_2 were constructed as contiguous sets, this need not necessarily be the case. The fact that both critical regions have size $\alpha = 0.1$ means that their powers can be compared on an equal footing. Now that the two potential critical regions have been identified, their powers will be calculated and compared. The power associated with the first potential critical region is

$$P(X \in C_1 \,|\, \theta = 3) = \int_{1.5174}^{\infty} \frac{2}{9} x e^{-(x/3)^2} \, dx = 10^{-1/9} \cong 0.7743.$$

The power associated with the second potential critical region is

$$P(X \in C_2 \,|\, \theta = 3) = \int_{0.8715}^{1} \frac{2}{9} x e^{-(x/3)^2} \, dx \cong 0.02423.$$

These two probabilities are plotted as areas under the probability density function associated with $H_1 : \theta = 3$ in Figure 4.43. The power associated with C_1 is much larger than the power associated with C_2, so C_1 is the *better critical region* of size α of the two, although we cannot confirm that it is the *best critical region* of size α. The infinite

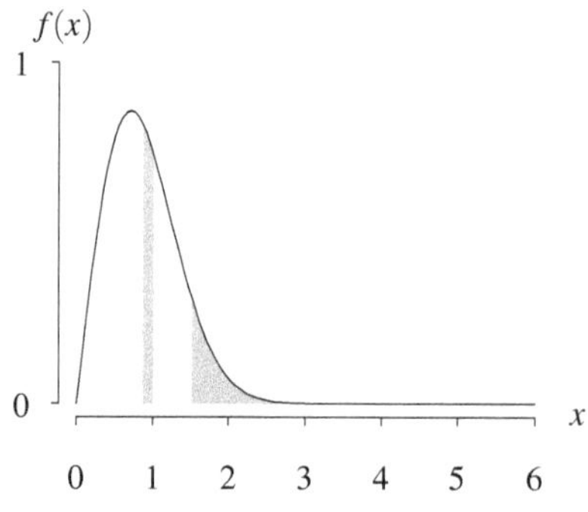

Figure 4.42: Areas under $f(x)$ when $\theta = 1$ (associated with H_0) for C_1 and C_2.

number of potential critical regions make this conclusion impossible to draw with the tools we have presently in place. This is where the next result, the Neyman–Pearson theorem, also known as the Neyman–Pearson lemma, comes to the rescue. It allows you to select a best critical region without enumerating all potential critical regions of size α.

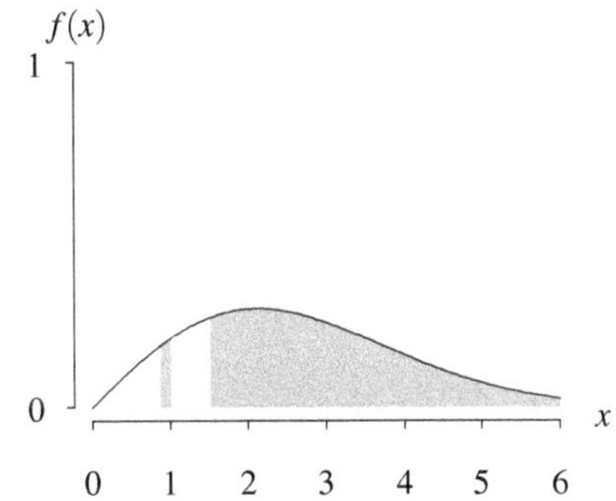

Figure 4.43: Areas under $f(x)$ when $\theta = 3$ (associated with H_1) for C_1 and C_2.

The Neyman–Pearson theorem stated below provides a method for determining the best critical region of size α without enumerating all of the potential critical regions of size α. It can be implemented for testing a simple null hypothesis H_0 against a simple alternative hypothesis H_1. Even though its assumption regarding simple hypotheses is rather restrictive, its outcome is remarkable: it is able to identify a best critical region C of size α. At least in terms of power—there is no better hypothesis test than that identified by the Neyman–Pearson theorem.

Theorem 4.2 (Neyman–Pearson) Let $X_1, X_2, \ldots, X_n$ denote a random sample drawn from a probability distribution described by $f(x)$ with an unknown parameter θ. The associated likelihood function is denoted by $L(\theta)$. The hypothesis test

$$H_0 : \theta = \theta_0$$

versus

$$H_1 : \theta = \theta_1,$$

where θ_0 and θ_1 are constants, is to be conducted at significance level α. A critical region C that satisfies

- $P\big((X_1, X_2, \ldots, X_n) \in C \,|\, \theta = \theta_0\big) = \alpha$,
- $\dfrac{L(\theta_0)}{L(\theta_1)} \le c$ for any $(x_1, x_2, \ldots, x_n) \in C$, and
- $\dfrac{L(\theta_0)}{L(\theta_1)} \ge c$ for any $(x_1, x_2, \ldots, x_n) \notin C$,

for some positive constant c, is a best critical region of size α, and the test associated with this critical region is known as a *most powerful test* of size α.

Proof This proof assumes that $X_1, X_2, \ldots, X_n$ are drawn from a continuous population. If instead $X_1, X_2, \ldots, X_n$ were drawn from a discrete population, then the integrals in the proof would be replaced by summations. Let C be a critical region of size α that satisfies the three conditions in the Neyman–Pearson theorem. If there is no other critical region of size α, then C must be a best critical region. If there is another critical region of size α, denote it as C_*. Let $\mathcal{A}$ denote the support of $X_1, X_2, \ldots, X_n$. To make the notation more compact, for any set $A \subset \mathcal{A}$,

$$P\big((X_1, X_2, \ldots, X_n) \in A\big) = \underset{A}{\int\int \cdots \int} L(\theta)\, dx_n \ldots dx_2\, dx_1 = \int_A L(\theta).$$

Using this compact notation, we want to show that C has the highest power among all critical regions of size α, that is,

$$\int_C L(\theta_1) - \int_{C_*} L(\theta_1) \ge 0$$

for any critical region C_* of size α. The second condition in the theorem implies that $L(\theta_1) \ge L(\theta_0)/c$ for every $(x_1, x_2, \ldots, x_n) \in C$ (and therefore this condition is true for every $(x_1, x_2, \ldots, x_n)$ in any subset of C). The third condition in the theorem implies that $L(\theta_1) \le L(\theta_0)/c$ for every $(x_1, x_2, \ldots, x_n) \in C'$ (and therefore this condition is true for every $(x_1, x_2, \ldots, x_n)$ in any subset of C'). Since the critical region C can be written as $C = (C \cap C_*) \cup (C \cap C_*')$ and the critical region C_* can be written as $C_* = (C_* \cap C) \cup (C_* \cap C')$,

$$\begin{aligned}
\int_C L(\theta_1) - \int_{C_*} L(\theta_1) &= \int_{C\cap C_*} L(\theta_1) + \int_{C\cap C_*'} L(\theta_1) - \left[\int_{C_*\cap C} L(\theta_1) + \int_{C_*\cap C'} L(\theta_1)\right] \\
&= \int_{C\cap C_*'} L(\theta_1) - \int_{C_*\cap C'} L(\theta_1) \\
&\ge \frac{1}{c}\int_{C\cap C_*'} L(\theta_0) - \frac{1}{c}\int_{C_*\cap C'} L(\theta_0) \\
&= \frac{1}{c}\left[\int_{C\cap C_*} L(\theta_0) + \int_{C\cap C_*'} L(\theta_0) - \int_{C_*\cap C} L(\theta_0) - \int_{C_*\cap C'} L(\theta_0)\right] \\
&= \frac{1}{c}\left[\int_C L(\theta_0) - \int_{C_*} L(\theta_0)\right] \\
&= \frac{1}{c}[\alpha - \alpha] \\
&= 0
\end{aligned}$$

because C and C_* are both critical regions of size α. This shows that

$$\int_C L(\theta_1) - \int_{C_*} L(\theta_1) \ge 0$$

for any critical region C_* of size α, so C must be a best critical region of size α. □

The Neyman–Pearson theorem, established in 1933 by Jerzy Neyman and Egon Pearson, is used to determine a best critical region (and the associated most powerful test) without enumerating all of the critical regions of size α. For sampling from discrete populations, the first criterion in Theorem 4.2 is typically written as $P\big((X_1, X_2, \ldots, X_n) \in C \,|\, \theta = \theta_0\big) \le \alpha$. The focus now shifts to the behavior of the likelihood ratio, as illustrated by a series of examples.

Example 4.24 Returning to Hayeon's coin flipping experiment from Example 4.22 to determine whether the coin that she is holding is fair, that is,

$$H_0 : \theta = \frac{1}{2},$$

or biased, that is,

$$H_1 : \theta = \frac{2}{3},$$

she will now use the Neyman–Pearson theorem (because H_0 and H_1 are both simple hypotheses) to determine a best critical region and associated most powerful test of size $\alpha = 5/16$ based on the number of heads X produced in four flips of the coin.

Even though there were four flips, there is only a single data value X, that is, the sample size is $n = 1$, the likelihood function $L(\theta)$ is the same as the population probability mass function $f(x)$. The table used in the enumeration approach in Example 4.22 for finding the best critical region of size $\alpha = 5/16$ is reproduced in Table 4.7, with slightly different row labels to match the notation in the Neyman–Pearson theorem.

	$x=0$	$x=1$	$x=2$	$x=3$	$x=4$
$L(\theta_0) = L\left(\frac{1}{2}\right) = f\left(x \,\middle\vert\, \theta = \frac{1}{2}\right)$	$\frac{1}{16}$	$\frac{1}{4}$	$\frac{3}{8}$	$\frac{1}{4}$	$\frac{1}{16}$
$L(\theta_1) = L\left(\frac{2}{3}\right) = f\left(x \,\middle\vert\, \theta = \frac{2}{3}\right)$	$\frac{1}{81}$	$\frac{8}{81}$	$\frac{24}{81}$	$\frac{32}{81}$	$\frac{16}{81}$
$\frac{L(1/2)}{L(2/3)}$	$\frac{81}{16}$	$\frac{81}{32}$	$\frac{81}{64}$	$\frac{81}{128}$	$\frac{81}{256}$

Table 4.7: Conditional distribution of X and likelihood ratio.

Hayeon recognizes that the entries in the bottom row of Table 4.7 are monotone decreasing, so the Neyman–Pearson theorem indicates that rejecting H_0 for

$$\frac{L(1/2)}{L(2/3)} \le \frac{81}{128},$$

that is, $c = 81/128$, gives a best critical region of size $\alpha = 5/16$. This is equivalent to the best critical region

$$C = \{x \mid x = 3, 4\}$$

as determined previously by enumeration.

We now shift to Graham's dilemma established in Example 4.23 with an infinite number of potential critical regions in which enumeration was not possible. Invoking the Neyman–Pearson theorem allows him to determine a best critical region of size α.

Example 4.25 Graham uses a single observation, X, that is collected to test the hypothesis

$$H_0 : \theta = 1$$

versus

$$H_1 : \theta = 3,$$

where θ is a positive unknown parameter from the continuous Rayleigh population distribution described by the probability density function

$$f(x) = 2\theta^{-2} x e^{-(x/\theta)^2} \qquad x > 0.$$

Find the best critical region and associated most powerful test of size $\alpha = 0.1$.

The null and alternative hypotheses are simple, so the Neyman–Pearson theorem can be applied. Since there is only a single ($n = 1$) observation X, the likelihood function $L(\theta)$ is identical to the probability density function $f(x)$. The ratio of the likelihood functions in the Neyman–Pearson theorem is

$$\frac{L(\theta_0)}{L(\theta_1)} = \frac{L(1)}{L(3)} = \frac{f(x \mid \theta = 1)}{f(x \mid \theta = 3)} = \frac{2xe^{-x^2}}{(2/9)xe^{-(x/3)^2}} = 9e^{-8x^2/9}$$

for $x > 0$. The ratio of the likelihood functions is a monotone decreasing function of x. Using the Neyman–Pearson theorem, a best critical region of size α has the form

$$\frac{L(\theta_0)}{L(\theta_1)} = 9e^{-8x^2/9} \le c$$

for some $c > 0$ which yields a significance level α. Working with the test statistic $9e^{-8x^2/9}$ can be a bit clumsy. The following sequence of inequalities shows that using $9e^{-8x^2/9}$ as a test statistic is equivalent to using just x as a test statistic:

$$\begin{aligned}
9e^{-8x^2/9} \le c \quad &\Longleftrightarrow \quad e^{-8x^2/9} \le c/9 \\
&\Longleftrightarrow \quad -8x^2/9 \le \ln(c/9) \\
&\Longleftrightarrow \quad 8x^2/9 \ge -\ln(c/9) \\
&\Longleftrightarrow \quad x^2 \ge -9\ln(c/9)/8 \\
&\Longleftrightarrow \quad x \ge \sqrt{-9\ln(c/9)/8}.
\end{aligned}$$

Since the right-hand side of the last inequality is also a constant, the best critical region of size α is of the form $x \ge c'$ for some constant c'. In order to establish a best critical region of size α, this constant c' is just the 90th percentile (because $\alpha = 0.1$) of the distribution of X under H_0, which, from Example 4.23, is $c' = \sqrt{-\ln(0.1)} \cong 1.5174$. This value of c' is consistent with $c = 9 \cdot 10^{-8/9} \cong 1.1624$. So the best critical region of size $\alpha = 0.1$ can be written in terms of the likelihood ratio as

$$C = \left\{ x \,\middle|\, \frac{L(\theta_0)}{L(\theta_1)} = 9e^{-8x^2/9} \le 1.1624 \right\}$$

or in terms of just x as

$$C = \{x \mid x \ge 1.5174\}.$$

The two forms are equivalent. Graham's most powerful test collects a single observation x from the population and rejects H_0 if $9e^{-8x^2/9} \le 1.1624$, or equivalently, if $x \ge 1.5174$. The likelihood ratio function is plotted in Figure 4.44. The best critical region of size $\alpha = 0.1$ can be expressed as the likelihood ratio (plotted on the vertical axis) being less than or equal to 1.1624 or the x value (plotted on the horizontal axis) observed being greater than or equal to 1.5174.

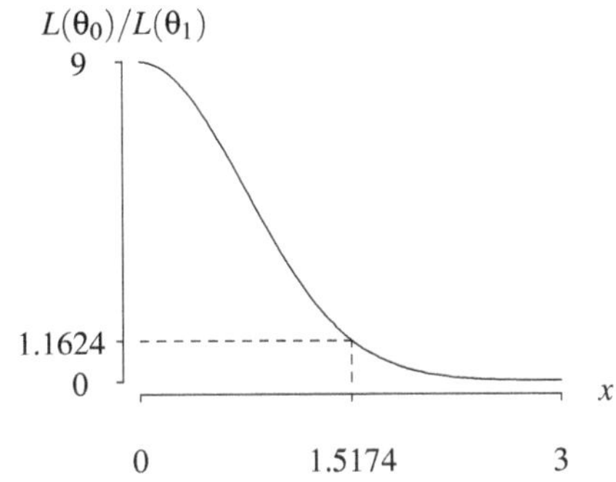

Figure 4.44: Likelihood ratio for a single observation X from a Rayleigh population.

We now apply the Neyman–Pearson theorem to the second example from this chapter, which involved a single observation drawn from a special case of the beta distribution.

Example 4.26 A single observation X is collected to test the simple null hypothesis

$$H_0 : \theta = 1/2$$

versus the simple alternative hypothesis

$$H_1 : \theta = 3,$$

where θ is a positive unknown parameter from the continuous population distribution described by the probability density function

$$f(x) = \theta x^{\theta - 1} \qquad 0 < x < 1.$$

Find the most powerful test of level $\alpha = 0.2$.

The null and alternative hypotheses are simple, so the Neyman–Pearson theorem can be applied. As in the previous example, there is only a single ($n = 1$) observation, so the likelihood function is equivalent to the probability density function. The ratio of the likelihood functions is

$$\frac{L(\theta_0)}{L(\theta_1)} = \frac{L(1/2)}{L(3)} = \frac{f(x \,|\, \theta = 1/2)}{f(x \,|\, \theta = 3)} = \frac{x^{-1/2}/2}{3x^2} = \frac{1}{6}x^{-5/2}$$

for $0 < x < 1$. This ratio of the likelihood functions is a monotone decreasing function of x. Using the Neyman–Pearson theorem, a best critical region of size α has the form

$$\frac{L(\theta_0)}{L(\theta_1)} = \frac{1}{6}x^{-5/2} \le c$$

for some $c > 0$ which yields a significance level α. Since working with the test statistic $x^{-5/2}/6$ can be clumsy, the following sequence of inequalities shows that it is equivalent to work with the test statistic x:

$$x^{-5/2}/6 \le c \qquad \Longleftrightarrow \qquad x^{-5/2} \le 6c$$
$$\qquad\qquad\qquad \Longleftrightarrow \qquad x \ge (6c)^{-2/5}.$$

Denote the right-hand side of the last inequality as the constant c'. In order to achieve a best critical region of size $\alpha = 0.2$, the critical value c' is the 80th percentile of the probability distribution of X under H_0. This is found by solving

$$\int_0^{c'} \frac{1}{2\sqrt{x}}\, dx = 0.8$$

for c', yielding the critical value $c' = 0.64$. This value of c' is consistent with the critical value $c = (5/4)^5/6 = 3125/6144 = (0.64)^{-5/2}/6 \cong 0.5086$ associated with the likelihood ratio. A graph of the ratio of the likelihood functions, including these two critical values, is given in Figure 4.45. So the decision rule for the most powerful test is to collect a single observation x and reject H_0 if the likelihood ratio $x^{-5/2}/6$ falls in the best critical region of size $\alpha = 0.2$

$$C = \{x \,|\, x^{-5/2}/6 \le 0.5086\}.$$

Alternatively, a simpler decision rule for the most powerful test is to collect a single observation x and reject H_0 if the test statistic x falls in the best critical region of size $\alpha = 0.2$

$$C = \{x \,|\, x \ge 0.64\}.$$

The two decision rules are equivalent in the sense that they will always draw the same conclusion for the same data value, as illustrated in Figure 4.45.

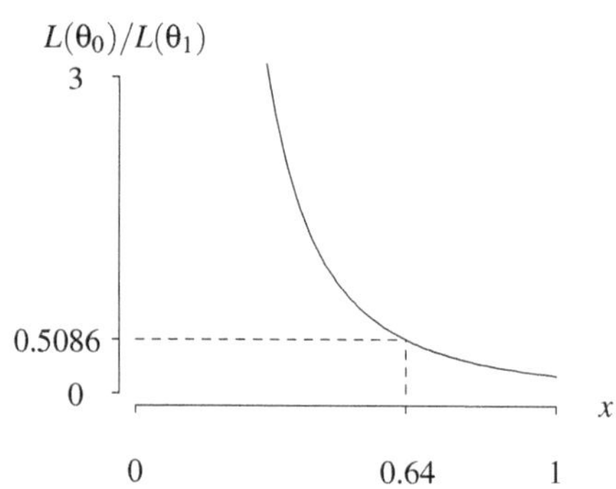

Figure 4.45: Likelihood ratio for a single observation X from a beta population.

The next example highlights the fact that the Neyman–Pearson theorem can be applied in a fairly general setting.

Example 4.27 Advertising a product has grown more complex over time because of the increasing number of potential advertising outlets. A company wants to get the word out that it is having a buy-one-get-one-free sale during the next week. They intend to use the following advertising outlets:

- print: newspapers, magazines, trade journals, billboards, etc.
- electronic: radio, television

- internet: web pages, search engine advertisement, e-mail, etc.
- social media: Facebook, Twitter, etc.

From past advertising efforts of this kind, the company knows that customers under 40 years of age are reached in a different manner than customers age 40 and over. The associated probabilities, classified by age and advertising venue, are given in Table 4.8.

	print	electronic	internet	social media
under 40	5%	5%	10%	80%
40 or over	50%	25%	20%	5%

Table 4.8: Advertising venues by age.

The company asks each customer how they heard about the sale, then tries to classify the customer by age based on their response. (Everyone knows it is never polite to ask someone their age.) Find the most powerful test for

$$H_0 : \text{customer under 40}$$

versus

$$H_1 : \text{customer 40 or over}$$

at $\alpha = 0.1$.

At first glance, the Neyman–Pearson theorem does not seem to apply. There are four parameters, rather than just the single parameter θ as stated in the theorem. But a review of the proof indicates that the theorem does not require that there be a single parameter. The key element is that both hypotheses are simple, and in this case they are. Conceptually, the null and alternative hypotheses can be thought of in terms of a vector of probabilities $\boldsymbol{\theta}$, so they can be restated as

$$H_0 : \boldsymbol{\theta} = (0.05, 0.05, 0.10, 0.80)$$

versus

$$H_1 : \boldsymbol{\theta} = (0.50, 0.25, 0.20, 0.05)$$

at $\alpha = 0.1$. So Table 4.8 is augmented in the usual fashion in Table 4.9 in order to implement the Neyman–Pearson theorem. Recognizing that the entries in the third row in Table 4.9 increase, the Neyman–Pearson theorem tells us that rejecting H_0 when $L(\boldsymbol{\theta}_0)/L(\boldsymbol{\theta}_1) \le 0.20$ results in a best critical region of size $\alpha = 0.1$. So the most powerful test of size $\alpha = 0.1$ rejects H_0 (thus concluding that the customer is 40 or over) when the customer claims that they heard of the sale by print or electronic media. The power of this test is $0.50 + 0.25 = 0.75$. This most powerful test is consistent with the most powerful test of size $\alpha = 0.1$ obtained by enumeration.

The previous four examples have all involved the collection of just a single ($n = 1$) data value X. The next example revisits the third example from this chapter which involves a statistical test concerning a single parameter θ by drawing a random sample of size $n = 2$.

	print	electronic	internet	social media
$L(\theta_0)$	0.05	0.05	0.10	0.80
$L(\theta_1)$	0.50	0.25	0.20	0.05
$\dfrac{L(\theta_0)}{L(\theta_1)}$	0.10	0.20	0.50	16

Table 4.9: Likelihood ratios by advertising venue.

Example 4.28 Let X_1 and X_2 be a random sample of size $n = 2$ from a $N(\theta, 1)$ population, where θ is an unknown parameter. Find the most powerful test of

$$H_0 : \theta = 5$$

versus

$$H_1 : \theta = 3$$

at $\alpha = 0.1$.

One of the advantages of using the Neyman–Pearson theorem will become apparent in this example. There is no discussion in the problem statement of an appropriate test statistic for this particular hypothesis test. The process of applying the Neyman–Pearson theorem will result in the development of a test statistic that results in a most powerful test.

The population variance has been set to the constant $\sigma^2 = 1$ in the problem statement so that there is just one unknown parameter θ, the population mean. The population probability density function associated with a $N(\theta, 1)$ population is

$$f(x) = \frac{1}{\sqrt{2\pi}} e^{-(x-\theta)^2/2} \qquad -\infty < x < \infty.$$

Since X_1 and X_2 are independent, the joint probability density function of X_1 and X_2, which is the likelihood function, is the product of the marginal probability density functions

$$L(\theta) = f(x_1, x_2) = f(x_1)f(x_2) = \frac{1}{2\pi} e^{-(x_1-\theta)^2/2-(x_2-\theta)^2/2}$$

for $-\infty < x_1 < \infty$ and $-\infty < x_2 < \infty$. The ratio of the likelihood functions is

$$\frac{L(\theta_0)}{L(\theta_1)} = \frac{L(5)}{L(3)} = \frac{\frac{1}{2\pi} e^{-(x_1-5)^2/2-(x_2-5)^2/2}}{\frac{1}{2\pi} e^{-(x_1-3)^2/2-(x_2-3)^2/2}} = e^{2x_1+2x_2-16}$$

for $-\infty < x_1 < \infty$ and $-\infty < x_2 < \infty$. This ratio of the likelihood functions associated with H_0 and H_1 is monotone increasing in x_1 and x_2. Using the Neyman–Pearson theorem, a best critical region of size α has the form

$$\frac{L(\theta_0)}{L(\theta_1)} = e^{2x_1+2x_2-16} \le c$$

for some $c > 0$ that yields a significance level α. The following sequence of inequalities gives an alternative test statistic:

$$e^{2x_1+2x_2-16} \le c \quad \Longleftrightarrow \quad 2x_1 + 2x_2 - 16 \le \ln c$$
$$\Longleftrightarrow \quad 2x_1 + 2x_2 \le 16 + \ln c$$
$$\Longleftrightarrow \quad x_1 + x_2 \le 8 + \frac{1}{2}\ln c.$$

Since the right-hand side of the last inequality is also a constant, it can be denoted by c'. Implementing the Neyman–Pearson theorem indicates that $x_1 + x_2$ is an appropriate test statistic and the best critical region of size α is of the form $x_1 + x_2 \le c'$ for some constant c'. Since $\alpha = 0.1$ and the sum $X_1 + X_2 \sim N(10,\, 2)$ under H_0 (that is, when $\theta = 5$), the critical value c' is the 10th percentile of the $N(10,\, 2)$ distribution. This percentile can be calculated with the R statement `qnorm(0.1, 10, sqrt(2))` which yields $c' = 8.1876$. The associated value of the critical value c is $c = 1.4553$. In addition to the two usual test statistics and associated critical values that are generated by the Neyman–Pearson theorem, Table 4.4 indicates that

$$Z = \frac{\bar{X} - \theta_0}{\sigma/\sqrt{n}} = \frac{\bar{X} - 5}{1/\sqrt{2}} = \sqrt{2}(\bar{X} - 5)$$

is also an appropriate test statistic. The associated critical value is the 10th percentile of a standard normal random variable, which is $z_{0.9} = -1.2816$ as calculated with the R statement `qnorm(0.1)`. So to summarize, the two observational values, x_1 and x_2, of the random variables X_1 and X_2 are collected. There are three different test statistics and associated best critical regions for the test, which means that all three correspond to most powerful tests of size $\alpha = 0.1$. All are equivalent in the sense that they will each draw the same conclusion for the same data set.

- H_0 is rejected when the test statistic $e^{2x_1+2x_2-16}$ is less than or equal to $c = 1.4553$.
- H_0 is rejected when the test statistic $x_1 + x_2$ is less than or equal to $c' = 8.1876$, or, equivalently, if $\bar{x}$ is less than or equal to 4.0938.
- H_0 is rejected when the test statistic $\sqrt{2}\,(\bar{x} - 5)$ is less than or equal to $z_{0.9} = -1.2816$.

A figure like those in the previous examples is not possible here because there are two data values x_1 and x_2. A similar graph can be drawn, however, with $\bar{x}$ on the horizontal axis and

$$\frac{L(\theta_0)}{L(\theta_1)} = e^{2x_1+2x_2-16} = e^{4\bar{x}-16}$$

on the vertical axis. This graph is shown in Figure 4.46. The fact that $L(\theta_0)/L(\theta_1)$ is a monotone increasing function of $\bar{x}$ is why the three decision rules stated above all involve rejecting the null hypothesis when the test statistic is *less than or equal to* a critical value.

In the previous four examples, the critical value associated with the most powerful test has been fairly easy to calculate analytically. In the next example, that will not be the case.

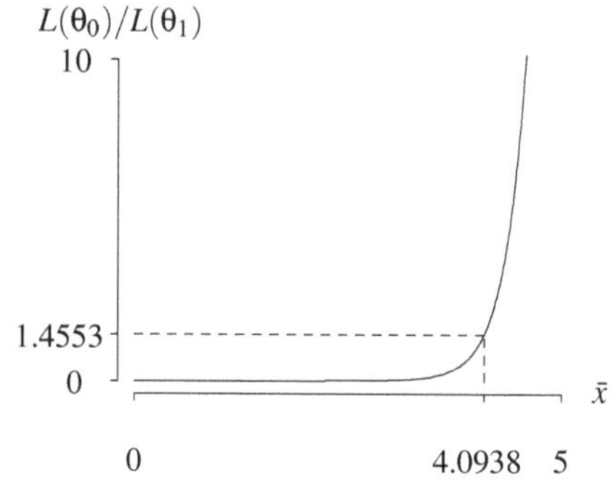

Figure 4.46: Likelihood ratio for a random sample X_1, X_2 from a $N(\theta,\, 1)$ population.

Example 4.29 Does the geometric or the Poisson distribution provide a better fit for the horse kick data from Example 2.9? Assume that the fits are via maximum likelihood estimation. Conduct an appropriate hypothesis test at $\alpha = 0.05$.

The $n = 200$ horse kick data values, ranging from 0 to 4 deaths per corps-year, are given in Table 4.10. The probability mass function associated with the geometric(p) distribution, which is parameterized beginning at $x = 0$, is

$$f(x) = p(1-p)^x \qquad\qquad x = 0,\, 1,\, 2,\, \ldots\, .$$

The probability mass function associated with the Poisson(λ) distribution is

$$f(x) = \frac{\lambda^x e^{-\lambda}}{x!} \qquad\qquad x = 0,\, 1,\, 2,\, \ldots\, .$$

Let $X_1, X_2, \ldots, X_n$ be the random sample. The maximum likelihood estimator of the unknown parameter p in a geometric(p) distribution, parameterized beginning at $x = 0$, is

$$\hat{p} = \frac{n}{n + \sum_{i=1}^{n} x_i} = \frac{200}{200 + 122} = \frac{200}{322} = \frac{100}{161} \cong 0.6211.$$

(The details are left as an exercise.) From Example 2.9, the maximum likelihood estimator of the unknown parameter λ in a Poisson(λ) distribution is the sample mean:

$$\hat{\lambda} = \frac{1}{n} \sum_{i=1}^{n} x_i = \frac{122}{200} = 0.61.$$

The next question is to determine which distribution corresponds to the null hypothesis and which distribution corresponds to the alternative hypothesis. This is an important

number of deaths per corps per year	0	1	2	3	4
number of observed values	109	65	22	3	1

Table 4.10: Observed horse kick deaths.

consideration because it determines which distribution will carry the burden of proof in the hypothesis test. We choose to place the geometric($\hat{p}$) distribution in the null hypothesis and the Poisson($\hat{\lambda}$) distribution in the alternative hypothesis. So the hypothesis test is

$$H_0 : \text{geometric}\,(\hat{p}\,)\,\text{population distribution}$$

versus

$$H_1 : \text{Poisson}\big(\hat{\lambda}\big)\text{population distribution.}$$

The hypothesis test can also be constructed as a choice between two population probability mass functions:

$$H_0 : f(x) = \frac{100}{161}\left(1 - \frac{100}{161}\right)^x \qquad\qquad x = 0, 1, 2, \ldots$$

versus

$$H_1 : f(x) = \frac{0.61^x e^{-0.61}}{x!} \qquad\qquad x = 0, 1, 2, \ldots .$$

Both H_0 and H_1 completely define a probability distribution (although they are two different probability distributions unlike most of the previous examples), so they are simple hypotheses. This allows us to invoke the Neyman–Pearson theorem to conduct the hypothesis test.

The fact that we are comparing *fitted*, rather than *hypothesized*, distributions induces some additional error into the hypothesis test. The error is small, however, because the sample size $n = 200$ is large.

The likelihood ratio $L(\theta_0)/L(\theta_1)$ in this case is

$$\frac{\prod_{i=1}^{n} \frac{100}{161}\left(1 - \frac{100}{161}\right)^{x_i}}{\prod_{i=1}^{n} \frac{0.61^{x_i} e^{-0.61}}{x_i!}} = \frac{\left(\frac{100}{161}\right)^{200}\left(1 - \frac{100}{161}\right)^{\sum_{i=1}^{n} x_i} \prod_{i=1}^{n} x_i!}{0.61^{\sum_{i=1}^{n} x_i} e^{-(200)(0.61)}}.$$

The notion of somehow simplifying or finding the distribution of this horrific test statistic under H_0 is out of the question. It can, however, still be computed. The R code below calculates the value of the likelihood ratio for the horse kick data, which yields 0.0005288.

```
x = c(rep(0, 109), rep(1, 65), rep(2, 22), rep(3, 3), rep(4, 1))
n = length(x)
p = n / (n + sum(x))
l = mean(x)
r = p ^ n * (1 - p) ^ sum(x) * prod(factorial(x)) /
    (l ^ sum(x) * exp(-n * l))
```

So the Neyman–Pearson theorem applied in this setting gives a best critical region that rejects H_0 when the test statistic, which in this case is 0.0005288, is less than or equal to a critical value c. But since there is no hope of finding the probability distribution of the test statistic under H_0, how is a critical value to be determined? In this case we can turn to Monte Carlo simulation to estimate the critical value. The *additional* R statements given below generate one million test statistics under H_0. Within the `for` loop, $n = 200$ random geometric variates are generated from the fitted distribution under H_0, then the test statistic is computed and stored. The last statement prints the estimate of the 5th percentile of the distribution of the test statistic, which is the critical value associated with $\alpha = 0.05$.

```
nrep = 1000000
r = numeric(nrep)
for (i in 1:nrep) {
  x = rgeom(n, p)
  r[i] = p ^ n * (1 - p) ^ sum(x) * prod(factorial(x)) /
         (1 ^ sum(x) * exp(-n * 1))
}
sort(r)[50000]
```

After a call to `set.seed(6)` to initialize the random number stream, five calls to this code yield the following estimates of the critical value associated with $\alpha = 0.05$:

$$5.06 \qquad 5.11 \qquad 5.09 \qquad 5.11 \qquad 5.12.$$

We conclude from the Monte Carlo simulation that the critical value, to two significant digits, is $c = 5.1$. The most powerful test of size $\alpha = 0.05$ rejects H_0 whenever the test statistic is less than or equal to 5.1. For the horse kick data, the value of the test statistic, 0.0005288, is much less than the critical value $c = 5.1$, so we overwhelmingly reject H_0. We conclude that the horse kick data set is better modeled by the Poisson distribution than the geometric distribution. Stated another way based on the Poisson process interpretation, we conclude the deaths due to horse kicks occur randomly over time. Is this conclusion supported by the data? The bottom two rows in Table 4.11 give the expected number of observations associated with the fitted geometric and Poisson models, rounded to the nearest integer. The last column corresponds to four or more horse kick deaths per corps-year. The last two rows can be calculated with the following R statements.

```
200 * c(dgeom(0:3, 100 / 161), 1 - pgeom(3, 100 / 161))
200 * c(dpois(0:3, 0.61), 1 - ppois(3, 0.61))
```

number of deaths per corps per year	$x=0$	$x=1$	$x=2$	$x=3$	$x \geq 4$
number of observed data values	109	65	22	3	1
geometric($\hat{p}$): $200 \cdot \frac{100}{161}\left(1 - \frac{100}{161}\right)^x$	124	47	18	7	4
Poisson$\left(\hat{\lambda}\right)$: $200 \cdot \frac{0.61^x e^{-0.61}}{x!}$	109	66	20	4	1

Table 4.11: Observed and fitted horse kick deaths.

Based on this table, our conclusion to reject the fitted geometric distribution in favor of the fitted Poisson distribution seems justified. The Poisson distribution provides a far superior fit to the data values.

The next example highlights a general sample size n, a general level of significance α, and simple hypotheses that are more generic in nature than in the previous examples.

Example 4.30 Let $X_1, X_2, \ldots, X_n$ be a random sample of size n from a $N(0, \theta)$ population, where θ is the positive unknown population variance. Find the most powerful test of size α to test

$$H_0 : \theta = \theta_0$$

versus

$$H_1 : \theta = \theta_1,$$

where θ_0 is a fixed value of θ and $\theta_1 < \theta_0$ is a different fixed value of θ.

The population mean has been set to the constant $\mu = 0$ in the problem statement so that there is just one unknown parameter θ, the population variance. The population probability density function associated with a $N(0, \theta)$ population is

$$f(x) = \frac{1}{\sqrt{2\pi\theta}} e^{-x^2/(2\theta)} \qquad -\infty < x < \infty.$$

Since $X_1, X_2, \ldots, X_n$ are mutually independent random variables, the joint probability density function of $X_1, X_2, \ldots, X_n$, which is the likelihood function, is the product of the marginal probability density functions:

$$L(\theta) = \prod_{i=1}^{n} f(x_i) = (2\pi\theta)^{-n/2} e^{-\sum_{i=1}^{n} x_i^2/(2\theta)}$$

for $-\infty < x_i < \infty$, $i = 1, 2, \ldots, n$. The ratio of the likelihood functions is

$$\frac{L(\theta_0)}{L(\theta_1)} = \frac{(2\pi\theta_0)^{-n/2} e^{-\sum_{i=1}^{n} x_i^2/(2\theta_0)}}{(2\pi\theta_1)^{-n/2} e^{-\sum_{i=1}^{n} x_i^2/(2\theta_1)}} = \left(\frac{\theta_0}{\theta_1}\right)^{-n/2} e^{-\left(\frac{1}{2\theta_0} - \frac{1}{2\theta_1}\right)\sum_{i=1}^{n} x_i^2}$$

for $-\infty < x_i < \infty$, $i = 1, 2, \ldots, n$. This ratio of the likelihood functions associated with H_0 and H_1 is monotone increasing in $\sum_{i=1}^{n} x_i^2$. Using the Neyman–Pearson theorem, a best critical region of size α has the form

$$\frac{L(\theta_0)}{L(\theta_1)} = \left(\frac{\theta_0}{\theta_1}\right)^{-n/2} e^{-\left(\frac{1}{2\theta_0} - \frac{1}{2\theta_1}\right)\sum_{i=1}^{n} x_i^2} \le c$$

for some $c > 0$ yielding a significance level α. The following sequence of inequalities gives an alternative test statistic:

$$\left(\frac{\theta_0}{\theta_1}\right)^{-n/2} e^{\frac{\theta_0-\theta_1}{2\theta_0\theta_1}\sum_{i=1}^{n} x_i^2} \le c \quad \Longleftrightarrow \quad e^{\frac{\theta_0-\theta_1}{2\theta_0\theta_1}\sum_{i=1}^{n} x_i^2} \le c\left(\frac{\theta_0}{\theta_1}\right)^{n/2}$$

$$\Longleftrightarrow \quad \frac{\theta_0-\theta_1}{2\theta_0\theta_1}\sum_{i=1}^{n} x_i^2 \le \ln\left[c\left(\frac{\theta_0}{\theta_1}\right)^{n/2}\right]$$

$$\Longleftrightarrow \quad \sum_{i=1}^{n} x_i^2 \le \frac{2\theta_0\theta_1}{\theta_0-\theta_1}\ln\left[c\left(\frac{\theta_0}{\theta_1}\right)^{n/2}\right].$$

The right-hand side of the last inequality is also a constant, which can be denoted by c'. The Neyman–Pearson theorem indicates that $\sum_{i=1}^{n} x_i^2$ is an appropriate test statistic and the best critical region of size α is of the form $\sum_{i=1}^{n} x_i^2 \le c'$ for some constant c'. The critical value c' can be determined based on the following sequence of implications:

$$X_i \sim N(0, \theta_0) \quad \Rightarrow \quad \frac{X_i}{\sqrt{\theta_0}} \sim N(0, 1) \quad \Rightarrow \quad \frac{X_i^2}{\theta_0} \sim \chi^2(1) \quad \Rightarrow \quad \frac{\sum_{i=1}^{n} X_i^2}{\theta_0} \sim \chi^2(n)$$

under H_0 because $X_1, X_2, \ldots, X_n$ are mutually independent and identically distributed random variables. This implies that

$$P\left(\sum_{i=1}^{n} X_i^2 \le c'\right) = P\left(\frac{1}{\theta_0}\sum_{i=1}^{n} X_i^2 \le \frac{c'}{\theta_0}\right) = P\left(\chi^2(n) \le \frac{c'}{\theta_0}\right) = \alpha$$

under H_0. The final equation results in $c'/\theta_0 = \chi^2_{n,1-\alpha}$ or $c' = \theta_0\chi^2_{n,1-\alpha}$. So a decision rule for a most powerful test of size α is to calculate the test statistic $\sum_{i=1}^{n} x_i^2$, and reject H_0 when

$$\sum_{i=1}^{n} x_i^2 \le \theta_0\chi^2_{n,1-\alpha}.$$

If one would rather work with the likelihood ratio as a statistic, this critical value is equivalent to

$$c = \left(\frac{\theta_0}{\theta_1}\right)^{-n/2} e^{\chi^2_{n,1-\alpha}(\theta_0-\theta_1)/(2\theta_1)}.$$

The decision rule for the equivalent most powerful test of size α based on the likelihood ratio is to reject H_0 when

$$\left(\frac{\theta_0}{\theta_1}\right)^{-n/2} e^{-\frac{\theta_1-\theta_0}{2\theta_0\theta_1}\sum_{i=1}^{n} x_i^2} \le \left(\frac{\theta_0}{\theta_1}\right)^{-n/2} e^{\chi^2_{n,1-\alpha}(\theta_0-\theta_1)/(2\theta_1)}.$$

Obviously, the first approach gives a much more compact expression for the decision rule. The probability of a Type II error for this test is

$$\begin{aligned}
\beta &= P(\text{fail to reject } H_0 \,|\, H_1 \text{ true}) \\
&= P\left(\sum_{i=1}^{n} X_i^2 > \theta_0\chi^2_{n,1-\alpha} \,\middle|\, \theta = \theta_1\right) \\
&= P\left(\frac{1}{\theta_1}\sum_{i=1}^{n} X_i^2 > \frac{\theta_0}{\theta_1}\chi^2_{n,1-\alpha} \,\middle|\, \theta = \theta_1\right) \\
&= P\left(\chi^2(n) > \frac{\theta_0}{\theta_1}\chi^2_{n,1-\alpha}\right).
\end{aligned}$$

To determine numerical values for c and c' requires specific values for n, α, θ_0, and θ_1. Arbitrarily setting $n = 19$, $\alpha = 0.05$, $\theta_0 = 9$, and $\theta_1 = 4$, we can determine the critical values c and c' associated with a best critical region of size α. Using $\sum_{i=1}^{n} x_i^2$ as a test statistic, the critical value is $c' = \theta_0\chi^2_{n,1-\alpha} = 9\chi^2_{19,0.95} = 91.0531$. This critical value is calculated with the R statement

```
9 * qchisq(0.05, 19)
```

The null hypothesis is to be rejected when $\sum_{i=1}^{n} x_i^2 \le 91.0531$. The corresponding value of c is $c = 0.2514$. A graph of the likelihood ratio $L(\theta_0)/L(\theta_1) = L(9)/L(4)$ as a function of $\sum_{i=1}^{n} x_i^2$ is shown in Figure 4.47 with $c' = 91.0531$ highlighted on the horizontal axis and $c = 0.2514$ highlighted on the vertical axis.

The probability of a Type II error is

$$\beta = P\left(\chi^2(n) > \frac{\theta_0}{\theta_1}\chi^2_{n,1-\alpha}\right) = P\left(\chi^2(19) > \frac{9}{4}\chi^2_{19,0.95}\right) \cong 0.2479,$$

which is calculated with the R statement

```
1 - pchisq(9 / 4 * qchisq(0.05, 19), 19)
```

The power function associated with this test can also be easily calculated. Using a similar approach to the calculation of β, the power of the test is

$$\begin{aligned} q(\theta) &= P(\text{reject } H_0 \mid \theta) \\ &= P\left(\sum_{i=1}^{n} X_i^2 \le c' \,\middle|\, \theta\right) \\ &= P\left(\frac{1}{\theta}\sum_{i=1}^{n} X_i^2 \le \frac{c'}{\theta} \,\middle|\, \theta\right) \\ &= P\left(\chi^2(n) \le \frac{c'}{\theta}\right) \end{aligned}$$

for any value of $\theta > 0$. The power function is calculated using the R statements that follow, and is graphed in Figure 4.48 for $n = 19$, $\alpha = 0.05$, $\theta_0 = 9$, and $\theta_1 = 4$. As expected, the power function passes through the points

$$(\theta_0, \alpha) = (9, 0.05) \qquad \text{and} \qquad (\theta_1, 1-\beta) = (4, 0.7521)$$

and is not particularly steep because of the small sample size of $n = 19$.

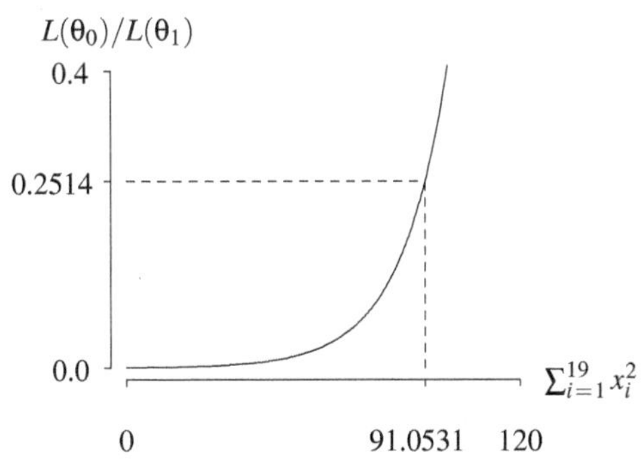

Figure 4.47: Likelihood ratio for a random sample of size $n = 19$ from a $N(0, \theta)$ population.

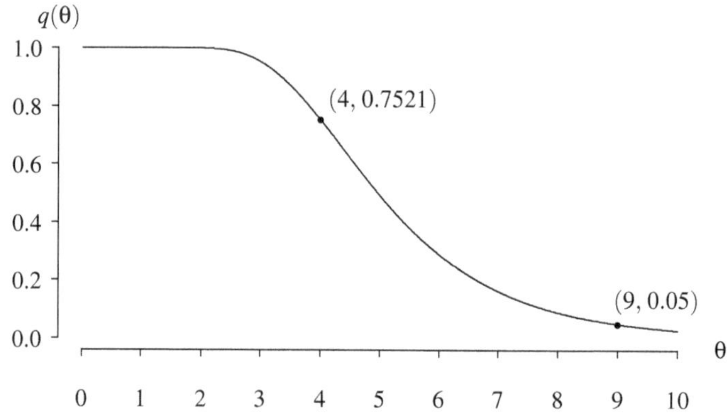

Figure 4.48: Power function for a random sample of $n = 19$ from a $N(0, \theta)$ population.

```
n      = 19
theta0 = 9
crit   = theta0 * qchisq(0.05, 19)
theta  = seq(0.01, 10, by = 0.01)
power  = pchisq(crit / theta, n)
plot(theta, power, type = "l")
```

This has been a long example with a lot of calculations. This means that there is plenty of opportunity for a mathematical mistake. The correctness of the calculations can be supported by using Monte Carlo simulation. The R code below conducts one million hypothesis tests using the parameters given above, and prints the fraction of times that H_0 is rejected when H_0 is true.

```
nrep   = 1000000
count  = 0
n      = 19
theta0 = 9
crit   = theta0 * qchisq(0.05, 19)
for (i in 1:nrep) {
  x = rnorm(n, 0, sqrt(theta0))
  if (sum(x ^ 2) <= crit) count = count + 1
}
print(count / nrep)
```

After a call to `set.seed(3)` to initialize the random number stream, five runs of this Monte Carlo simulation yield the following estimates of the size of the test α.

$$0.0499 \qquad 0.0500 \qquad 0.0501 \qquad 0.0498 \qquad 0.0503.$$

These values hover around $\alpha = 0.05$, which provides us with support that we have calculated the critical value c' correctly. A similar Monte Carlo experiment can be conducted to estimate the value of β for the test. The code below conducts one million hypothesis tests using the parameters given above, and prints the fraction of times that the test fails to reject H_0 when H_1 is true.

```
nrep   = 1000000
count  = 0
n      = 19
theta0 = 9
theta1 = 4
crit   = theta0 * qchisq(0.05, n)
for (i in 1:nrep) {
  x = rnorm(n, 0, sqrt(theta1))
  if (sum(x ^ 2) > crit) count = count + 1
}
print(count / nrep)
```

After a call to `set.seed(3)` to initialize the random number stream, five runs of this Monte Carlo simulation yield the following estimates of the probability of a Type II error β.

$$0.2480 \qquad 0.2484 \qquad 0.2479 \qquad 0.2478 \qquad 0.2490.$$

These values hover around $\beta = 0.2479$, which provides us with support that we have calculated the probability of a Type II error correctly.

The previous example provides an opportunity for an insight regarding best critical regions and most powerful tests of size α. Both the definition of the best critical region of size α and the Neyman–Pearson theorem require simple null and alternative hypotheses, which is rather restrictive. In practice, very few hypothesis tests have simple null and alternative hypotheses. In the previous example, we constructed the most powerful test of size α for

$$H_0 : \theta = \theta_0$$

versus

$$H_1 : \theta = \theta_1.$$

But reviewing the derivation of the most powerful test of size α reveals that the only requirement for θ_1 is that it is less than θ_0. As a specific instance, for testing

$$H_0 : \theta = 9$$

versus

$$H_1 : \theta = 4,$$

we were able to construct a most powerful test of size $\alpha = 0.05$. But the same best critical region would have resulted for the alternative hypotheses

$$H_1 : \theta = 2, \qquad \text{or} \qquad H_1 : \theta = 7, \qquad \text{or} \qquad H_1 : \theta = 8.9.$$

So in fact, *any* θ_1 value less than $\theta_0 = 9$ would also have produced a most powerful test of size $\alpha = 0.05$ in this setting. When a situation of this nature arises, statisticians refer to a hypothesis test with a two-sided alternative, such as

$$H_0 : \theta = 9$$

versus

$$H_1 : \theta < 9,$$

as a *uniformly most powerful test of size* α. The formal definition of such a test is given next. The definition extends the notions of a best critical region of size α and a most powerful test of size α beyond just simple hypotheses.

Definition 4.12 Consider a hypothesis test of a simple null hypothesis H_0 and a composite alternative hypothesis H_1. The critical region C is a *uniformly most powerful critical region of size* α if C is the best critical region of size α for testing H_0 against each simple alternative hypothesis consistent with H_1. The associated hypothesis test is known as a *uniformly most powerful test of size* α.

With this definition in place, it is easy to revisit the derivations associated with many of the examples presented in the section thus far and see that there was a uniformly most powerful test lurking in the background.

- In Example 4.24, with a single observation X from a binomial(4, θ) distribution, testing the simple null hypothesis $H_0 : \theta = 1/2$ versus the composite alternative hypothesis $H_1 : \theta > 1/2$ using the test statistic $X \sim \text{binomial}(4, 1/2)$ under H_0 results in a uniformly most powerful critical region of size $\alpha = 5/16$ of $C = \{x \,|\, x = 3, 4\}$.

- In Example 4.25, with a single observation X from a Rayleigh(θ) distribution, testing the simple null hypothesis $H_0 : \theta = 1$ versus the composite alternative hypothesis $H_1 : \theta > 1$ using the test statistic $X \sim \text{Rayleigh}(1)$ under H_0 results in a uniformly most powerful critical region of size $\alpha = 0.1$ of $C = \{x \,|\, x \ge 1.5174\}$.

- In Example 4.26, with a single observation X from a beta(θ, 1) distribution, testing the simple null hypothesis $H_0 : \theta = 1/2$ versus the composite alternative hypothesis $H_1 : \theta > 1/2$ using the test statistic $X \sim \text{beta}(1/2, 1)$ under H_0 results in a uniformly most powerful critical region of size $\alpha = 0.2$ of $C = \{x \,|\, x \ge 0.64\}$.

- In Example 4.28, with a random sample X_1 and X_2 from a $N(\theta, 1)$ distribution, testing the simple null hypothesis $H_0 : \theta = 5$ versus the composite alternative hypothesis $H_1 : \theta < 5$ using the test statistic $X_1 + X_2 \sim N(10, 2)$ under H_0 results in a uniformly most powerful critical region of size $\alpha = 0.1$ of $C = \{(x_1, x_2) \,|\, x_1 + x_2 \le 8.1876\}$.

- In Example 4.30, with a random sample $X_1, X_2, \ldots, X_n$ from a $N(0, \theta)$ distribution, testing the simple null hypothesis $H_0 : \theta = \theta_0$ versus the composite alternative hypothesis $H_1 : \theta < \theta_0$ using the test statistic $X_1^2 + X_2^2 + \cdots + X_n^2$, where $\left(X_1^2 + X_2^2 + \cdots + X_n^2\right)/\theta_0 \sim \chi^2(n)$ under H_0, results in a uniformly most powerful critical region of size α of

$$C = \left\{ (x_1, x_2, \ldots, x_n) \,\middle|\, \sum_{i=1}^{n} x_i^2 \le \theta_0 \chi^2_{n,1-\alpha} \right\}.$$

The question arises as to whether all most powerful tests of size α are indeed uniformly most powerful tests of size α. A simple counterexample will show that this is not the case. This will be accomplished by revisiting Example 4.28. As indicated in the fourth bullet from the list above, the hypothesis test of the simple null hypothesis

$$H_0 : \theta = 5$$

versus the one-tailed composite alternative hypothesis

$$H_1 : \theta < 5$$

is based on the random sample X_1 and X_2 drawn from a $N(\theta,\, 1)$ distribution. Using the test statistic $X_1 + X_2 \sim N(10,\, 2)$ under H_0 results in a uniformly most powerful critical region of size $\alpha = 0.1$ of $C = \{(x_1, x_2)\,|\,x_1 + x_2 \le 8.1876\}$. But if you return to the derivation and think about simply switching the direction of the alternative hypothesis, that is, the same simple null hypothesis

$$H_0 : \theta = 5$$

versus the one-tailed composite alternative hypothesis in the opposite direction,

$$H_1 : \theta > 5,$$

you realize that the test statistic $X_1 + X_2$ still has the same $N(10,\, 2)$ distribution under H_0. A nearly-identical derivation will conclude that a uniformly most powerful test of size $\alpha = 0.1$ has critical region $C = \{(x_1, x_2)\,|\,x_1 + x_2 \ge 11.8124\}$. (This critical value is 1.8124 above 10, and the previous critical value was 1.8124 below 10.) Finally, consider the same simple null hypothesis

$$H_0 : \theta = 5$$

versus the two-tailed composite alternative hypothesis

$$H_1 : \theta \ne 5$$

at level of significance $\alpha = 0.1$. The critical region will necessarily be split into two distinct regions. If one considers a simple alternative to the left of $\theta = 5$, for example, $H_1 : \theta = 4$, then the test with the split critical region is outperformed by the uniformly most powerful test associated with $H_1 : \theta < 5$. Likewise, if one considers a simple alternative to the right of $\theta = 5$, for example, $H_1 : \theta = 17$, then the test with the split critical region is outperformed by the uniformly most powerful test associated with $H_1 : \theta > 5$. Since the two-tailed test can be outperformed by another test of size α, a uniformly most powerful test does not exist for this particular hypothesis test with a two-tailed alternative.

The notion of a sufficient statistic was first introduced back in Section 2.4. Recall that a sufficient statistic for an unknown parameter θ captures all of the information contained in a data set with respect to θ. The next result shows the role that sufficient statistics play in hypothesis testing.

Theorem 4.3 Let $X_1, X_2, \ldots, X_n$ denote a random sample drawn from a probability distribution described by $f(x)$ with an unknown parameter θ. The associated likelihood function is denoted by $L(\theta)$. Let $Y = g(X_1, X_2, \ldots, X_n)$ be a sufficient statistic for θ. A best test or a uniformly most powerful test of size α concerning θ must be based on the sufficient statistic Y.

Proof Using the Fisher–Neyman factorization criterion (Theorem 2.7), the likelihood ratio for a best test or a uniformly most powerful test of size α concerning θ is

$$\frac{L(\theta_0)}{L(\theta_1)} = \frac{h_1\left(g(x_1, x_2, \ldots, x_n), \theta_0\right) \cdot h_2(x_1, x_2, \ldots, x_n)}{h_1\left(g(x_1, x_2, \ldots, x_n), \theta_1\right) \cdot h_2(x_1, x_2, \ldots, x_n)} = \frac{h_1\left(g(x_1, x_2, \ldots, x_n), \theta_0\right)}{h_1\left(g(x_1, x_2, \ldots, x_n), \theta_1\right)},$$

where h_1 and h_2 are nonnegative functions. Since the likelihood ratio depends on the sample values $x_1, x_2, \ldots, x_n$ only through the sufficient statistic $y = g(x_1, x_2, \ldots, x_n)$, any best test or uniformly most powerful test of size α concerning θ must be based on the sufficient statistic $y = g(x_1, x_2, \ldots, x_n)$. □

Theorem 4.3 has a very important practical application—the search for a test statistic for a best test or a uniformly most powerful test of size α can be limited to a sufficient statistic for θ, or some function of a sufficient statistic.

We can now return to the problem of the competing test statistics from the three brothers Chico, Harpo, and Groucho in Example 4.5. The hypothesis test

$$H_0 : \theta = 5$$

versus

$$H_1 : \theta < 5$$

was tested at $\alpha = 0.08$ for independent observations X_1 and X_2 drawn from a $U(0, \theta)$ population, where θ is a positive unknown parameter. The three test statistics used to conduct the test are

- Chico: $X_1 + X_2$,
- Harpo: $\min\{X_1, X_2\}$,
- Groucho: $\max\{X_1, X_2\}$.

Figure 4.12 displayed the power functions for the three test statistics, revealing that Groucho's test statistic, $\max\{X_1, X_2\}$, had the highest power of the three. But is there another test statistic that could have been devised that would have performed even better than Groucho's test statistic? The answer is provided by Theorem 4.3. Since Groucho's test statistic is a sufficient statistic for θ (as was established in Example 2.26), we know that Theorem 4.3 established that no other test statistic will outdo Groucho's in terms of power and that there is no need to look further for another test statistic in the hopes of devising a more powerful test. One can do no better than Groucho's test statistic.

This discussion concerning the relationship between sufficient statistics and hypothesis testing can be extended. Certain parametric families of probability distributions possess a characteristic known as the *monotone likelihood ratio*. This is the case when the likelihood ratio $L(\theta_0)/L(\theta_1)$ is a monotone function in some statistic. When the statistic that possesses the monotone likelihood ratio property happens to be a sufficient statistic, then a hypothesis test with a one-sided alternative hypothesis can be constructed based on the sufficient statistic which is a uniformly most powerful test of size α. This result is known as the *Karlin–Rubin theorem*, and is stated and proved in most advanced statistics textbooks.

4.7 Likelihood Ratio Tests

The likelihood ratio statistic was defined in Definition 3.8 in conjunction with asymptotically exact confidence intervals for an unknown parameter. We present a slight modification of the likelihood ratio statistic in this section to allow us to develop *likelihood ratio tests* which are used in hypothesis testing.

The most powerful tests and uniformly most powerful tests from the previous section are helpful in that they can identify, in some limited settings, a hypothesis test with the highest possible power. But uniformly most powerful tests do not always exist, as was illustrated in the two-tailed hypothesis test given in the previous section. So we desire a more general procedure for hypothesis testing, which is provided by likelihood ratio tests. These tests are designed for simple and composite hypotheses and for one or several unknown parameters. Likelihood ratio tests can be applied when there are unknown parameters that are not of interest to the experimenter. These parameters are appropriately known as *nuisance parameters*.

The notation remains the same as before: $X_1, X_2, \ldots, X_n$ are the observations that are drawn as a random sample from a population with a single unknown parameter θ or k unknown parameters $\theta_1, \theta_2, \ldots, \theta_k$. The parameter(s) are defined on a parameter space that is denoted by Ω. It is desired to test

$$H_0 : \theta \in \Omega_0$$

versus

$$H_1 : \theta \in \Omega_1,$$

where $\Omega_0 \cap \Omega_1 = \emptyset$ and $\Omega_0 \cup \Omega_1 = \Omega$. The assumption that $\Omega_0 \cup \Omega_1 = \Omega$ might force a slight modification of some null and alternative hypotheses in some instances. The more general statement of H_0 and H_1 emphasizes the generality of likelihood ratio tests, which are also known as *generalized likelihood ratio tests*.

In the Neyman–Pearson theorem (Theorem 4.2), the likelihood ratio was defined as

$$\frac{L(\theta_0)}{L(\theta_1)},$$

where θ_0 and θ_1 correspond to the simple null and simple alternative hypotheses. A natural extension of the likelihood ratio from the Neyman–Pearson theorem to the more general setting here is

$$\Lambda = \frac{\sup_{\Omega_0} L(\theta)}{\sup_{\Omega_1} L(\theta)}.$$

But this definition of Λ is typically not used in practice. Instead, the following definition of the likelihood ratio test statistic is used in practice because it yields equivalent results and is often easier to evaluate. For simplicity of notation, the definition considers just a single unknown parameter θ; a similar definition applies to k unknown parameters $\theta_1, \theta_2, \ldots, \theta_k$.

Definition 4.13 Let $X_1, X_2, \ldots, X_n$ denote a random sample from a population whose probability distribution is described by $f(x)$ with a single unknown parameter θ which is defined on a parameter space Ω. Let $L(\theta)$ denote the associated likelihood function. Consider the hypothesis test

$$H_0 : \theta \in \Omega_0$$

versus

$$H_1 : \theta \in \Omega_1,$$

where $\Omega_0 \cap \Omega_1 = \emptyset$ and $\Omega_0 \cup \Omega_1 = \Omega$. The *likelihood ratio test statistic* is

$$\Lambda = \frac{\sup_{\Omega_0} L(\theta)}{\sup_{\Omega} L(\theta)}.$$

The upper case Λ is used here because the ratio of the likelihood functions is a random variable. It varies from one sample to the next. This is the notation used when referring to Λ as a random variable. The observed value of Λ calculated from the data values $x_1, x_2, \ldots, x_n$ is denoted by λ.

Some authors define Λ as the reciprocal of this quantity, but most authors use the formulation in Definition 4.13. The fact that the likelihood ratio test statistic Λ is defined explicitly is an advantage associated with likelihood ratio tests; likelihood ratio tests suggest an appropriate test statistic.

Notice that the denominator of Λ is equivalent to the problem of finding the value of the likelihood function at the maximum likelihood estimator because $L(\theta)$ is being maximized over the parameter space Ω. The numerator is known as the restricted maximization problem; the denominator is known as the unrestricted maximization problem. A close match between the numerator and denominator is evidence supporting H_0. Since the likelihood function $L(\theta)$ is a joint probability density function (when sampling is from a continuous population) or a joint probability mass function (when sampling is from a discrete population), we know that Λ must be nonnegative. Furthermore, since the denominator is the largest possible value of the likelihood function on the parameter space Ω, and $\Omega_0 \subset \Omega$, we can also conclude that $\Lambda \leq 1$. To summarize, Λ is a random variable with support $\{\lambda \,|\, 0 \leq \lambda \leq 1\}$ and, since small values of Λ lead to rejecting H_0, the critical region is of the form

$$C = \{\Lambda \,|\, \Lambda \leq c\}$$

for some real critical value c that is determined from the probability distribution of Λ under H_0. More specifically, when H_0 is simple, the critical value c is determined by

$$P(\Lambda \leq c \,|\, H_0 \text{ true}) = P(\Lambda \leq c \,|\, \theta = \theta_0) = \alpha.$$

When H_0 is composite, the critical value c is determined by

$$\sup_{\theta} P(\Lambda \leq c \,|\, H_0 \text{ true}) = \sup_{\theta} P(\Lambda \leq c \,|\, \theta \in \Omega_0) = \alpha.$$

The usual modification is made for sampling from a discrete population.

The procedure for constructing and executing a likelihood ratio test of size α concerning θ consists of the following steps. The steps are stated in terms of a single parameter θ, but the same steps apply if there are k unknown parameters $\theta_1, \theta_2, \ldots, \theta_k$.

1. Develop a hunch, or theory, concerning a problem of interest.
2. Translate the theory into a question about an unknown parameter θ on a parameter space Ω.
3. State the null hypothesis H_0 as $\theta \in \Omega_0$.
4. State the alternative hypothesis H_1 as $\theta \in \Omega_1$.
5. If necessary, restate H_0 and H_1 so that $\Omega_0 \cup \Omega_1 = \Omega$.
6. Determine the significance level α of the test.
7. Determine the sample size n.
8. Find the maximum likelihood estimator $\hat{\theta}$, by solving the unrestricted maximization problem.
9. Find the maximum value of $L(\theta)$ under H_0, which solves the restricted maximization problem.
10. Find the probability distribution of the likelihood ratio test statistic Λ under H_0.
11. Find the critical value c, which is the α percentile of the distribution of Λ under H_0.
12. Collect the data values $x_1, x_2, \ldots, x_n$.
13. Compute the observed value of the test statistic λ.
14. Reject H_0 if $\lambda \leq c$.

So the mathematical challenges associated with implementing a likelihood ratio test are to (a) calculate the numerator and denominator of Λ, (b) calculate the probability distribution of Λ under H_0, and (c) calculate an appropriate critical value c based on the probability distribution of Λ under H_0. Sometimes the probability distribution of Λ under H_0 is easy to find, and sometimes it is not. Not surprisingly, we start with an easy case. The first example of conducting a likelihood ratio test concerns a random sample of data values drawn from a normal population distribution.

Example 4.31 Let $X_1, X_2, \ldots, X_n$ be a random sample from a $N\left(\mu, \sigma^2\right)$ distribution, where μ is an *unknown* parameter and $\sigma^2 > 0$ is a *known* parameter. Construct a likelihood ratio test of

$$H_0 : \mu = \mu_0$$

versus

$$H_1 : \mu \neq \mu_0$$

at significance level α, where μ_0 is a real-valued constant.

Since the parameter σ^2 is known, it will be treated as a constant throughout the derivation of the likelihood ratio test. The restricted parameter spaces associated with H_0 and H_1 are $\Omega_0 = \{\mu \,|\, \mu = \mu_0\}$ and $\Omega_1 = \{\mu \,|\, \mu \neq \mu_0\}$. The parameter space is $\Omega = \Omega_0 \cup \Omega_1 = \{\mu \,|\, -\infty < \mu < \infty\}$. We begin with the denominator of the likelihood ratio test statistic Λ, which involves finding the maximum likelihood estimator of μ. The population probability density function is

$$f(x) = \frac{1}{\sqrt{2\pi\sigma^2}} e^{-\frac{1}{2}\left(\frac{x-\mu}{\sigma}\right)^2} \qquad -\infty < x < \infty.$$

The likelihood function is

$$L(\mu) = \prod_{i=1}^{n} f(x_i) = \prod_{i=1}^{n} \frac{1}{\sqrt{2\pi\sigma^2}} e^{-\frac{1}{2}\left(\frac{x_i-\mu}{\sigma}\right)^2} = (2\pi\sigma^2)^{-n/2} e^{-\frac{1}{2}\sum_{i=1}^{n}\left(\frac{x_i-\mu}{\sigma}\right)^2}.$$

The log likelihood function is

$$\ln L(\mu) = -\frac{n}{2}\ln(2\pi\sigma^2) - \frac{1}{2\sigma^2}\sum_{i=1}^{n}(x_i-\mu)^2.$$

The score is

$$\frac{\partial \ln L(\mu)}{\partial \mu} = \frac{1}{\sigma^2}\sum_{i=1}^{n}(x_i-\mu).$$

Equating the score to zero and solving for μ gives the maximum likelihood estimator

$$\hat{\mu} = \frac{1}{n}\sum_{i=1}^{n} x_i,$$

which is the sample mean $\hat{\mu} = \bar{x}$. The second derivative of the log likelihood function evaluated at the maximum likelihood estimator is negative, which confirms that $\hat{\mu}$ maximizes both the log likelihood function and the likelihood function. So the solution to the unrestricted maximization problem is

$$\sup_{\Omega} L(\mu) = L(\hat{\mu}) = \left(2\pi\sigma^2\right)^{-n/2} e^{-\frac{1}{2}\sum_{i=1}^{n}\left(\frac{x_i-\bar{x}}{\sigma}\right)^2}.$$

We now turn to the restricted maximization problem associated with the numerator of Λ. Since the null hypothesis restricts the parameter space to just $\mu = \mu_0$, the restricted maximization problem is solved by simply using μ_0 as an argument in $L(\mu)$. So the solution to the restricted maximization problem is

$$\sup_{\Omega_0} L(\mu) = L(\mu_0) = \left(2\pi\sigma^2\right)^{-n/2} e^{-\frac{1}{2}\sum_{i=1}^{n}\left(\frac{x_i-\mu_0}{\sigma}\right)^2}.$$

Putting the numerator and denominator together, the likelihood ratio statistic is

$$\Lambda = \frac{\sup_{\Omega_0} L(\mu)}{\sup_{\Omega} L(\mu)} = \frac{L(\mu_0)}{L(\hat{\mu})} = \frac{\left(2\pi\sigma^2\right)^{-n/2} e^{-\frac{1}{2}\sum_{i=1}^{n}\left(\frac{x_i-\mu_0}{\sigma}\right)^2}}{\left(2\pi\sigma^2\right)^{-n/2} e^{-\frac{1}{2}\sum_{i=1}^{n}\left(\frac{x_i-\bar{x}}{\sigma}\right)^2}} = e^{-\frac{1}{2\sigma^2}\sum_{i=1}^{n}\left[(x_i-\mu_0)^2-(x_i-\bar{x})^2\right]}.$$

Some notational sloppiness is present here in presenting a random variable Λ on the left-hand side of the equation and using lower-case x_i values on the right-hand side of the equation. After some algebra, the summation in Λ can be eliminated, yielding

$$\Lambda = e^{-\frac{n}{2\sigma^2}(\bar{x}-\mu_0)^2}.$$

Determining the probability distribution of Λ is nontrivial. But there is a glimmer of hope. With a little rearrangement, Λ can be rewritten as

$$\Lambda = e^{-\frac{1}{2}\left(\frac{\bar{x}-\mu_0}{\sigma/\sqrt{n}}\right)^2}.$$

Since the sample mean $\bar{x}$ is normally distributed with expected value μ_0 and population standard deviation $\sigma/\sqrt{n}$ under H_0, the quantity in the exponent in parentheses has a standard normal distribution under H_0. Since the square of a standard normal random variable has the chi-square distribution with one degree of freedom, the following APPL statements can be used to find the probability density function of Λ under H_0.

```
Y := ChiSquareRV(1);
g := [[y -> exp(-y / 2)], [0, infinity]];
L := Transform(Y, g);
c := IDF(L, 1 / 10);
```

The last APPL statement calculates the 10th percentile of the distribution of Λ, which is approximately $c = 0.2585$. The probability density function of Λ under H_0 is

$$f_\Lambda(\lambda) = \frac{1}{\sqrt{-\pi \ln \lambda}} \qquad 0 < \lambda < 1.$$

As shown in Figure 4.49, the probability density function of Λ under H_0 has a vertical asymptote at $\lambda = 1$. The next step is to find a critical value c which satisfies

$$P(\Lambda \le c \,|\, H_0 \text{ true}) = P(\Lambda \le c \,|\, \mu = \mu_0) = \alpha.$$

This cannot be solved for c in closed form, so numerical methods are required. The procedure for the likelihood ratio test in this form is to collect the data values $x_1, x_2, \ldots, x_n$

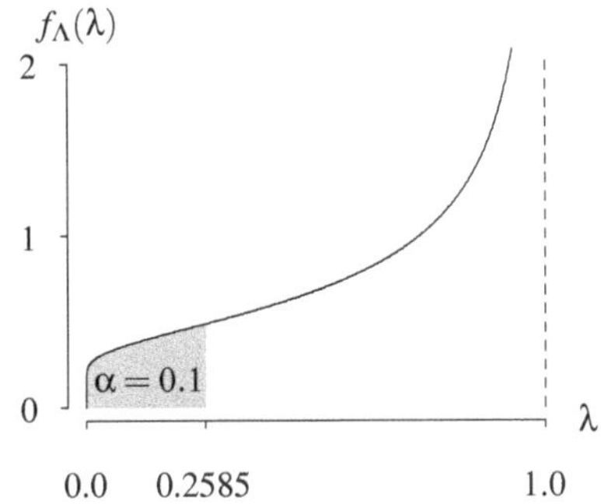

Figure 4.49: Probability density function of Λ under H_0.

and reject $H_0 : \mu = \mu_0$ in favor of $H_1 : \mu \neq \mu_0$ if the test statistic $\lambda = e^{-\frac{1}{2}\left(\frac{\bar{x}-\mu_0}{\sigma/\sqrt{n}}\right)^2}$ is less than or equal to a critical value c that is determined by numerical methods. Figure 4.49 shows the critical value $c = 0.2585$ associated with a test of size $\alpha = 0.1$.

The probability distribution of Λ under H_0 is a bit unsatisfying for two reasons. First, it requires numerical methods to determine a critical value c associated with a particular value of α. Second, the probability distribution of Λ is not one of the well-known distributions from probability theory. The first weakness cannot be overcome—but the second weakness can. We need not necessarily use λ as a test statistic. Consider the following sequence of inequalities.

$$\begin{aligned}
\lambda \le c \qquad & \Longleftrightarrow \qquad e^{-\frac{1}{2}\left(\frac{\bar{x}-\mu_0}{\sigma/\sqrt{n}}\right)^2} \le c \\
& \Longleftrightarrow \qquad -\frac{1}{2}\left(\frac{\bar{x}-\mu_0}{\sigma/\sqrt{n}}\right)^2 \le \ln c \\
& \Longleftrightarrow \qquad \left(\frac{\bar{x}-\mu_0}{\sigma/\sqrt{n}}\right)^2 \ge -2\ln c \\
& \Longleftrightarrow \qquad \left|\frac{\bar{x}-\mu_0}{\sigma/\sqrt{n}}\right| \ge \sqrt{-2\ln c}.
\end{aligned}$$

Since the right-hand side of the last inequality is also a constant (denote it as c') and the quantity in the absolute value is a standard normal random variable under H_0, one can see that $c' = \sqrt{-2\ln c} = z_{\alpha/2}$. As an illustration, when $\alpha = 0.1$,

$$c' = \sqrt{-2\ln c} = \sqrt{-2\ln 0.2585} = 1.6449,$$

which is the same as $c' = z_{\alpha/2} = z_{0.05} = 1.6449$, and can be calculated with the R statement `qnorm(0.95)`.

So the procedure for the equivalent likelihood ratio test in this form is to collect the data values $x_1, x_2, \ldots, x_n$ and reject $H_0 : \mu = \mu_0$ in favor of $H_1 : \mu \neq \mu_0$ if the test statistic $\left|\frac{\bar{x}-\mu_0}{\sigma/\sqrt{n}}\right|$ is greater than or equal to $z_{\alpha/2}$. This is equivalent to the hypothesis test that was given in the first row of Table 4.4 associated with Theorem 1.3.

The power function for this test is

$$
\begin{aligned}
q(\mu) &= P(\text{reject } H_0 \mid \mu) \\
&= P\left(\left|\frac{\bar{X}-\mu_0}{\sigma/\sqrt{n}}\right| \ge z_{\alpha/2} \,\Big|\, \mu\right) \\
&= 1 - P\left(\left|\frac{\bar{X}-\mu_0}{\sigma/\sqrt{n}}\right| < z_{\alpha/2} \,\Big|\, \mu\right) \\
&= 1 - P\left(-z_{\alpha/2} < \frac{\bar{X}-\mu_0}{\sigma/\sqrt{n}} < z_{\alpha/2} \,\Big|\, \mu\right) \\
&= 1 - P\left(-z_{\alpha/2} + \frac{\mu_0-\mu}{\sigma/\sqrt{n}} < \frac{\bar{X}-\mu_0}{\sigma/\sqrt{n}} + \frac{\mu_0-\mu}{\sigma/\sqrt{n}} < z_{\alpha/2} + \frac{\mu_0-\mu}{\sigma/\sqrt{n}} \,\Big|\, \mu\right) \\
&= 1 - P\left(-z_{\alpha/2} + \frac{\mu_0-\mu}{\sigma/\sqrt{n}} < \frac{\bar{X}-\mu}{\sigma/\sqrt{n}} < z_{\alpha/2} + \frac{\mu_0-\mu}{\sigma/\sqrt{n}} \,\Big|\, \mu\right) \\
&= 1 - P\left(-z_{\alpha/2} + \frac{\mu_0-\mu}{\sigma/\sqrt{n}} < Z < z_{\alpha/2} + \frac{\mu_0-\mu}{\sigma/\sqrt{n}} \,\Big|\, \mu\right)
\end{aligned}
$$

for $-\infty < \mu < \infty$, where $Z \sim N(0, 1)$. Power functions for generic μ_0 and α and for several values of n are plotted in Figure 4.50. All of these power functions pass through the point (μ_0, α). The power functions are the same regardless of the choice of the two test statistics suggested in this example. As usual, the power functions become steeper as the sample size n increases.

Finally, we conclude this example by conducting a likelihood ratio test using the two test statistics for $\mu_0 = 100$, $\sigma = 10$, $\alpha = 0.1$, and the $n = 4$ data values

106.3	98.6	116.9	112.2.

The hypothesis test is

$$H_0 : \mu = 100$$

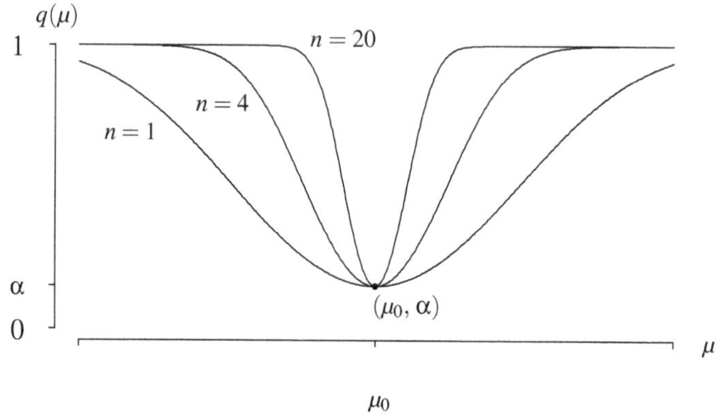

Figure 4.50: Power functions for the likelihood ratio test of $H_0 : \mu = \mu_0$ versus $H_1 : \mu \ne \mu_0$.

versus

$$H_1 : \mu \neq 100$$

at significance level $\alpha = 0.1$, so the critical value for the first test statistic is $c = 0.2585$ and the critical value for the second test statistic is $c' = 1.6449$. Since the sample mean is $\bar{x} = (106.3 + 98.6 + 116.9 + 112.2)/4 = 108.5$, the first test statistic assumes the value

$$\lambda = e^{-\frac{1}{2}\left(\frac{\bar{x}-\mu_0}{\sigma/\sqrt{n}}\right)^2} = e^{-\frac{1}{2}\left(\frac{108.5-100}{10/\sqrt{4}}\right)^2} = e^{-\frac{1}{2}(1.7)^2} = e^{-1.445} = 0.2357.$$

This test statistic can be calculated with the R statements given below.

```
x = c(106.3, 98.6, 116.9, 112.2)
lambda = exp(-((mean(x) - 100) / (10 / sqrt(4))) ^ 2 / 2)
```

Since this test statistic is (just barely) below the associated critical value $c = 0.2585$, H_0 is rejected. There is enough statistical evidence in the sample at $\alpha = 0.1$ to conclude that the population mean differs from $\mu_0 = 100$.

The second test statistic assumes the value

$$\left|\frac{\bar{x}-\mu_0}{\sigma/\sqrt{n}}\right| = \left|\frac{108.5-100}{10/\sqrt{4}}\right| = \left|\frac{8.5}{5}\right| = |1.7| = 1.7.$$

Since this test statistic is (again, barely) above the associated critical value $c' = 1.6449$, H_0 is rejected. There is again enough statistical evidence in the sample at $\alpha = 0.1$ to conclude that the population mean differs from $\mu_0 = 100$. The two tests using the different test statistics are equivalent, but which is better? The second test is preferred because the critical value is calculated from the standard normal distribution, which is much more familiar than $f_\Lambda(\lambda)$. This allows the test to be coded up into a statistical language such as R. The R statements listed below calculate the test statistic and critical value c' at $\alpha = 0.1$.

```
x = c(106.3, 98.6, 116.9, 112.2)
alpha = 0.1
test.statistic = abs((mean(x) - 100) / (10 / sqrt(4)))
critical.value = qnorm(1 - alpha / 2)
```

The next example of a likelihood ratio test involves a small, seemingly harmless tweak (a one-sided alternative hypothesis) to the hypothesis test in the previous example. As will be seen, this tweak leads to a much bumpier road to arrive at a test statistic and associated critical value. The resulting hypothesis test, however, is still one of the tests in Table 4.4 associated with classical hypothesis testing involving random samples drawn from normal populations.

Example 4.32 Let $X_1, X_2, \ldots, X_n$ be a random sample from a $N(\mu, \sigma^2)$ distribution, where μ is an *unknown* parameter and $\sigma^2 > 0$ is a *known* parameter. Construct a likelihood ratio test of

$$H_0 : \mu = \mu_0$$

versus

$$H_1 : \mu > \mu_0$$

at significance level α, where μ_0 is a real-valued constant.

Since the parameter σ^2 is known, it will again be treated as a constant throughout the derivation of the likelihood ratio test. The restricted parameter spaces associated with H_0 and H_1 are $\Omega_0 = \{\mu \,|\, \mu = \mu_0\}$ and $\Omega_1 = \{\mu \,|\, \mu > \mu_0\}$ for some real constant value μ_0. The parameter space is $\Omega = \{\mu \,|\, -\infty < \mu < \infty\}$. In this case, $\Omega_0 \cup \Omega_1 \neq \Omega$, so the hypotheses must be restated in order to meet the condition $\Omega_0 \cup \Omega_1 = \Omega$ as

$$H_0 : \mu \leq \mu_0$$

versus

$$H_1 : \mu > \mu_0.$$

So the restricted parameter spaces associated with the restated hypotheses H_0 and H_1 are now $\Omega_0 = \{\mu \,|\, \mu \leq \mu_0\}$ and $\Omega_1 = \{\mu \,|\, \mu > \mu_0\}$. This means that $\Omega_0 \cup \Omega_1 = \Omega$. The likelihood ratio test that will be formulated under these updated H_0 and H_1 is identical to the likelihood ratio test that would result under the original H_0 and H_1.

We begin with the denominator of the likelihood ratio test statistic Λ, which involves finding the maximum likelihood estimator of μ. This progresses in exactly the same fashion as in Example 4.31. The maximum likelihood estimator is

$$\hat{\mu} = \frac{1}{n} \sum_{i=1}^{n} x_i,$$

which is the sample mean $\hat{\mu} = \bar{x}$. So the solution to the unrestricted maximization problem is exactly the same as in the previous question:

$$\sup_{\Omega} L(\mu) = L(\hat{\mu}) = \left(2\pi\sigma^2\right)^{-n/2} e^{-\frac{1}{2}\sum_{i=1}^{n}\left(\frac{x_i - \bar{x}}{\sigma}\right)^2}.$$

We now turn to the restricted maximization problem associated with the numerator of the likelihood ratio statistic Λ, that is, we want to find the value of μ that maximizes $L(\mu)$ over Ω_0. This is more complicated than in the previous example. Consider the generic likelihood function shown in Figure 4.51. The likelihood function is maximized at $\bar{x}$. In the restricted maximization problem, however, $L(\mu)$ is maximized over μ values that satisfy the null hypothesis, that is, $\mu \leq \mu_0$. Consider two cases. First, referring to Figure 4.51, if $\mu_0 \geq \bar{x}$, then the likelihood function is maximized over $\mu \leq \mu_0$ at $\bar{x}$. Second, if $\mu_0 < \bar{x}$, then the likelihood function is maximized over $\mu \leq \mu_0$ at μ_0. To summarize, over Ω_0, the likelihood function is maximized at

$$\hat{\mu} = \begin{cases} \bar{x} & \mu_0 \geq \bar{x} \\ \mu_0 & \mu_0 < \bar{x}. \end{cases}$$

So the solution to the restricted maximization problem in the numerator of Λ is

$$\sup_{\Omega_0} L(\mu) = \begin{cases} \left(2\pi\sigma^2\right)^{-n/2} e^{-\frac{1}{2}\sum_{i=1}^{n}\left(\frac{x_i - \bar{x}}{\sigma}\right)^2} & \mu_0 \geq \bar{x} \\ (2\pi\sigma^2)^{-n/2} e^{-\frac{1}{2}\sum_{i=1}^{n}\left(\frac{x_i - \mu_0}{\sigma}\right)^2} & \mu_0 < \bar{x}. \end{cases}$$

Finally, the likelihood ratio test statistic is

$$\Lambda = \frac{\sup_{\Omega_0} L(\mu)}{\sup_{\Omega} L(\mu)} = \begin{cases} 1 & \mu_0 \geq \bar{x} \\ e^{-\frac{1}{2}\left(\frac{\bar{x} - \mu_0}{\sigma/\sqrt{n}}\right)^2} & \mu_0 < \bar{x} \end{cases}$$

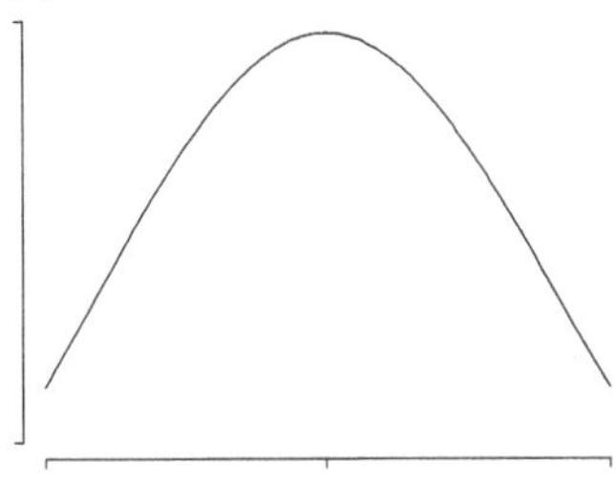

Figure 4.51: Likelihood function for a random sample from a $N(\mu, \sigma^2)$ population.

using the algebraic simplifications from the previous example. So the probability distribution of Λ in this case is a mixed distribution (part discrete, part continuous). Under H_0, the probability that $\Lambda = 1$ is

$$P(\Lambda = 1) = P(\mu_0 \geq \bar{X}) = \frac{1}{2}$$

because $\bar{X} \sim N(\mu_0, \sigma^2/n)$ under H_0. Thus, using a slight modification from the previous example, the probability distribution of Λ under H_0 is described by

$$f_\Lambda(\lambda) = \begin{cases} \dfrac{1}{2\sqrt{-\pi \ln \lambda}} & 0 < \lambda < 1 \\ \dfrac{1}{2} & \lambda = 1. \end{cases}$$

Since all examples in the chapter thus far have used a significance level α less than $1/2$, it is reasonable to assume that this is the case for this example as well. So it is fine to work with λ as a test statistic, or instead use the string of equivalent inequalities from the previous example and work with the test statistic

$$z = \frac{\bar{x} - \mu_0}{\sigma/\sqrt{n}}$$

which has the standard normal distribution under H_0. Since large values of this test statistic lead to rejecting H_0, the null hypothesis should be rejected whenever $z > z_\alpha$. This is the hypothesis test from the first row of Table 4.4 associated with Theorem 1.3 and is also a uniformly most powerful test.

This example concludes by calculating p-values associated with significance tests for $\mu_0 = 100$, $\sigma = 10$, and the $n = 4$ data values from the previous example, that is,

106.3	98.6	116.9	112.2,

for each of the two possible test statistics. The hypothesis test is

$$H_0 : \mu \leq 100$$

versus

$$H_1 : \mu > 100.$$

From the previous example, the likelihood ratio statistic is $\lambda = e^{-1.445} \cong 0.2357$. The p-value is the area under $f_\Lambda(\lambda)$ to the left of $\lambda = e^{-1.445} \cong 0.2357$, which can be calculated with the Maple statement

```
int(1 / (2 * sqrt(-Pi * ln(lambda))), lambda = 0 .. exp(-1.445));
```

which returns $p = 0.04457$. The value of the second test statistic is again

$$z = \frac{\bar{x} - \mu_0}{\sigma/\sqrt{n}} = \frac{108.5 - 100}{10/\sqrt{4}} = 1.7.$$

The p-value is the area under the probability density function of a standard normal random variable to the right of 1.7, which can be calculated with the R statement

```
1 - pnorm(1.7)
```

This also gives $p = 0.04457$. The fact that the two p-values are equal lends support to the notion that the likelihood ratio tests based on the two statistics are equivalent. Furthermore, since the two-tailed test from the previous example barely rejected H_0 at $\alpha = 0.1$, it is not surprising that the p-value for the associated one-tailed test for the same data set is just slightly less than $\alpha = 0.05$.

The previous two examples have considered random sampling from a normal population. The next example considers random sampling from a uniform population.

Example 4.33 Let $X_1, X_2, \ldots, X_n$ be a random sample from a $U(0, \theta)$ distribution, where θ is a positive unknown parameter. Construct a likelihood ratio test of

$$H_0 : \theta = \theta_0$$

versus

$$H_1 : \theta < \theta_0$$

at significance level α, where θ_0 is a positive, real-valued constant.

In order to satisfy the constraint that $\Omega = \Omega_0 \cup \Omega_1$, rewrite H_0 and H_1 as

$$H_0 : \theta \geq \theta_0$$

versus

$$H_1 : 0 < \theta < \theta_0.$$

With these modifications, the restricted parameter spaces associated with the updated H_0 and H_1 are now $\Omega_0 = \{\theta \,|\, \theta \geq \theta_0\}$ and $\Omega_1 = \{\theta \,|\, 0 < \theta < \theta_0\}$. The parameter space in this problem is $\Omega = \Omega_0 \cup \Omega_1 = \{\theta \,|\, \theta > 0\}$. The first step is to solve the unrestricted maximization problem in the denominator of Λ. Recall from Example 2.10 that the population probability density function is

$$f(x) = \frac{1}{\theta} \qquad 0 < x < \theta$$

and the likelihood function is

$$L(\theta) = \left(\frac{1}{\theta}\right)^n$$

for $0 < x_i < \theta$, $i = 1, 2, \ldots, n$, and 0 otherwise. Recall also from Example 2.10 that the maximum likelihood estimator of θ does not exist, but with some small tweaks to the problem statement, the maximum likelihood estimator is the largest value in the data set:

$$\hat{\theta} = \max\{x_1, x_2, \ldots, x_n\} = x_{(n)}.$$

Since the supremum function (sup) is used in the unrestricted maximization problem in the denominator of Λ in Definition 4.13, this establishes the denominator of Λ as

$$\sup_{\Omega} L(\theta) = L\left(\hat{\theta}\right) = \left(\frac{1}{x_{(n)}}\right)^n.$$

We now turn to the restricted maximization problem in the numerator of Λ. As in the previous example, two cases need to be considered when maximizing $L(\theta)$ over Ω_0. Using the likelihood function graphed in Figure 4.52, it is clear that the likelihood function is maximized over Ω_0 at $x_{(n)}$ when $\theta_0 < x_{(n)}$, and the likelihood function is maximized over Ω_0 at θ_0 when $\theta_0 \geq x_{(n)}$. Putting these two pieces together, the likelihood function is maximized at

$$\hat{\theta} = \begin{cases} x_{(n)} & \theta_0 < x_{(n)} \\ \theta_0 & \theta_0 \geq x_{(n)} \end{cases}$$

over Ω_0. This means that the solution to the restricted maximization problem is

$$\sup_{\Omega_0} L(\theta) = \begin{cases} \left(1/x_{(n)}\right)^n & \theta_0 < x_{(n)} \\ (1/\theta_0)^n & \theta_0 \geq x_{(n)}. \end{cases}$$

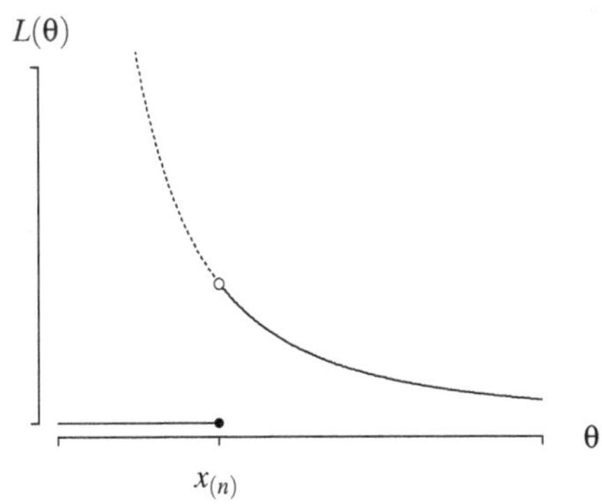

Figure 4.52: The likelihood function $L(\theta) = 1/\theta^n$ for $\theta > x_{(n)}$.

Finally, the likelihood ratio test statistic is

$$\Lambda = \frac{\sup_{\Omega_0} L(\theta)}{\sup_{\Omega} L(\theta)} = \begin{cases} 1 & \theta_0 < x_{(n)} \\ \frac{(1/\theta_0)^n}{\left(1/x_{(n)}\right)^n} & \theta_0 \geq x_{(n)} \end{cases}$$

or

$$\Lambda = \frac{\sup_{\Omega_0} L(\theta)}{\sup_{\Omega} L(\theta)} = \begin{cases} 1 & \theta_0 < x_{(n)} \\ \left(x_{(n)}/\theta_0\right)^n & \theta_0 \geq x_{(n)}. \end{cases}$$

Once again, Λ is a mixed (part discrete, part continuous) random variable. For now, we will ignore the discrete portion of Λ associated with $\Lambda = 1$, which is justified under the original statement of H_0. The next step is to find the probability distribution of Λ under H_0. Recall from Example 2.26 that the probability density function of $X_{(n)}$ under H_0 is

$$f_{X_{(n)}}(x) = \frac{nx^{n-1}}{\theta_0^n} \qquad 0 < x < \theta_0.$$

The probability density function of $\Lambda = g(X_{(n)}) = \left(X_{(n)}/\theta_0\right)^n$ is found using the transformation technique. The function $\lambda = g(x_{(n)}) = \left(x_{(n)}/\theta_0\right)^n$ is a one-to-one transformation from the support of $X_{(n)}$, which, under H_0, is $\mathcal{A} = \{x \,|\, 0 < x < \theta_0\}$ to the support of Λ, which, under H_0, is $\mathcal{B} = \{\lambda \,|\, 0 < \lambda < 1\}$, with inverse

$$x = g^{-1}(\lambda) = \theta_0 \lambda^{1/n}$$

and

$$\frac{dx}{d\lambda} = \frac{\theta_0}{n}\lambda^{1/n-1}.$$

So by the transformation technique, the probability density function of Λ under H_0 is

$$f_\Lambda(\lambda) = f_{X_{(n)}}\left(g^{-1}(\lambda)\right)\left|\frac{dx}{d\lambda}\right| = \frac{n\left(\theta_0\lambda^{1/n}\right)^{n-1}}{\theta_0^n}\left|\frac{\theta_0}{n}\lambda^{1/n-1}\right| = 1 \qquad 0 < \lambda < 1.$$

So for this hypothesis test, the probability distribution of Λ under H_0 has a particularly tractable form: $\Lambda \sim U(0, 1)$. A critical value can be determined by solving

$$P(\Lambda \leq c \,|\, H_0 \text{ true}) = \alpha.$$

This is easily calculated as

$$P(\Lambda \leq c \,|\, H_0 \text{ true}) = P(\Lambda \leq c \,|\, \theta = \theta_0) = \int_0^c 1\, d\lambda = \left[\lambda\right]_0^c = c = \alpha$$

as illustrated in Figure 4.53.

So the procedure for the likelihood ratio test is to collect the data values $x_1, x_2, \ldots, x_n$ and reject $H_0: \theta = \theta_0$ in favor of $H_1: \theta > \theta_0$ if the test statistic $\lambda = (x_{(n)}/\theta_0)^n \leq \alpha$. This is an ideal likelihood ratio test because the probability density function of Λ under H_0 and the critical value c are both mathematically tractable in that they can be expressed in closed form. Notice that the likelihood ratio test is equivalent to that which was derived for Groucho's test statistic $X_{(2)}$ in the case of $\theta_0 = 5$, $n = 2$, and $\alpha = 0.08$ in Example 4.5. Since the distribution of Λ is tractable, the power function is

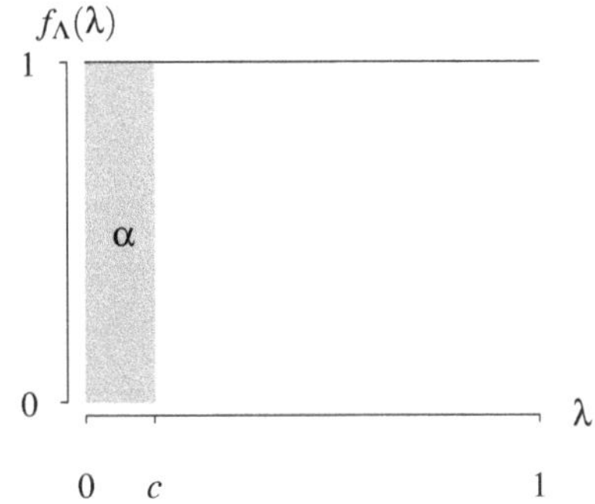

Figure 4.53: Probability density function of Λ under H_0.

$$
\begin{aligned}
q(\theta) &= P(\text{reject } H_0 \,|\, \theta) \\
&= P\big((X_{(n)}/\theta_0)^n \le \alpha \,|\, \theta\big) \\
&= P\big(X_{(n)} \le \theta_0 \alpha^{1/n} \,|\, \theta\big) \\
&= \begin{cases} 1 & \theta \le \theta_0 \alpha^{1/n} \\ \displaystyle\int_0^{\theta_0 \alpha^{1/n}} n x^{n-1}/\theta^n \, dx & \theta > \theta_0 \alpha^{1/n} \end{cases} \\
&= \begin{cases} 1 & \theta \le \theta_0 \alpha^{1/n} \\ \theta_0^n \alpha/\theta^n & \theta > \theta_0 \alpha^{1/n}. \end{cases}
\end{aligned}
$$

Generic power functions for several values of n are plotted in Figure 4.54. As usual, all of these power functions pass through the point (θ_0, α) and become steeper as n increases.

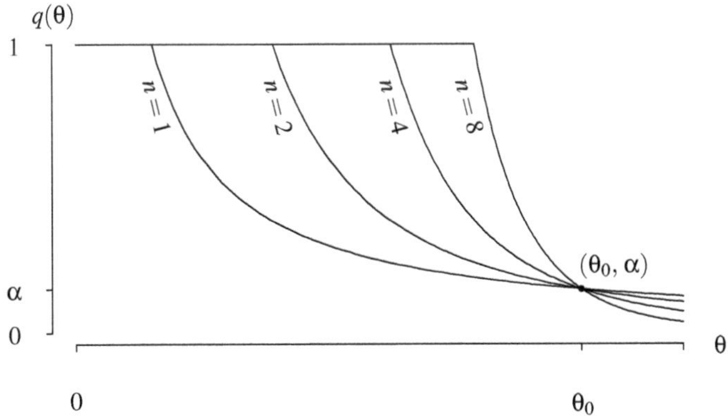

Figure 4.54: Power functions for the likelihood ratio test statistic $\Lambda = (X_{(n)}/\theta_0)^n$.

All three of the examples of likelihood ratio tests presented thus far have involved a single unknown parameter (the unknown population mean μ of a normally distributed population for the first two examples and the unknown population maximum θ for a uniform distribution for the third example). But situations arise in which there can be extraneous unknown parameters that are not of interest to the experimenter. These parameters are known as *nuisance parameters*, and the next example shows how they are dealt with in a likelihood ratio test.

There are four rectangles shown in Figure 4.55. Which rectangle do you consider the most handsome of the four? The notion of intrinsic beauty is a difficult one. Beauty is in the eye of the beholder. I asked a group of students to choose among the four, and the majority picked the rectangle at the far right. All of the rectangles have the same width, but they have varying heights. Consider the ratio of the height to the width for each rectangle. The rectangle at the far left is exactly half as high as it is wide, so the ratio of the height to the width is $1/2$. These are the dimensions used, for example, in a cinder block. The next rectangle is nearly a square, so its height-to-width ratio is nearly 1. The next rectangle is stretched very thin, so its height-to-width ratio is nearly 0. The last rectangle has a ratio of its width to its height of

$$\frac{\sqrt{5}-1}{2} \cong 0.618,$$

which is a famous proportion known as the "golden ratio." This ratio has been used by the Greeks in architectural design, and subsequent cultures have adopted it as well. Ratios near the golden ratio are currently used on business cards, credit cards, television sets, computer monitors, artwork, etc. Did the group of students that I asked to rate the four rectangles choose the golden rectangle because it is more intrinsically beautiful than the others, or have they been conditioned over their lifetimes to favor it because of their exposure to the golden rectangle?

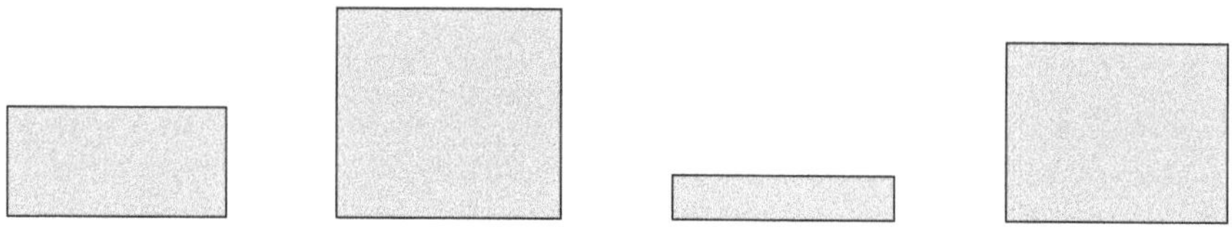

Figure 4.55: Four rectangles.

The next example considers a group of rectangles consisting of beads sewn by the Shoshone Indians as decorations on their clothes, blankets, and leather goods. They would not have had any exposure to golden rectangles, so they could choose any height-to-width ratio. If the ratios that they chose for their rectangles are near the golden ratio, this would be evidence supporting its reputed intrinsic beauty.

Example 4.34 Let $X_1, X_2, \ldots, X_n$ be a random sample from a $N\left(\mu, \sigma^2\right)$ distribution, where μ and $\sigma^2 > 0$ are unknown parameters. Construct a likelihood ratio test of

$$H_0 : \mu = \mu_0$$

versus

$$H_1 : \mu \neq \mu_0$$

at significance level α, where μ_0 is a real-valued constant. Determine a p-value associated with this likelihood ratio test for the golden ratio $\mu_0 = 0.618$ for the height-to-width ratios of $n = 20$ rectangles sewn by the Shoshone Indians given below.

0.693	0.662	0.690	0.606	0.570
0.749	0.672	0.628	0.609	0.844
0.654	0.615	0.668	0.601	0.576
0.670	0.606	0.611	0.553	0.933

The parameter of interest in this hypothesis test is μ. Since σ^2 is an unknown parameter that is not of interest, it is a nuisance parameter. It cannot be treated as a constant as in the first two examples in this section.

As shown in the shaded region in Figure 4.56, the unrestricted parameter space is

$$\Omega = \left\{ \left(\mu, \sigma^2\right) \mid -\infty < \mu < \infty, \sigma^2 > 0 \right\}.$$

The restricted parameter spaces associated with H_0 and H_1 are

$$\Omega_0 = \left\{ \left(\mu, \sigma^2\right) \mid \mu = \mu_0, \sigma^2 > 0 \right\} \qquad \text{and} \qquad \Omega_1 = \left\{ \left(\mu, \sigma^2\right) \mid \mu \neq \mu_0, \sigma^2 > 0 \right\}.$$

The ray pointing upwards (excluding the point on the line $\sigma^2 = 0$) in Figure 4.56 is Ω_0. Since $\Omega = \Omega_0 \cup \Omega_1$, the null and alternative hypotheses do not need to be restated.

We begin with the denominator of the likelihood ratio test statistic Λ, which involves finding the maximum likelihood estimators of μ and σ^2 over Ω. This problem was already encountered in Chapter 2. From Example 2.13, the likelihood function is

$$L\left(\mu, \sigma^2\right) = \left(2\pi\sigma^2\right)^{-n/2} e^{-\frac{1}{2\sigma^2}\sum_{i=1}^{n}(x_i-\mu)^2},$$

which is maximized at the maximum likelihood estimators

$$\hat{\mu} = \bar{x} = \frac{1}{n}\sum_{i=1}^{n} x_i \qquad \text{and} \qquad \hat{\sigma}^2 = \frac{1}{n}\sum_{i=1}^{n}(x_i - \bar{x})^2.$$

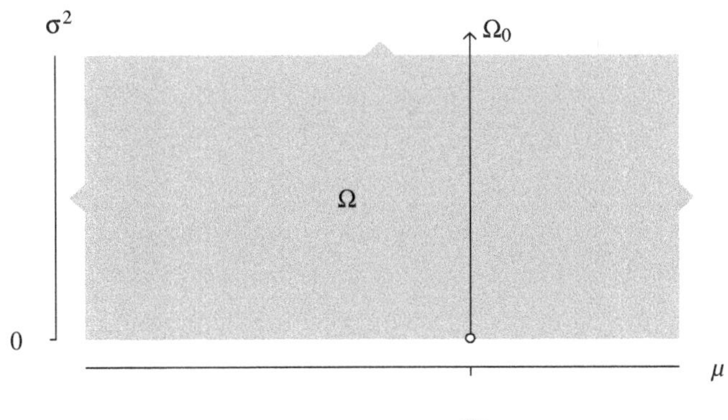

Figure 4.56: Parameter spaces Ω and Ω_0.

So the solution to the unrestricted maximization problem is

$$\begin{aligned}
\sup_{\Omega} L\left(\mu, \sigma^2\right) &= L\left(\hat{\mu}, \hat{\sigma}^2\right) \\
&= \left(2\pi\hat{\sigma}^2\right)^{-n/2} e^{-\frac{1}{2\hat{\sigma}^2}\sum_{i=1}^{n}(x_i-\bar{x})^2} \\
&= \left(2e\pi\hat{\sigma}^2\right)^{-n/2}.
\end{aligned}$$

We now turn to the restricted maximization problem over Ω_0 in the numerator of Λ. Since $\mu = \mu_0$ on Ω_0, the likelihood function is

$$L\left(\mu_0, \sigma^2\right) = \left(2\pi\sigma^2\right)^{-n/2} e^{-\frac{1}{2\sigma^2}\sum_{i=1}^{n}(x_i-\mu_0)^2}.$$

The log likelihood function is

$$\ln L(\mu_0, \sigma^2) = -\frac{n}{2}\ln\left(2\pi\sigma^2\right) - \frac{1}{2\sigma^2}\sum_{i=1}^{n}\left(x_i - \mu_0\right)^2.$$

Differentiating the log likelihood function with respect to σ^2 gives the score

$$\frac{\partial \ln L\left(\mu_0, \sigma^2\right)}{\partial \sigma^2} = -\frac{n}{2\sigma^2} + \frac{1}{2\sigma^4}\sum_{i=1}^{n}\left(x_i - \mu_0\right)^2.$$

Equating the score to zero and solving for σ^2 gives

$$\hat{\sigma}_0^2 = \frac{1}{n}\sum_{i=1}^{n}\left(x_i - \mu_0\right)^2.$$

The subscript 0 is attached to the estimator in order to signify that the likelihood function has been maximized over Ω_0. So the solution to the restricted maximization problem on Ω_0 is

$$\begin{aligned}
\sup_{\Omega_0} L\left(\mu, \sigma^2\right) &= L\left(\mu_0, \hat{\sigma}_0^2\right) \\
&= \left(2\pi\hat{\sigma}_0^2\right)^{-n/2} e^{-\frac{1}{2\hat{\sigma}_0^2}\sum_{i=1}^{n}(x_i-\mu_0)^2} \\
&= \left(2e\pi\hat{\sigma}_0^2\right)^{-n/2}.
\end{aligned}$$

Putting the restricted and unrestricted maximization problems together, the likelihood ratio statistic is

$$\Lambda = \frac{\sup_{\Omega_0} L\left(\mu, \sigma^2\right)}{\sup_{\Omega} L\left(\mu, \sigma^2\right)} = \frac{L\left(\mu_0, \hat{\sigma}_0^2\right)}{L\left(\hat{\mu}, \hat{\sigma}^2\right)} = \frac{\left(2e\pi\sigma_0^2\right)^{-n/2}}{\left(2e\pi\hat{\sigma}^2\right)^{-n/2}} = \left(\hat{\sigma}_0^2/\hat{\sigma}^2\right)^{-n/2}.$$

The next step is to determine the probability distribution of Λ under H_0. Although the form of the likelihood ratio statistic $\left(\hat{\sigma}_0^2/\hat{\sigma}^2\right)^{-n/2}$ is compact and tidy, its distribution under H_0 is not mathematically tractable. Consider the following sequence of inequalities which lead to a familiar probability distribution from probability theory:

$$
\begin{aligned}
\lambda \le c \quad & \Longleftrightarrow \quad \left(\hat{\sigma}_0^2/\hat{\sigma}^2\right)^{-n/2} \le c \\
& \Longleftrightarrow \quad \hat{\sigma}_0^2/\hat{\sigma}^2 \ge c^{-2/n} \\
& \Longleftrightarrow \quad \frac{(1/n)\sum_{i=1}^n (x_i-\mu_0)^2}{(1/n)\sum_{i=1}^n (x_i-\bar{x})^2} \ge c^{-2/n} \\
& \Longleftrightarrow \quad \frac{\sum_{i=1}^n (x_i-\bar{x})^2 + n(\bar{x}-\mu_0)^2}{\sum_{i=1}^n (x_i-\bar{x})^2} \ge c^{-2/n} \\
& \Longleftrightarrow \quad 1+\frac{1}{n-1}\left(\frac{\bar{x}-\mu_0}{s/\sqrt{n}}\right)^2 \ge c^{-2/n} \\
& \Longleftrightarrow \quad \left(\frac{\bar{x}-\mu_0}{s/\sqrt{n}}\right)^2 \ge (n-1)\left(c^{-2/n}-1\right) \\
& \Longleftrightarrow \quad \left|\frac{\bar{x}-\mu_0}{s/\sqrt{n}}\right| \ge \sqrt{(n-1)\left(c^{-2/n}-1\right)},
\end{aligned}
$$

where s is the sample standard deviation. The right-hand side of the last inequality is also a constant (which can be denoted by c', although it is a function of n). In addition, the quantity in the absolute value bars on the left-hand side of the last inequality has the t distribution with $n-1$ degrees of freedom under H_0 by Theorem 1.7. The critical value for the likelihood ratio test is $c' = t_{n-1,\alpha/2}$. So the procedure of the likelihood ratio test in this form is to collect the data values $x_1, x_2, \ldots, x_n$ and reject $H_0 : \mu = \mu_0$ in favor of $H_1 : \mu \ne \mu_0$ if the test statistic $\left|\frac{\bar{x}-\mu_0}{s/\sqrt{n}}\right|$ is greater than or equal to $t_{n-1,\alpha/2}$. This is equivalent to the hypothesis test that was given in the fourth row of Table 4.4 associated with Theorem 1.6.

We now turn to the Shoshone rectangle data set. The R code given below stores the data values $x_1, x_2, \ldots, x_{20}$ in the R vector shoshone, calculates the test statistic, and calculates the associated p-value.

```
shoshone = c(0.693, 0.662, ... , 0.933)
n = length(shoshone)
xbar = mean(shoshone)
golden = (sqrt(5) - 1) / 2
test.statistic = abs((xbar - golden) / sqrt(var(shoshone) / n))
p = 2 * (1 - pt(test.statistic, n - 1))
```

Of course the same p-value can be calculated with fewer keystrokes using the R function named t.test illustrated below.

```
shoshone = c(0.693, 0.662, ... , 0.933)
golden = (sqrt(5) - 1) / 2
t.test(shoshone, mu = golden)
```

In either case, the p-value is $p = 0.054$. There is some, but not overwhelming, statistical evidence that the height-to-width ratio for the Shoshone rectangles differs from the golden ratio.

One thing that has been noticeably missing in the analysis of the rectangles is checking for normality. A histogram of the $n = 20$ values in Figure 4.57 reveals that the probability distribution might not be symmetric; the right-hand tail appears to be longer than the left-hand tail. A second factor working against the normality assumption is that the height-to-width ratios lie between zero and one. This is inconsistent with the infinite-length tails of the normal distribution. The normal distribution does not cut off in this fashion. While it is *possible* that these values could have been drawn from an approximately normal population, it does not look likely. So with the normal assumption suspect, it would be prudent to analyze the data set with either another parametric model or use a nonparametric approach. We will use the latter.

Bootstrapping can be used to determine a p-value for the significance test. It makes no parametric assumptions about the population distribution. The following steps constitute the bootstrap experiment.

- Calculate the sample mean of the $n = 20$ data values.
- Generate B bootstrap samples by sampling $n = 20$ values with replacement from the data set.
- For each bootstrap sample, calculate and store the bootstrap sample mean.
- The p-value is twice the proportion of times that the difference between the bootstrap mean and the sample mean exceeds the difference between the sample mean and $\mu_0 = 0.618$.

The R code below generates nine million bootstrap samples in order to estimate the p-value for the significance test.

```
golden   = (sqrt(5) - 1) / 2
shoshone = c(0.693, 0.662, ... , 0.933)
xbar     = mean(shoshone)
B        = 9000000
bootmean = numeric(B)
for (i in 1:B) {
```

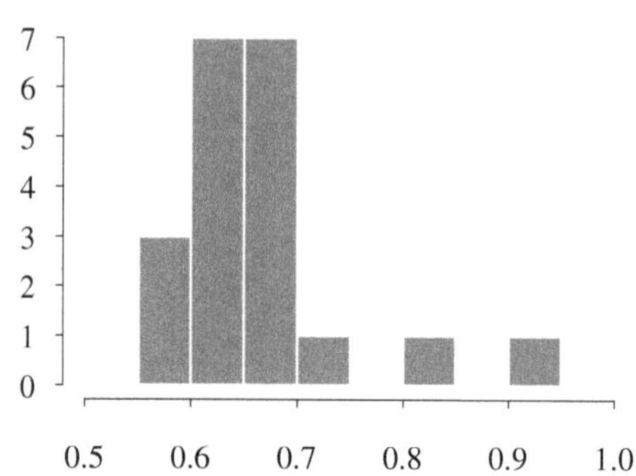

Figure 4.57: Histogram of the height-to-width ratios for $n = 20$ Shoshone rectangles.

```
    bootsample   = sample(shoshone, replace = TRUE)
    bootmean[i] = mean(bootsample)
}
p = 2 * mean(bootmean - xbar > xbar - golden)
```

After a call to `set.seed(1)` to initialize the random number stream, five runs of this bootstrap analysis yield

0.05158 0.05159 0.05165 0.05145 0.05146.

So the bootstrap p-value is approximately $p = 0.052$, which is not far from the value from the t test. The parametric and nonparametric tests yield essentially the same results. This supports the notion from Example 4.17 that tests concerning the population mean are robust to violation of the normality assumption. The power function for this test is based on the noncentral t distribution.

In the four illustrations of likelihood ratio tests so far, we have been able to compute the likelihood ratio test statistic Λ and to determine the associated critical value c (or c' if Λ is transformed) in closed form or by using numerical methods. But sometimes the probability distribution of Λ under H_0 is so mathematically intractable that other techniques must be used to conduct a likelihood ratio test. One such technique relies on the asymptotic distribution of Λ when n is large using a result known as Wilks's theorem, named after Samuel S. Wilks (1906–1964). It can be used to construct an approximate likelihood ratio test on a wide variety of population distributions. Wilks's theorem will be stated in the case of a probability distribution with k unknown parameters $\theta_1, \theta_2, \ldots, \theta_k$, of which there is a subset of r of the parameters that are of interest to the experimenter.

Theorem 4.4 (Wilks's theorem) Let $X_1, X_2, \ldots, X_n$ be a random sample from a population distribution described by $f(x)$ with unknown parameters $\boldsymbol{\theta} = (\theta_1, \theta_2, \ldots, \theta_k)$. Let

$$\Omega_0 = \{(\theta_1, \theta_2, \ldots, \theta_r) \,|\, \theta_1 = \theta_{10}, \theta_2 = \theta_{20}, \ldots, \theta_r = \theta_{r0}\}$$

define a subset of the k-dimensional parameter space Ω associated with the unknown parameters that are of interest to the experimenter. Consider the likelihood ratio test concerning

$$H_0 : \boldsymbol{\theta} \in \Omega_0$$

versus

$$H_1 : \boldsymbol{\theta} \in \Omega_1$$

where $\Omega_0 \cap \Omega_1 = \emptyset$ and $\Omega_0 \cup \Omega_1 = \Omega$. Under certain regularity conditions concerning the population distribution,

$$-2\ln\Lambda = -2\ln\left(\frac{\sup_{\Omega_0} L(\boldsymbol{\theta})}{\sup_{\Omega} L(\boldsymbol{\theta})}\right) \xrightarrow{D} \chi^2(r)$$

under H_0.

The regularity conditions are similar to the regularity conditions associated with results presented previously concerning maximum likelihood estimators. The population distribution has k unknown parameters, but only r of these parameters are of interest to the experimenter, so there are $k-r$ nuisance parameters. The proof of Wilks's theorem follows along the same lines as the proof to Theorem 3.8.

Wilks's theorem provides an asymptotic approximation to the probability distribution of $-2\ln\Lambda$ under H_0. This in turn leads to a critical value associated with the test statistic $-2\ln\lambda$ calculated from the data values $x_1, x_2, \ldots, x_n$. Notice that because the support of Λ is the unit interval, the support of $-2\ln\Lambda$ is the positive real axis, which is consistent with the asymptotic $\chi^2(r)$ distribution. In terms of determining a critical value for the test, consider the following sequence of inequalities:

$$\lambda \le c \qquad \Longleftrightarrow \qquad \ln\lambda \le \ln c \qquad \Longleftrightarrow \qquad -2\ln\lambda \ge -2\ln c.$$

Since $-2\ln c$ is a constant, the procedure for conducting a likelihood ratio test of size α based on Wilks's theorem is to collect the data values $x_1, x_2, \ldots, x_n$ and reject H_0 if the test statistic $-2\ln\lambda$ is greater than or equal to the critical value $\chi^2_{r,\alpha}$. The asymptotic result based on Wilks's theorem is

$$\lim_{n\to\infty} P(\text{reject } H_0 \,|\, H_0 \text{ true}) = \alpha.$$

Note in passing that in the first example in this section, that is, Example 4.31, that the distribution of

$$-2\ln\Lambda = -2\ln e^{-\frac{1}{2}\left(\frac{\bar{X}-\mu_0}{\sigma/\sqrt{n}}\right)^2} = \left(\frac{\bar{X}-\mu_0}{\sigma/\sqrt{n}}\right)^2 \sim \chi^2(1)$$

under H_0 for all values of n because

$$\frac{\bar{X}-\mu_0}{\sigma/\sqrt{n}} \sim N(0,1)$$

and the square of a standard normal random variable has the chi-square distribution with one degree of freedom. So for that particular likelihood ratio test, the asymptotic result is exact for all values of n. This, of course, is a rare case. In general, the degree to which Wilks's theorem is satisfied for finite values of n is determined by the population distribution.

The next example illustrates the application of Wilks's theorem for a population that is a special case of the beta distribution.

Example 4.35 Let $X_1, X_2, \ldots, X_n$ be a random sample from a population distribution described by the probability density function

$$f(x) = \theta x^{\theta-1} \qquad 0 < x < 1,$$

where θ is a positive unknown parameter. Use Wilks's theorem to construct a likelihood ratio test of

$$H_0 : \theta = \theta_0$$

versus

$$H_1 : \theta \ne \theta_0$$

at significance level α, where θ_0 is a positive real-valued constant.

There is just $k = 1$ unknown parameter and it is the parameter of interest, so $r = 1$. With a slight modification to H_1, the restricted parameter spaces associated with H_0 and H_1 are $\Omega_0 = \{\theta \,|\, \theta = \theta_0\}$ and $\Omega_1 = \{\theta \,|\, \theta \ne \theta_0, \theta > 0\}$. The parameter space is $\Omega = \Omega_0 \cup \Omega_1 = \{\theta \,|\, \theta > 0\}$. We begin with the denominator of the likelihood ratio test statistic Λ, which involves finding the maximum likelihood estimator of θ. The likelihood function is

$$L(\theta) = \prod_{i=1}^{n} f(x_i) = \prod_{i=1}^{n} \theta x_i^{\theta-1} = \theta^n \prod_{i=1}^{n} x_i^{\theta-1}.$$

The log likelihood function is

$$\ln L(\theta) = n \ln \theta + (\theta - 1) \sum_{i=1}^{n} \ln x_i.$$

The score is

$$\frac{\partial \ln L(\theta)}{\partial \theta} = \frac{n}{\theta} + \sum_{i=1}^{n} \ln x_i.$$

Equating the score to zero and solving for θ gives the maximum likelihood estimator

$$\hat{\theta} = -\frac{n}{\sum_{i=1}^{n} \ln x_i},$$

which is positive because all of the data values lie on the support $0 < x < 1$. The second derivative of the log likelihood function evaluated at the maximum likelihood estimator is negative, which confirms that $\hat{\theta}$ maximizes both the log likelihood function and the likelihood function. So the solution to the unrestricted maximization problem is simply the likelihood function evaluated at the maximum likelihood estimator:

$$\sup_{\Omega} L(\theta) = L\left(\hat{\theta}\right) = \hat{\theta}^n \prod_{i=1}^{n} x_i^{\hat{\theta}-1}.$$

We now turn to the restricted maximization problem associated with the numerator of Λ. Since the null hypothesis restricts the parameter space to just $\theta = \theta_0$, the restricted maximization problem is solved by simply using θ_0 as an argument in $L(\theta)$. So the solution to the restricted maximization problem is

$$\sup_{\Omega_0} L(\theta) = L(\theta_0) = \theta_0^n \prod_{i=1}^{n} x_i^{\theta_0 - 1}.$$

Finally, the likelihood ratio statistic is

$$\Lambda = \frac{\sup_{\Omega_0} L(\theta)}{\sup_{\Omega} L(\theta)} = \frac{L(\theta_0)}{L\left(\hat{\theta}\right)} = \frac{\theta_0^n \prod_{i=1}^{n} x_i^{\theta_0 - 1}}{\hat{\theta}^n \prod_{i=1}^{n} x_i^{\hat{\theta}-1}} = \left(\frac{\theta_0}{\hat{\theta}}\right)^n \prod_{i=1}^{n} x_i^{\theta_0 - \hat{\theta}}.$$

This is a random variable because $\hat{\theta}$ is a random variable. Finding the probability density function of Λ under H_0 is hopeless in this case, so we turn to Wilks's theorem to construct an approximate test. The procedure is to collect the data values $x_1, x_2, \ldots, x_n$ and reject H_0 when the test statistic

$$-2 \ln \Lambda = -2 \left[n \ln \left(\frac{\theta_0}{\hat{\theta}}\right) + \left(\theta_0 - \hat{\theta}\right) \sum_{i=1}^{n} \ln x_i \right]$$

is greater than or equal to the critical value $\chi^2_{r,\alpha}$.

For a numerical illustration, consider the starting five players on a basketball team whose season free-throw percentages are

$$0.51 \qquad 0.59 \qquad 0.77 \qquad 0.81 \qquad 0.89.$$

Although the data set is fairly small, a likelihood ratio test will be conducted to see if this data set is well modeled by the assumed population distribution with $\theta_0 = 3$. So the hypothesis test is

$$H_0 : \theta = 3$$

versus

$$H_1 : \theta \neq 3.$$

Figure 4.58 shows the probability density function of the population distribution under H_0 and the five data values as points along the horizontal axis.

The R code below calculates the value of the maximum likelihood estimator $\hat{\theta} = 2.8$, the likelihood ratio statistic $\lambda = 0.99$, the test statistic $-2\ln\lambda = 0.026$, and the associated p-value $p = 0.87$.

```
x           = c(0.51, 0.59, 0.77, 0.81, 0.89)
n           = length(x)
mle         = -n / sum(log(x))
theta0      = 3
numerator   = theta0 ^ n * prod(x ^ (theta0 - 1))
denominator = mle ^ n * prod(x ^ (mle - 1))
lambda      = numerator / denominator
p           = 1 - pchisq(-2 * log(lambda), 1)
```

This p-value is very high. We fail to reject H_0. There is not enough statistical evidence here to reject the conclusion that the free throw percentages are well modeled by the population with probability density function associated with $\theta_0 = 3$. Keep in mind that an asymptotic result is being used on a sample of just $n = 5$ observations.

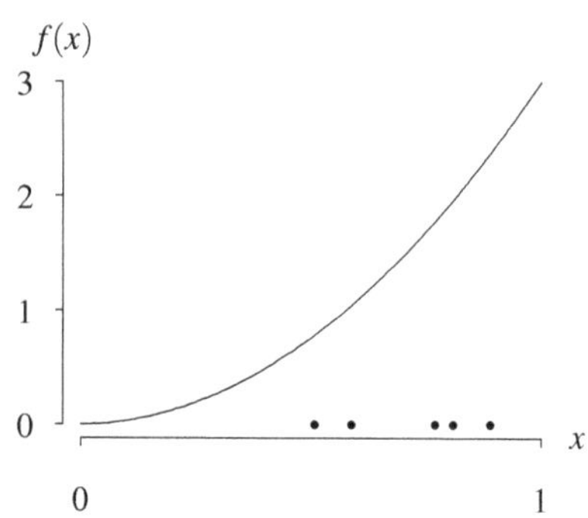

Figure 4.58: Population probability density function for $\theta = 3$ and five data values.

The numerical illustration in the previous example used the asymptotic distribution with only $n = 5$ observations. How close is the distribution of $-2\ln\Lambda$ to the chi-square distribution with one degree of freedom under H_0? The R code below conducts 500 Monte Carlo simulation experiments of generating $n = 5$ observations from the population distribution under H_0 and calculating $-2\ln\lambda$ under H_0. The population cumulative distribution function is $F(x) = \int_0^x \theta w^{\theta-1} = x^\theta$ for $0 < x < 1$.

This means that random variates from this particular distribution can be generated by inversion via the inverse cumulative distribution function $F^{-1}(u) = u^{1/\theta}$ for $0 < u < 1$.

```
nrep      = 500
test.stat = numeric(nrep)
theta0    = 3
n         = 5
for (i in 1:nrep) {
  x              = runif(n) ^ (1 / theta0)
  mle            = -n / sum(log(x))
  numerator      = theta0 ^ n * prod(x ^ (theta0 - 1))
  denominator    = mle ^ n * prod(x ^ (mle - 1))
  test.stat[i]   = -2 * log(numerator / denominator)
}
```

The empirical cumulative distribution function of these values is plotted against the cumulative distribution function of a chi-square distribution with one degree of freedom in Figure 4.59. The smooth curve corresponds to the cumulative distribution function of the chi-square distribution with one degree of freedom. The step function with 500 steps corresponds to the simulated values of Λ. Clearly, the distribution of $-2\ln\Lambda$ is remarkably close to the chi-square distribution with one degree of freedom for this particular population distribution, so we can have faith in the p-value from the previous example.

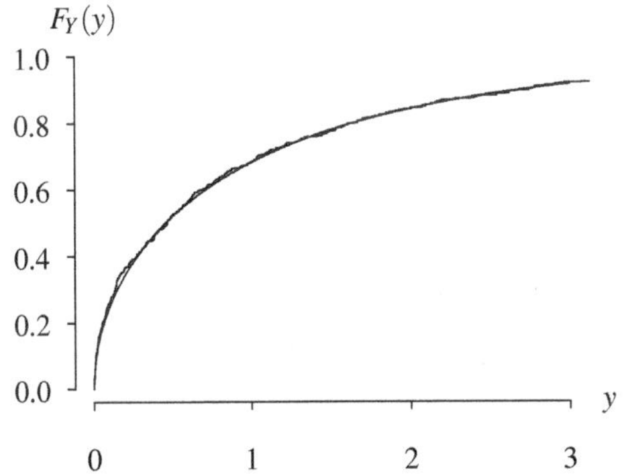

Figure 4.59: Cumulative distribution functions for $y = -2\ln\lambda$ and the $\chi^2(1)$ distribution.

This ends the introduction to hypothesis testing given in this chapter. A summary of the first five sections is given at the end of Section 4.5. The subsequent section on most powerful tests provides a mechanism for determining whether a hypothesis test has the highest power among all tests of size α. This leads to uniformly most powerful tests for a composite alternative hypothesis. In this final section, likelihood ratio tests were introduced. They are often used in practice because they actually suggest a test statistic. In addition, when that test statistic does not have a mathematically tractable probability distribution under H_0, Wilks's theorem provides a mechanism for computing an asymptotically exact critical value.

4.8 Exercises

4.1 Let $X_1, X_2, \ldots, X_n$ be a random sample from a population with probability density function

$$f(x) = \theta x^{\theta-1} \qquad 0 < x < 1,$$

where θ is a positive unknown parameter. What is the critical value associated with the test H_0: $\theta = 7$ versus H_1: $\theta = 3$ using the test statistic $X_{(n)} = \max\{X_1, X_2, \ldots, X_n\}$ and significance level $\alpha = 0.01$?

4.2 Let X_1, X_2, X_3 be a random sample from a Poisson(λ) population, where λ is a positive unknown parameter. The null hypothesis H_0: $\lambda = 4$ is tested against H_1: $\lambda \neq 4$ using the test statistic $\bar{X}$ and critical region $|\bar{X} - 4| > 2$. Find the level of significance α.

4.3 Let $X_1, X_2, \ldots, X_{20}$ be a random sample from a population with probability density function

$$f(x) = \theta x^{\theta-1} \qquad 0 < x < 1,$$

where θ is a positive unknown parameter. To test H_0: $\theta = 1$ versus H_1: $\theta > 1$, reject H_0 if more than half of the observations exceed 0.75. Find α, the level of significance of the test.

4.4 Let $X \sim$ binomial$(6, p)$, where p is an unknown parameter satisfying $0 < p < 1$. Test the null hypothesis H_0: $p = 1/2$ versus the alternative hypothesis H_1: $p = 1/5$ using the critical region $X \leq 1$. What is the level of significance α for the test?

4.5 Let $X \sim$ binomial$(6, p)$, where p is an unknown parameter satisfying $0 < p < 1$. Test the null hypothesis H_0: $p = 1/2$ versus the alternative hypothesis H_1: $p = 1/5$ using the critical region $X \leq 1$. What is the probability of Type II error for the test?

4.6 Julia conducts an *acceptance test* on a large batch of microprocessors in order to test

$$H_0 : p \leq 0.2$$

versus

$$H_1 : p > 0.2,$$

where p is the probability that a microprocessor is defective. A random sample of three microprocessors is drawn from the batch. If all three microprocessors in the initial sample are defective, H_0 is rejected immediately. If one or two of the microprocessors in the initial sample of three are defective, then a fourth microprocessor is randomly sampled from the batch, and H_0 is rejected if the fourth microprocessor is defective. If none of the three microprocessors in the initial sample is defective, then the test immediately fails to reject H_0. A hypothesis test, like this one, that might require two samples to draw a conclusion is often referred to as a *double sampling plan* by quality assurance engineers. What is the power function for this hypothesis test?

4.7 A study was conducted to determine if heart attacks were equally likely to occur throughout the week. Dr. Eliot (*American Medical News*, May 15, 1981) found that 43% of the sudden cardiac deaths occur on Monday. He has concluded that stress is the major cause of coronary heart disease. Assuming that he collected $n = 1000$ data values to determine this percentage, perform a test to determine if heart attacks occur randomly throughout the week at $\alpha = 0.01$.

4.8 In order to test the fairness of a coin at significance level $\alpha = 1/16$, Pedro tosses the coin five times and records the number of heads. What is the power of this test when the true probability of tossing a head on a single toss is $p = 3/5$?

4.9 Barney is a statistician at the corporate headquarters of the Burger Barn. The Colonial Burger Barn, located in historic Williamsburg, Virginia is being threatened with extinction by upper management due to lack of customers. The manager of the Colonial Burger Barn has contacted Barney to prove to the corporate executives that the arrival rate of customers is above the national Barn average. Since there are thousands of Barns nationwide, Barney knows that Burger Barn arrivals are well-modeled by a Poisson process during the "lunch rush" with an arrival rate of 37.7 customers per hour. The manager agrees to collect the number of arrivals during five one-hour lunch rush periods. He counts a total of 208 customer arrivals over the five hours. Can Barney conclude that the Colonial Burger Barn is busier than average at significance level $\alpha = 0.1$?

4.10 Let X_1, X_2, X_3 be a random sample from a $U(0, \theta)$ population, where θ is a positive unknown parameter.

(a) Find the appropriate critical value c for testing

$$H_0 : \theta = 2$$

versus

$$H_1 : \theta = 1$$

based on the test statistic $X_{(3)}$ at significance level $\alpha = 1/64$.

(b) Perform a Monte Carlo simulation experiment to support your critical value c determined in part (a).

4.11 Let X be a single observation from a population with probability density function

$$f(x) = \theta x^{\theta-1} \qquad 0 < x < 1,$$

where θ is a positive unknown parameter. Find the critical value c that minimizes $\alpha + \beta$ for the hypothesis test with null hypothesis

$$H_0 : \theta = \frac{1}{2}$$

and alternative hypothesis

$$H_1 : \theta = 4.$$

4.12 Let X be a single observation from a population with probability density function

$$f(x) = 1 + \theta^2 \qquad 0 < x < \frac{1}{1+\theta^2},$$

where θ is a nonnegative unknown parameter. The critical region for testing

$$H_0 : \theta = 0$$

versus

$$H_1 : \theta = 1$$

is $X < 0.4$. Find the values of the α and β for this test. Also, sketch a diagram which indicates the geometry associated with this hypothesis test.

4.13 Let $X_1, X_2, \ldots, X_{10}$ be a random sample from a Bernoulli(p) population, where p is an unknown parameter which satisfies $0 < p < 1$. Danielle would like to test

$$H_0 : p = 1/3$$

versus

$$H_1 : p > 1/3$$

using the test statistic $X = X_1 + X_2 + \cdots + X_{10}$. Her decision rule rejects H_0 when $X \geq 9$.

(a) Find the significance level α.

(b) Find β when $p = 3/5$.

4.14 Let $X_1, X_2, \ldots, X_n$ be a random sample from a Pareto population with probability density function

$$f(x) = \frac{\theta}{(1+x)^{\theta+1}} \qquad x > 0,$$

where θ is a positive unknown parameter.

(a) Find the Cramér–Rao lower bound on the population variance of any unbiased estimator of θ.

(b) Find an exact two-sided $100(1-\alpha)\%$ confidence interval for θ from a *single* ($n = 1$) observation X.

(c) A single observation X is used to test

$$H_0 : \theta = 2$$

versus

$$H_1 : \theta < 2.$$

Find the critical value at $\alpha = 0.01$.

(d) Plot the power function for the hypothesis test in part (c).

4.15 A company is concerned about the quality of a particular item that it produces. Liz tests H_0: $p = 1/20$ versus H_1: $p > 1/20$, where p is the probability that an item is defective, in the following fashion. An initial random sample of 5 items is drawn. If all 5 items are nondefective, she fails to reject H_0. If one or more of the items in the initial sample are defective, she takes a second sample of 10 items. If all 10 items in the second sample are nondefective, she fails to reject H_0. If one or more items are defective in the second sample, she rejects H_0. If the test result for each item sampled can be regarded as an independent and identically distributed Bernoulli trial, what is the power of the test as a function of p?

4.16 Allison would like to test the hypothesis

$$H_0 : \lambda = 17$$

versus

$$H_1 : \lambda > 17$$

using a single value X from an exponential population with probability density function

$$f(x) = \lambda e^{-\lambda x} \qquad x > 0,$$

where λ is a positive unknown population rate parameter. The null hypothesis is rejected if $X < 0.01$. Find the significance level α for the test.

4.17 Let $X_1, X_2, \ldots, X_n$ be a random sample from a distribution with probability density function

$$f(x) = \theta x^{\theta-1} \qquad 0 < x < 1,$$

where θ is a positive unknown parameter.

(a) Find the maximum likelihood estimator of θ.

(b) Find the method of moments estimator of θ.

(c) Find a sufficient statistic for θ using the Fisher–Neyman factorization criterion.

(d) Is your sufficient statistic an unbiased estimator of θ?

(e) Find the Cramér–Rao lower bound.

(f) Test

$$H_0 : \theta = 2$$

versus

$$H_1 : \theta < 2$$

at $\alpha = 0.04$ using a single ($n = 1$) observation X from the population. Find the critical value and find the power function for the test.

(g) Find the maximum likelihood estimator of the population standard deviation.

(h) Find an exact two-sided $100(1-\alpha)\%$ confidence interval for θ from a single ($n = 1$) observation X from the population.

(i) Consider the popular "major vote" proportions (which include only the two major parties) of the winners in the U.S. presidential elections between 1924 and 2024.

Year	President	Party	Proportion	Year	President	Party	Proportion
2024	Trump	R	0.508	1972	Nixon	R	0.618
2020	Biden	D	0.523	1968	Nixon	R	0.504
2016	Trump	R	0.489	1964	Johnson	D	0.613
2012	Obama	D	0.520	1960	Kennedy	D	0.501
2008	Obama	D	0.537	1956	Eisenhower	R	0.578
2004	Bush	R	0.512	1952	Eisenhower	R	0.555
2000	Bush	R	0.497	1948	Truman	D	0.524
1996	Clinton	D	0.547	1944	Roosevelt	D	0.538
1992	Clinton	D	0.535	1940	Roosevelt	D	0.550
1988	Bush	R	0.539	1936	Roosevelt	D	0.625
1984	Reagan	R	0.592	1932	Roosevelt	D	0.591
1980	Reagan	R	0.553	1928	Hoover	R	0.588
1976	Carter	D	0.511	1924	Coolidge	R	0.652

Apply parts (a), (b), and (g) to this data set. Do you think that this is a good distribution to use for this particular data set? Why?

4.18 Ron is holding a coin. He would like to conduct a statistical test to determine whether the coin is fair or biased toward tails. Let p be the probability of tossing a head on a single toss. Ron will toss the coin repeatedly and will record X, the number of tosses required until the first head appears. The null hypothesis H_0 is that the coin is fair. The null hypothesis is rejected when $X \geq 3$. State H_0 and H_1 in terms of p and find the significance level α.

4.19 Consider the following *randomized hypothesis test*. Let $X_1, X_2, \ldots, X_9$ be a random sample from a Bernoulli(p) population with unknown parameter p satisfying $0 < p < 1$. Consider the hypothesis test:

$$H_0 : p = 3/8$$

versus

$$H_1 : p > 3/8.$$

Consider the following decision rule: reject H_0 if

- $\sum_{i=1}^{9} X_i \geq 8$ or
- $\sum_{i=1}^{9} X_i = 7$ and two flips of a fair coin come up heads.

Find the significance level α for this test.

4.20 Let X be an observation from an exponential population with positive unknown population mean θ. This observation is used to test

$$H_0 : \theta = 6$$

versus

$$H_1 : \theta = 2.$$

(a) Find the critical value for the test for a fixed significance level α.

(b) Find β for a fixed significance level α.

4.21 Let X be an observation from a $U(0, \theta)$ population, where θ is a positive unknown parameter. Consider testing

$$H_0 : \theta = 10$$

versus

$$H_1 : \theta \neq 10$$

by rejecting H_0 when $X < 1$ or $X > 9$.

(a) Find the significance level of the test.

(b) Plot the power function for the test for $0 < \theta < 16$.

4.22 Let $X_1, X_2, \ldots, X_n$ be a random sample from a population with probability density function

$$f(x) = \theta x^{\theta - 1} \qquad 0 < x < 1,$$

where θ is a positive unknown parameter.

(a) Find the maximum likelihood estimator of the population median of this distribution.

(b) Consider the hypothesis test

$$H_0 : \theta = 5$$

versus

$$H_1 : \theta > 5.$$

If you use the test statistic $X_1 + X_2$ at $\alpha = 0.05$ when $n = 2$, find the critical value c.

4.23 Let $X_1, X_2, \ldots, X_n$ be a random sample from an exponential population with probability density function

$$f(x) = \frac{1}{\theta} e^{-x/\theta} \qquad x > 0,$$

where θ is a positive unknown parameter.

(a) Show that the sample mean, $\bar{X}$, and n times the first order statistic, $nX_{(1)}$, are both unbiased estimators of θ.

(b) Calculate the relative efficiency of the sample mean, $\bar{X}$, versus n times the first order statistic, $nX_{(1)}$. Indicate the preferred estimator.

(c) Determine the probability density functions of the sample mean, $\bar{X}$, and n times the first order statistic, $nX_{(1)}$.

(d) Graph the two probability density functions determined in part (c) for $n = 5$ and $\theta = 2$.

(e) Consider testing

$$H_0 : \theta = 7$$

versus

$$H_1 : \theta < 7$$

based on a single observation X ($n = 1$). Find the critical value corresponding to significance level $\alpha = 0.03$.

(f) Graph the power curve for the hypothesis test

$$H_0 : \theta = 7$$

versus

$$H_1 : \theta < 7$$

based on a single observation X ($n = 1$) at significance level $\alpha = 0.03$.

4.24 Ivy is not sure whether a particular coin is fair. She flips the coin 100 times and decides that the coin is not fair if she observes 40 or fewer heads, or 60 or more heads. State the null and alternative hypotheses H_0 and H_1, calculate the significance level α, and plot the power function associated with her test.

4.25 Let $X_1, X_2, \ldots, X_n$ denote a random sample of light bulb lifetimes from an exponential population with probability density function

$$f(x) = \lambda e^{-\lambda x} \qquad x > 0,$$

where λ is a positive unknown rate parameter. The bulbs are stamped with "1000 hour MTTF," indicating that the *mean time to failure* equals 1000 hours. Determine whether there is statistical evidence that indicates the bulbs last *longer* than 1000 hours.

(a) State the appropriate H_0 and H_1.

(b) Daniel uses the test statistic $\bar{X}$ and Katie uses $nX_{(1)}$ to test the hypothesis. Find the critical values for their test statistics when $\alpha = 0.05$ and $n = 10$.

(c) Draw the power curves associated with each of the test statistics from part (b) on the same set of axes using a computer. Again assume that $\alpha = 0.05$ and $n = 10$.

4.26 Let X_1 and X_2 be a random sample from a population with probability density function

$$f(x) = \theta x^{\theta-1} \qquad 0 < x < 1,$$

where θ is a positive unknown parameter. It is desired to test

$$H_0 : \theta = 2$$

versus

$$H_1 : \theta < 2$$

at significance level $\alpha = 0.04$. Kevin, Nancy, Jordan, Greta, and Emma each choose a different test statistic in order to conduct the test. For each test statistic, determine the appropriate critical value and the associated power function. *Carefully* plot all five power functions on a single set of axes over $0 < \theta < 3$ and determine which test statistic is preferred. The test statistics are

(a) Kevin: X_2

(b) Nancy: $1/X_1$

(c) Jordan: $X_1 \cdot X_2$

(d) Greta: $X_1 + X_2$

(e) Emma: $\min\{X_1, X_2\}$.

4.27 Xin is a sociologist. She has a theory. She believes that people are depressed more often during the 30 days prior to Christmas. She has several years of data concerning admission dates to a mental health clinic where depressed individuals can receive counseling and/or treatment. State the appropriate H_0 and H_1 to test Xin's theory, and indicate the appropriate test statistic.

4.28 Let $X_1, X_2, \ldots, X_{15}$ be a random sample from a Poisson population with positive unknown population mean λ, denoting the number of major hurricanes in the United States per decade from the 1850s through the 1990s given in Example 4.13.

6 1 7 5 8 4 7 5 8 10 8 6 4 5 5.

(a) Find the maximum likelihood estimator of the probability that there will be no hurricanes in a decade.

(b) Find the significance level α associated with the hypothesis test

$$H_0 : \lambda = 5$$

versus

$$H_1 : \lambda > 5$$

using the decision rule: reject H_0 if the test statistic $\sum_{i=1}^{15} X_i \geq 87$. What is the conclusion for the data set given above?

(c) Plot the power function for the hypothesis test from part (b).

4.29 Let $X_1, X_2, \ldots, X_n$ be a random sample from a population with probability density function

$$f(x) = \theta x^{\theta-1} \qquad 0 < x < 1,$$

where θ is a positive unknown parameter. Test

$$H_0 : \theta = 1$$

versus

$$H_1 : \theta > 1$$

using the test statistic $X_{(1)} = \min\{X_1, X_2, \ldots, X_n\}$.

(a) Find the critical value c in terms of the significance level α and the sample size n.

(b) Plot the power function when $n = 2$ and $\alpha = 0.04$.

4.30 Consider a single observation X from an exponential population with a positive unknown population mean θ. You would like to test

$$H_0 : \theta = 1$$

versus

$$H_1 : \theta > 1$$

at $\alpha = 0.07$ using X as a test statistic.

(a) Find the critical value c for the test.

(b) Plot the power function for the test.

4.31 Let X be a single observation from a population with probability density function

$$f(x) = \frac{\theta}{(1+x)^{\theta+1}} \qquad x > 0,$$

where θ is a positive unknown parameter. This is a special case of the Lomax distribution. Test

$$H_0 : \theta = 1$$

versus

$$H_1 : \theta = 2$$

at $\alpha = 0.1$ using X as a test statistic.

(a) Give the decision rule for the test.

(b) Calculate $\beta = P(\text{Type II error})$.

4.32 Let X be a single observation from a $U(\theta, 5\theta)$ population, where θ is a positive unknown parameter. Test

$$H_0 : \theta = 1$$

versus

$$H_1 : \theta = 2$$

using the critical region $X > c$, where $2 < c < 5$. Find the value of the power function evaluated at $\theta = 2$, that is, find $q(2)$.

4.33 Let X be a single observation from a $U(\theta, 5\theta)$ population, where θ is a positive unknown parameter. Test

$$H_0 : \theta = 1$$

versus

$$H_1 : \theta = 2$$

using the critical region $X > 3$. Find the power function.

4.34 Let X_1 and X_2 be observations from a population which is a special case of the bivariate normal distribution defined by

$$\mu_{X_1} = \theta + 10 \qquad \mu_{X_2} = 2\theta + 15 \qquad \sigma_{X_1}^2 = 100 \qquad \sigma_{X_2}^2 = 144 \qquad E[X_1 X_2] = 150,$$

where θ is an unknown parameter.

(a) Find the parameter space for θ.

(b) The critical region for testing the null hypothesis $H_0 : \theta = 0$ versus the alternative hypothesis $H_1 : \theta > 0$ is $X_1 + 2X_2 > 50$. Find the significance level α for this test.

4.35 Let X_1 and X_2 be a random sample from a population with probability density function

$$f(x) = \frac{\theta e^x}{(1 + \theta e^x)^2} \qquad -\infty < x < \infty,$$

where θ is a positive unknown parameter. This is a special case of the logistic distribution. Find the appropriate critical region for testing

$$H_0 : \theta = 1$$

versus

$$H_1 : \theta = 2$$

using the test statistic $X_{(2)} = \max\{X_1, X_2\}$ at significance level $\alpha = 0.05$.

4.36 Let X be a single observation from a population with probability density function

$$f(x) = \frac{1}{2} e^{-|x-\theta|} \qquad -\infty < x < \infty,$$

where θ is an unknown parameter. The is a special case of the Laplace or double exponential distribution. The hypothesis test

$$H_0 : \theta = 0$$

versus

$$H_1 : \theta \neq 0$$

is conducted at level $\alpha = 0.1$ using X as a test statistic. If H_0 is rejected for $X < -c$ or $X > c$, find the critical value c.

4.37 Jutta tosses a biased coin repeatedly until the third head appears. Let X be the number of tosses required for her to see the third head. Let p be the probability of heads appearing on an individual toss, $0 < p < 1$. The null hypothesis $H_0 : p = 1/3$ is tested against the alternative hypothesis $H_1 : p < 1/3$. If H_0 is rejected when $X \geq 7$, find the power of the test when $p = 1/10$, that is, find $q(1/10)$.

4.38 A pond contains 50 fish. A total of θ of these fish are tagged and the remaining fish are untagged. Yogi would like to test

$$H_0 : \theta = 10$$

versus

$$H_1 : \theta < 10$$

by catching 12 of the fish (without replacement) and counting X, the number of tagged fish in his sample. He rejects H_0 if zero or one of the fish in his sample are tagged. Find the significance level α for this test.

4.39 Olivia is performing a hypothesis test. She obtains a p-value of 0.07 for the test. Should she reject H_0 at $\alpha = 0.05$?

4.40 Let $X_1, X_2, \ldots, X_n$ be a random sample from a population with a single unknown parameter θ. A significance test procedure is developed to test the null hypothesis $H_0 : \theta = 2$ versus the alternative hypothesis $H_1 : \theta = 1$, resulting in a p-value of 0.02. What is the resulting p-value if this same statistical test procedure is applied to testing the null hypothesis $H_0 : \theta = 2$ versus the alternative hypothesis $H_1 : \theta < 2$.

4.41 Aastha is crashing cars going 45 miles per hour into a wall. She is interested in whether an unbelted test dummy will go through a windshield made of a new product known as "flexi-glass." She crashes five cars and one dummy breaks through the windshield. Assume that p is the true probability that a dummy crashes through the windshield.

(a) Give an approximate two-sided Wald 95% confidence interval for p using large-sample theory (which is clearly inappropriate here since n is only 5).

(b) Give the p-value for the hypothesis test

$$H_0 : p = 1/2$$

versus

$$H_1 : p < 1/2.$$

4.42 Let X_1, X_2, X_3 be a random sample from a Poisson(λ) population with positive unknown population mean λ. In order to test

$$H_0 : \lambda = 2$$

versus

$$H_1 : \lambda < 2,$$

the null hypothesis is rejected if $X_1 + X_2 + X_3 < 4$.

(a) Find α, the significance level for this test.

(b) Find β when $\lambda = 1$.

(c) Plot the power function.

(d) Find the p-value associated with the data set: $x_1 = 1, x_2 = 0, x_3 = 1$.

4.43 Joel is concerned that he is holding a die that is loaded (that is, unfair) such that it produces fewer fours than a fair die. He rolls this die 20 times and gets only one four. He hires you as a statistical consultant. State the appropriate H_0 and H_1 and give the p-value for his test. Is there enough statistical evidence here for him to conclude that the die is loaded?

4.44 Let $X_1, X_2, \ldots, X_n$ be a random sample from an exponential population with probability density function

$$f(x) = \lambda e^{-\lambda x} \qquad x > 0,$$

where λ is a positive unknown rate parameter.

(a) Show that $2\lambda \sum_{i=1}^{n} X_i \sim \chi^2(2n)$.

(b) Using the result from part (a), find an exact two-sided $100(1-\alpha)\%$ confidence interval for λ.

(c) Determine an exact two-sided 95% confidence interval for λ for the data set: 6, 24, 58, 72.

(d) Give the p-value for the hypothesis test

$$H_0 : \lambda = 0.01$$

versus

$$H_1 : \lambda > 0.01$$

for the data set: 6, 24, 58, 72.

4.45 Let $X_1, X_2, \ldots, X_{10}$ be a random sample of light bulb lifetimes from an exponential(λ) population, where λ is a positive unknown failure rate. Adrienna is a reliability engineer. She is confident from previous test results that the time to failure for these light bulbs is exponentially distributed. She is interested in testing whether a manufacturer's claim that the population mean time to failure for the bulbs is 1000 hours. So she would like to test

$$H_0 : \lambda = 0.001$$

versus

$$H_1 : \lambda > 0.001.$$

But she is in a hurry. She places ten bulbs on test and only observes the first bulb fail at $x_{(1)} = 14$ hours. Give the p-value for the test based on the value of this single order statistic.

4.46 Let $X_1, X_2, \ldots, X_7$ be a random sample from a Rayleigh population with probability density function

$$f(x) = 2\theta^{-2} x e^{-(x/\theta)^2} \qquad x > 0,$$

where θ is a positive unknown parameter. Gexin tests

$$H_0 : \theta = 10$$

versus

$$H_1 : \theta > 10$$

using the test statistic $X_{(1)} = \min\{X_1, X_2, \ldots, X_7\}$, which assumes the value $x_{(1)} = 6$. Find the p-value for his test.

4.47 Lindsey rents out a room using Airbnb. She determines that she will need to rent the room 63% of the evenings in order to achieve profitability. Letting H_0 correspond to break-even at 63% rentals and H_1 correspond to profitability, find the p-value for the appropriate hypothesis test if she has rented the room on 25 of the last 30 evenings. Even though it is suspect in this case, assume that a rental on each of the 30 evenings is a mutually independent Bernoulli trial.

4.48 Previous testing indicates that the number of cycles to failure for a particular type of switch are well modeled by a Geometric(p) population distribution, where p is the probability of failure on a single cycle, which satisfies $0 < p < 1$. Testing $n = 3$ such switches to failure results in the random sample X_1, X_2, X_3. Find the p-value associated with the hypothesis test

$$H_0 : p = 0.001$$

versus

$$H_1 : p > 0.001$$

for the experimental values $x_1 = 243$, $x_2 = 155$, $x_3 = 606$.

4.49 Let $X_1, X_2, \ldots, X_{18}$ be a random sample from a $N\left(\mu, \sigma^2\right)$ population, where μ and $\sigma^2 > 0$ are unknown parameters. Find the critical value for testing

$$H_0 : \sigma^2 = 16$$

versus

$$H_1 : \sigma^2 > 16$$

at $\alpha = 0.05$ using the test statistic $\sum_{i=1}^{18} \left(X_i - \bar{X}\right)^2$.

4.50 Find the critical value for a paired t-test involving $n = 4$ pairs of observations and the one-tailed alternative hypothesis H_1: $\mu_X < \mu_Y$ at significance level $\alpha = 0.05$.

4.51 Let $X_1, X_2, \ldots, X_8$ be a random sample from a $N\left(\mu, \sigma^2\right)$ population, where μ and σ^2 are unknown parameters. Find the p-value for testing H_0: $\sigma^2 = 10$ versus H_1: $\sigma^2 < 10$ when $s^2 = 6$.

4.52 Let $X_1, X_2, \ldots, X_{10}$ be a random sample from a $N(\mu_X, 2)$ population, where μ_X is an unknown parameter. Let $Y_1, Y_2, \ldots, Y_{25}$ be a random sample from a $N(\mu_Y, 4)$ population, where μ_Y is an unknown parameter. The two random samples are independent. The null hypothesis H_0: $\mu_X = \mu_Y$ is tested against the alternative hypothesis H_1: $\mu_X > \mu_Y$ using the test statistic $\bar{X} - \bar{Y}$. If H_0 is rejected when $\bar{X} - \bar{Y} > 3.2$, find the power of the test when $\mu_X - \mu_Y = 2$.

4.53 Let $X_1, X_2, \ldots, X_7$ be a random sample from a $N(\mu, 100)$ population, where μ is an unknown parameter. Consider the hypothesis test

$$H_0 : \mu = 12$$

versus

$$H_1 : \mu \neq 12.$$

Find the critical values for conducting this test at significance level α associated with the test statistics

(a) $\frac{\bar{X}-12}{10/\sqrt{7}}$,

(b) $\bar{X}$,

(c) $\sum_{i=1}^{7} X_i$.

4.54 Let −4, 2, 2 be a random sample from a normal population with unknown population mean μ_X and positive unknown population variance σ^2. Let −6, 1, 1, 4 be a random sample from a normal population with unknown population mean μ_Y and positive unknown population variance σ^2. The two random samples are independent. Give the p-value for the hypothesis test

$$H_0 : \mu_X = \mu_Y$$

versus

$$H_1 : \mu_X < \mu_Y.$$

4.55 Let −4, 2, 2 be a random sample from a normal population with unknown population mean μ_X and positive unknown population variance σ_X^2. Let −6, 1, 1, 4 be a random sample from a normal population with unknown population mean μ_Y and positive unknown population variance σ_Y^2. The two random samples are independent. At significance level $\alpha = 0.05$, test

$$H_0 : \sigma_X^2 = \sigma_Y^2$$

versus

$$H_1 : \sigma_X^2 < \sigma_Y^2.$$

4.56 The IQ scores of four sixth-grade children randomly sampled from a school are:

94 126 110 114.

Assume that these values are drawn from a normal population with unknown population mean μ and unknown population variance $\sigma^2 > 0$.

(a) Find the maximum likelihood estimator of σ^2.

(b) Find the p-value for testing

$$H_0 : \mu = 100$$

versus

$$H_1 : \mu > 100.$$

4.57 The data in the table below are the prices of ten different types of food, in pence, bought in a particular area of the UK at a particular time in 1987, and again exactly one year later in 1988 (Hand et al., 1994, *Small Data Sets*, Chapman and Hall, page 49). Run a hypothesis test to see if the prices have risen over the year. You may assume that the two sets of prices are independent random samples from two normal populations.

Food (quantity)	1987 price	1988 price
Fish (5 lb)	856	943
Milk (20 pints)	496	516
Cheese (5 lb)	663	732
Bread (20 loaves)	478	511
Breakfast cereal (10 lb)	753	802
Potatoes, old (70 lb)	659	617
Apples (20 lb)	685	711
Frozen peas (20 lb)	750	891
Beef (5 lb)	823	896
Margarine (5 lb)	407	427

4.58 Many theories about the causation of schizophrenia involve changes in the activity of a substance called dopamine in the central nervous system. In one study, 25 hospitalized schizophrenic patients were treated with antipsychotic medication, and after a period of time were classified as psychotic or nonpsychotic by hospital staff (Hand et al., 1994, *Small Data Sets*, Chapman and Hall, page 37). Samples of cerebrospinal fluid were taken from each patient and assayed for the dopamine b-hydroxylase (DBH) activity. The units are nmol/(ml)(h)/(mg) of protein. The DBH activity levels for those patients judged nonpsychotic were:

$$0.0104\ \ 0.0105\ \ 0.0112\ \ 0.0116\ \ 0.0130\ \ 0.0145\ \ 0.0154\ \ 0.0156$$
$$0.0170\ \ 0.0180\ \ 0.0200\ \ 0.0200\ \ 0.0210\ \ 0.0230\ \ 0.0252.$$

The DBH activity levels for those patients judged psychotic were:

$$0.0150\ \ 0.0204\ \ 0.0208\ \ 0.0222\ \ 0.0226\ \ 0.0245\ \ 0.0270\ \ 0.0275\ \ 0.0306\ \ 0.0320.$$

Assuming normal sampling for the two sets of patients:

(a) find an exact two-sided 90% confidence interval for the ratio of the population variances, and

(b) report the p-value for a hypothesis test to determine whether there is a statistically significant difference in the population mean DBH activity between the two populations.

4.59 Let $X_1, X_2, \ldots, X_{15}$ be a random sample from a $N(\mu_X, \sigma^2)$ population, and $Y_1, Y_2, \ldots, Y_{12}$ be a random sample from a $N(\mu_Y, \sigma^2)$ population, where μ_X, μ_Y, and $\sigma^2 > 0$ are unknown parameters. The two random samples are independent. Give the test statistic and critical value for a test of

$$H_0 : \mu_X = \mu_Y$$

versus

$$H_1 : \mu_X > \mu_Y$$

at $\alpha = 0.01$.

4.60 Professor Hunt gives a statistics exam in two parts: in-class, worth 70% of the grade, and take-home, worth 30% of the grade. There are $n = 20$ students in the class. He posits that there is a positive correlation between the in-class and take-home scores. Are his instincts correct? The sample correlation coefficient between the in-class scores and take-home scores is $r = 0.546$. Test the appropriate hypothesis at $\alpha = 0.05$. Base your test statistic on the following approximate relationship (known as Fisher's transformation after Sir Ronald Fisher):

$$\frac{1}{2}\ln\left(\frac{1+R}{1-R}\right) \overset{a}{\sim} N\left(\frac{1}{2}\ln\left[\frac{1+\rho}{1-\rho}\right], \frac{1}{n-3}\right),$$

where

$$R = \frac{1}{n-1}\sum_{i=1}^{n}\left(\frac{X_i - \bar{X}}{S_X}\right)\left(\frac{Y_i - \bar{Y}}{S_Y}\right)$$

is the sample correlation coefficient and

$$\rho = E\left[\left(\frac{X - \mu_X}{\sigma_X}\right)\left(\frac{Y - \mu_Y}{\sigma_Y}\right)\right]$$

is the population correlation coefficient.

4.61 Hans and Franz are working out in their gym. In order to determine who is the strongest, they devise ten feats of strength. The performance of the two weight lifters, in pounds, is given in the table below. Is there statistical evidence (at $\alpha = 0.05$) that one of the weight lifters is stronger than the other?

Feat	Hans	Franz
Clean and Jerk	216	224
Jerk and Clean	186	202
Snatch and Jerk	360	401
Clean and Nip	87	83
Nip and Tuck	410	440
Nip and Press	160	170
Pump–U–Up	224	226
Tuck and Knupp	180	170
Bric and Brac	316	338
Snatch and Tuck	228	252

4.62 Let $X_1, X_2, \ldots, X_{16}$ be a random sample from a $N(\mu, 25)$ population, where μ is an unknown parameter. In order to test

$$H_0 : \mu = 100$$

versus

$$H_1 : \mu > 100,$$

H_0 is rejected if the test statistic $\sum_{i=1}^{16} X_i$ is greater than 1650.

(a) Find the significance level α.

(b) Find β when $\mu = 105$.

4.63 Let X_1, X_2, X_3, X_4 be a random sample from a normal population with unknown population mean μ and known population variance $\sigma^2 = 25$. You intend to test

$$H_0 : \mu = 68$$

versus

$$H_1 : \mu > 68$$

using the test statistic $X_1 + X_2 + X_3 + X_4$ at level of significance $\alpha = 0.08$.

(a) Find the critical region associated with this test statistic.

(b) Draw a graph of the probability density function of the test statistic when H_0 is true. Indicate the critical value and α on your graph.

(c) Draw the power function for this test.

(d) Draw the appropriate conclusion for this data set: 71, 73, 64, 69.

4.64 Dr. Reynolds believes that adult heights can be predicted by doubling a toddler's height on his or her second birthday. Given that X_i is the height of toddler i on their second birthday, and Y_i is the associated adult height, for $i = 1, 2, \ldots, n$, give the appropriate H_0, H_1, test statistic and critical value(s) for Dr. Reynolds's test. You may assume that $X_1, X_2, \ldots, X_n$ denote a random sample from a normal distribution and $Y_1, Y_2, \ldots, Y_n$ also denote a random sample from a normal distribution.

4.65 Let $X_1, X_2, \ldots, X_n$ be a random sample from a $N(\mu, 16)$ population, where μ is an unknown parameter. Test

$$H_0 : \mu = 10$$

versus

$$H_1 : \mu \neq 10$$

at $\alpha = 0.05$ using the test statistic $\bar{X}$.

(a) Find the decision rule for the test.

(b) Draw the power curves for $n = 9$ and $n = 100$ on the same set of axes.

4.66 Jace is interested in testing

$$H_0 : \mu = 70$$

versus

$$H_1 : \mu > 70$$

using the test statistic $\bar{X}$. Assume that $X_1, X_2, \ldots, X_n$ are data values from a random sample from a $N\left(\mu, \sigma_0^2\right)$ population with unknown population mean μ and known population variance $\sigma_0^2 > 0$. If the level of significance for the test is 0.01, find the critical value for the test.

4.67 Let $X_1, X_2, \ldots, X_9$ be a random sample from a $N\left(\mu, \sigma^2\right)$ population, where μ is an unknown parameter and σ^2 is a known parameter. Consider the hypothesis test of

$$H_0 : \mu = 0$$

versus

$$H_1 : \mu \neq 0,$$

which rejects H_0 when $|\bar{X}| > 2$.

(a) Find the value of σ associated with a level of significance $\alpha = 0.01$.

(b) Conduct a Monte Carlo simulation experiment that provides convincing numerical evidence that your solution to part (a) is correct.

4.68 Let X_1, X_2, X_3 be a random sample from a $N\left(\mu_X, \sigma_X^2\right)$ population, where μ_X and $\sigma_X^2 > 0$ are unknown parameters. Let $Y_1, Y_2, \ldots, Y_5$ be a random sample from a $N\left(\mu_Y, \sigma_Y^2\right)$ population, where μ_Y and $\sigma_Y^2 > 0$ are unknown parameters. The two random samples are independent.

(a) Find the critical value for the hypothesis test

$$H_0 : 3\sigma_X^2 = \sigma_Y^2$$

versus

$$H_1 : 3\sigma_X^2 > \sigma_Y^2$$

associated with the test statistic S_X^2 / S_Y^2 at $\alpha = 0.01$.

(b) Conduct a Monte Carlo simulation experiment to provide convincing numerical evidence that your critical value from part (a) is correct.

4.69 Iris collects a random sample X_1, X_2, X_3, X_4 from a $N(\mu, 1)$ population, where μ is an unknown parameter. She tests the simple hypotheses

$$H_0 : \mu = 5$$

versus

$$H_1 : \mu = 6$$

and rejects H_0 when $\bar{X} > 5.8$. Find α and β for her hypothesis test.

4.70 Let $X_1, X_2, \ldots, X_n$ be a random sample from a $N(\mu, 100)$ population, where μ is an unknown parameter. Test H_0: $\mu = 20$ versus H_1: $\mu = 40$ using the test statistic $\bar{X}$ and the critical region $\bar{X} > 32$. Find the smallest sample size n to assure that the probability of a Type II error is less than 0.025.

4.71 Let $X_1, X_2, \ldots, X_n$ be a random sample from a $N(\mu, 9)$ population, where μ is an unknown parameter. In testing

$$H_0 : \mu = 68$$

versus

$$H_1 : \mu = 71,$$

H_0 is rejected if $\bar{X} > 70$. Find the smallest sample size n such that the probability of a Type II error is less than 0.1.

4.72 Gergana is a statistician for the Environmental Protection Agency. She is testing whether an automobile manufacturer's claim that their economy car averages 45 miles per gallon (MPG) is exaggerated. She will perform a statistical hypothesis test. She wants the probability of a Type I error to be exactly 0.01 and the probability of a Type II error to be 0.05 or smaller when the true MPG is $\mu = 40$. Assuming that her test statistic is $\bar{X}$ and the MPGs for the cars are $N(\mu, 9)$, find

(a) the smallest sample size n to achieve her goals for the probabilities of Type I and Type II error, and

(b) the decision rule for the sample size determined in part (a).

4.73 Jamie is a quality control engineer for a microprocessor manufacturing facility. The facility has set a goal of a 70% yield on the manufacture of its microprocessors. Let p be the proportion of microprocessors that undergo the manufacturing process successfully. A hypothesis test of interest is

$$H_0 : p = 0.7$$

versus

$$H_1 : p < 0.7$$

at $\alpha = 0.05$. To test this hypothesis, Jamie collects the condition of n microprocessors $X_1, X_2, \ldots, X_n$, where $X_i = 0$ denotes a failed microprocessor and $X_i = 1$ denotes a functioning microprocessor, for $i = 1, 2, \ldots, n$. Find the smallest sample size n that achieves a Type II error that is less than 0.2 when $p = 0.6$.

4.74 Let $X_1, X_2, \ldots, X_n$ be a random sample from an exponential(λ) population, where λ is a positive unknown parameter. Consider the hypothesis test

$$H_0 : \lambda = 0.001$$

versus

$$H_1 : \lambda > 0.001$$

at level of significance $\alpha = 0.05$. Example 4.20 showed that the smallest sample size to achieve a Type II error of less than $\beta = 0.1$ when $\lambda_1 = 0.0025$ was $n = 11$. Find the smallest sample size n for the following three *individual* alterations to the original problem.

(a) Change $\alpha = 0.05$ to $\alpha = 0.01$.

(b) Change $\beta = 0.1$ to $\beta = 0.01$.

(c) Change $\lambda_1 = 0.0025$ to $\lambda_1 = 0.0015$.

Comment on the effect of these changes on the resulting minimal sample sizes.

4.75 Let $X_1, X_2, \ldots, X_n$ be a random sample from an exponential(λ) distribution, where λ is a positive unknown rate parameter. Randa conducts a significance test of

$$H_0 : \lambda = 1$$

versus

$$H_1 : \lambda \neq 1,$$

which achieves a p-value of $p = 0.07$ for a particular data set. If she then computes an exact two-sided 95% confidence interval for λ for this particular data set, will the confidence interval contain 1?

4.76 Let X_1, X_2, X_3 be a random sample from a $N\left(\mu, \sigma^2\right)$ population, where μ and $\sigma^2 > 0$ are unknown parameters. The experimental values of X_1, X_2, X_3 are

$$x_1 = 2.53, x_2 = 3.11, x_3 = 1.29.$$

(a) Find an exact two-sided 95% confidence interval for μ.

(b) Conduct the hypothesis test

$$H_0 : \mu = 0$$

versus

$$H_1 : \mu \neq 0$$

at $\alpha = 0.05$.

(c) Find the p-value for the significance test

$$H_0 : \mu = 0$$

versus

$$H_1 : \mu \neq 0.$$

(d) Comment on the relationship between the confidence interval, hypothesis test, and significance test.

4.77 In order to test the fairness of a coin, Cerry tosses the coin 32 times and observes 22 heads. Let p be the probability of obtaining a head on a single toss of the coin, where $0 < p < 1$.

(a) Use the R `binom.test` function to compute an approximate two-sided 95% confidence interval for p using the Clopper–Pearson method for computing a confidence interval for a binomial proportion.

(b) Conduct the appropriate hypothesis test at $\alpha = 0.05$.

(c) Use the R `binom.test` function to compute the p-value for the appropriate significance test.

(d) Comment on the relationship between the confidence interval, hypothesis test, and significance test.

4.78 Let $X_1, X_2, \ldots, X_{25}$ be a random sample from a $N(\mu, 1)$ population, where μ is an unknown parameter. Wenying would like to test

$$H_0 : \mu = 0$$

versus

$$H_1 : \mu = 1$$

at $\alpha = 0.05$.

(a) Give the critical region for the most powerful test.

(b) Calculate the probability of Type II error.

4.79 Let X_1 and X_2 be a random sample from a population probability distribution with probability density function $f(x)$ defined on positive support. Owen tests

$$H_0 : f(x) = e^{-x} \qquad x > 0$$

versus

$$H_1 : f(x) = xe^{-x} \qquad x > 0$$

at $\alpha = 0.05$. Find a test statistic and associated critical region for the most powerful test. *Hint*: Determining the critical value might require the use of numerical methods.

4.80 Let $X_1, X_2, \ldots, X_n$ be a random sample from a Rayleigh population with probability density function

$$f(x) = 2\theta^{-2} x e^{-(x/\theta)^2} \qquad x > 0,$$

where θ is positive unknown parameter. Find the critical region for the most powerful test

$$H_0 : \theta = 1$$

versus

$$H_1 : \theta = 3$$

at level of significance α.

4.81 Let X be a single observation from a population with probability density function $f(x)$. Consider the hypothesis test

$$H_0 : X \sim N(0, 1)$$

versus

$$H_1 : X \sim U\left(-\sqrt{3}, \sqrt{3}\right)$$

at $\alpha = 0.1$.

(a) Find the critical region for the most powerful test.

(b) Find the probability of Type II error associated with the most powerful test.

4.82 Let X be a single observation from a population with probability density function $f(x)$. Consider the hypothesis test

$$H_0 : X \sim U\left(-\sqrt{3}, \sqrt{3}\right)$$

versus

$$H_1 : X \sim N(0, 1)$$

at $\alpha = 0.1$.

(a) Find the critical region for the most powerful test.

(b) Find the probability of Type II error associated with the most powerful test.

4.83 Let X be a single observation from a population with probability density function $f(x)$ used to test

$$H_0 : f(x) = 2x \qquad 0 < x < 1$$

versus

$$H_1 : f(x) = 3x^2 \qquad 0 < x < 1$$

at significance level α. Find the critical region for the most powerful test.

4.84 Let $X_1, X_2, \ldots, X_n$ be a random sample from a population with probability density function

$$f(x) = \frac{\theta}{x^{\theta+1}} \qquad x > 1,$$

where θ is a positive unknown parameter.

(a) Use the Fisher–Neyman factorization criterion to suggest a sufficient statistic for θ.

(b) Is the sufficient statistic an unbiased estimator of θ?

(c) Find the Cramér–Rao lower bound on the population variance of any unbiased estimator of θ.

(d) Find the method of moments estimator of θ.

(e) Find the maximum likelihood estimator of θ.

(f) Use the invariance property of maximum likelihood estimators to find the maximum likelihood estimator of the population mean of the distribution.

(g) Find an exact two-sided $100(1-\alpha)\%$ confidence interval for θ.

(h) Find the likelihood ratio test statistic Λ for testing $H_0 : \theta = \theta_0$ versus $H_1 : \theta \neq \theta_0$.

(i) Apply parts (d), (e) and (f) to the data set of $n = 3$ observations: 1.2, 1.7, 3.1.

4.85 Let $X_1, X_2, \ldots, X_n$ be a random sample from a *discrete uniform* population with probability density function

$$f(x) = \frac{1}{\theta} \qquad x = 1, 2, \ldots, \theta,$$

where θ is a positive unknown integer parameter.

(a) Find the method of moments estimator of θ.

(b) Determine whether the method of moments estimator of θ is an unbiased estimator of θ.

(c) Find the population variance of the method of moments estimator of θ.

(d) Discuss any problems that you can see with the method of moments estimator. *Hint*: Consider the data set 1, 2, 9.

(e) Find the maximum likelihood estimator of θ.

(f) Give a formula for the likelihood ratio test statistic for testing

$$H_0 : \theta = 20$$

versus

$$H_1 : \theta < 20.$$

(g) Find the critical value for this hypothesis test using the maximum likelihood estimator as a test statistic and $n = 1$ observation at level of significance $\alpha = 0.05$.

(h) Plot the power function for this hypothesis test when $n = 1$.

4.86 Let $X_1, X_2, \ldots, X_n$ be a random sample from a Poisson(λ) population with positive unknown parameter λ. Compute the likelihood ratio test statistic for testing

$$H_0 : \lambda = \lambda_0$$

versus

$$H_1 : \lambda < \lambda_0.$$

Use the likelihood ratio test statistic to suggest the form of the decision rule for such a test.

4.87 Let $X_1, X_2, \ldots, X_n$ be a random sample from a Rayleigh(λ) population with cumulative distribution function

$$F(x) = 1 - e^{-(\lambda x)^2} \qquad x > 0,$$

where λ is a positive unknown parameter.

(a) Find the maximum likelihood estimator of λ.

(b) Show that the log likelihood function is maximized at the maximum likelihood estimator $\hat{\lambda}$.

(c) Given that $E[X] = \sqrt{\pi}/(2\lambda)$, find the method of moments estimator of λ.

(d) Use the Fisher–Neyman factorization criterion to suggest a sufficient statistic for λ.

(e) Find the likelihood ratio test statistic Λ when $H_0 : \lambda = \lambda_0$.

(f) For a single observation X from a Rayleigh(λ) population, find an exact two-sided $100(1 - \alpha)\%$ confidence interval for λ.

4.88 Let $X_1, X_2, \ldots, X_n$ be a random sample from a population with probability density function

$$f(x) = \theta x \qquad 0 < x < \sqrt{\frac{2}{\theta}},$$

where θ is a positive unknown parameter.

(a) Find the method of moments estimator of θ.

(b) Find the maximum likelihood estimator of θ.

(c) Give the likelihood ratio test statistic for testing

$$H_0 : \theta = 8$$

versus

$$H_1 : \theta > 8.$$

(d) Grace calls five people at random and asks them their hourly wages. The numbers she collects are:

$$22.18, \quad 62.92, \quad 35.07, \quad 44.33, \quad 86.20.$$

(Note that the sample mean of these numbers is *exactly* 50.14.) Find the method of moments estimator of θ and the maximum likelihood estimator of θ for this data set.

4.89 Let $X_1, X_2, \ldots, X_n$ be a random sample from a population with probability density function

$$f(x) = \frac{\theta}{(1+x)^{\theta+1}} \qquad x > 0,$$

where θ is a positive unknown parameter.

(a) Find the maximum likelihood estimator $\hat{\theta}$.

(b) Show that the maximum likelihood estimator $\hat{\theta}$ is a biased estimator of θ. Show that $\hat{\theta}$ is asymptotically unbiased. Find an unbiasing constant c_n, which is a function of n, such that $E\left[c_n\hat{\theta}\right] = \theta$.

(c) Determine whether the unbiased estimate $c_n\hat{\theta}$ from part (c) is efficient.

(d) Calculate the maximum likelihood estimator $\hat{\theta}$ and give an exact two-sided 93% confidence interval for θ for the wooden toy price data set (in pounds, from Hand, et al., 1994, *Small Data Sets*, Chapman and Hall, page 48):

4.20 1.12 1.39 2.00 3.99 2.15 1.74 5.81 1.70 2.85 0.50
0.99 11.50 5.12 0.90 1.99 6.24 2.60 3.00 12.20 7.36 4.75
11.59 8.69 9.80 1.85 1.99 1.35 10.00 0.65 1.45.

(e) Find the p-value associated with the test

$$H_0 : \theta = 1$$

versus

$$H_1 : \theta \neq 1$$

using the likelihood ratio statistic for the wooden toy prices given in part (d). Use Wilks's theorem to arrive at the p-value.

4.90 Let $X_1, X_2, \ldots, X_n$ be a random sample from a $U(\theta_1, \theta_2)$ population, where $\theta_1 < 0$ and $\theta_2 > 0$ are unknown parameters. What is the form of the critical region of the likelihood ratio test of H_0: $\theta_2 = -\theta_1$ versus H_1: $\theta_2 \neq -\theta_1$?

4.91 Let $X_1, X_2, \ldots, X_n$ be a random sample from an exponential population with positive unknown population mean θ_X. Let $Y_1, Y_2, \ldots, Y_m$ be a random sample from an exponential population with positive unknown population mean θ_Y. The two random samples are independent. Find the likelihood ratio test for

$$H_0 : \theta_X = \theta_Y$$

versus

$$H_1 : \theta_X \neq \theta_Y$$

at significance level α.

Index

T

www.ingramcontent.com/pod-product-compliance
Ingram Content Group UK Ltd.
Pitfield, Milton Keynes, MK11 3LW, UK
UKHW061932100926
8556IPUK00007B/880

9 789819 832927